T0178379

Lecture Notes in Computer Science   14334

Founding Editors

Gerhard Goos
Juris Hartmanis

Editorial Board Members

Elisa Bertino, *Purdue University, West Lafayette, IN, USA*
Wen Gao, *Peking University, Beijing, China*
Bernhard Steffen, *TU Dortmund University, Dortmund, Germany*
Moti Yung, *Columbia University, New York, NY, USA*

The series Lecture Notes in Computer Science (LNCS), including its subseries Lecture Notes in Artificial Intelligence (LNAI) and Lecture Notes in Bioinformatics (LNBI), has established itself as a medium for the publication of new developments in computer science and information technology research, teaching, and education.

LNCS enjoys close cooperation with the computer science R & D community, the series counts many renowned academics among its volume editors and paper authors, and collaborates with prestigious societies. Its mission is to serve this international community by providing an invaluable service, mainly focused on the publication of conference and workshop proceedings and postproceedings. LNCS commenced publication in 1973.

Xiangyu Song · Ruyi Feng · Yunliang Chen ·
Jianxin Li · Geyong Min
Editors

Web and Big Data

7th International Joint Conference, APWeb-WAIM 2023
Wuhan, China, October 6–8, 2023
Proceedings, Part IV

Springer

Editors
Xiangyu Song
Peng Cheng Laboratory
Shenzhen, China

Yunliang Chen (ID)
China University of Geosciences
Wuhan, China

Geyong Min (ID)
University of Exeter
Exeter, UK

Ruyi Feng (ID)
China University of Geosciences
Wuhan, China

Jianxin Li (ID)
Deakin University
Burwood, VIC, Australia

ISSN 0302-9743   ISSN 1611-3349 (electronic)
Lecture Notes in Computer Science
ISBN 978-981-97-2420-8   ISBN 978-981-97-2421-5 (eBook)
https://doi.org/10.1007/978-981-97-2421-5

Preface

This volume (LNCS 14334) and its companion volumes (LNCS 14331, LNCS 14332, and LNCS 14333) contain the proceedings of the 7th Asia-Pacific Web (APWeb) and Web-Age Information Management (WAIM) Joint Conference on Web and Big Data, called APWeb-WAIM 2023. Researchers and practitioners from around the world came together at this leading international forum to share innovative ideas, original research findings, case study results, and experienced insights in the areas of the World Wide Web and big data, thus covering web technologies, database systems, information management, software engineering, knowledge graphs, recommend systems and big data.

The 7th APWeb-WAIM conference was held in Wuhan during 6–8 October 2023. As an Asia-Pacific flagship conference focusing on research, development, and applications in relation to Web information management, APWeb-WAIM builds on the successes of APWeb and WAIM. Previous APWeb conferences were held in Beijing (1998), Hong Kong (1999), Xi'an (2000), Changsha (2001), Xi'an (2003), Hangzhou (2004), Shanghai (2005), Harbin (2006), Huangshan (2007), Shenyang (2008), Suzhou (2009), Busan (2010), Beijing (2011), Kunming (2012), Sydney (2013), Changsha (2014), Guangzhou (2015), and Suzhou (2016); and WAIM was held in Shanghai (2000), Xi'an (2001), Beijing (2002), Chengdu (2003), Dalian (2004), Hangzhou (2005), Hong Kong (2006), Huangshan (2007), Zhangjiajie (2008), Suzhou (2009), Jiuzhaigou (2010), Wuhan (2011), Harbin (2012), Beidaihe (2013), Macau (2014), Qingdao (2015), and Nanchang (2016). The APWeb-WAIM conferences were held in Beijing (2017), Macau (2018), Chengdu (2019), Tianjin (2020), Guangzhou (2021), Nanjing (2022), and Wuhan (2023). With the ever-growing importance of appropriate methods in these data-rich times and the fast development of web-related technologies, APWeb-WAIM will become a flagship conference in this field.

The high-quality program documented in these proceedings would not have been possible without the authors who chose APWeb-WAIM for disseminating their findings. APWeb-WAIM 2023 received a total of 434 submissions and, after the double-blind review process (each paper received at least three review reports), the conference accepted 133 regular papers (including research and industry track) (acceptance rate 31.15%), and 6 demonstrations. The contributed papers address a wide range of topics, such as big data analytics, advanced database and web applications, data mining and applications, graph data and social networks, information extraction and retrieval, knowledge graphs, machine learning, recommender systems, security, privacy and trust, and spatial and multi-media data. The technical program also included keynotes by Jie Lu, Qing-Long Han, and Hai Jin. We are grateful to these distinguished scientists for their invaluable contributions to the conference program.

We would like to express our gratitude to all individuals, institutions, and sponsors that supported APWeb-WAIM2023. We are deeply thankful to the Program Committee members for lending their time and expertise to the conference. We also would like to acknowledge the support of the other members of the Organizing Committee. All of them helped to make APWeb-WAIM2023 a success. We are grateful for the guidance of the honorary chair (Lizhe Wang), the steering committee representative (Yanchun Zhang) and the general co-chairs (Guoren Wang, Schahram Dustdar, Bruce Xuefeng Ling, and Hongyan Zhang) for their guidance and support. Thanks also go to the program committee chairs (Yunliang Chen, Jianxin Li, and Geyong Min), local co-chairs (Chengyu Hu, Tao Lu, and Jianga Shang), publicity co-chairs (Bohan Li, Chang Tang, and Xin Bi), proceedings co-chairs (David A. Yuen, Ruyi Feng, and Xiangyu Song), tutorial co-chairs (Ye Yuan and Rajiv Ranjan), CCF TCIS liaison (Xin Wang), CCF TCDB liaison (Yueguo Chen), Ph.D. consortium co-chairs (Pablo Casaseca, Xiaohui Huang, and Yanan Li), Web co-chairs (Wei Han, Huabing Zhou, and Wei Liu), and industry co-chairs (Jun Song, Wenjian Qin, and Tao Yu).

We hope you enjoyed the exciting program of APWeb-WAIM 2023 as documented in these proceedings.

October 2023

Yunliang Chen
Jianxin Li
Geyong Min
David A. Yuen
Ruyi Feng
Xiangyu Song

Organization

General Chairs

Guoren Wang   BIT, China
Schahram Dustdar   TU Wien, Austria
Bruce Xuefeng Ling   Stanford University, USA
Hongyan Zhang   China University of Geosciences, China

Program Committee Chairs

Yunliang Chen   China University of Geosciences, China
Jianxin Li   Deakin University, Australia
Geyong Min   University of Exeter, UK

Steering Committee Representative

Yanchun Zhang   Guangzhou University & Pengcheng Lab, China;
Victoria University, Australia

Local Co-chairs

Chengyu Hu   China University of Geosciences, China
Tao Lu   Wuhan Institute of Technology, China
Jianga Shang   China University of Geosciences, China

Publicity Co-chairs

Bohan Li   Nanjing University of Aeronautics and
Astronautics, China
Chang Tang   China University of Geosciences, China
Xin Bi   Northeastern University, China

Proceedings Co-chairs

David A. Yuen	Columbia University, USA
Ruyi Feng	China University of Geosciences, China
Xiangyu Song	Swinburne University of Technology, Australia

Tutorial Co-chairs

Ye Yuan	BIT, China
Rajiv Ranjan	Newcastle University, UK

CCF TCIS Liaison

Xin Wang	Tianjin University, China

CCF TCDB Liaison

Yueguo Chen	Renmin University of China, China

Ph.D. Consortium Co-chairs

Pablo Casaseca	University of Valladolid, Spain
Xiaohui Huang	China University of Geosciences, China
Yanan Li	Wuhan Institute of Technology, China

Web Co-chairs

Wei Han	China University of Geosciences, China
Huabing Zhou	Wuhan Institute of Technology, China
Wei Liu	Wuhan Institute of Technology, China

Industry Track Co-chairs

Jun Song	China University of Geosciences, China
Wenjian Qin	Shenzhen Institute of Advanced Technology CAS, China
Tao Yu	Tsinghua University, China

Program Committee Members

Alex Delis	University of Athens, Greece
Amr Ebaid	Google, USA
An Liu	Soochow University, China
Anko Fu	China University of Geosciences, China
Ao Long	China University of Geosciences, Wuhan, China
Aviv Segev	University of South Alabama, USA
Baoning Niu	Taiyuan University of Technology, China
Bin Zhao	Nanjing Normal University, China
Bo Tang	Southern University of Science and Technology, China
Bohan Li	Nanjing University of Aeronautics and Astronautics, China
Bolong Zheng	Huazhong University of Science and Technology, China
Cai Xu	Xidian University, China
Carson Leung	University of Manitoba, Canada
Chang Tang	China University of Geosciences, China
Chen Shaohao	China University of Geosciences, China
Cheqing Jin	East China Normal University, China
Chuanqi Tao	Nanjing University of Aeronautics and Astronautics, China
Dechang Pi	Nanjing University of Aeronautics and Astronautics, China
Dejun Teng	Shandong University, China
Derong Shen	Northeastern University, China
Dong Li	Liaoning University, China
Donghai Guan	Nanjing University of Aeronautics and Astronautics, China
Fang Wang	Hong Kong Polytechnic University, China
Feng Yaokai	Kyushu University, Japan
Giovanna Guerrini	University of Genoa, Italy
Guanfeng Liu	Macquarie University, Australia

Guoqiong Liao	Jiangxi University of Finance & Economics, China
Hailong Liu	Northwestern Polytechnical University, USA
Haipeng Dai	Nanjing University, China
Haiwei Pan	Harbin Engineering University, China
Haoran Xu	China University of Geosciences, Wuhan, China
Haozheng Ma	China University of Geosciences, Wuhan, China
Harry Kai-Ho Chan	University of Sheffield, UK
Hiroaki Ohshima	University of Hyogo, Japan
Hongzhi Wang	Harbin Institute of Technology, China
Hua Wang	Victoria University, Australia
Hui Li	Xidian University, China
Jiabao Li	China University of Geosciences, Wuhan, China
Jiajie Xu	Soochow University, China
Jiali Mao	East China Normal University, China
Jian Chen	South China University of Technology, China
Jian Yin	Sun Yat-sen University, China
Jianbin Qin	Shenzhen University, China
Jiannan Wang	Simon Fraser University, Canada
Jianqiu Xu	Nanjing University of Aeronautics and Astronautics, China
Jianxin Li	Deakin University, Australia
Jianzhong Qi	University of Melbourne, Australia
Jianzong Wang	Ping An Technology (Shenzhen) Co., Ltd., China
Jinguo You	Kunming University of Science and Technology, China
Jizhou Luo	Harbin Institute of Technology, China
Jun Gao	Peking University, China
Jun Wang	China University of Geosciences, Wuhan, China
Junhu Wang	Griffith University, Australia
K. Selçuk Candan	Arizona State University, USA
Krishna Reddy P.	IIIT Hyderabad, India
Ladjel Bellatreche	ISAE-ENSMA, France
Le Sun	Nanjing University of Information Science and Technology, China
Lei Duan	Sichuan University, China
Leong Hou U	University of Macau, China
Li Jiajia	Shenyang Aerospace University, China
Liang Hong	Wuhan University, China
Lin Xiao	China University of Geosciences, Wuhan, China
Lin Yue	University of Newcastle, UK

Lisi Chen	University of Electronic Science and Technology of China, China
Lizhen Cui	Shandong University, China
Long Yuan	Nanjing University of Science and Technology, China
Lu Chen	Zhejiang University, China
Lu Qin	UTS, Australia
Luyi Bai	Northeastern University, China
Miaomiao Liu	Northeast Petroleum University, China
Min Jin	China University of Geosciences, Wuhan, China
Ming Zhong	Wuhan University, China
Mirco Nanni	CNR-ISTI Pisa, Italy
Mizuho Iwaihara	Waseda University, Japan
Nicolas Travers	Pôle Universitaire Léonard de Vinci, France
Peiquan Jin	University of Science and Technology of China, China
Peng Peng	Hunan University, China
Peng Wang	Fudan University, China
Philippe Fournier-Viger	Shenzhen University, China
Qiang Qu	SIAT, China
Qilong Han	Harbin Engineering University, China
Qing Xie	Wuhan University of Technology, China
Qiuyan Yan	China University of Mining and Technology, China
Qun Chen	Northwestern Polytechnical University, China
Rong-Hua Li	Beijing Institute of Technology, China
Rui Zhu	Shenyang Aerospace University, China
Runyu Fan	China University of Geosciences, China
Sanghyun Park	Yonsei University, South Korea
Sanjay Madria	Missouri University of Science & Technology, USA
Sara Comai	Politecnico di Milano, Italy
Shanshan Yao	Shanxi University, China
Shaofei Shen	University of Queensland, Australia
Shaoxu Song	Tsinghua University, China
Sheng Wang	China University of Geosciences, Wuhan, China
ShiJie Sun	Chang'an University, China
Shiyu Yang	Guangzhou University, China
Shuai Xu	Nanjing University of Aeronautics and Astronautics, China
Shuigeng Zhou	Fudan University, China
Tanzima Hashem	Bangladesh University of Engineering and Technology, Bangladesh

Tianrui Li	Southwest Jiaotong University, China
Tung Kieu	Aalborg University, Denmark
Vincent Oria	NJIT, USA
Wee Siong Ng	Institute for Infocomm Research, Singapore
Wei Chen	Hebei University of Environmental Engineering, China
Wei Han	China University of Geosciences, Wuhan, China
Wei Shen	Nankai University, China
Weiguo Zheng	Fudan University, China
Weiwei Sun	Fudan University, China
Wen Zhang	Wuhan University, China
Wolf-Tilo Balke	TU Braunschweig, Germany
Xiang Lian	Kent State University, USA
Xiang Zhao	National University of Defense Technology, China
Xiangfu Meng	Liaoning Technical University, China
Xiangguo Sun	Chinese University of Hong Kong, China
Xiangmin Zhou	RMIT University, Australia
Xiao Pan	Shijiazhuang Tiedao University, China
Xiao Zhang	Shandong University, China
Xiao Zheng	National University of Defense Technology, China
Xiaochun Yang	Northeastern University, China
Xiaofeng Ding	Huazhong University of Science and Technology, China
Xiaohan Zhang	China University of Geosciences, Wuhan, China
Xiaohui (Daniel) Tao	University of Southern Queensland, Australia
Xiaohui Huang	China University of Geosciences, Wuhan, China
Xiaowang Zhang	Tianjin University, China
Xie Xiaojun	Nanjing Agricultural University, China
Xin Bi	Northeastern University, China
Xin Cao	University of New South Wales, Australia
Xin Wang	Tianjin University, China
Xingquan Zhu	Florida Atlantic University, USA
Xinwei Jiang	China University of Geosciences, Wuhan, China
Xinya Lei	China University of Geosciences, Wuhan, China
Xinyu Zhang	China University of Geosciences, Wuhan, China
Xujian Zhao	Southwest University of Science and Technology, China
Xuyun Zhang	Macquarie University, Australia
Yajun Yang	Tianjin University, China
Yanfeng Zhang	Northeastern University, China

Yanghui Rao	Sun Yat-sen University, China
Yang-Sae Moon	Kangwon National University, South Korea
Yanhui Gu	Nanjing Normal University, China
Yanjun Zhang	University of Technology Sydney, Australia
Yaoshu Wang	Shenzhen University, China
Ye Yuan	China University of Geosciences, Wuhan, China
Yijie Wang	National University of Defense Technology, China
Yinghui Shao	China University of Geosciences, Wuhan, China
Yong Tang	South China Normal University, China
Yong Zhang	Tsinghua University, China
Yongpan Sheng	Southwest University, China
Yongqing Zhang	Chengdu University of Information Technology, China
Youwen Zhu	Nanjing University of Aeronautics and Astronautics, China
Yu Liu	Huazhong University of Science and Technology, China
Yuanbo Xu	Jilin University, China
Yue Lu	China University of Geosciences, Wuhan, China
Yuewei Wang	China University of Geosciences, Wuhan, China
Yunjun Gao	Zhejiang University, China
Yunliang Chen	China University of Geosciences, Wuhan, China
Yunpeng Chai	Renmin University of China, China
Yuwei Peng	Wuhan University, China
Yuxiang Zhang	Civil Aviation University of China, China
Zhaokang Wang	Nanjing University of Aeronautics and Astronautics, China
Zhaonian Zou	Harbin Institute of Technology, China
Zhenying He	Fudan University, China
Zhi Cai	Beijing University of Technology, China
Zhiwei Zhang	Beijing Institute of Technology, China
Zhixu Li	Soochow University, China
Ziqiang Yu	Yantai University, China
Zouhaier Brahmia	University of Sfax, Tunisia

Contents – Part IV

YOLO-SA: An Efficient Object Detection Model Based on Self-attention Mechanism

Ang Li[1], Xiangyu song[2], ShiJie Sun[1], Zhaoyang Zhang[1(✉)], Taotao Cai[3],
and Huansheng Song[1]

[1] Chang'an University, Xi'an, China
{2021124113,shijieSun,zhaoyang_zh,hshsong}@chd.edu.cn
[2] Swinburne University of Technology, Melbourne, Australia
x.song@deakin.edu.au
[3] Macquarie University, Sydney, Australia
taotao.cai@mq.edu.au

Abstract. Object detector based on CNN structure has been widely used in object detection, object classification and other tasks. The traditional CNN module usually adopts complex multi-branch design, which reduces the reasoning speed and memory utilization. Moreover, in many works, attention mechanism is usually added to the object detector to extract rich features in spatial information, which are usually used as additional modules of convolution without fundamental improvement from the limitations of convolution operation. Finally, traditional object detectors often have coupled detection heads, which can compromise model performance. To solve the above problems, we propose a new object detection model, YOLO-SA, based on the current popular object detector model YOLOv5. We introduce a new reparameterized module RepVGG to replace the original DarkNet53 structure of YOLOv5 model, which greatly reduces the complexity of the model and improves the detection accuracy. We introduce a self-attention mechanism module in the feature fusion part of the model, which is independent from other convolutional layers and has higher performance than other mainstream attention mechanism modules. We replace the coupled detection head in YOLOv5 model with an anchor-based decoupled detection head, which greatly improved the convergence speed in the training process. Experiments show that the detection accuracy of the YOLO-SA model proposed by us reaches 71.2% and 75.8% on COCO2014 and VOC2012 dataset respectively, which is superior to the YOLOv5s model as the baseline and other mainstream object detection models, showing certain superiority.

Keywords: Object detection · CNN architecture · Attention mechanism · Decoupled detection head

1 Introduction

The architecture of Deep Convolutional Neural Networks (CNN) has undergone a period of significant evolution over the years, resulting in increased accuracy and speed. This evolution has been marked by the milestone work of LeNet [1] and the subsequent improvement of ImageNet [2] classification accuracy through novel structures such as AlexNet [3], ZFNet [4], VGG [5], GoogLeNet [6], ResNet [7], and DenseNet [8], among others.

Object detection is a key component in many vision-based intelligent applications, and thanks to the strong ability of CNN, numerous models based on this technology have been proposed. These models can be divided into two categories: one-stage and two-stage object detectors, which differ in their detection processes. Two-stage object detectors, such as the Region-based Convolutional Neural Network (R-CNN) series [9–11], have been instrumental in significantly improving the accuracy of object detection. These detectors first select possible object regions and then classify them. Furthermore, due to their region-level features, these detectors for still images can be easily adapted to more complex tasks, such as segmentation and video object detection. Although the two-stage object detector has excellent performance, its efficiency is a major limitation for practical use. In contrast, one-stage object detectors allow for the joint and direct production of location and classification from the dense predictions of feature maps. One-stage detectors, such as the YOLO series [12–15] and SSD [16], have superior speed compared to two-stage approaches, making them suitable for scenarios with real-time requirements. These detectors do not involve region proposals directly, thus providing an efficient alternative for object detection.

The YOLO series of object detectors have been designed to optimize the speed-accuracy trade-off for real-time applications, drawing on the most advanced detection technologies available and optimizing the implementation for optimal practice. The YOLOv5 model is renowned for its outstanding performance amongst the YOLO series models, and is widely utilised for object detection tasks. Currently, YOLOv5 has demonstrated a trade-off performance of 48.2% average precision (AP) on the COCO dataset at 13.7 ms. The structure of the YOLOv5 model is shown in Fig. 1.

In this paper, based on the YOLOv5 model, we propose an efficient object detection model based on the self-attention mechanism, YOLO-SA, which occupies a small computing overhead but has a high detection accuracy. Our contributions are as follows:

1. The complex multi-branch design of traditional convolutional neural networks will greatly reduce the reasoning speed and memory utilization. We introduce an improved RepVGG [17] module, whose performance is better than many complex models, which can greatly improve the reasoning speed while maintaining good detection accuracy.
2. Traditional attention mechanisms are introduced as additional modules of convolution without fundamental improvement from the limitations of convolution operation. We introduce a new self-attention mechanism [18] in the

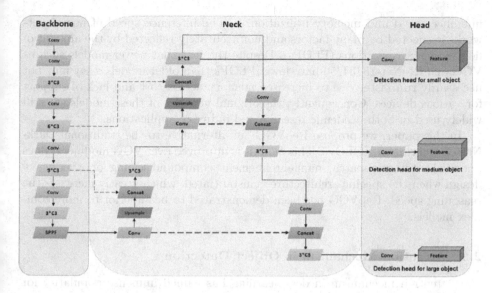

Fig. 1. The structure of the YOLOv5 model.

feature fusion part, which makes it a separate layer of the visual model and achieves good results.

3. The detection head of YOLO series models adopts anchor-based design, which will increase the complexity of the detection head and there is a potential bottleneck in the calculation process. And their detection heads are usually coupled, which compromises detection performance. We introduced a decoupling head [19] based on anchor-free, which greatly reduce parameters and FLOPs in the detection process.

2   Related Work

2.1   Traditional CNN Module in Object Detection

The YOLO series model utilizes a classic deep neural network, DarkNet [12], which is combined with the feature-rich characteristics of ResNet to ensure strong feature expression while avoiding the gradient problem that can be caused by a network that is too deep. There are two main architectures of DarkNet, namely DarkNet19 and DarkNet53. DarkNet53 in the YOLOv5 model, achieved huge success in image recognition with a simple architecture composed of a stack of Conv, Batch Normalization, and LeakyReLu. The emergence of object detection models has generated a heightened interest in the creation of well-designed architectures, resulting in a marked increase in complexity.

Although many ConvNets with complex architectures are capable of achieving higher accuracy than their simpler counterparts, they are often accompanied with several drawbacks. 1. The implementation and customization of the multi-branch model is challenging due to its complex design, which can lead to reduced inference speed and memory utilization. 2. The inference speed of neural networks is affected by many factors and not accurately reflected by the amount of floating-point operations (FLOPs). Despite the fact that newer models such as VGG and ResNet-18/34/50 have lower FLOPs than older models, they may not necessarily run faster due to increased memory access cost and lack of support for various devices. Consequently, the original versions of these models are still widely used in both academic research and industry applications.

In this paper, we propose RepVGG, an alternative to the traditional DarkNet53 structure in YOLOv5. This module is improved over VGG module. It does not need automatic search, manual refinement, compound scaling, or other heavy design when the specific architecture is instantiated, which greatly increases the reasoning speed. RepVGG has been demonstrated to be superior to more complex models.

2.2   Attention Mechanism in Object Detection

The attention mechanism in deep learning has gained immense popularity for its ability to extract more accurate information from a large number of data points, similar to the way the human visual attention mechanism works. This has become a subject of intense academic research in recent years. The utilization of attention mechanisms has been seen to have a considerable amount of success in numerous tasks, such as language modeling, speech recognition, and neural captioning. More recently, attention modules have been applied in discriminative computer vision models to augment the performance of traditional Convolutional Neural Networks.

Squeeze-and-Excitation Networks (SENet) [20] is a deep learning architecture that integrates spatial information into the feature response in terms of channels via two multilayer perceptron (MLP) layers to compute the corresponding attention. Coordinate Attention (CA) [21] is a further development of SENet which embeds position information into channel attention, allowing long-range dependencies to be captured in one spatial direction while preserving accurate position information in the other. Convolutional Block Attention Module (CBAM) [22] provides a solution which embeds channel and spatial attention submodules in a sequential manner. These traditional attention mechanisms modules use the global attention layer as an additional module of convolution without fundamentally improving the limitations of convolution operations. Moreover, due to the need to pay attention to all positions of the input, the size of the input feature map is relatively small, otherwise the computational cost would be too high.

To solve this problem, we propose a novel local self-attention layer. The proposed stand-alone local self-attention layer can be used to construct a fully attentional network instead of being used as an enhancement of convolution, which has demonstrated competitive predictive performance on object detection tasks while requiring fewer parameters and floating point operations than the corresponding convolution baselines.

2.3 Detection Head in Object Detection

YOLOv4 and YOLOv5 follow the anchor-based pipeline of YOLOv3, which has several known issues. For optimal detection performance, one must conduct clustering analysis to determine an appropriate set of anchors before training. However, this approach yields domain-specific anchors which are not generalised. Besides, the utilization of anchor mechanisms in detection models leads to an increased complexity of the detection heads and a greater number of predictions per image. This causes a potential bottleneck in terms of latency on some edge AI systems, as a large number of predictions must be transferred between devices.

In the past two years, the development of anchor-free detectors [23–25] has seen rapid progress. Studies have demonstrated that the performance of anchor-free detectors can be comparable to that of anchor-based detectors. The use of an anchor-free mechanism significantly reduces the number of design parameters that require heuristic tuning, thus simplifying the training and decoding phases of the detector. Moreover, the use of such a mechanism eliminates the need for many tricks that are usually necessary for achieving good performance.

The conflict between classification and regression tasks in object detection is a widely-acknowledged problem [26,27]. As a result, most one- and two-stage detectors utilize a decoupled head for classification and localization [24,27,28]. However, despite the evolution of YOLO series' backbones and feature pyramids, its detection head remains coupled, the coupled detection head may harm the performance.

To solve these problems, we replace the YOLOv5 detection head with compact decoupled head and converted to an anchor-free form. This modification reduces the detector's parameters and GFLOPs, which is faster but performs better.

3 YOLO-SA

3.1 Overview of YOLO-SA

YOLOv5 consists of five models: YOLOv5n, YOLOv5s, YOLOv5m, YOLOv5l, and YOLOv5x. Each model utilizes a CSPDarknet53 architecture with an SPPF layer as the backbone, FPN+PAN as the neck, and YOLO detection head. The overall structure of our model still follows this design, our YOLO-SA model is based on the YOLOv5s model, YOLOv5s model achieves the balance of detection accuracy and detection speed, and can adapt to most complex object detection tasks. The structure of the YOLO-SA model is shown in Fig. 2.

3.2 RepVGG Module

The multi-branch topology is known to be memory-inefficient due to the requirement of storing the results of each branch until the addition or concatenation is executed, thus significantly increasing the peak value of memory occupation. In

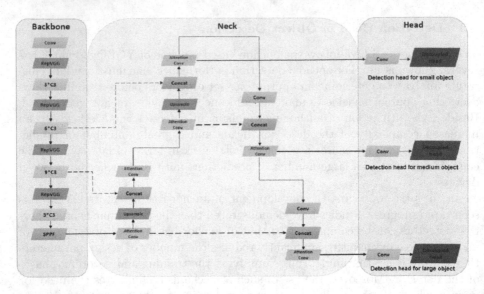

Fig. 2. The structure of the YOLO-SA model.

contrast to other topologies, a plain topology allows for the memory occupied by the inputs of a specific layer to be released immediately following the completion of the operation. This facilitates efficient utilization of the available memory and ensures that the system does not become overburdened.

RepVGG is a convolutional neural network composed of five stages. Each stage begins with a stride-2 convolution operation, which downsamples the input. The structure of RepVGG is shown in Fig. 3. This is inspired by the ResNet architecture, which also uses identity and 1×1 branches, however, these are only used during the training phase. Here, we will only show the first four layers of a particular stage. A VGG-like plain topology is adopted in this model, which means that each layer takes the output of its preceding layer as input and feeds it into its succeeding layer. Consequently, there are no branches present in this model. The body of an inference-time RepVGG is composed of a single type of operator: a 3×3 convolutional layer followed by a ReLU non-linearity. This uniformity in architecture makes RepVGG particularly efficient when implemented on generic computing devices such as GPUs.

3.3 Self-attention Mechanism

This paper presents a self-attention layer that can be used as a substitute for spatial convolution in order to construct a model of full attention. The self-attention layer is stand-alone, implying that it can be implemented independently. The independent self-attention layer defines three concepts: query, key, value. The self-attention layer operates locally, so there is no constraint on the size of the input. The number of parameters of the self-attention layer is independent of the size of the receptive field, and the number of parameters of the convolution

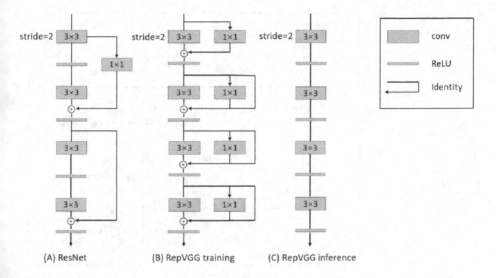

Fig. 3. Part of the RepVGG model structure.

is squared with the size of the receptive field. The amount of computation also grows more slowly than that of convolution.

The training efficiency and computational requirements of this self-attention-based structure are more favorable than traditional convolution. The ability of self-attention module to directly simulate remote interaction and its parallelism make use of the advantages of modern hardware and bring great performance improvement for various tasks.

3.4   Decoupled Head

The traditional coupled head has been shown to have detrimental effects on performance. To address this issue, we replace the YOLOv5 detection head with compact decoupled head, as shown in Fig. 4. At each level of the Feature Pyramid Network (FPN), a 1×1 convolutional layer is used to reduce the feature channel to 256. Two parallel branches are then added, each comprised of two 3×3 convolutional layers. The first branch is employed for the classification task, while the second branch is for regression and includes an Intersection over Union (IoU) branch.

We also converted YOLO's anchor-based detection head to anchor-free detection head. By reducing the number of predictions for each location from three to one and directly predicting four values, namely two offsets of the left-top corner of the grid, and the height and width of the predicted box, our proposed method improves the efficiency of the object detection process. The center location of each object is designated as the positive sample, and a scale range is pre-defined to designate the FPN level for each object. The modification of the detector

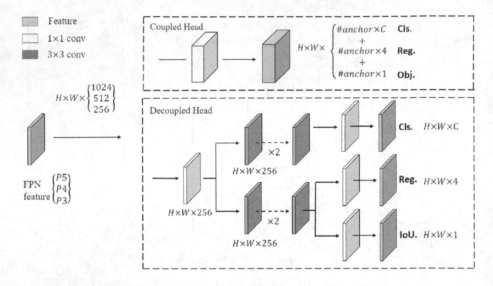

Fig. 4. The difference between traditional detection head and decoupled head.

results in a reduction of parameters and GFLOPs, leading to increased speed; however, the performance of the detector is improved.

4   Experiments

4.1   Implementation Details

We use the COCO2014 and VOC2012 dataset to evaluate our model. We implement YOLO-SA on Pytorch 1.9.1, cuda 11.3. All of our models use an NVIDIA RTX3080ti GPU for training and testing. During the training process, we employ part of the pre-trained model from YOLOv5s due to the fact that YOLO-SA and YOLOv5 share the majority of the backbone and some components of the head. This results in many weights that can be transferred from YOLOv5s to YOLO-SA, thereby saving a considerable amount of training time.

4.2   Comparison Experiment of RepVGG and Mainstream CNN Modules

We compare RepVGG with the classic and state-of-art models including VGG-16, ResNet-101/152, ResNeXt-101 [29], EfficientNet [30] on the COCO2014 and VOC2012 dataset, and use the YOLOv5s model of DarkNet53 backbone as the baseline to verify the performance of these CNN modules in feature extraction. The experimental results are shown in Tables 1 and 2.

It can be seen from the experimental results that on COCO2014 dataset, the mAP@0.5 of YOLOv5s model with improved RepVGG reaches 67.2%, the highest among the compared models. Precision and recall reach 71.2% and

Table 1. Detection results of YOLOv5s models after using different CNN modules in feature extraction part on COCO2014 dataset.

Model	mAP@0.5	Precision	Recall	Params (M)	FLOPs	Speed(ms)
YOLOv5s-DarkNet53	65.9	70.6	60.3	7.2	16.5	6.5
YOLOv5s-VGG-16	59.8	66.5	58.4	6.9	16.0	7.2
YOLOv5s-ResNet-101	61.7	67.9	59.7	7.5	16.9	6.8
YOLOv5s-ResNet-152	62.2	70.1	63.6	7.7	17.2	8.3
YOLOv5s-ResNeXt-101	64.2	65.2	70.6	7.4	16.7	6.7
YOLOv5s-EfficientNet	65.3	69.7	62.3	6.8	15.8	6.2
YOLOv5s-RepVGG(ours)	**67.2**	**71.2**	65.9	**6.7**	**15.7**	**5.8**

Table 2. Detection results of YOLOv5s models after using different CNN modules in feature extraction part on VOC2012 dataset.

Model	mAP@0.5	Precision	Recall	Params (M)	FLOPs	Speed(ms)
YOLOv5s-DarkNet53	68.7	72.6	65.3	7.2	16.5	6.5
YOLOv5s-VGG-16	60.1	67.1	59.2	6.9	16.0	7.2
YOLOv5s-ResNet-101	63.9	70.4	61.5	7.5	16.9	6.8
YOLOv5s-ResNet-152	65.8	73.1	65.7	7.7	17.2	8.3
YOLOv5s-ResNeXt-101	67.2	68.9	72.6	7.4	16.7	6.7
YOLOv5s-EfficientNet	69.3	70.7	68.1	6.8	15.8	6.2
YOLOv5s-RepVGG(ours)	**70.7**	**78.4**	**73.5**	**6.7**	**15.7**	**5.8**

65.9% respectively. On VOC2012 dataset, the mAP@0.5 of YOLOv5s model with improved RepVGG reaches 70.7%, the highest among the compared models. Precision and recall reach 78.4% and 73.5% respectively, the highest among the compared models. Params and FLOPs reach 6.7M and 15.7 respectively, occupying the least parameter and calculation quantity among the compared models. The operation speed of the model reaches 5.8 ms, which is the fastest among the compared models. The experiment proves that our improvement has certain superiority.

4.3 Comparison Experiment of Self-attention Mechanism and Mainstream Attention Mechanism

We compare the self-attention mechanism with the mainstream attention mechanism SE, CA, ECA [31], CBAM on COCO2014 and VOC2012 dataset. We take the YOLOv5s model without adding attention mechanism as the baseline. The experimental results are shown in Tables 3 and 4.

It can be seen from the experimental results that on COCO2014 dataset, the mAP@0.5 of YOLOv5s model with improved self-attention mechanism reaches 69.6%, the highest among the compared models. Precision and recall reach 75.4% and 74.3% respectively, the highest among the compared models. On VOC2012

Table 3. Detection results of YOLOv5s models after adding different attention mechanisms in feature fusion part on COCO2014 dataset.

Model	mAP@0.5	Precision	Recall	Params (M)	FLOPs	Speed(ms)
YOLOv5s	65.9	70.6	60.3	7.2	16.5	6.5
YOLOv5s-SE	66.7	74.1	68.2	8.5	21.3	8.1
YOLOv5s-CA	67.2	70.4	66.9	8.9	22.7	7.8
YOLOv5s-ECA	66.8	69.3	72.7	8.5	20.1	7.6
YOLOv5s-CBAM	67.8	68.2	70.1	11.1	25.6	9.2
YOLOv5s-SA(ours)	**69.6**	**75.4**	**74.3**	9.2	23.0	7.5

Table 4. Detection results of YOLOv5s models after adding different attention mechanisms in feature fusion part on VOC2012 dataset.

Model	mAP@0.5	Precision	Recall	Params (M)	FLOPs	Speed(ms)
YOLOv5s	68.7	72.6	65.3	7.2	16.5	6.5
YOLOv5s-SE	69.8	76.6	70.2	8.5	21.3	8.1
YOLOv5s-CA	70.2	75.4	68.9	8.9	22.7	7.8
YOLOv5s-ECA	68.9	71.3	70.7	8.5	20.1	7.6
YOLOv5s-CBAM	72.6	77.9	74.7	11.1	25.6	9.2
YOLOv5s-SA(ours)	**74.6**	**80.4**	**77.9**	9.2	23.0	7.5

dataset, the mAP@0.5 of YOLOv5s model with improved self-attention mechanism reaches 74.6%, the highest among the compared models. Precision and recall reach 80.4% and 77.9% respectively, the highest among the compared models. Params and FLOPs reach 9.2M and 23.0 respectively. The operation speed of the model reaches 7.5 ms. The experiment proves that our improvement has certain superiority.

4.4   Comparison Experiment of Decoupled Detection Head and Coupled Detection Head

We replace the coupled head of the YOLOv5s model with decoupled head, and test their performance on COCO2014 and VOC2012 dataset. The experimen-

Table 5. Detection results of YOLOv5s models after improving decoupled head in detection head part on COCO2014 dataset.

Model	mAP@0.5	Precision	Recall	Params (M)	FLOPs	Speed(ms)
Coupled Head	65.9	70.6	60.3	7.2	16.5	6.5
Decoupled Head(ours)	**68.7**	**77.1**	**66.7**	7.0	**15.7**	**4.8**

tal results are shown in Tables 5 and 6. The loss function of the model after improving the decoupled detection head is shown in the Fig. 5.

Table 6. Detection results of YOLOv5s models after improving decoupled head in detection head part on VOC2012 dataset.

Model	mAP@0.5	Precision	Recall	Params (M)	FLOPs	Speed(ms)
Coupled Head	68.7	72.6	65.3	7.2	16.5	6.5
Decoupled Head(ours)	**70.6**	**78.2**	**67.9**	**7.0**	**15.7**	**4.8**

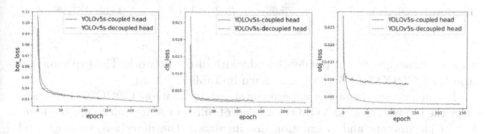

Fig. 5. Loss functions for YOLOv5s models with coupled head and decoupled head.

It can be seen from the experimental results that on COCO2014 dataset, the mAP@0.5 of YOLOv5s model with improved decoupled head reaches 68.7%, it is 2.8% higher than the original coupled head structure model. Precision and recall reach 77.1% and 66.7% respectively, it is 6.5% and 6.4% higher than the original coupled head structure model respectively. On VOC2012 dataset, the mAP@0.5 of YOLOv5s model with improved decoupled head reaches 70.6%, it is 1.9% higher than the original coupled head structure model. Precision and recall reach 78.2% and 67.9% respectively, it is 5.6% and 2.6% higher than the original coupled head structure model respectively. Params and FLOPs reach 7.0 M and 15.7 respectively, it is 0.2 M and 0.8 less than the original coupling head structure model respectively. The operation speed of the model reaches 4.8 ms, it is 1.7 ms less than the original coupling head structure model. The experiment proves that our improvement has certain superiority.

4.5 Comparison Experiments of Improved YOLOv5 Models at Different Scales

In order to verify the effect of our improvement on YOLOv5 models at different scales, we apply our improvement to YOLOv5n, YOLOv5s, YOLOv5m, YOLOv5l, and YOLOv5x models respectively. There is no difference in their

Table 7. Detection results of improved YOLOv5 models at different scales on COCO2014 dataset.

Model	mAP@0.5	Precision	Recall	Params (M)	FLOPs	Speed(ms)
YOLOv5n	59.7	61.2	65.7	1.9	4.5	6.4
YOLOv5s	65.9	70.6	60.3	7.2	16.5	6.5
YOLOv5m	66.2	69.8	72.6	21.2	49.0	8.4
YOLOv5l	69.6	72.1	65.4	46.5	109.1	10.5
YOLOv5x	72.3	66.1	74.1	86.7	205.7	12.4
Improved YOLOv5n	62.9	60.1	65.7	2.7	6.5	5.5
Improved YOLOv5s	71.2	79.6	69.7	9.0	21.4	5.7
Improved YOLOv5m	72.8	77.4	69.5	25.7	55.6	8.0
Improved YOLOv5l	72.7	70.8	77.6	55.7	125.1	9.6
Improved YOLOv5x	74.5	75.6	70.2	101.8	234.9	11.8

network structure except the depth and width of the network. The experimental results on COCO2014 dataset are shown in Table 7.

It can be seen from the experimental results that on COCO2014 dataset, after the improvement of YOLOv5 models of different scales, although the number of parameters and calculation are increased, the detection accuracy and detection speed are significantly improved, which proves the effectiveness of our improvement.

4.6   Comparison Experiment of YOLO-SA and Mainstream Object Detection Methods

In order to verify the validity of the YOLO-SA model proposed in this paper, we conduct comparative experiments on COCO2014 and VOC2012 dataset with the current mainstream object detection algorithm models, such as YOLOv3, YOLOv4, YOLOv4-tiny, SSD, Faster R-CNN, YOLOX-s, YOLOv6s, YOLOv7s. The experimental results are shown in Tables 8 and 9.

It can be seen from the experimental results that on COCO2014 dataset, the mAP@0.5 of YOLO-SA model reaches 71.2%, the highest among the compared models. Precision and recall reach 79.6% and 69.7% respectively, the highest among the compared models. On VOC2012 dataset, the mAP@0.5 of YOLO-SA reaches 75.8%, the highest among the compared models. Precision and recall reach 77.6% and 67.1% respectively, the highest among the compared models. Params and GFLOPs reach 9.0M and 21.4 respectively. The operation speed of the model reaches 5.7 ms. Experimental results show that our proposed YOLO-SA model maintains the highest detection accuracy among the current mainstream object detection models, while maintaining light weight, which has certain advantages. Detection results of YOLO-SA on COCO2014 and VOC2012 dataset are shown in Fig. 6.

Table 8. Detection result of YOLO-SA and mainstream methods on COCO2014 dataset.

Model	mAP@0.5	Precision	Recall	Params (M)	FLOPs	Speed(ms)
YOLOv3	47.8	59.1	51.6	62.0	147.0	23.2
YOLOv4	54.4	66.7	57.8	62.1	128.4	17.8
YOLOv4-tiny	58.6	65.7	60.3	6.1	3.4	3.7
SSD	41.2	58.6	47.9	50.4	114.2	10.7
Faster R-CNN	49.3	69.4	54.2	67.9	147.2	12.5
YOLOX-s	61.2	72.6	58.7	8.9	26.6	7.2
YOLOv5s	65.9	70.6	60.3	7.2	16.5	6.5
YOLOv6s	69.7	77.4	66.9	17.2	44.1	6.8
YOLOv7s	67.4	72.5	64.6	36.5	103.5	6.0
YOLO-SA(ours)	**71.2**	**79.6**	**69.7**	9.0	21.4	5.7

Table 9. Detection result of YOLO-SA and mainstream methods on VOC2012 dataset.

Model	mAP@0.5	Precision	Recall	Params (M)	FLOPs	Speed(ms)
YOLOv3	51.7	60.5	56.6	62.0	147.0	23.2
YOLOv4	57.4	70.1	55.8	62.1	128.4	17.8
YOLOv4-tiny	57.7	69.5	59.8	6.1	3.4	3.7
SSD	43.4	59.7	48.8	50.4	114.2	10.7
Faster R-CNN	50.6	71.5	56.6	67.9	147.2	12.5
YOLOX-s	70.2	77.6	68.7	8.9	26.6	7.2
YOLOv5s	68.7	72.6	65.3	7.2	16.5	6.5
YOLOv6s	70.9	78.4	67.6	17.2	44.1	6.8
YOLOv7s	69.6	75.5	70.1	36.5	103.5	6.0
YOLO-SA(ours)	**75.8**	**77.6**	**67.1**	9.0	21.4	5.7

Fig. 6. Detection results of YOLO-SA on COCO2014 and VOC2012 dataset.

5   Conclusion

In this paper, we add some cutting-edge techniques to the current popular object detector YOLOv5 model. We introduce a new RepVGG reparameterized module, which simplifies the traditional CNN module while maintaining high performance. We introduce a new self-attention mechanism in the feature fusion part of the model, which is independent of other convolution layers to form a separate self-attention visual model. We also replace the traditional coupled head in the YOLOv5 model with the decoupled head of anchor-free. Comparative experiments show that the detection accuracy of the YOLO-SA model proposed in this paper on COCO2014 and VOC2012 dataset is higher than that of the original YOLOv5s model, reaching 71.2% and 75.8% respectively, while maintaining good real-time performance, which also has certain advantages compared with mainstream object detection models.

References

1. LeCun, Y., Bottou, L., Bengio, Y., Haffner, P.: Gradient-based learning applied to document recognition. Proc. IEEE **86**, 2278–2324 (1998)
2. Deng, J., Dong, W., Socher, R., Li, L.J., Li, K., Fei-Fei, L.: ImageNet: a large-scale hierarchical image database. In: Computer Vision and Pattern Recognition (2009)
3. Krizhevsky, A., Sutskever, I., Hinton, G.E.: ImageNet classification with deep convolutional neural networks. In: Neural Information Processing Systems (2012)
4. Zeiler, M.D., Fergus, R.: Visualizing and understanding convolutional networks. In: Fleet, D., Pajdla, T., Schiele, B., Tuytelaars, T. (eds.) ECCV 2014. LNCS, vol. 8689, pp. 818–833. Springer, Cham (2014). https://doi.org/10.1007/978-3-319-10590-1_53
5. Simonyan: Very deep convolutional networks for large-scale image recognition. (No Title) (2015)
6. Szegedy, C., et al.: Going deeper with convolutions (2023)
7. He, K., Zhang, X., Ren, S., Sun, J.: Deep residual learning for image recognition. Cornell University - arXiv (2015)
8. Huang, G., Liu, Z., van der Maaten, L., Weinberger, K.Q.: Densely connected convolutional networks. Computer Vision and Pattern Recognition. arxiv (2016)
9. Girshick, R., Donahue, J., Darrell, T., Malik, J.: Rich feature hierarchies for accurate object detection and semantic segmentation (2014)
10. Girshick, R.: Fast R-CNN. Computer Vision and Pattern Recognition. arxiv (2015)
11. Ren, S., He, K., Girshick, R., Sun, J.: Faster R-CNN: towards real-time object detection with region proposal networks. Cornell University - arXiv (2015)
12. Redmon, J., Divvala, S.K., Girshick, R., Farhadi, A.: You only look once: unified, real-time object detection. In: Computer Vision and Pattern Recognition (2016)
13. Redmon, J., Farhadi, A.: Yolo9000: better, faster, stronger. In: Computer Vision and Pattern Recognition (2017)
14. Redmon, J., Farhadi, A.: Yolov3: an incremental improvement. Computer Vision and Pattern Recognition. arxiv (2018)
15. Bochkovskiy, A., Wang, C.Y., Liao, H.Y.M.: Yolov4: optimal speed and accuracy of object detection. Computer Vision and Pattern Recognition. arxiv (2020)

16. Liu, W., et al.: SSD: single shot multibox detector. In: Leibe, B., Matas, J., Sebe, N., Welling, M. (eds.) ECCV 2016. LNCS, vol. 9905, pp. 21–37. Springer, Cham (2016). https://doi.org/10.1007/978-3-319-46448-0_2
17. Ding, X., Zhang, X., Ma, N., Han, J., Ding, G., Sun, J.: RepVGG: making VGG-style convnets great again. In: Computer Vision and Pattern Recognition (2021)
18. Ramachandran, P., Parmar, N., Vaswani, A., Bello, I., Levskaya, A., Shlens, J.: Stand-alone self-attention in vision models. arxiv (2019)
19. Ge, Z., Liu, S., Wang, F., Li, Z., Sun, J.: Yolox: exceeding yolo series in 2021 (2021)
20. Hu, J., Shen, L., Albanie, S., Sun, G., Wu, E.: Squeeze-and-excitation networks. In: IEEE Transactions on Pattern Analysis and Machine Intelligence (2018)
21. Hou, Q., Zhou, D., Feng, J.: Coordinate attention for efficient mobile network design. Cornell University - arXiv (2021)
22. Woo, S., Park, J., Lee, J.Y., Kweon, I.S.: CBAM: convolutional block attention module. Computer Vision and Pattern Recognition. arxiv (2018)
23. Law, H., Deng, J.: CornerNet: detecting objects as paired keypoints. Computer Vision and Pattern Recognition. arxiv (2018)
24. Tian, Z., Shen, C., Chen, H., He, T.: FCOS: fully convolutional one-stage object detection. Cornell University - arXiv (2019)
25. Zhou, X., Wang, D., Krähenbühl, P.: Objects as points. Computer Vision and Pattern Recognition. arxiv (2019)
26. Song, G., Liu, Y., Wang, X.: Revisiting the sibling head in object detector. Cornell University - arXiv (2020)
27. Wu, Y., et al.: Rethinking classification and localization for object detection. Cornell University - arXiv (2020)
28. Lin, T.Y., Goyal, P., Girshick, R., He, K., Dollár, P.: Focal loss for dense object detection. Cornell University - arXiv (2017)
29. Xie, S., Girshick, R., Dollár, P., Tu, Z., He, K.: Aggregated residual transformations for deep neural networks. Cornell University - arXiv (2016)
30. Tan, M., Le, Q.V.: EfficientNet: rethinking model scaling for convolutional neural networks (2019)
31. Wang, Q., Wu, B., Zhu, P., Li, P., Zuo, W., Hu, Q.: ECA-Net: efficient channel attention for deep convolutional neural networks. Computer Vision and Pattern Recognition. arxiv (2019)

Retrieval-Enhanced Event Temporal Relation Extraction by Prompt Tuning

Rong Luo[1,2,3] and Po Hu[1,2,3]($\boxtimes$)

[1] Hubei Provincial Key Laboratory of Artificial Intelligence and Smart Learning,
Central China Normal University, Wuhan, Hubei, China
`rl@mails.ccnu.edu.cn, phu@mail.ccnu.edu.cn`
[2] School of Computer Science, Central China Normal University,
Wuhan, Hubei, China
[3] National Language Resources Monitoring and Research Center for Network Media,
Central China Normal University, Wuhan, Hubei, China

Abstract. Event temporal relation extraction aims to automatically identify the temporal order between a pair of events, which is an essential step towards event-oriented natural language understanding and generation. For this task, impressive improvements have been made in neural network-based approaches. However, they typically treat it as a supervised classification task and inevitably suffer from under-annotated data and label imbalance problems. In this paper, we propose a new retrieval-enhanced event temporal relation extraction model, called PRetrieval, which addresses these issues by making full use of the golden relation labels to learn potential temporal knowledge in a pre-trained language model and to perform prompt tuning. Specifically, PRetrieval first generates an initial prediction using the prompt-based input and constructs a datastore using this prediction result and the associated golden labels, which has the advantage of not relying on any external datasets or additional knowledge enhancement. PRetrieval then retrieves examples with outputs similar to the preliminary prediction results as the new prompt to improve the model's relation extraction ability under few-shot settings. In addition, it also retrieves examples with similar input to further enhance the model performance. The experimental results on two benchmark datasets (i.e., MATRES and TB-Dense) show that our proposed approach significantly outperforms the state-of-the-art methods.

Keywords: Prompt Tuning · Retrieval Enhanced Model · Event Temporal Relation Extraction

1 Introduction

Event temporal relation extraction (ETRE) is an important task in natural language processing, which aims to identify and classify the temporal relationships among event mentions within a given text. Figure 1 shows a specific example sentence and the temporal relationships within it. There are four events in this sentence, i.e., "**allows**", "**say**", "**wanted**" and "**said**". A good ETRE model is expected to extract all pair-wise temporal relations in it by understanding the semantics of the sentence, which can benefit many downstream tasks such as text summarization [20], question answering [15], reading comprehension [22] and so on.

"It **allows** us to spread his name far and wide across Canada and **say** that he is **wanted** for attempted murder , " **said** police Inspector Dave Bowen .

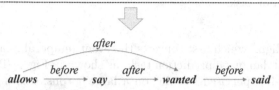

Fig. 1. An example sentence. All pair-wise temporal relations within it are "*allows before say*", "*allows after wanted*", "*say after wanted*" and "*wanted before said*".

The traditional feature-based machine learning methods [5,17] and the recent deep learning-based methods [31,32] mostly regard ETRE as a classification task, using supervised learning to train models on annotated data. However, existing ETRE datasets tend to be small because annotating events and their temporal relationships from texts are time-consuming. Thus, although existing methods have made significant progress, they are still challenged by data scarcity and label imbalance problems due to limited training data. In addition, the label imbalance problem will also cause the ETRE model to perform much worse than the others in few-shot settings.

To alleviate these problems, various methods have been proposed to leverage either the augmented dataset [1,3] or external knowledge [28,32]. However, indexing a large external knowledge base and performing information retrieval can often lead to a substantial computational burden. The introduction of external knowledge or the generation of pseudo data via data augmentation may also introduce new noise into the model. Thanks to the rapid development of pre-trained language models (PLMs) such as BERT [8] and RoBERTa [19], prompt-based methods have been proposed and demonstrated promising performance on many NLP tasks. Existing work [24,29] has shown that retrieving relevant training samples (i.e. examples) and appending the retrieved input-output pairs as prompts to the sample inputs can improve model performance in low-resource settings.

Inspired by such work, in this paper, we propose a novel approach for extracting temporal relationships between events based on prompt retrieval enhancement. Given a test instance, a group of similar examples from the training set are selected for prompt tuning.

Specifically, we propose a new prompt tuning method with retrieval enhancement named PRetrieval, which can identify different temporal relationships between events without introducing any external knowledge or data. We adopt a

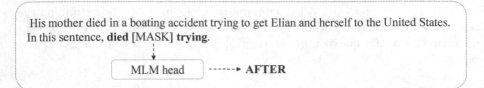

Fig. 2. An example of converting the event temporal relation extraction task into a mask language prediction task.

prompt-based paradigm, which first converts the event temporal relation extraction task into a mask language prediction task, as shown in Fig. 2. Through pre-training based on self-supervised learning on a large corpus, existing PLMs are often already able to learn a large amount of knowledge, including knowledge about the temporal relationships between events. The prompt-based learning paradigm, on the other hand, facilitates the ability of PLMs to identify the temporal relationships of events in few-shot settings and significantly reduces the need for large amounts of labeled data for existing models to perform this task. Furthermore, considering that the text with the most similar distribution to the test data is often the training data, we propose to retrieve some examples with similar inputs and outputs directly from the training set as prompts so that the semantic information of the golden temporal relational labels (i.e., the real temporal relation provided in the dataset) can be more fully utilized.

More specifically, to retrieve examples with similar output, we first use the prompt-based prediction for each instance in the training set to construct a datastore. Then, during inference, we retrieve Top-k nearest instances as prompt examples from the datastore. Since a specific pattern of temporal relationships may exist between similar event pairs, to retrieve examples with similar input, we retrieve relevant examples based on the content similarity of the event pairs. Citing similar instances from the training set and treating them as evidence allows to memorize rare patterns explicitly and not implicitly in the model parameters, which can make our model focus on few-shot relationships that are difficult to train under traditional settings, thus improving the robustness of our model when it encounters label imbalance problems. Extensive experimental results on two widely used benchmark datasets have indicated that PRetrieval significantly outperforms previous state-of-the-art methods.

Our contributions are summarized as follows:

- To make better use of the temporal knowledge in the PLMs, we propose a novel prompt tuning method with retrieval enhancement strategy that effectively alleviates label imbalance and data scarcity problems. As far as we know, this is the first time to use a prompt-based method for ETRE.

- We retrieve relevant examples from two dimensions, both examples with similar input and examples with similar output from the training set, to yield improvements for our model's performance on few-shot relations.
- The experimental results on two datasets further verify the superiority of our method.

2   Related Work

Event Temporal Relation Extraction. Event Temporal Relation Extraction (ETRE) is a crucial information extraction task. Most of the earlier studies on ETRE [2,16,27] were based on feature classifiers that relied on hand-crafted features or rules to extract temporal relationships. Recently, methods based on PLMs have also made great progress [1,7,9,12].

Despite the advanced results achieved by these models, they still encounter problems with data scarcity and label imbalance. Several studies have attempted to eliminate these issues through data augmentation. Ballesteros et al. (2020) [1] utilize complementary datasets for multi-task learning to supplement both temporal and non-temporal information in the model. Cao et al. (2021) [3] employ a semi-supervised learning approach to ETRE and used the latest advancement in Bayesian deep learning to measure the model uncertainty. Although data augmentation can mitigate the lack of data to some extent, the pseudo-data generated by self-training will also inevitably introduce new noise into the model.

Similarly, to alleviate the limitation of small-scale imbalanced training data, researchers have proposed new methods to improve ETRE by using external knowledge resources. Wang et al. (2020) [28] suggest using joint constraint learning and incorporate event background knowledge. Zhang et al. (2022) [32] integrate temporal commonsense knowledge, data enhancement and focus loss functions to enhance the model. However, indexing and retrieving large-scale external knowledge bases incurs a considerable computational cost, and may also introduce noise.

Retrieval-Enhanced Methods. Recently, with the development of pre-trained language models like BERT [8] and RoBERTa [19], retrieval-enhanced language models have been applied to various tasks, such as machine translation [14], text classification [18], semantic analysis [25], question answering [13], and relation extraction [6]. A retrieval-enhanced language model is a non-parametric approach that employs training examples to enhance the predictions of the language model during the inference stage. Chen et al. (2022) [6] have developed a semi-parametric paradigm for relation extraction, which involves referencing relations by interpolating the base output of pre-trained language model (PLM) with the non-parametric example distribution in a linear fashion.

In contrast, we treat the retrieval examples as prompts to fully utilize the golden temporal relation label information. We employ a mask language prediction paradigm to effectively elicit potential temporal knowledge in PLM. To the best of our knowledge, ours is the first retrieval-enhanced prompt learning approach for ETRE.

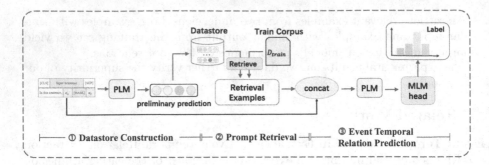

Fig. 3. The architecture of our proposed PRetrieval.

3 Methodology

3.1 Prompt Tuning for ETRE

The prompt tuning method was proposed to bridge the significant gap between pre-training objectives and fine-tuning objectives. This method converts the downstream task into the same form as the pre-training task. In this study, we denote $\mathcal{M}$, $\mathcal{T}$ and $\mathcal{V}$ as the PLM, prompt template and label words, respectively. We also define a verbalizer function $f : \mathcal{R} \longrightarrow \mathcal{V}$, which maps the relational label space to the label word space.

Regarding ETRE with prompt tuning, for each instance $x = (S, (e_s, e_t))$, where S represents an input sentence and (e_s, e_t) is an event pair in S. The prompt template $\mathcal{T}$ map x to x_{prompt} by adding some additional words to x. Concretely, in addition to retaining the original tokens in x, one or more [MASK] is placed into x_{prompt}. $\mathcal{M}$ then processes x_{prompt} to generate the hidden vector, $h_{[mask]}$, at the position of the [MASK] token. The probability distribution $P(y|x)$ over the relation set is calculated as:

$$P(y|x) = P_{MLM\_head}([MASK] = \mathcal{M}(v|T(x))) \tag{1}$$

where $v \in \mathcal{V}$ and P_{MLM_head} represents the probability distribution predicted by MLM head layer. The ETRE task can be converted to a masking language modeling problem by populating the input with [MASK] tokens. To simplify the process, we utilize a soft verbalizer that maps the relation label to the label word via a one-to-one connection.

3.2 Model Overview

The overall architecture of our model is shown in Fig. 3, which consists of three major components: (1) *Datastore Construction* (Sect. 3.3), which provides prompt-based preliminary predictions and constructs a datastore containing preliminary predictions and corresponding golden temporal relation labels as key-value pairs. (2) *Prompt Retrieval* (Sect. 3.4), which retrieves the examples related to the given test instance x from two perspectives. (3) *Event Temporal*

Relation Prediction (Sect. 3.5), which augments the input x with the retrieved examples. The augmented input sequence, x_{aug}, is encoded using the PLM before utilizing its MLM head to predict the relation.

3.3 Datastore Construction

As the true output of a sample is usually unknown, we convert the ETRE task into a mask language prediction task. This involves considering the embedding of the "[MASK]" token as the initial prediction for retrieving examples with a similar output. In this subsection, we will describe how to construct the datastore, and the detailed construction process is shown in Fig. 4.

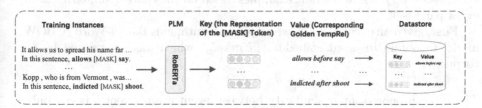

Fig. 4. Datastore Construction

Our datastore is composed of a set of key-value pairs. To encode prompt-based instances, we use a PLM and represent the entire instance using the embedding of the "[MASK]" token. Utilizing the embedding of the "[MASK]" token enables us to leverage the pertinent knowledge in the language model for our task. To be specific, we establish the prompt template like this:

$\mathcal{T}_1(\cdot)$: In this sentence, e_s *[MASK]* e_t.

where e_s and e_t are the representations of the source event and target event in the given text, respectively. For each instance in the training set q, we splice the prompt template $\mathcal{T}_1$ after the sentence and feed it into the RoBERTa-based encoder $\mathcal{M}_1(\cdot)$ to obtain the hidden features of "[MASK]" tokens h_q as the preliminary prediction,

$$h_q = \mathcal{M}_1(\mathcal{T}_1(q) + q)_{[MASK]} \qquad (2)$$

where '+' means sequence splicing. We use h_q as the key of instance q in the datastore, and the value is the corresponding golden event relational triple r_i. The complete datastore is defined as follows:

$$(\mathcal{K}, \mathcal{V}) = \{h_{q_i}, r_i | (q_i, r_i) \in Train\} \qquad (3)$$

where $Train$ denotes the training set, r_i represents the event relational triple $(e_s, TempRel, e_t)$ in the i-th instance in the training set.

3.4 Prompt Retrieval

In order to leverage the golden temporal relation label information to stimulate potential temporal knowledge in PLMs, we retrieve novel prompts from the training data by selecting examples with similar input and output.

Retrieve Examples with Similar Input. The conventional method usually involves creating a similarity measure that takes into account the similarity between the vector representations of two instances. These vector representations are typically provided by the hidden features of the [CLS] token in the output of a PLM, which contains the semantic information of the entire sentence. However, for the task of ETRE, a sentence usually contains multiple event pairs, and the relation between each event pair is not necessarily the same, as shown in Fig. 1. Hence if the semantic information of the whole sentence is considered when measuring similarity, it is likely that other temporal relations in the same sentence will be retrieved. Considering that the existing approaches may introduce more noise to our model and a specific pattern of relations may exist between similar event pairs, we retrieve relevant examples based on the content similarity of the event pair itself.

First, given an event word w, we employ a Continuous Bag-of-Word (CBOW) model to obtain its word embedding $Embed_w$, which was pre-trained by our corpus,

$$Embed_w = CBOW(w) \tag{4}$$

Then, the proposed input similarity between event a and event b is computed by the cosine similarity between the $Embed_a$ and $Embed_b$,

$$cos(a, b) = \frac{Embed_a \cdot Embed_b}{|Embed_a| \times |Embed_b|} \tag{5}$$

We improve on the input-similarity baseline by creating a hybrid similarity measure that considers not only the similarity between the word embeddings of the source events but also that of the target events. Finally, PRetrieval retrieves Top-n most similar event pairs to the event pair in the given test instance x, based on the hybrid similarity measure and uses the corresponding event relational triples, consisting of event pairs and the golden temporal relations, as examples e_{input} with similar input.

Retrieve Examples with Similar Output. Given a test instance x, we first append the prompt template $T_1(x)$ to x, and obtain its representation h_x in the preliminary prediction by Eq. 2. Then, PRetrieval retrieves the k-nearest neighbours K of h_x by querying the datastore (K, V) constructed in Sect. 3.3 according to the Euclidean distance function $d(\cdot, \cdot)$:

$$d(h_{x_i}, h_{x_j}) = \sqrt{\sum_{l=1}^{m} \left| h_{x_i}^{(l)} - h_{x_j}^{(l)} \right|^2} \tag{6}$$

where h_{x_i} and h_{x_j} are two instances' representations, and m denotes the dimension of h_x. We take the k retrieved examples as the examples e_{output} with similar output of x.

3.5 Event Temporal Relation Prediction

Converting the ETRE task into a mask language prediction task enables us to utilize contextual information more effectively to predict the temporal relation between the pairs of events. After retrieving the examples with similar input as well as output, we can obtain an augmented input of each sample. Specifically, assume that $(e_{s_j}, relation_j, e_{t_j})$ represents the retrieved examples in e_{input} and e_{output}, We design the prompt template like this:

$T_2(\cdot)$: *According to* $(e_{s_1}, relation_1, e_{t_1})$ ... $(e_{s_{k+n}}, relation_{k+n}, e_{t_{k+n}})$, *we can know that* e_s *[MASK]* e_t. *[SEP]*.

For each sample x, we concatenate $T_2(x)$ and x together to get the augmented sample x_{aug}, which will be fed into the PLM RoBERTa. The MLM head of RoBERTa is used to predict the temporal relation between the event pair. The probability of its temporal label $p(y|x_{aug})$ is:

$$p(y|x_{aug}) = P_{MLM\_head}([MASK] = RoBERTa(y|x_{aug})) \qquad (7)$$

where P_{MLM_head} represents the probability distribution predicted by the MLM head layer. During training, we use cross-entropy objective for temporal relation prediction, as shown in Eq. 8.

$$\mathcal{L} = -log(p(y|x_{aug})) \qquad (8)$$

4   Experiments

4.1   Dataset and Metrics

We conducted experiments on two public benchmark datasets:

1) **TB-Dense** [4] is a densely annotated dataset, which contains 36 documents from TimeBank that have been re-annotated by a dense event ordering framework. TB-Dense is labeled with six types of relations: *BEFORE, AFTER, SIMULTANEOUS, INCLUDES, IS_INCLUDED,* and *VAGUE*.
2) **MATRES** [23] dataset, which consists of 3 subsections: TimeBank (TB), AQUAINT (AQ) and Platinum (PT), is labeled with four types of temporal relationships: *BEFORE, AFTER, EQUAL,* and *VAGUE*. For MATRES dataset, we followed the previous work [21], using TimeBank and Aquaint for training and Platinum as the test set. Table 1 shows the detailed data distribution.

Following previous work [3,30], we evaluate performance based on Precision (P), Recall (R), and F1-score (F1). For the TB-Dense dataset, we regard all classes as positive classes. Thus, it has the same precision, recall and F1 score on this dataset. For the MATRES dataset, since we seen *VAGUE* as negative class, the precision, recall and F1 score are different.

4.2   Experimental Settings

To demonstrate the effectiveness of our proposed model, we compared PRetrieval with various representative baselines that also use RoBERTa as their pre-trained language model, including:

Table 1. Data statistics for MATRES and TB-Dense.

	MATRES		TB-Dense	
	Docs	Relations	Docs	Relations
Train	260	10,888	22	4,032
Dev	21	1,852	5	629
Test	20	840	9	1,427

Data augmentation based methods (1) **KJETE** [32] leverages both data enhancement and temporal commonsense knowledge enhancement to extract event temporal relation. (2) **UAST** [3] employs semi-supervised learning and counteracts pseudo-labeling errors by quantifying the uncertainty of the model.

External knowledge enhanced methods (1) **HGRU+knowledge** [26] maps events into a hyperbolic space via incorporating event contextual knowledge. (2) **Joint Constrained Learning** [28] extracts temporal and hierarchical relationships through joint constraint learning. (3) **Domain-K** [11] utilizes TEMPROB, which contains the knowledge of probability.

Other recent representative methods (1) **ECONET** [10] equips PLMs with knowledge about event temporal relations. (2) **Syntax-T** [31] captures temporal cues between events through a syntactically guided attention mechanism.

In addition, we also compared with **RoBERTa-base**, which uses a linear classifier after RoBERTa for ETRE, the input of the classifier is the hidden feature of the source event and target event. Since some studies used only F1-score as an evaluation metric, there are some Precision and Recall rates with "–" instead.

Table 2. Temporal relation extraction results on MATRES dataset (%). Here * denotes the results are statistically significant ($p < 0.05$).

Model	P	R	F1
RoBERTa-base	77.3	79.0	78.9
Joint Constrained Learning (Wang et al., 2020)	73.4	85.0	78.8
ECONET (Han et al., 2021)	–	–	80.3
HGRU+knowledge (Tan et al., 2021)	79.2	81.7	80.5
UAST (Cao et al., 2021)	76.6	84.9	80.5
Syntax-T (Zhang et al., 2022a)	–	–	80.3
KJETE (Zhang et al., 2022b)	–	–	76.6
PRetrieval (ours)	77.2	**85.9**	**81.4***

Table 3. Temporal relation extraction results on TB-Dense dataset (%). Here * denotes the results are statistically significant ($p < 0.05$).

Model	P	R	F1
RoBERTa-base	59.4	59.4	59.4
ECONET (Han et al., 2021)	–	–	66.8
Domain-K (Han et al., 2021)	–	–	50.5
UAST (Cao et al., 2021)	64.3	64.3	64.3
Syntax-T (Zhang et al., 2022a)	–	–	67.1
KJETE (Zhang et al., 2022b)	–	–	65.6
PRetrieval (ours)	**67.6**	**67.6**	**67.6***

4.3   Main Results

The overall experimental results are shown in Table 2 and Table 3. Based on the results, it is evident that our proposed PRetrieval outperforms all baselines. PRetrieval shows a significant improvement in the F1 score on the MATRES dataset, increasing it from 80.5% to 81.4% compared to the best baseline. Similarly, PRetrieval achieves a 0.5% absolute improvement in the F1 score for the TB-Dense dataset.

It is worth noting that PRetrieval largely outperforms the method enhanced by external knowledge and data. This demonstrates that adequately stimulating temporal knowledge in a PLM is more beneficial for the model to understand temporal relations in sentences than introducing an external temporal common-sense knowledge base or expanding original labeled data with pseudo-labeled data. The reason for this may be that neither the knowledge base nor pseudo-labeled data cover "include" or "simultaneous" relations, which are difficult to obtain for these methods. As a result, they often struggle to deal with class imbalance.

In short, the prompt tuning method with retrieval enhancement can more effectively mine temporal knowledge in PLMs, so as to improve the performance of the ETRE model. The extensive experiments also demonstrate the superiority of our model.

4.4   Ablation Study

To assess the effect of prompt retrieval, we conduct several ablation studies, which further illustrate the importance of the examples we retrieved. We designed three variants of the model to verify the effectiveness of our model: **RoBERTa_mlm** represents a prompt-based baseline, we use a basic prompt $T_1(\cdot)$ mentioned in Sect. 3.3. **PRetrieval w/o similar_output** only use the examples with similar input and **PRetrieval w/o similar_input** only use the examples with similar output.

The results of the ablation studies are shown in Table 4. From the results, we can observe that adding examples with similar input to the prompt helps improve the performance, and adding examples with similar output further

Table 4. Experimental results of ablation study on MATRES and TB-Dense datasets (%).

	MATRES			TB-Dense		
	P	R	F1	P	R	F1
RoBERTa_mlm	76.0	84.9	80.2	65.2	65.2	65.2
PRetrieval w/o similar_output	77.2	84.2	80.5	66.6	66.6	66.6
PRetrieval w/o similar_input	76.2	86.6	81.1	66.9	66.9	66.9
PRetrieval	**77.2**	85.9	**81.4**	**67.6**	**67.6**	**67.6**

boosts the performance. The explanation is that the retrieval examples can effectively exploit the potential temporal knowledge in the PLM to better guide the model to optimize the weights.

4.5 Single Temporal Relation Performance

From Fig. 5 we can see a serious label imbalance in the corpus for the ETRE task. For example, the TB-Dense dataset is dominated by the *BEFORE*, *AFTER* and *VAGUE* labels, while the *SIMULTANEOUS*, *INCLUDES* and *IS_INCLUDED* labels are severely underrepresented. As a result, the model struggles to learn temporal relationships involving these minority classes during training.

From Table 5, we can find that our model PRetrieval outperforms KJETE, a method that is both enhanced by knowledge and data, on all label relations. Of particular interest is the fact that PRetrieval breaks through the dilemma of previous studies in the identification of *SIMULTANEOUS* relations and gets an F1 score of 20.7%. According to our observations, the improvement on the *BEFORE*, *AFTER* and *VAGUE* relations are due to the conversion of the ETRE task into a mask language prediction task. In addition, the improvement on the *INCLUDES*, *IS_INCLUDED* and *SIMULTANEOUS* relations are mainly attributable to the retrieved examples, which provides more sufficient evidence that PRetrieval can effectively mitigate the class imbalance problem.

Table 5. Model performance breakdown on TB-Dense dataset (%). Label abbreviations are explained below *BEFORE*(B), *AFTER*(A), *INCLUDES*(I), *IS INCLUDED*(II), *SIMULTANEOUS*(S), *VAGUE*(V).

Model	KJETE			PRetrieval		
	P	R	F1	P	R	F1
B	79.9	58	67.2	73.2	**72.9**	**73.1**
A	70.1	66.7	68.4	**80.0**	59.8	**68.5**
I	38.9	12.5	18.9	**50.0**	**35.7**	**41.7**
II	46.7	26.4	33.7	40.4	**32.0**	**35.8**
S	–	–	–	**42.8**	**13.6**	**20.7**
V	61.4	78.4	68.9	**63.8**	75.0	**69.0**

Table 6. Performance comparison with various amounts of labeled data on the TB-Dense and MATRES datasets (%).

	TB-Dense						MATRES					
	30%			40%			30%			40%		
	P	R	F1	P	R	F1	P	R	F1	P	R	F1
RoBERTa_mlm	60.4	60.4	60.4	64.1	64.1	64.1	75.1	78.4	76.7	78.6	77.4	78.0
UAST	58.2	58.2	58.2	60.8	60.8	60.8	74.4	76.2	75.3	73.7	79.1	76.3
PRetrieval	**66.8**	**66.8**	**66.8**	**67.1**	**67.1**	**67.1**	**76.4**	**81.3**	**78.8**	**79.4**	**79.8**	**79.6**

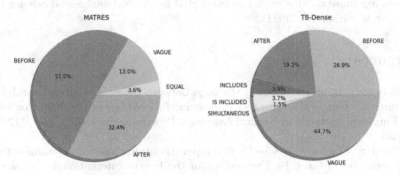

Fig. 5. Statistical analysis of labels on test sets.

4.6 Effectiveness in Low-Resource Scenarios

We further study the effectiveness of PRetrieval in low resource scenarios by utilizing limited labeled data as UAST [30] have done. For two datasets, we randomly sample 30% and 40% training data to train our model. We compare PRetrieval with **RoBERTa_mlm** and a representative data augmentation approach **UAST**, which sample an additional 50% training data as unlabeled set.

Table 6 shows the experimental results on MATRES and TB-Dense. It can be observed that RoBERTa_mlm outperforms UAST, which illustrates the superiority of prompt-based paradigm and indicates that prompt-based methods can elicit the potential of PLMs to solve ETRE better than data augmentation. Comparing PRetrieval and RoBERTa_mlm, we notice that PRetrieval is significantly better than RoBERTa_mlm. It proves that prompt retrieval can obtain more useful temporal knowledge from PLMs.

5 Conclusion

In this paper, we introduce a new retrieval-enhanced prompt tuning method. This method enables PLMs to reference examples with similar input and output from the training set. It augments the test input with the retrieved examples to generate a final prediction that better stimulates the potential for temporal reasoning in PLMs. Extensive experimental results have demonstrated that our proposed approach achieves state-of-the-art performance and the retrieval-enhanced prompt tuning method is very effective. We also demonstrate that using retrieval examples yields the model's improvements on few-shot relations.

In the future, we intend to improve our work from several perspectives. Firstly, we aim to enhance our strategy for retrieving relevant examples. Secondly, we plan to develop a more effective prompt template for integrating information from the retrieved examples. Thirdly, we will explore the effectiveness of the prompt retrieval method in other tasks.

Acknowledgements. This work was supported by the National Social Science Fund of China under Grant No. 20BTQ068.

References

1. Ballesteros, M., et al.: Severing the edge between before and after: neural architectures for temporal ordering of events. In: Proceedings of the 2020 Conference on Empirical Methods in Natural Language Processing (EMNLP), pp. 5412–5417 (2020)
2. Bethard, S., Martin, J.H.: CU-TMP: temporal relation classification using syntactic and semantic features. In: Proceedings of the Fourth International Workshop on Semantic Evaluations (SemEval), pp. 129–132 (2007)
3. Cao, P., Zuo, X., Chen, Y., Liu, K., Zhao, J., Bi, W.: Uncertainty-aware self-training for semi-supervised event temporal relation extraction. In: Proceedings of the 30th ACM International Conference on Information & Knowledge Management (CIKM), pp. 2900–2904 (2021)
4. Cassidy, T., McDowell, B., Chambers, N., Bethard, S.: An annotation framework for dense event ordering. In: Proceedings of the 52nd Annual Meeting of the Association for Computational Linguistics (ACL), pp. 501–506 (2014)
5. Chambers, N., Cassidy, T., McDowell, B., Bethard, S.: Dense event ordering with a multi-pass architecture. Trans. Assoc. Comput. Linguist. (TACL) 2, 273–284 (2014)
6. Chen, X., et al.: Relation extraction as open-book examination: retrieval-enhanced prompt tuning. In: Proceedings of the 45th International ACM Conference on Research and Development in Information Retrieval (SIGIR), pp. 2443–2448 (2022)
7. Cheng, F., Asahara, M., Kobayashi, I., Kurohashi, S.: Dynamically updating event representations for temporal relation classification with multi-category learning. In: Findings of the Association for Computational Linguistics (EMNLP), pp. 1352–1357 (2020)
8. Devlin, J., Chang, M.W., Lee, K., Toutanova, K.: BERT: pre-training of deep bidirectional transformers for language understanding. In: Proceedings of the 2019 Conference of the North American Chapter of the Association for Computational Linguistics: Human Language Technologies (NAACL), pp. 4171–4186 (2019)
9. Han, R., Ning, Q., Peng, N.: Joint event and temporal relation extraction with shared representations and structured prediction. In: Proceedings of the 2019 Conference on Empirical Methods in Natural Language Processing and the 9th International Joint Conference on Natural Language Processing (EMNLP-IJCNLP), pp. 434–444 (2019)

10. Han, R., Ren, X., Peng, N.: ECONET: effective continual pretraining of language models for event temporal reasoning. In: Proceedings of the 2021 Conference on Empirical Methods in Natural Language Processing (EMNLP), pp. 5367–5380 (2021)
11. Han, R., Zhou, Y., Peng, N.: Domain knowledge empowered structured neural net for end-to-end event temporal relation extraction. In: Proceedings of the 2020 Conference on Empirical Methods in Natural Language Processing (EMNLP), pp. 5717–5729 (2020)
12. Hwang, E., Lee, J.Y., Yang, T., Patel, D., Zhang, D., McCallum, A.: Event-event relation extraction using probabilistic box embedding. In: Proceedings of the 60th Annual Meeting of the Association for Computational Linguistics (ACL), pp. 235–244 (2022)
13. Kassner, N., Schütze, H.: BERT-kNN: adding a KNN search component to pre-trained language models for better QA. In: Findings of the Association for Computational Linguistics (EMNLP), pp. 3424–3430 (2020)
14. Khandelwal, U., Levy, O., Jurafsky, D., Zettlemoyer, L., Lewis, M.: Generalization through memorization: nearest neighbor language models. In: Proceedings of International Conference on Learning Representations (ICLR) (2020)
15. Khashabi, D., Khot, T., Sabharwal, A., Roth, D.: Question answering as global reasoning over semantic abstractions. In: Proceedings of the AAAI Conference on Artificial Intelligence (AAAI), vol. 32 (2018)
16. Laokulrat, N., Miwa, M., Tsuruoka, Y., Chikayama, T.: UTTime: temporal relation classification using deep syntactic features. In: Proceedings of the Seventh International Workshop on Semantic Evaluation (SemEval), pp. 88–92 (2013)
17. Leeuwenberg, A., Moens, M.F.: Temporal information extraction by predicting relative time-lines. In: Proceedings of the 2018 Conference on Empirical Methods in Natural Language Processing (EMNLP), pp. 1237–1246 (2018)
18. Li, L., Song, D., Ma, R., Qiu, X., Huang, X.: KNN-BERT: fine-tuning pre-trained models with KNN classifier. arXiv preprint arXiv:2110.02523 (2021)
19. Liu, Y., et al.: RoBERTa: a robustly optimized BERT pretraining approach. arXiv preprint arXiv:1907.11692 (2019)
20. Ng, J.P., Chen, Y., Kan, M.Y., Li, Z.: Exploiting timelines to enhance multi-document summarization. In: Proceedings of the 52nd Annual Meeting of the Association for Computational Linguistics (ACL), pp. 923–933 (2014)
21. Ning, Q., Subramanian, S., Roth, D.: An improved neural baseline for temporal relation extraction. In: Proceedings of the 2019 Conference on Empirical Methods in Natural Language Processing and the 9th International Joint Conference on Natural Language Processing (EMNLP-IJCNLP), pp. 6203–6209 (2019)
22. Ning, Q., Wu, H., Han, R., Peng, N., Gardner, M., Roth, D.: TORQUE: a reading comprehension dataset of temporal ordering questions. In: Proceedings of the 2020 Conference on Empirical Methods in Natural Language Processing (EMNLP), pp. 1158–1172 (2020)
23. Ning, Q., Wu, H., Roth, D.: A multi-axis annotation scheme for event temporal relations. arXiv preprint arXiv:1804.07828 (2018)
24. Pasupat, P., Zhang, Y., Guu, K.: Controllable semantic parsing via retrieval augmentation. In: Proceedings of the 2021 Conference on Empirical Methods in Natural Language Processing (EMNLP), pp. 7683–7698 (2021)
25. Rubin, O., Herzig, J., Berant, J.: Learning to retrieve prompts for in-context learning. In: Proceedings of the 2022 Conference of the North American Chapter of the Association for Computational Linguistics (NAACL), pp. 2655–2671 (2022)

26. Tan, X., Pergola, G., He, Y.: Extracting event temporal relations via hyperbolic geometry. In: Proceedings of the 2021 Conference on Empirical Methods in Natural Language Processing (EMNLP), pp. 8065–8077 (2021)
27. Verhagen, M., Pustejovsky, J.: Temporal processing with the TARSQI toolkit. In: Proceedings of the 22nd International Conference on Computational Linguistics (COLING), pp. 189–192 (2008)
28. Wang, H., Chen, M., Zhang, H., Roth, D.: Joint constrained learning for event-event relation extraction. In: Proceedings of the 2020 Conference on Empirical Methods in Natural Language Processing (EMNLP), pp. 696–706 (2020)
29. Wang, S., et al.: Training data is more valuable than you think: a simple and effective method by retrieving from training data. In: Proceedings of the 60th Annual Meeting of the Association for Computational Linguistics (ACL), pp. 3170–3179 (2022)
30. Wen, H., Ji, H.: Utilizing relative event time to enhance event-event temporal relation extraction. In: Proceedings of the 2021 Conference on Empirical Methods in Natural Language Processing (EMNLP), pp. 10431–10437 (2021)
31. Zhang, S., Ning, Q., Huang, L.: Extracting temporal event relation with syntax-guided graph transformer. In: Findings of the Association for Computational Linguistics (NAACL), pp. 379–390 (2022)
32. Zhang, X., Zang, L., Cheng, P., Wang, Y., Hu, S.: A knowledge/data enhanced method for joint event and temporal relation extraction. In: ICASSP 2022 - 2022 IEEE International Conference on Acoustics, Speech and Signal Processing (ICASSP), pp. 6362–6366 (2022)

MICA: Multi-channel Representation Refinement Contrastive Learning for Graph Fraud Detection

Guifeng Wang[1]([✉])[iD], Disheng Tang[2][iD], Anatoli Shatsila[3], and Xuecang Zhang[1]

[1] Huawei Technologies Co., Ltd., Hangzhou, Zhejiang, China
wgf1109@mail.ustc.edu.cn, {wangguifeng4,zhangxuecang}@huawei.com
[2] School of Life Science, Tsinghua University, Beijing, China
dstang@mail.tsinghua.edu.cn
[3] Jagiellonian University, Kraków, Poland
anatoli.shatsila@student.uj.edu.pl

Abstract. Detecting fraudulent nodes from topological graphs is important in many real applications, such as financial fraud detection. This task is challenging due to both the class imbalance issue and the camouflaged behaviors of anomalous nodes. Recently, some graph contrastive learning (GCL) methods have been proposed to solve the above issue. However, local aggregation-based GNN encoders can not consider the long-distance nodes, leading to over-smoothing and false negative samples. Also, random perturbation data augmentation hinders separately considering camouflaged behaviors at the topological and feature levels. To address that, this paper proposes a novel contrastive learning architecture for enhancing the performance of graph fraud detection. Specifically, a context generator and a representation refinement module are embraced for mitigating the limitation of local aggregation in finding long-distance fraudsters, as well as the introduction of false negative samples in GCL. Further, a multi-channel fusion module is designed to adaptively defend against diverse camouflaged behaviors. The experimental results on real-world datasets show a significant performance improvement over baselines, which demonstrates its effectiveness.

Keywords: Graph Fraud Detection · Graph Contrastive Learning

1 Introduction

While enjoying the benefits of the surge in users under the convenience of the Internet, all walks of life have also incubated various fraudulent activities. For instance, fraudsters may create malicious accounts on payment platforms [18], spread rumors (e.g., by posting fake reviews) on e-commerce platforms [11] or make fictitious claims in the insurance industry [12]. Since many fraudulent activities are performed by multiple entities, graph-based fraud detection methods which are able to discover and incorporate structural patterns have naturally become the target of attention in both academia and industry.

© The Author(s), under exclusive license to Springer Nature Singapore Pte Ltd. 2024
X. Song et al. (Eds.): APWeb-WAIM 2023, LNCS 14334, pp. 31–46, 2024.
https://doi.org/10.1007/978-981-97-2421-5_3

The mainstream graph fraud detection methods are Graph Neural Networks (GNNs) based learning models, which have leveraged the power of message passing to learn node representations with the goal of identifying the fraudsters in the embedding space [17,30]. They work well when features and topology keep consistent with the labels [26]. However, this assumption is broken because the graph-based detectors have motivated perpetrators to artificially disturb networks to camouflage their activities. For example, fraudsters disguise themselves among a group of benign users by modifying features (e.g., camouflaged features) or adding connections (e.g., camouflaged links) to many normal users. Hence, the diverse camouflaged behaviors pose challenges to graph fraud detection.

In addition, to tackle the class imbalance issue, there are also some graph contrastive learning (GCL) attempts for fraud detection [27], which show a good performance and are becoming the trend in this area. However, there are three main limitations that severely hinder obtaining significant node representations and degrade the performance of fraud detection. First, most GCL methods employ GNNs in the form of local aggregation as encoders, which makes the representation of samples averaged or smooth in a local scope. Second, negative samples are typically randomly selected from distant nodes, and it is challenging for GNNs to capture long-distance yet similar nodes, resulting in false negative samples. Third, random perturbation way commonly used in data augmentation may lose some useful structure or attribute information, which hinders the analysis of diverse camouflaged behaviors.

Motivated by the above gaps, the purpose of this paper is to alleviate the above issues thereby improving the performance of fraud detection. To address that, based on a general contrastive learning framework, a **M**ulti-channel representation ref**I**nement **C**ontrastive learning method for fr**A**ud detection (**MICA** for abbreviation) is innovatively proposed. Specifically, we first embrace a context generator for capturing global information, hoping to learn similar camouflage patterns of fraudsters even in a long-range situation. Based on it, an augmentation-agnostic representation refinement module works on representation space, for the purpose of reducing false negative samples and mitigating over-smoothing. Then, considering that the random perturbation way may disturb the analysis of where the camouflaged behaviors come from, an adaptive and fine-grained multi-channel fusion module is designed to defend against diverse attacks. The contributions can be summarized as follows:

- To our best knowledge, this is the first paper to simultaneously explore the problems of over-smoothing, false negative samples, and the conflicts of random perturbation data augmentation under camouflage behaviors in the GCL framework for graph fraud detection.
- We proposed a general MICA model which solved the above limitations of existing GCL solutions, the adequate experiments show that our method achieves significant improvements for the fraud detection tasks.

2   Related Work

2.1   GNN-Based Fraud Detection

There are some survey works about graph anomaly or fraud detection recently [19], the key challenges referring to fraud detection include the existence of camouflage behaviors, the class imbalance issue, and data scarcity. The camouflage behavior makes the graph vulnerable to topology and features, resulting in smoother node representations obtained by GNNs, and thus fraudsters are indistinguishable from normal nodes. To solve that, some previous studies have designed some strategies such as neighbor filtering [13], adversarial learning [28] or active generative learning [7] to achieve robustness and generalization in the presence of fraudsters. As for the class imbalance issue, some undersampling [5] and data augmentation [14] methods are employed. In addition, Some recent approaches like active learning [21], meta learning [4], and data augmentation [31] are proposed to solve the data scarcity problem. Although these methods have explored fraud detection from varied perspectives and achieved effective results, few of them consider these issues simultaneously.

2.2   Graph Contrastive Learning Way

Due to the advantages of graph contrastive learning (GCL) models in various fields, some methods have also been utilized to learn node representation for fraud/anomaly detection. Such methods generally assume that abnormal users can be distinguished by structural patterns. However, there is a problem in that the structural patterns and the label semantics are not consistent, which is a vital factor that impacts the representation learning found by DCI [27] method. Hence, the DCI method injects a clustering step in the GCL scheme to reduce data inconsistency. In addition, a graph contrastive coding-based method GCCAD [2] is proposed for contrasting abnormal nodes with normal ones in terms of their distances to the global context, with scarce labels in a self-supervised way. The above methods do not consider the attributed networks, so CoLA [15] was designed to learn informative embedding from high-dimensional attributes and local structure, and measure the anomaly score for each instance pair. Although the above GCL-based studies have been proposed, the basic limitation of GCL methods for fraud detection has not been explored.

3   Methodology

In this section, we first formulate the problem and introduce the overview of the proposed MICA framework, then systematically explore each module for the fraud detection task.

3.1   Problem Definition

In fraud detection problem on graph $\mathbf{G} = (\mathcal{V}, \mathcal{E}, \mathcal{X})$ with n nodes, each node $v_i \in \mathcal{V}$ with features $x_i \in \mathcal{X} \in \mathbb{R}^{n \times d}$ represents the target entity whose suspiciousness needs to be justified. For example, to detect fake reviews on the

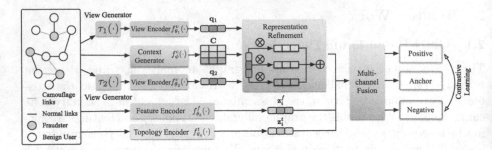

Fig. 1. The architecture of our MICA model. Based on the general GCL framework, our MICA additionally designed the context generator, representation refinement module, and multi-channel fusion module.

e-commerce platform, the target entities are genuine and fake reviews. Correspondingly, nodes have labels $\mathbf{Y} \in \{0,1\}$, where 0 denotes benign and 1 represents suspicious. The edge $(i,j) \in \mathcal{E}$ links node v_i and v_j due to some certain relationships or shared attributes, e.g., two reviews from the same user or posted from the same devices. Hence, the fraud detection problem is a binary node classification task on the graph.

3.2 Overview of Proposed MICA

As explained above, the false negative samples and over-smoothing caused by the GNN-based view encoder, and the conflicts of random perturbation way with existing diverse camouflaged attacks pose certain challenges to GCL methods. Thanks to the superiority of the CL model for this problem [27], we make improvements based on the GCL framework. As shown in Fig. 1, we additionally design three components in our MICA model, namely context generator, representation refinement, and multi-channel fusion. The context generator is to learn a mapping function $f_\phi^c(\cdot)$ from a global scope with notable normal and fraudster patterns (Sect. 3.3). Based on it, the augmentation-agnostic representation refinement module is proposed to map each view (e.g., $\mathbf{q}_1, \mathbf{q}_2$) into a unified space and refine their distances for mitigating the averaged embedding and false negative samples (Sect. 3.4). Besides, to explore different camouflaged behaviors under random perturbation, an adaptive and fine-grained multi-channel fusion module is designed (Sect. 3.5). Finally, a supervised contrastive loss is applied on the anchor, positive and negative samples for learning downstream relevant representations (Sect. 3.6).

3.3 Context Generator

In order to alleviate the false negative samples and the over-smoothing brought by local aggregation-based GNNs, we put forward a context generator to obtain the information from a global scope, then local feature differences are easier to be distinguished under its transformation. For example, for a fraud node that intentionally connects many normal nodes, it is easier to recognize the camouflage nature of this behavior by searching for similar cheating patterns from distant abnormal nodes.

Context Definition. The *context* represents a representation space with a global view, which could be sub-graphs of m nodes with d-dimensional node feature $\mathbf{X}^c \in \mathbb{R}^{m \times d}$ and topological information $\mathbf{A}^c \in \mathbb{R}^{m \times m}$. In our experiments, we consider context as a whole graph for a stable and global expression, which means $n = m$ in such a case. Inspired from attention mechanism [1], we calculate it via the keys $\mathbf{K} \in \mathbb{R}^{m \times k}$ and values $\mathbf{V} \in \mathbb{R}^{m \times v}$, k and v are the query/key depth and value depth respectively:

$$\mathbf{K} = \mathcal{F}_K(\mathbf{X}^c, \mathbf{A}^c) \tag{1}$$

$$\mathbf{V} = \mathcal{F}_V(\mathbf{X}^c, \mathbf{A}^c) \tag{2}$$

Here the function $\mathcal{F}_K(\cdot)$ and $\mathcal{F}_V(\cdot)$ are keys and values generators, which can be linear/non-linear functions only considering the feature information or graph-based algorithms like graph convolutional networks (GCN). We implemented several solutions for the generator, whose results could be seen in Table 2.

Generating the Context Map. With the generated keys and values, we wish to generate a linear function $f_\phi^c(\cdot) : \mathbb{R}^k \to \mathbb{R}^v$, i.e., a matrix $\mathbf{C} \in \mathbb{R}^{k \times v}$. We start by normalizing the keys across the nodes in a context via a softmax operation, which is $\overline{\mathbf{K}} = \text{softmax}(\mathbf{K}, axis = m)$, then the matrix $\mathbf{C} = \overline{\mathbf{K}}^{\mathrm{T}} \mathbf{V}$ is obtained by using the normalized keys $\overline{\mathbf{K}} \in \mathbb{R}^{m \times k}$ to aggregate the values $\mathbf{V}$.

Note that we choose to use different model parameters to generate keys and values, which can represent different levels of context. After the above operation, it is equivalent to performing screening in the context and retaining the expressive information. That is to say, some notable patterns of benign users and fraudsters remain in the context map.

3.4 Representation Refinement

With the notable patterns in the context map, to capture the long-distance but similar nodes, it is natural to refine the distance of nodes by transforming their hidden representations. Similarly, we denote n nodes as the *queries* $\mathcal{Q} = \{\mathbf{q}_1, \mathbf{q}_2, ..., \mathbf{q}_n\} \in \mathbb{R}^{n \times k}$. In a graph, although each query can be calculated by topology and features, here we think that only considering node features is more conducive to interacting with the context. Therefore, a query is generated by linearly mapping the node features to a specific space via function $\mathcal{F}_Q(\cdot)$:

$$\mathbf{Q} = \mathcal{F}_Q(\mathbf{X}) = \mathbf{X} \mathbf{W}_Q \tag{3}$$

Here $\mathbf{W}_Q \in \mathbb{R}^{d \times k}$ is the parameter transforming node features $\mathbf{X} \in \mathbb{R}^{n \times d}$ into queries. For a graph, the interactions between nodes and subgraphs as well as the whole graph can be useful. Essentially, we can treat each row of matrix $\mathbf{C}$ as a basis, that is $\mathbf{C} = \{\mathbf{C}_1, \mathbf{C}_2, ..., \mathbf{C}_k\}$, and each basis maps each dimension of the query $\mathbf{q}_i = \{q_1, q_2, ..., q_k\}$ to get a unified representation from a global scope. Based on that, we apply the context matrix to the query $\mathbf{q}_i$ for obtaining the context-aware embedding of node v_i, which is denoted as $\mathbf{z}_i^c \in \mathbb{R}^v$:

$$\mathbf{z}_i^c = \mathbf{C}^{\mathrm{T}} \mathbf{q}_i = \mathbf{C}_1 q_1 + \mathbf{C}_2 q_2 + ... + \mathbf{C}_k q_k \tag{4}$$

The context matrix $\mathbf{C}$ is shared across all queries and is invariant to the permutation of the context elements, which helps distinct nodes with similar patterns to be close to each other in the representation space. Especially for fraudsters, their embeddings are closer even if they are not directly connected.

3.5 Multi-channel Fusion

Although the context matrix maps each node to a global space, the random permutation way of data augmentation in GCL usually randomly masks the node attributes or edges, which hinders the exploration of analyzing the effect of different elusive camouflaged behaviors, therefore we further conduct an adaptive multi-channel way to better represent the contrastive views.

Concretely, apart from the context-aware embedding $\mathbf{z}_i^c \in \mathbb{R}^v$ of node v_i, we also obtain the feature-aware embedding $\mathbf{z}_i^f$ and topology-aware embedding $\mathbf{z}_i^t$ with the same dimension via MLP operation:

$$\mathbf{z}_i^f = f_{\theta_3}^f(\mathbf{x}_i) = \mathrm{MLP}(\mathbf{x}_i) \tag{5}$$

$$\mathbf{z}_i^t = f_{\theta_4}^t(\mathbf{A}) = \mathrm{MLP}(\tilde{\mathbf{u}}_i) \tag{6}$$

For the feature encoder $f_{\theta_3}^f(\cdot)$ and topology encoder $f_{\theta_4}^t(\cdot)$, the MLP consists of three linear layers with the exponential linear unit as the activation functions. Here $\tilde{\mathbf{u}}_i$ represents the top r dimensions of the eigen vector of node v_i, which is calculated by the eigen decomposition [10] based on adjacency matrix $\mathbf{A}$, i.e., $\mathbf{A} = \mathbf{U}\mathbf{\Lambda}\mathbf{U}^{-1}$. Except for the eigen decomposition way to obtain the topological information, other graph embedding methods such as random walks can be tried.

To explore their contributions in the final representation, we employ a gated way to adaptively adjust their weights. Concretely, we calculate the activation vectors $\boldsymbol{\alpha}_i^c, \boldsymbol{\alpha}_i^f, \boldsymbol{\alpha}_i^t$ for the context gate, feature gate, and topology gate respectively based on the embedding $\mathbf{z}_i^c, \mathbf{z}_i^f, \mathbf{z}_i^t$. For space limitation, we give the calculation process of $\boldsymbol{\alpha}_i^c$ as an example, the same process but with linear function $\mathcal{F}_\alpha^f(\cdot)$ and $\mathcal{F}_\alpha^t(\cdot)$ for the other two gated vectors:

$$\boldsymbol{\alpha}_i^c = \sigma\left(\mathcal{F}_\alpha^c\left(\mathbf{z}_i^c \| \mathbf{z}_i^f \| \mathbf{z}_i^t\right)\right) \tag{7}$$

This fine-grained way allows a good trade-off between the refined representation and original information, so as to automatically explore the camouflaged attack under random perturbation. Then the final embedding $\mathbf{z}_i$ is the weighted sum of the above three, which serves as the contrastive samples.

$$\mathbf{z}_i = \boldsymbol{\alpha}_i^c \circ \mathbf{z}_i^c + \boldsymbol{\alpha}_i^f \circ \mathbf{z}_i^f + \boldsymbol{\alpha}_i^t \circ \mathbf{z}_i^t \tag{8}$$

Here $\mathcal{F}_\alpha^c(\cdot), \mathcal{F}_\alpha^f(\cdot), \mathcal{F}_\alpha^t(\cdot)$ are linear functions for each channel, operator $\|$ means concatenation, operator $\circ$ represents Hadamard product, and $\sigma(\cdot)$ is the Sigmoid function. After the pre-processing phase, we load the learned parameters to initialize the encoder in fine-tuning process, and the classifier is trained with a cross-entropy loss.

3.6   Supervised Contrastive Loss

On the basis of node representation, we can construct anchors and positive as well as negative samples for contrastive learning. In general, self-supervised loss functions like InfoNCE are adopted for training in CL. In our case, since the camouflage behavior disturbs the original input data, we strongly need the label information to perform some bias correction in the representation space. In addition, according to the conclusions of some recent studies [22], incorporating supervised information helps to reduce the mutual information between contrastive views while keeping task-relevant information intact.

The representation of the anchor, the positive and negative sample is separately defined as $\mathbf{z}_o, \mathbf{z}_p, \mathbf{z}_n$, then the supervised contrastive loss [8] with temperature parameter τ in the pre-training process, adapting it to the fully supervised setting for leveraging label information is formulated as:

$$\mathcal{L}_{SupCon} = -\sum_{o \in \mathcal{V}} \frac{1}{|P(o)|} \sum_{p \in P(o)} \log \frac{\exp(\mathbf{z}_o \cdot \mathbf{z}_p / \tau)}{\sum_{j \in \mathcal{V} \setminus \{o\}} \exp(\mathbf{z}_o \cdot \mathbf{z}_n / \tau)} \tag{9}$$

Here $P(o) \equiv \{p \in \mathcal{V} \setminus \{o\} : y_p = y_o\}$ is the set of indices of all positives distinct from node v_o but sharing the same label, and $|P(o)|$ is its cardinality. If the anchor is graph embedding without label information, then positive and negative nodes are selected in terms of their labels.

3.7   Model Discussion

It is worth noting that our framework is compatible with the major components in CL, whether it is data augmentation strategies, pretext tasks (e.g., the same scale views like node-node pair or cross-scale views like node-graph pair) or contrastive objectives. We explore the varied strategies in the experiments (see Sect. 4.4). Besides, the method of generating context based on the keys and values has a small number of parameters. Concretely, in regular attention mechanism, the attention map calculated by $\mathbf{Q}\mathbf{K}^\top$ has $\mathcal{O}(n * m)$ space complexity with n inputs (which is used to generate queries) and m context (which is used to generate the keys and values). While in our model, we obtain the context map by $\mathbf{K}^\top\mathbf{V}$ with $\mathcal{O}(k * v)$ space complexity. Generally, it is called local attention

Table 1. The Statistics of Datasets.

Dataset	#Nodes	#Edges	Fraud	HR
Amazon	11,944	4,404,364	6.87%	0.91
YelpChi	45,954	3,869,956	14.53%	0.77

when the context length m is smaller than the number of samples n $(m \ll n)$, while global attention is when $n = m$. Since the key depth k will be set to a small value in practice, the final output dimension v is also smaller than the number of inputs, we have a much smaller memory cost. That is to say, it is an efficient calculation way compared to regular attention mechanisms.

4 Experiments

In the experiments, we mainly focus on verifying the effectiveness (in Sect. 4.2 and Sect. 4.3), expandability (in Sect. 4.4) and explainability (in Sect. 4.5).

4.1 Experimental Settings

Datasets. For the sake of fully verifying the above questions, we conduct experiments on two widely used real-world datasets (i.e., Amazon and YelpChi) whose statistics can be found in Table 1. Here we calculate the node-level homophily ratio (HR) for a better understanding of the camouflage links in the datasets.

Baselines. In order to fully analyze the performance of the model, we compare it with the following models from three categories:

Message passing GNNs have proved to be powerful in a variety of tasks on graphs. We select three architectures that are trained in an end-to-end manner. **GCN** [9], **GAT** [23] and **GPR-GNN** [3]. Specifically, GPR-GNN is a more powerful GNN that utilizes adaptive multi-hop aggregation to avoid over-smoothing and learn difficult label patterns.

GCL-based models show great potential in handling labels related issues, which is crucial in anomaly detection tasks. We choose **Deep Graph Infomax (DGI)** [24] and recently proposed **Deep Cluster Infomax (DCI)** [27] (which also falls into the Fraud Detection category). Besides, we also design a general GCL method called **GCN+SupCL** as a baseline, it uses GCN as the encoder in a supervised contrastive loss-based CL.

Fraud detection schemes include models which were proven to perform well in anomaly detection tasks. In this category we consider **CARE-GNN** [5], **GeniePath** (with its variant **GeniePathLazy**) [17], **FRAUDRE** [29] and **PC-GNN** [13]. CARE-GNN leverage reinforcement learning (RL) to effectively deal with camouflaged fraudsters and GeniePath utilizes multi-hop attention to learn receptive paths and propagate signals in the graph in a more effective fashion.

Table 2. Overall evaluation on YelpChi and Amazon datasets.

Method		YelpChi				Amazon			
		AUC (%)	AP (%)	Recall (%)	Acc. (%)	AUC (%)	AP (%)	Recall (%)	Acc. (%)
Baselines	GCN	61.31	25.32	50.02	85.47	74.40	20.72	50.02	93.12
	GAT	63.31	26.72	50.81	85.59	78.19	26.77	51.77	93.12
	GPR-GNN	83.65	52.77	63.07	87.48	96.57	84.54	88.48	97.99
	GCN+SupCL	59.99	22.37	50.07	85.47	85.35	39.35	54.11	93.36
	DGI	66.88	29.64	53.71	85.71	83.24	40.47	58.36	93.60
	DCI	66.72	29.44	53.42	85.63	83.29	42.08	64.64	93.32
	CARE-GNN	79.21	42.24	72.00	71.16	95.20	86.30	88.47	96.19
	GeniePath	76.90	39.78	56.70	86.13	77.40	33.33	57.75	94.11
	GeniePathLazy	84.93	56.61	64.64	88.05	96.65	85.82	88.56	**98.10**
	FRAUDRE	83.82	52.35	75.00	72.72	95.88	87.68	89.86	96.23
	PC-GNN	83.76	55.05	70.00	68.43	94.96	81.86	88.58	86.86
Variants	ICA(GPR)	77.81	42.90	62.54	85.97	94.30	82.15	87.76	97.72
	ICA(MLP)	83.47	51.69	63.26	87.33	95.53	83.13	86.92	97.72
	ICA(GCN)	86.67	62.28	69.53	89.03	97.60	87.58	87.46	97.88
	MICA(topo.)	86.92	62.28	69.74	88.84	97.76	87.51	87.46	97.76
	MICA(feat.)	88.90	66.97	74.52	**89.57**	97.79	88.07	**91.22**	97.75
	MICA	**89.36**	**67.94**	**75.33**	89.56	**98.01**	**88.89**	90.95	97.84

FRAUDRE considered the multi-relation among users and unified the graph-agnostic embedding and fraud-aware graph convolution module into a GNN framework, while PC-GNN proposed a pick-and-choose step to sample neighborhoods and getting the final node embeddings by aggregating neighbor information under different relations. Although there are more architectures designed for anomaly detection (e.g., Player2Vec [30], SemiGNN [25] and GraphConsis [16]), we decided not to include them in our study because these recently published methods outperform them significantly on both datasets.

Evaluation Metrics. In fraud detection problems we are naturally more interested in correctly identifying fraudsters (positive instances). Moreover, as we mentioned before, the classes are naturally imbalanced, which is reflected in both datasets. For the above reasons, we utilize four metrics: ROC-AUC (AUC), Average precision score (AP), Recall and Accuracy (Acc.).

Parameters Settings. For fair comparison, we separately adopt 40% and 60% of whole data as training set and testing set, and use the following unified setup for all models: hidden dimensions = 64, number of epochs = 1000 on YelpChi and 300 on Amazon, batch size = 1024 for YelpChi and 256 for Amazon, learning rate = 0.002 and L2 regularization weight = $5e^{-4}$. In our MICA model, the representation dimension of query/key $k = 16$ and the representation dimension of value $v = 64$. All of the models are trained with Adam optimizer, the results are averaged on 20 run times. We implement code based on PyTorch Geometric [6]. The code is available at https://github.com/goiter/anomaly_detection.git.

4.2   Overall Evaluation

In addition to conducting experiments on the baselines, we also implement some variants of our models from two perspectives.

- **ICA**: In order to focus on verifying the performance of the context generator and representation refinement module, we name a method as ICA which is our MICA model without the multi-channel fusion module. Besides, we implement the same function but not sharing parameters for keys and values generator (i.e., the function $\mathcal{F}_K(\cdot)$ and $\mathcal{F}_V(\cdot)$ in Eq. (1)) in three ways: GPR-GNN, MLP and GCN. We refer to these models as **ICA(GPR)**, **ICA(MLP)** and **ICA(GCN)** respectively.
- **MICA**: This is our proposed model which includes all modules in Fig. 1 with the keys (and values) generator implemented as GCN. While **MICA(topo.)** and **MICA(feat.)** are the variants of MICA fused with only the topology-aware embedding and the feature-aware embedding respectively.

According to the results shown in Table 2, we summarize our conclusions as follows: (1) **The proposed model MICA contributes to fraud detection significantly.** Even without the multi-channel fusion module, our ICA(GCN) outperforms the baselines on almost all evaluation metrics. The performance improvement between ICA methods and GCL-based methods like DCI indicates that the context generator and representation refinement module have a certain effect on solving the problem of false negative samples. (2) **The multi-channel fusion module is more advantageous in heterophilic datasets.** By comparing the results of ICA(GCN) and the variants of MICA, it is obvious that the multi-channel module plays an important role in the overall performance of YelpChi. We hypothesize that the main reason for that is the dataset being relatively heterophilic (with 0.77 homophily ratio [20]). That is to say, there is a considerable portion of connected nodes belong to different classes. In this situation, it may be necessary to pay more attention to the context and feature information. (3) **The results under different context generators reflect that the context should retain as much information as possible from a global scope.** By comparing the results under the variants of ICA, we can see that the GCN-based context generator performs best. Although the GPR-GNN method achieves impressive results in a supervised manner, the performance of ICA(GPR) is not as good as that of ICA(MLP), indicating that the context in the pre-training phrase requires as comprehensive information as possible.

Besides, according to the results of the baselines, we can draw some observations: (1) The widely used GNN methods (i.e., GCN and GAT) fail to achieve good results on fraud detection datasets, while the results of GPR-GNN are impressive due to its ability to adaptively learn the weights in order to optimize nodes features and topological information extraction process. (2) The performance of the GCL-based models (i.e., GCN+SupCL, DGI, and DCI) on these two datasets is not very satisfactory. The results of GCN+SupCL on the YelpChi dataset are worse than the GCN method but better on the Amazon dataset, which illustrates that graph contrastive learning methods are not necessarily better than GNNs, and more powerful view encoders are needed. As for DGI

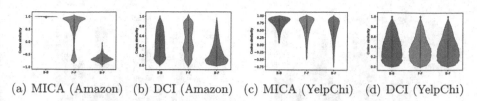

(a) MICA (Amazon)   (b) DCI (Amazon)   (c) MICA (YelpChi)   (d) DCI (YelpChi)

Fig. 2. Comparison of cosine similarity of MICA and DCI between benign-benign (B-B), fraudster-fraudster (F-F) and benign-fraudster (B-F) users. Sub-figure (a) and (c) are the results of MICA while Sub-figure (b) and (d) are from DCI.

and DCI, we believe that for larger datasets, contrasting each node to the full graph or some feature-based clusters is not good enough to separate them with camouflaged behaviors. (3) CARE-GNN, GeniePathLazy, FRAUDRE, and PC-GNN also perform well. By selecting neighbors and considering the multi-relation on edges, CARE-GNN and PC-GNN deeply mitigate the attack brought by the topological camouflage behavior. FRAUDRE which investigated aspects of the features, topologies, and relations proved the ability in solving the heterophily issue. GeniePathLazy with the individual feature map in the multi-hop attention has achieved better performance than GeniePath, which also verifies the necessity of our proposed multi-channel fusion module.

4.3 Visualization on Distinguishable Representations

In order to verify the motivation and intuitively prove that our MICA model learned better representations and mitigated the over-smoothing, we compared the pairwise cosine similarity of node embeddings with DCI methods. Figure 2 shows the violin plot of inner-class (B-B, F-F) and inter-class (B-F) similarities, which is a box plot with the addition of a rotated probability density and the average value represented as white dots. On one hand, according to Fig. 2(a) and Fig. 2(b), we can clearly observe that the inner-class similarity of MICA on the Amazon dataset is higher than that of DCI, and the opposite for inter-class. That is to say, MICA has distinct distributions for inner- and inter-class similarities, while the distributions of DCI are closer and more difficult to distinguish. On the other hand, the YelpChi dataset has a lower homophily ratio, implying the prevalence of structural anomaly compared with the Amazon dataset. According to Fig. 2(c) and Fig. 2(d), though with overlapped distributions, for MICA there's a clear difference between the means of inner- and inter-class distributions, while we can barely observe any difference between the counterparts from DCI. In a word, our MICA model makes node representation more distinguishable on inner- and inter-class, which proved that MICA alleviated the over-smoothing issue and is more beneficial to detect camouflaged behaviors.

4.4 Expandability and Ablation Study on ICA

In Sect. 4.2 we analyze the performance of our ICA model and its variants with different context generators, which is one of the ablation studies. Besides, due to

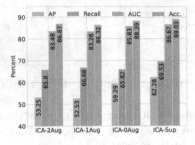

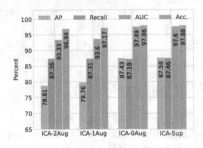

(a) Variants of ICA (YelpChi)   (b) Variants of ICA (Amazon)

ICA Variants	Contrastive Modes	Contrastive Losses	Augmentation Way		
			Anchor	Positives	Negatives
ICA-2Aug	node-graph pair	self-supervised	aug1	aug1	aug2
ICA-1Aug	node-graph pair	self-supervised	-	-	aug
ICA-0Aug	node-graph pair	supervised	-	-	-
ICA-Sup	node-node pair	supervised	-	-	-

Fig. 3. Comparison of our ICA model on varied contrastive modes, training losses and data augmentation strategies for contrastive views. The different settings of the ICA variants are listed in the table.

the expandability of our proposed framework, it supports different contrastive modes, contrastive losses and data augmentation strategies for contrastive views. Therefore, in this section we analyze the effects of them. For fair comparison, we merely conduct variant experiments in our ICA model.

As the table in Fig. 3 shows, we implement the ICA method with varied contrastive modes, contrastive losses and data augmentation strategies. Concretely, the node-graph pair contrastive mode is to treat the graph embedding as the anchor and node embeddings as positive/negative samples. The graph embedding is obtained by sum pooling on node embeddings. Similarly, the anchor and contrastive samples in node-node pair contrastive mode are all base on node embeddings. The samples in ICA-2Aug are all obtained by two kinds of data augmentation methods (i.e., aug1 and aug2). Differently, the data augmented nodes as negative samples in ICA-1Aug, while the original nodes and their corresponding graph representation are used as positive samples and the anchor. Note that we implement the data augmentation by randomly removing the edges with a 0.2 dropout rate and masking the features with a 0.1 probability.

By observing Fig. 3 we can draw the following conclusions. Firstly, the methods training with supervised contrastive loss (i.e., ICA-0Aug and ICA-Sup) outperforms the ones with self-supervised loss (i.e., ICA-2Aug and ICA-1Aug), which verifies that leveraging label information in pre-training makes representations keeping task-relevant information. Secondly, compared with the results of ICA-Sup and ICA-0Aug, it shows that the node-node pair contrastive mode may be more suitable for this node-level fraud detection task than the node-graph

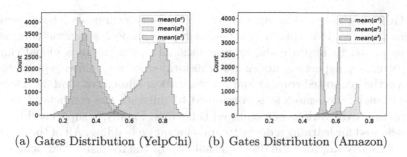

(a) Gates Distribution (YelpChi)   (b) Gates Distribution (Amazon)

Fig. 4. Distributions of the context gate, feature gate, and topology gate in MICA. Each data point in the x-axis is obtained by averaging over the feature dimension.

pair. In addition, it may also be due to the design of the context generator and representation refinement module enriching the node samples with notable pattern information, making it more effective and intuitive when doing the sample contrastive learning. Thirdly, by comparing the results on simply designed data augmentation strategies, using different data augmentation strategies in ICA-2Aug and ICA-1Aug does not seem to have much effect on the results, which implies that the representation refinement module is augmentation-agnostic. In a summary, all the above ablation studies show the expandability and effectiveness of our proposed ICA architecture.

4.5 Explainability of Multi-channel Fusion Module

Table 2 indicates that the feature-aware embedding alone (MICA(feat.)) brings the most significant improvement to our ICA model, and it will continue to advance with the addition of topology-aware embedding (MICA). However, it is impossible to directly measure the contributions of each module solely depending on Table 2, since the gating module relies upon the coupling of the involved embeddings. Therefore, we draw Fig. 4 to analyze which part of information is more crucial for fraud detection. Specifically, we draw the distribution of three gates (i.e., $\alpha^c, \alpha^f, \alpha^t$) based on their average values over feature dimensions.

From the distribution figures, we can infer that the contribution of context-aware embedding is the highest on YelpChi dataset, and the role of feature-aware embedding is slightly more important than that of topology-aware embedding. While on Amazon dataset, the contributions of them are less different, with the topology-aware embedding being the most important component. These results imply that the designed multi-channel module enables our MICA model to be applicable to varied datasets, and be able to interpret the results.

5   Conclusion

In this paper, we aim to design a new solution for enhancing prediction performance in fraud detection tasks. Considering the node representations learned from GCL methods are hampered by the over-smoothing and false negative samples under various camouflage behaviors, we propose a general contrastive learning model to improve the representational power. From a global camouflage patterns perspective, nodes are transformed into a unified representation space via the generated context and representation refinement module. Then the multi-channel fusion module is considered to mitigate the conflicts of random perturbation with various camouflaged behaviors. Finally, a supervised training loss is adopted for learning a downstream relevant embedding. All of the designed modules greatly improve the distinguishability of the fraudster and benign user in the representation space. The implemented experiments fully verify the effectiveness of the proposed solution in fraud detection. Due to the generality of our proposed framework, we believe that in the future, designing a different context map for each application scenario is a worthy direction to explore.

References

1. Bello, I.: LambdaNetworks: modeling long-range interactions without attention. In: International Conference on Learning Representations (2020)
2. Chen, B., et al.: GCCAD: graph contrastive coding for anomaly detection. arXiv: abs/2108.07516 (2021)
3. Chien, E., Peng, J., Li, P., Milenkovic, O.: Adaptive universal generalized PageRank graph neural network. In: International Conference on Learning Representations (2020)
4. Ding, K., Zhou, Q., Tong, H., Liu, H.: Few-shot network anomaly detection via cross-network meta-learning. In: Proceedings of the Web Conference (2021)
5. Dou, Y., Liu, Z., Sun, L., Deng, Y., Peng, H., Yu, P.S.: Enhancing graph neural network-based fraud detectors against camouflaged fraudsters. In: Proceedings of the 29th ACM International Conference on Information & Knowledge Management (2020)
6. Fey, M., Lenssen, J.E.: Fast graph representation learning with PyTorch Geometric. In: ICLR Workshop on Representation Learning on Graphs and Manifolds (2019)
7. Jiang, Z., et al.: Camouflaged Chinese spam content detection with semi-supervised generative active learning. In: Proceedings of the 58th Annual Meeting of the Association for Computational Linguistics, pp. 3080–3085 (2020)
8. Khosla, P., et al.: Supervised contrastive learning. In: Advances in Neural Information Processing Systems, vol. 33, pp. 18661–18673 (2020)
9. Kipf, T., Welling, M.: Semi-supervised classification with graph convolutional networks. arXiv: abs/1609.02907 (2017)
10. Kreuzer, D., Beaini, D., Hamilton, W.L., L'etourneau, V., Tossou, P.: Rethinking graph transformers with spectral attention. In: Advances in Neural Information Processing Systems, vol. 34 (2021)
11. Li, A., Qin, Z., Liu, R., Yang, Y., Li, D.: Spam review detection with graph convolutional networks. In: Proceedings of the 28th ACM International Conference on Information & Knowledge Management (2019)
12. Liang, C., et al.: Uncovering insurance fraud conspiracy with network learning. In: Proceedings of the 42nd International ACM SIGIR Conference on Research and Development in Information Retrieval (2019)

13. Liu, Y., et al.: Pick and choose: a GNN-based imbalanced learning approach for fraud detection. In: Proceedings of the Web Conference (2021)
14. Liu, Y., Ao, X., Zhong, Q., Feng, J., Tang, J., He, Q.: Alike and unlike: resolving class imbalance problem in financial credit risk assessment. In: Proceedings of the 29th ACM International Conference on Information & Knowledge Management (2020)
15. Liu, Y., Li, Z., Pan, S., Gong, C., Zhou, C., Karypis, G.: Anomaly detection on attributed networks via contrastive self-supervised learning. IEEE Trans. Neural Netw. Learn. Syst. **33**(6), 2378–2392 (2021)
16. Liu, Z., Dou, Y., Yu, P.S., Deng, Y., Peng, H.: Alleviating the inconsistency problem of applying graph neural network to fraud detection. In: Proceedings of the 43nd International ACM SIGIR Conference on Research and Development in Information Retrieval (2020)
17. Liu, Z., Chen, C., Li, L., Zhou, J., Li, X., Song, L.: GeniePath: graph neural networks with adaptive receptive paths. In: Proceedings of the 33rd AAAI Conference on Artificial Intelligence (2019)
18. Liu, Z., Chen, C., Yang, X., Zhou, J., Li, X., Song, L.: Heterogeneous graph neural networks for malicious account detection. In: Proceedings of the 27th ACM International Conference on Information & Knowledge Management (2018)
19. Ma, X., Wu, J., Xue, S., Yang, J., Sheng, Q.Z., Xiong, H.: A comprehensive survey on graph anomaly detection with deep learning. IEEE Trans. Knowl. Data Eng. **35**(12), 12012–12038 (2021)
20. Pei, H., Wei, B., Chang, K.C.C., Lei, Y., Yang, B.: Geom-GCN: geometric graph convolutional networks. In: International Conference on Learning Representations (2020)
21. Ren, Y., Wang, B., Zhang, J., Chang, Y.: Adversarial active learning based heterogeneous graph neural network for fake news detection. In: 2020 IEEE International Conference on Data Mining (ICDM), pp. 452–461 (2020)
22. Tian, Y., Sun, C., Poole, B., Krishnan, D., Schmid, C., Isola, P.: What makes for good views for contrastive learning. In: Advances in Neural Information Processing Systems, vol. 33, pp. 6827–6839 (2020)
23. Velickovic, P., Cucurull, G., Casanova, A., Romero, A., Lio', P., Bengio, Y.: Graph attention networks. In: International Conference on Learning Representations (2018)
24. Velickovic, P., Fedus, W., Hamilton, W.L., Lio', P., Bengio, Y., Hjelm, R.D.: Deep graph infomax. In: International Conference on Learning Representations (2018)
25. Wang, D., et al.: A semi-supervised graph attentive network for financial fraud detection. In: IEEE International Conference on Data Mining (ICDM), pp. 598–607 (2019)
26. Wang, X., Zhu, M., Bo, D., Cui, P., Shi, C., Pei, J.: AM-GCN: adaptive multi-channel graph convolutional networks. In: Proceedings of the 26th ACM SIGKDD International Conference on Knowledge Discovery & Data Mining (2020)
27. Wang, Y., Zhang, J., Guo, S., Yin, H., Li, C., Chen, H.: Decoupling representation learning and classification for GNN-based anomaly detection. In: Proceedings of the 44th International ACM SIGIR Conference on Research and Development in Information Retrieval (2021)
28. Yang, X., Lyu, Y., Tian, T., Liu, Y., Liu, Y., Zhang, X.: Rumor detection on social media with graph structured adversarial learning. In: Proceedings of the Twenty-Ninth International Conference on International Joint Conferences on Artificial Intelligence, pp. 1417–1423 (2021)

29. Zhang, G., et al.: FRAUDRE: fraud detection dual-resistant to graph inconsistency and imbalance. In: 2021 IEEE International Conference on Data Mining (ICDM), pp. 867–876 (2021)
30. Zhang, Y., Fan, Y., Ye, Y., Zhao, L., Shi, C.: Key player identification in underground forums over attributed heterogeneous information network embedding framework. In: Proceedings of the 28th ACM International Conference on Information & Knowledge Management (CIKM 2019), pp. 549–558 (2019)
31. Zhao, T., Ni, B., Yu, W., Guo, Z., Shah, N., Jiang, M.: Action sequence augmentation for early graph-based anomaly detection. In: Proceedings of the 30th ACM International Conference on Information & Knowledge Management (2021)

Detecting Critical Nodes in Hypergraphs via Hypergraph Convolutional Network

Zhuang Miao[1], Fuhui Sun[2(✉)], Xiaoyan Wang[2], Pengpeng Qiao[1], Kangfei Zhao[1], Yadong Wang[1], Zhiwei Zhang[1], and George Y. Yuan[3]

[1] Beijing Institute of Technology, Beijing, China
{mz,qpp,bit_wyd,zwzhang}@bit.edu.cn
[2] Information Technology Service Center of People's Court, Beijing, China
sunfh6732@163.com
[3] Thinvent Digital Technology Co., Ltd., Nanchang, China
yuanye@thinvent.com

Abstract. In many real-world networks, such as co-authorship, etc., relationships are complex and go beyond pairwise associations. Hypergraphs provide a flexible and natural modeling tool to model such complex relationships. Detecting the set of critical nodes that keeps the hypergraph structure cohesive and tremendous has great significance. At present, all the researches in detecting critical nodes area focus on traditional pairwise graphs, and how to extend these researches to hypergraphs is an urgent problem. In this paper, we propose a novel framework, called Hypercore Hypergraph Convolutional Network (H^2GCN) to detect the critical nodes. Specifically, we first transform the raw hypergraph into a k-hypercore, and then generate training samples using biased sampling. Furthermore, we adopt the hypergraph convolutional network to capture the complex hypergraph topological structure information for generating node embedding, and then generate a hypergraph embedding by aggregating node embedding using a self-attention mechanism. Finally, by feeding the hypergraph embedding to MLP, H^2GCN has the ability to predict the probability that the remaining nodes will be selected as critical nodes under the condition of existing partial solutions. Our experiments on many real-world hypergraphs show that H^2GCN significantly improves the quality of the detected critical nodes compared to existing traditional algorithms extended to hypergraphs.

Keywords: Hypergraph · Critical Nodes · HGCN

1 Introduction

The critical node detection problem has attracted many interests due to its large applicability in various fields, such as immunization [7], network vulnerability [8,12], economics [1], social network analysis [3,18], etc. In general, these applications model their data as a pairwise graph and measure the criticalness of a node by evaluating its impact after removing it from the graph.

However, many applications need to mine the information on the higher-order graph, in which the corresponding relationships can be modeled as a hypergraph. In a hypergraph, each hyperedge may contain a set of nodes instead of two nodes, leading to complex topology. Although there are many studies on the critical node set detection problem on pairwise graphs, less attention has been paid to introducing the formulation and solutions to the hypergraph domain.

Figure 1 illustrates how co-author relationships can be modeled differently using pairwise graphs and hypergraphs. In a co-author network, authors are represented as nodes and co-author relationships as edges. In pairwise graphs, each edge connects two authors who have collaborated on at least one paper, while in hypergraphs, each hyperedge includes all authors of a paper. While nodes in both graphs represent authors, the amount of information represented by edges varies greatly. Hypergraphs contain richer information because we can convert a hypergraph to a pairwise graph by connecting the nodes in the hyperedge in pairs, but not vice versa. For instance, in Fig. 1(b), we can discern a co-authorship relationship among v_1, v_2, and v_3, whereas no such relationship exists among v_1, v_2, and v_5. However, in Fig. 1(a), this relationship cannot be determined.

To evaluate the criticalness of a node set U in the graph, recent methods model it as the number of nodes to be removed under certain conditions after removing U from the graph. The left of the critical nodes usually deconstructs the engagement of the community and triggers the leave of a great number of nodes. Many works take the community as a k-core. That is, for each node in a k-core, its degree is at least k. Based on the definition of k-core, the collapsed k-core problem is to find a set U with b nodes that maximize the number of nodes to be deleted due to the removal of U under the conditions for k-core [18].

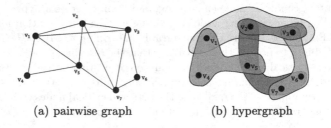

(a) pairwise graph   (b) hypergraph

Fig. 1. Comparison of a hypergraph and a pairwise graph in co-author domain.

However, most of the existing works for the critical nodes detection focus on pairwise graph, and these approaches cannot be applied to deal with hypergraphs easily. In order to detect critical nodes, some approaches are based on greedy heuristics. The difference between them lies in the heuristic-based function $f(v)$ for selecting node v in the community. In each iteration, greedy-based algorithms select the node v with the largest $f(v)$ repeatedly until the desired critical nodes are found. The function $f(v)$ is usually designed based on the degree or the number of nodes removed by deleting v only [18]. Recently, some works design the GNN model as $f(v)$, to select a candidate node in each iteration in the pairwise graph. However, for hypergraphs, applying these approaches have several drawbacks. First of all, in the hypergraph, the node connection is more complex than pairwise graph. Secondly, the criticalness of a node needs to consider not only the number of hyperedges that contain it, but also the whole structure. For learning-based approaches, they leverage the whole graph structure when finding critical nodes. [19] proposed a learning-based approach $SCGCN$ that not only consider the criticalness of every single node, but also the criticalness of node combinations based on the complex graph structure. However, applying $SCGCN$ to hypergraph still cannot fully utilized the structural information in hypergraphs.

To tackle the aforementioned issues, we present a solution by formulating the critical node detection problem in hypergraphs as a collapsed k-hypercore problem. However, when implementing the greedy algorithm for hypergraphs, it is found that the heuristic function is nonsubmodular, leading to poor performance on hypergraphs with complex topology. As a remedy, we propose a DL model called H^2GCN, which serves as a Hypergraph Convolutional Network (HGCN) based model to encode the criticalness of node sets for a specific hypergraph. The model uses self-attention to obtain the graph embedding for predicting a conditional probability distribution, namely the likelihood of selecting a node as a critical node given a partial solution. The DL-based heuristic can approximate complex combinatorial functions by jointly modeling the hypergraph structure and node set criticalness, while being tailored to the specific hypergraph instance and trained using a sampled node set with a criticalness bias. Overall, this paper makes significant contributions in addressing the problem at hand. The contributions of this paper are summarized as follows:

- We formulate the problem of finding critical nodes in hypergraphs, where the removal of the critical nodes cause the collapsion of k-hypercore. We adapt existing greedy algorithm of pairwise graphs to hypergraphs and prove that the algorithm is lack approximate guarantee due to the non-submodularity of the heuristic function.
- We propose a learning-based framework following the greedy paradigm. We design a DL model, H^2GCN, as the greedy heuristic that models the criticalness of node sets in hypergraph, by Hypergraph convolutional Network. H^2GCN is able to infer the critcalness of unseen node sets and guide the greedy search towards better solutions.
- We conduct substantial experiments to evaluate H^2GCN and the greedy algorithms on 5 real-world hypergraphs. Our experimental results reveal that effectiveness of different approaches depends on the inherent topology of the hypergraphs. H^2GCN outperforms traditional greedy algorithm in mining 'jumping points' that incur sudden collapsing.

Organization. The rest of this paper is organized as follows. Section 2 shows the related works, and Sect. 3 illustrates the problem formulation. We extend the existing traditional algorithms to hypergraphs in Sect. 4, and introduce our proposed H^2GCN in Sect. 5. Then, the experimental results are shown in Sect. 6. Finally, we conclude this work in Sect. 7.

2   Related Works

Critical Node Detection: For the dense subgraph, k-core is introduced in [15] and has been used in many graph analysis tasks. [18] first proposes the collapsed k-core problem and provides the greedy algorithm to detect the critical users in social communities. [5,22] studies the k-core minimization problem, which aims to find the critical edges that minimize the k-core by removing edges from the network. [17,21] studies the k-core maximization problem, which aims to find the critical edge that maximizes the k-core when edges are added to the network. These works are intended to find critical dense subgraph structures by cumulatively removing edges, and have been applied directly to find critical nodes. [19] proposes a learning-based approach for finding critical users to solve the collapsed k-core problem. It also adopts a greedy search framework like [18].

Graph Neural Network: In the literature, geometric deep learning [4] is an umbrella phrase for emerging techniques attempting to generalize deep neural network models to non-Euclidean domains such as graphs and manifolds. Graph convolutional network (GCN) [11] defines the convolution using a simple linear function of the graph Laplacian and is shown to be effective on semi-supervised classification on attributed graphs. The clique expansion of a hypergraph is introduced in a seminal work [20] and has become a popular approach for learning hypergraph-structured data. Hypergraph convolutional networks (HGCN) [6] and their variants [9,10] use the clique expansion to extend GCNs for hypergraphs.

Compared with the above methods, most current work is based on pairwise graphs for critical node detection, and we are the first work to use a hypergraph convolutional network for critical node detection on hypergraphs.

3   Preliminaries

We model a hypergraph as $G = (V, E)$, where $V(G) = \{v_1, \cdots, v_N\}$ is the set of nodes, $E(G) = \{\mathcal{H}_1, \cdots, \mathcal{H}_M\}$ is the set of hyperedges. A hyperedge $\mathcal{H}_i = \{v_{i_1}, \cdots, v_{i_k}\} \subset V$ is a set of nodes in G that interact beyond pairwise relationship. The incident hyperedges of a node v, denoted as $Adj(v, G) = \{\mathcal{H}_i | v \in \mathcal{H}_i\}$, is the set of hyperedges that contains the node v. The degree of node v is denoted as $deg(v, G) = |Adj(v, G)|$.

To model the community in a pairwise graph, subgraph cohesiveness metrics such as k-core, k-truss, k-clique are proposed, where k-core is the most widely used. Accordingly, for hypergraphs, we use the k-hypercore to model cohesive subgraphs in hypergraphs, which is defined as follows:

Definition 1. *(k-hypercore): For a given k, the k-hypercore of hypergraph G, $C_k(G)$, is the largest subgraph of G satisfies $\forall v \in C_k(G)$, $deg(v, C_k(G)) \geq k$.*

Given a hypergraph G with its k-hypercore $C_k(G)$, if a node v is removed from $C_k(G)$, to preserve the structure of the k-hypercore, other nodes may also need to be removed due to the degree constraint. Similar to [18] that formulates this collapsing phenomenon as the collapsed k-core problem in pairwise graphs, we adapt the collapsed k-core to hypergraphs, which is the collapsed k-hypercore as below.

Definition 2. *(Collapsed k-hypercore): Given a hypergraph G and a node set $U \subseteq V(G)$, the collapsed k-hypercore, denoted as $C_k(G_U)$, is the corresponding k-hypercore of G with nodes in U removed.*

In this paper, we study the problem to find critical nodes in a hypergraph, and such problem can be seen as the collapsed k-hypercore problem. To evaluate the criticalness for nodes in U in a k-hypercore $C_k(G)$, we define the *followers* of node set U as $Fll(U,G) = C_k(G) \setminus (C_k(G_U) \cup U)$ when U is removed from $C_k(G)$. Following the criticalness of node set regarding k-core [19], the *criticalness* value of node set U for a k-hypercore $C_k(G)$ is evaluated by U and its followers $Fll(U,G)$, i.e., $Cr(U) = |Fll(U,G) \cup U|$. The problem of this paper is to find a fixed-size node set U that maximizes the criticalness for a k-hypercore as the following.

Problem Statement: Given a hypergraph G, an integer k for a k-hypercore, and budget b for critical nodes, the problem of critical nodes detection in hypergraph is to find the set of nodes $U_{\max} = \arg\max_U Cr(U)$ where $|U_{\max}| = b$.

4   The Trivial Approaches for Hypergraphs

The problem to detect critical nodes in hypergraph is an NP-hard problem. The reason is that, finding the critical nodes in pairwise graph is proved to be NP-hard [18], and it is a special case for hypergraphs. Existing approaches, such as *Greedy*, *Degree*, and *SCGCN* [18,19], adopt a greedy strategy to search critical nodes in k-core. For *Greedy* algorithm, it selects the node with the greatest criticalness in each iteration, while *Degree* removes the node with the highest degree. For *SCGCN*, it uses a GCN-based model to infer the candidate node in each iteration as its greedy strategy.

We also adapt the greedy strategy to hypergraphs, as shown in Algorithm 1, which takes a hypergraph G, an integer of k for k-hypercore and budget b as inputs, and outputs the critical node set with size b.

Initially, the candidate node set U is empty. During every iteration, the greedy framework chooses a node v^* and adds it to U until $|U| = b$, where v^* is the one with the max value of $f(v)$ among the nodes in current G_U. Using this framework, we can extend the traditional *Greedy*, *Degree*, and *SCGCN* algorithms to hypergraphs by replacing the greedy strategy in lines 4 and 5 in Algorithm 1. Specifically, for the *Greedy* algorithm, we use the criticalness of the nodes in the

---

Algorithm 1. Greedy Strategy For Hypergraphs

Input: A hypergraph G, integer k for k-hypercore, budget b for critical node set U;
Output: Critical node set U in hypergraph G;
  1: $U \leftarrow \varnothing$
  2: **while** $|U| < b$ **do**
  3:     compute $f(v)$ for node $v \in C_k(G_U)$;
  4:     $v^* \leftarrow \arg\max_v f(v),\ U \leftarrow U \cup \{v^*\}$;
  5: **end while**
  6: **return** U;

---

hypergraph as $f(v)$ to replace the criticalness in pairwise graphs. For the *Degree* algorithm, we choose the node with the highest degree in the hypergraph in each iteration. For the *SCGCN* algorithm, in each iteration, we first perform clique expansion on each hyperedge to form a traditional pairwise graph structure. On the basis of the pairwise structure, we use the original model of *SCGCN* to compute $f(v)$.

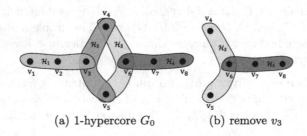

(a) 1-hypercore G_0 (b) remove v_3

Fig. 2. Example for detecting critical nodes and example for non-submodular

Table 1. The *Degree* And *Criticalness* Computed In Iteration

V	v_1	v_2	v_3	v_4	v_5	v_6	v_7	v_8
Degree	1	1	2	2	2	2	1	1
Criticalness	1	1	2	1	1	2	1	1
Followers	v_2	v_1	v_1, v_2	v_5	v_4	v_7, v_8	v_8	v_7

We demonstrate how critical node detection is performed on a specific hypergraph by an example in Fig. 2. Considering a 1-hypercore G_0 in Fig. 2(a). For the first iteration, among all nodes, the criticalness of v_3, v_6 is the largest. Because when deleting the node v_3, we need to delete the adjacent hyperedges $\mathcal{H}_1$ and $\mathcal{H}_2$ of the node v_3. Deleting the hyperedge means that the degrees of all vertices in the hyperedge are reduced by 1. Since the nodes v_1, v_2 no longer satisfy the 1-hypercore degree constraint, they are deleted to obtain the 1-hypercore subgraph with v_3 removed, as shown in Fig. 2(b). In this case, *Greedy* will select the node v_3 with the smallest node ID to delete. For the *Degree* algorithm, it will also choose v_3 because of v_3 has the largest degree and the smallest ID. We calculate the degree, criticalness and followers of each node in G_0 and put them in Table 1.

Since we transform the problem of critical nodes detection on hypergraphs into a collapsed k-hypercore problem, we prove the properties of monotone and non-submodular towards the collapsed k-hypercore problem in Theorem 1.

Theorem 1. *Given a hypergraph G, and an node set U, assume $f(U) = |Fll(U, C_k(G)) \cup U|$, for the given k-hypergraph, then f is monotone but not submodular for any k.*

Proof. Suppose there is a set $U' \supseteq U$. For every node v in $Fll(U, C_k(G))$, v will still be deleted in the collapsed k-hypercore with the collapsers set U', because removing vertices in U' cannot increase the degree of u. Thus $f(U') \geq f(U)$ and f is monotone. For two arbitrary collapsers sets U_1 and U_2, if f is submodular, it must hold that $f(U_1 \cup U_2) + f(U_1 \cap U_2) \leq f(U_1) + f(U_2)$. We show that the inequality does not hold using counterexamples. When $k=1$, we use the example shown in Fig. 2. Suppose $k = 1$, $U_1 = \{v_3\}$ and $U_2 = \{v_6\}$, we have $f(U_1) = \{v_1, v_2, v_3\}$, $f(U_2) = \{v_6, v_7, v_8\}$, $f(U_1 \cup U_2) = \{v_1, v_2, v_3, v_4, v_5, v_6, v_7, v_8\}$, $f(U_1 \cap U_2) = \emptyset$, so the inequation does not hold.

The greedy algorithm is not designed for a specific hypergraph structure and does not fully consider the importance of the structural characteristics of the hypergraph to the critical nodes detection. In the following, we introduce our approach. Our goal is to use a hypergraph convolutional network to learn a function that can better find the critical nodes in different types of hypergraphs.

5   The Hypergraph GCN Based H^2GCN Framework

As shown in Fig. 3, a new framework H^2GCN is proposed for critical nodes detection in hypergraphs. The architecture contains three main parts: k-hypercore computation, partial solution sampling, and critical nodes inference. First, we compute the k-hypercore $C_k(G)$ of hypergraph G in the k-hypercore computation module. Then, in the sampling module, the training data is generated by biased sampling on the $C_k(G)$. At last, a model based on hypergraph convolutional network fused with a self-attention mechanism is designed to infer the critical nodes. We will discuss these three modules in detail as follows.

5.1 k-Hypercore Computation

Since we define the problem of critical nodes detection on a hypergraph as a collapsed k-hypercore problem, we first need to identify the k-hypercore community. In this case, we use the hypercore decomposition algorithm [13] to obtain the k-hypercore of the graph by recursively removing nodes with degree less than k. In real applications, the value of k is determined by users based on their

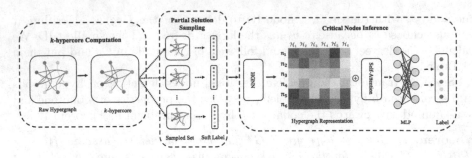

Fig. 3. The overview architecture of H^2GCN.

requirement for cohesiveness. Intuitively, the resulting k-hypercore will be more cohesive when the k value becomes larger. In the follow-up, we will model combinatorial criticalness on the generated k-hypercore community, and perform biased sampling based on the criticalness of nodes to generate partial solutions as training datasets.

5.2 Partial Solution Sampling

Because searching all the possible partial solutions S in a large graph is impossible, we employ a parametric hypergraph convolutional network learning the $f(v, S, G; \Theta)$ to estimate the probability $p(v|S)$ that node v ($v \notin S$) is selected as a critical node and then added it to partial solution S. The domain of partial solution S is a set of nodes, which is a subset of $V(C_k(G))$, with the size in the range of $[1, b-1]$. To train this network, we need to design a sampler that can sample batches of partial solutions of size in the range of $[1, b-1]$. The label of each training sample is a vector that represents the probability $p(v|S)$ of the remaining nodes being selected as critical nodes given the partial solution S. The probability $p(v|S)$ is defined below:

$$p(v|S) = \frac{p(v, S)}{p(S)} \propto p(S \cup \{v\}).$$ (1)

In Eq. (1), $p(S \cup \{v\})$ is the marginal probability of the comprehensive solution, which can be estimated by computing the criticalness of the extended set $Cr(S \cup \{v\})$ explicitly.

$$p(S \cup \{v\}) = \frac{Cr(S \cup \{v\})}{\sum_{v' \in V} Cr(S \cup \{v'\})}.$$ (2)

As for sampling partial solutions from hypergraph as the training data, the sampler must consider the quality of the single node and the node set sampled to ensure efficacy. Meanwhile, the sampler has to be efficient. In our sample method, instead of sampling node v uniformly from the node set $V(C_k(G))$, we sample each node with a sampling probability prob (v), i.e., the normalized criticalness of v (Eq. (3)).

$$\text{prob}(v) = \frac{Cr(v)}{\sum_{v' \in V} Cr(v')}. \tag{3}$$

During training, batches of sampled partial solutions are fed into the model, and the corresponding soft-label is calculated by $p(v|S)$ (Eq. (2)). Using the entropy as the training loss, as given in Eq. (4), the network has the ability to predict the most likely node v, given partial solution S.

$$L(S) = \sum_{v \in V} -p(v|S) \log f(v, S, G, \Theta). \tag{4}$$

Given any partial solution S, the output of $f(S, G, \Theta) \in \mathbb{R}^{|V(C_k(G))|}$ is a vector composed of all remaining nodes v as critical node probability. When generating the critical node set, we replace the greedy strategy in Algorithm 1 with trained $f(v, S, G, \Theta)$. Theoretically, the learned function $f(v, S, G, \Theta)$ can improve the greedy algorithm by leveraging the critical combinations of node sets, which are sampled from the hypergraph based on the node criticalness. The biased sampling enables the observation to concentrate on high potential critical nodes. To implement $f(v, S, G, \Theta)$, we next introduce a model based on a hypergraph convolutional network, which is a framework to handle combinatorial complexity and structural complexity in an end-to-end fashion.

5.3   Critical Node Inference

In H^2GCN, we adopt the hypergraph convolutional network (HGCN) to capture the complex hypergraph topological structure information for generating nodes representation in the sampled partial solution S. For the pooling solution, we adopt the self-attention mechanism to aggregate the node embedding to a fixed-length hypergraph representation, which encodes the importance of different nodes into the hypergraph representation. The overall architecture of our proposed critical node inference module is shown in Fig. 3.

Given a hypergraph $G = (V, E)$, the hypergraph can be denoted by a $|V| \times |E|$ incidence matrix $\mathbf{H}$ where V is the node set, E is the hyperedge set. If $v \in \mathcal{H}$, then $h(v, \mathcal{H})$ equals 1, otherwise equals 0. For a node $v \in V$, its degree is defined as $deg(v, G) = |Adj(v, G)| = \sum_{e \in E} h(v, \mathcal{H})$. For a hyperedge $\mathcal{H} \in E$, its degree is defined as $deg(\mathcal{H}, G) = \sum_{v \in V} h(v, \mathcal{H})$. We use $\mathbf{D}_e$ and $\mathbf{D}_v$ to denote the diagonal matrices of the edge degrees and node degrees. Given a hypergraph signal $\mathbf{X} \in \mathbb{R}^{n \times C_1}$ with n nodes and C_1 dimensional features, the HGCN hypergraph convolution can be formulated by:

$$\mathbf{Y} = \mathbf{D}_v^{-1/2} \mathbf{H} \mathbf{W} \mathbf{D}_e^{-1} \mathbf{H}^\top \mathbf{D}_v^{-1/2} \mathbf{X} \Theta, \tag{5}$$

where $\mathbf{W} = \text{diag}(w_1, ..., w_n)$ and $\Theta \in \mathbb{R}^{C_1 \times C_2}$ are the parameters to be learned during the training process. After the convolution, we can obtain $\mathbf{Y} \in \mathbb{R}^{n \times C_2}$, which can be regarded as special Laplacian smoothing on the node features for generating new features. The HGNN layer can perform node-edge-node transform, which can better refine the features using the hypergraph structure. More specifically, for each sampled partial solution S, the initial nodes feature $\mathbf{X}^{(0)}$ is built by assigning the $\mathbf{X}^{(0)}(i) = 1$ if the i-th node in S, otherwise $\mathbf{X}^{(0)}(i) = 0$. Auxiliary node features like degree, clustering coefficient of each node can be concatenated to $\mathbf{X}^{(0)}$ to form a more cohesive initial node feature. We also use a learnable filter matrix $\Theta^{(0)}$ to extract C_2-dimensional feature. Then, the node feature is gathered according to the hyperedge to form the hyperedge feature $\mathbb{R}^{E \times C_2}$, which is implemented by the multiplication of $\mathbf{H}^\top \in \mathbb{R}^{E \times n}$. Finally, the output node feature is obtained by aggregating their related hyperedge feature, which is achieved by multiplying matrix $\mathbf{H}$. Denote that $\mathbf{D}_v$ and $\mathbf{D}_e$ play a role of normalization in Eq. (5). Thus, the HGNN layer can efficiently extract the high-order correlation on hypergraph by the node-edge-node transform.

After utilizing the two-layer hypergraph convolution to smooth the node features over the hypergraph topology, we feed the node representations $\mathbf{Y}$ to a self-attention mechanism. The self-attention mechanism takes $\mathbf{Y} \in \mathbb{R}^{n \times C_2}$ as input and outputs the attention matrix A_{tt} as the attention weights. For the detailed calculation of A_{tt}, please refer to (6).

$$A_{tt} = \text{softmax}\left(W_{S2} \tanh\left(W_{S1}\mathbf{Y}^\top\right)\right), \tag{6}$$

where $W_{S1} \in \mathbb{R}^{d \times C_2}$, $W_{S2} \in \mathbb{R}^{r \times d}$ are two weight matrices. The weight W_{S1} transforms the node representation linearly from C_2-dim space to d-dim space, which is activated by the non-linear $tanh$ function. Subsequently, the weight W_{S2} learns the importance of each node on the k-hypercore, in r aspects. It serves as there are r experts rating the importance of nodes in independent perspectives.

By multiplying the annotation matrix A_{tt} with the node representation $\mathbf{Y}$ as Eq. (7), a final graph representation $\mathbf{Y}' \in \mathbb{R}^{r \times C_2}$ is computed. The attention mechanism acts as a pooling layer, which sums up the embedding of each node weighted by the attention learned in Eq. (6). It is worth noting that $\mathbf{Y}'$ is a fixed-length representation controlled only by the hyperparameters r and C_2.

$$\mathbf{Y}' = A_{tt}\mathbf{Y}. \tag{7}$$

Finally, flatting $\mathbf{Y}'$ to a vector e, we feed e to an MLP with sigmoid activation to estimate the probabilities $f(v, S, G; \Theta)$ for the given node u:

$$f(v, S, G; \Theta) = \hat{p}_\Theta(v|S) = \text{MLP}(e)[v], \tag{8}$$

where the output of the MLP is a n-dim vector, representing the probability of selecting a node v and adding it to critical node set given the partial solution S. As we mentioned before, the learned function $f(v, S, G; \Theta)$ can be used to replace the greedy strategy in Algorithm 1. In each iteration, we only need to select a node with the highest predicted probability that is not in partial solution S from all n nodes as the critical node.

6   Experimental Studies

6.1   Experiment Setup

Datasets: We use five real-world hypergraphs from three different domains [2] after removing duplicated or singleton hyperedges. Refer to Table 2 for some statistics of the hypergraphs. The specific description of the meaning of nodes and hyperedges in the datasets is as follows:

- **contact** (contact-primary [16] and contact-high [14]): Each node represents a person, and each hyperedge represents a group interaction among individuals.
- **drugs** (NDC-classes and NDC-substances): Each node represents a class label (in NDC-classes) or a substances (in NDC-substances) and each hyperedge represents a set of labels/substances of a drug.
- **threads** (threads-ubuntu): Each node represents a user, and each hyperedge represents a group of users participating in a thread.

Table 2. Summary statistics of five real-world hypergraphs: the number of nodes $|V|$, the number of hyperedges $|E|$.

Dataset	$|\mathbf{V}|$	$|\mathbf{E}|$	Network Type
contact-primary	242	12,704	Face-to-face communication
contact-high	327	7,818	
NDC-classes	1,149	1,049	Drug
NDC-substances	3,767	6,631	
threads-ubuntu	90,054	115,987	Online communities

Baselines: Since the previous works are all based on the traditional pairwise graphs, but we need to complete this task on the hypergraph, so we extend the current algorithms on the traditional pairwise graphs, such as *Greedy*, *Degree* and *SCGCN* to hypergraphs, as our experiment baselines. All of the above approaches are greedy strategies (Algorithm 1) with different heuristics function $f(u; U; G)$.

Implementation and Setting: The learning framework is built on Pytorch with Python 3, while sampling node set and the greedy algorithm are implemented by C++ in order to improve computational efficiency. While sampling the node set, we first randomly generate a number from $[1, b-1]$ as the size of the sample set, and then we sample the node set according to the probability prob in Eq. (3). During the training process, we use Adam optimizer with a learning rate decay to minimize our cross-entropy loss function. Here, we use a two-layer MLP to generate the final output. To achieve the best performance, we grid search for the dropout rates in $\{0.1, 0.2, 0.3, 0.4, 0.5, 0.6, 0.7, 0.8, 0.9\}$, the weight decay in

$\{10^{-4},\ 10^{-3},\ 10^{-2},\ 10^{-1}\}$, the learning rate in $\{10^{-4},\ 10^{-3},\ 10^{-2},\ 10^{-1}\}$ and the mini-batch size in $\{16, 32, 64, 128\}$. For the model-specific hyper-parameters, we fine-tune them based on the results in the validation set. Both training and inference are performed on a single RTX 6000.

6.2 Experimental Results

Exp-1: Effectiveness on Different Datasets. In Fig. 4, we list the overall results of different k-hypercore sizes on different datasets. These figures show the total number of nodes removed in k-hypercore after removing the critical nodes generated by different algorithms at each step, which is used as an indicator to compare different algorithms. From the macro point of view in the figure, the overall collapse of k-hypercore appears as a stable collapse with jumping points.

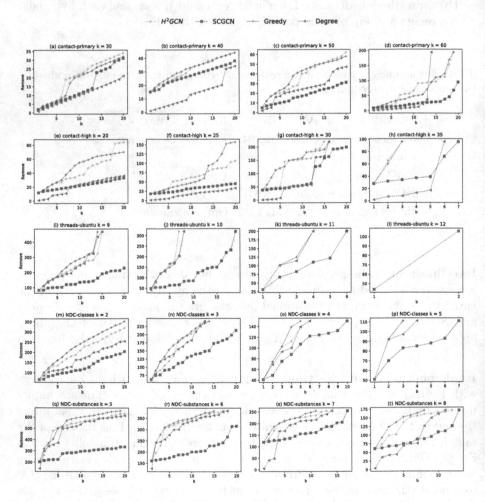

Fig. 4. Cumulative removals in various k.

The stable collapse process is due to the partial removal of each critical node, and the jumping point comes from the cumulative effect of the previous collapse of each node, which ultimately has a qualitative impact on the k-hypercore.

For different hypergraph structures, whether there is a structural critical node becomes the key to determining the quality of different algorithms. If the number of removed nodes is proportional to the budget size b, *Greedy* and *Degree* will perform well. However, if there are indeed some nodes that can change the graph structure qualitatively, the learning-based algorithm can find these nodes in advance compared to *Greedy* and *Degree*, so as to obtain the final good effect. For *SCGCN*, the reason for the poor performance is that the hypergraph has to be converted into a pairwise graph to fit the GCN, which results in the loss of structural information. This further illustrates the necessity of using hypergraph networks to capture structural information.

We further elaborate the results for 30-hypercore of contact-primary-school, as shown in Fig. 4(a). Our method finds a jumping point at $b = 7$, earlier than the one found by *Greedy* at $b = 8$. In addition, our method stays ahead until the last $b = 20$. For 60-core of contact-primary-school (Fig. 4(d)), our method first finds a jumping point at $b = 7$ and then another jump point at $b = 11$, resulting in an 80% improvement in Greedy. For 25-core of contact-high-school (Fig. 4(f)), our method finds a jumping point much earlier than *Greedy*. For 20-core of contact-high-school (Fig. 4(e)), our method finds a small jumping point at $b = 18$, which improves the number of collapsed followers by 20% for *Greedy*. Another interesting observation is that for 10-core of threads-ask-ubuntu (Fig. 4(j)), the whole community is collapsed by 7 critical users in our method, while *Greedy* and *Degree* generate a solution of 8 nodes for this complete collapse.

In most cases, our method successfully finds the jumping points earlier than other methods. Furthermore, we observe that before finding the jumping points, our model always selects an alternative critical candidate to remove, which could have slightly fewer local quantitative removing followers. But this selection leads to better final results, as shown in Fig. 4(c), Fig. 4(e) and Fig. 4(n). It indicates that our model successfully captures the long-term accumulative qualitative collapsing and overcomes the short-term deficiency. That is why our method can perform better than other baselines.

Exp-2: Gap to Optimal Solution. In order to further explore the effectiveness of our model, we conduct experiments on two relatively small datasets to compare the gap between *Greedy*, *Degree*, *SCGCN*, our proposed H^2GCN, and the optimal results by varying b on contact-high-school (Fig. 5(a)) and varying k on threads-ask-ubuntu (Fig. 5(b)). The optimal result is obtained by traversing all possible node combinations, where the time complexity is $O\left(n^b m\right)$. Here, m denotes the cost of computing the criticalness of a node set with b nodes and in the worst case, $m = O\left(|E|\right)$. From the Fig. 5(a), we can find that when $b = 1$, *Greedy*, *SCGCN* and H^2GCN can both find the optimal solution while *Degree* failed. That is because the node with the largest degree does not necessarily cause the collapse of k-hypercore. As b continues to increase, the optimal solution gradually opens up the gap with *Greedy* and H^2GCN, which also gives

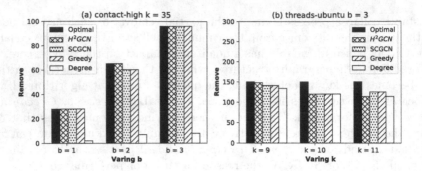

Fig. 5. Compare with optimal solution

us the possibility of surpassing the *Greedy* algorithm. When $b = 2$, the cumulative number of nodes removed by H^2GCN far exceeds the number of nodes removed by the *Greedy* and *SCGCN*, which means that our algorithm finds a potential structural critical node earlier than the *Greedy* and *SCGCN*. As for different k, it represents the density of different k-hypercore communities. Intuitively, the smaller the k value, the fewer the density of the k-hypercore community. Figure 5(b) shows that in a community with a small k value (like $k = 9$), our algorithm can find critical node more easily, while in a community with a large k value (like $k = 11$), it is more difficult to find structural critical node within a few steps due to the relatively large density.

Exp-3: Efficiency of Detection. Finally, we compare the running time of the prediction of H^2GCN, comparing with the other three baseline algorithms. Figure 6 shows the running time of H^2GCN in the prediction phase, and that of the baselines, where the time of loading the hypergraph and computing the corresponding k-hypercore is excluded because it is the same for all. In Fig. 6, the running time of the four algorithms is on the same magnitude, where in the vast majority of cases, *Degree* is the cheapest and *SCGCN* is the most expensive approach. The reason for this phenomenon can be attributed to the fact that

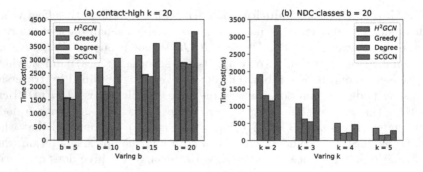

Fig. 6. Comparison on running time

the extended $SCGCN$ needs to perform an additional process of converting the hypergraph to a pairwise graph. With the competitive prediction time, H^2GCN is a promising option to improve the solution quality when sufficient computation resources are available.

7 Conclusions

In this paper, we propose a new approach for detecting critical nodes in hypergraphs using a hypergraph convolutional network. We analyze the limitations of existing pairwise algorithms and highlight the strengths of hypergraphs in representing associations between real-world data. Our new model, called H^2GCN, learns the criticalness of node sets generated by biased sampling of k-hypercore communities. Extensive experiments show that our hypergraph convolutional network-based approach outperforms traditional pairwise algorithms in detecting critical nodes. Our work demonstrates the potential of hypergraph convolutional networks in solving combinatorial optimization problems and provides a new method for detecting critical nodes in hypergraphs.

Acknowledgment. The work was supported by National Key Research and Development Program of China (Grant No. 2020YFB1707900, No. 2021YFB2700700), National Natural Science Foundation of China (Grant No. 62072035), Open Research Projects of Zhejiang Lab (Grant No. 2020KE0AB04), CCF-Huawei Database System Innovation Research Plan (Grant No. CCF-HuaweiDBIR2021007B).

References

1. Arulselvan, A., Commander, C.W., Pardalos, P.M., Shylo, O.: Managing network risk via critical node identification. In: Risk Management in Telecommunication Networks, pp. 79–92 (2011)
2. Benson, A.R., Abebe, R., Schaub, M.T., Jadbabaie, A., Kleinberg, J.: Simplicial closure and higher-order link prediction. PNAS **115**(48), E11221–E11230 (2018)
3. Borgatti, S.P.: Identifying sets of key players in a social network. Comput. Math. Organ. Theory **12**(1), 21–34 (2006)
4. Bronstein, M.M., Bruna, J., LeCun, Y., Szlam, A., Vandergheynst, P.: Geometric deep learning: going beyond Euclidean data. IEEE Signal Process. Mag. **34**(4), 18–42 (2017)
5. Chen, C., Zhu, Q., Sun, R., Wang, X., Wu, Y.: Edge manipulation approaches for K-core minimization: metrics and analytics. TKDE **35**(1), 390–403 (2021)
6. Feng, Y., You, H., Zhang, Z., Ji, R., Gao, Y.: Hypergraph neural networks. In: AAAI, vol. 33, pp. 3558–3565 (2019)
7. He, J., Liang, H., Yuan, H.: Controlling infection by blocking nodes and links simultaneously. In: Chen, N., Elkind, E., Koutsoupias, E. (eds.) WINE 2011. LNCS, vol. 7090, pp. 206–217. Springer, Heidelberg (2011). https://doi.org/10.1007/978-3-642-25510-6_18
8. Iyer, S., Killingback, T., Sundaram, B., Wang, Z.: Attack robustness and centrality of complex networks. PLoS ONE **8**(4), e59613 (2013)

9. Jiang, J., Wei, Y., Feng, Y., Cao, J., Gao, Y.: Dynamic hypergraph neural networks. In: IJCAI, pp. 2635–2641 (2019)
10. Jin, T., Cao, L., Zhang, B., Sun, X., Deng, C., Ji, R.: Hypergraph induced convolutional manifold networks. In: IJCAI, pp. 2670–2676 (2019)
11. Kipf, T.N., Welling, M.: Semi-supervised classification with graph convolutional networks. arXiv preprint arXiv:1609.02907 (2016)
12. Lozano, M., Garcia-Martinez, C., Rodriguez, F.J., Trujillo, H.M.: Optimizing network attacks by artificial bee colony. Inf. Sci. **377**, 30–50 (2017)
13. Luo, Q., Yu, D., Cai, Z., Lin, X., Cheng, X.: Hypercore maintenance in dynamic hypergraphs. In: 2021 IEEE 37th International Conference on Data Engineering (ICDE), pp. 2051–2056. IEEE (2021)
14. Mastrandrea, R., Fournet, J., Barrat, A.: Contact patterns in a high school: a comparison between data collected using wearable sensors, contact diaries and friendship surveys. PLoS ONE **10**(9), e0136497 (2015)
15. Seidman, S.B.: Network structure and minimum degree. Soc. Netw. **5**(3), 269–287 (1983)
16. Stehlé, J., et al.: High-resolution measurements of face-to-face contact patterns in a primary school. PLoS ONE **6**(8), e23176 (2011)
17. Sun, X., Huang, X., Jin, D.: Fast algorithms for core maximization on large graphs. Proc. VLDB Endow. **15**(7), 1350–1362 (2022)
18. Zhang, F., Zhang, Y., Qin, L., Zhang, W., Lin, X.: Finding critical users for social network engagement: the collapsed K-core problem. In: AAAI (2017)
19. Zhao, K., Zhang, Z., Rong, Y., Yu, J.X., Huang, J.: Finding critical users in social communities via graph convolutions. IEEE Trans. Knowl. Data Eng. **35**(1), 456–468 (2021)
20. Zhou, D., Huang, J., Schölkopf, B.: Learning with hypergraphs: clustering, classification, and embedding. Adv. Neural. Inf. Process. Syst. **19**, 1601–1608 (2006)
21. Zhou, Z., Zhang, F., Lin, X., Zhang, W., Chen, C.: K-core maximization: an edge addition approach. In: IJCAI, pp. 4867–4873 (2019)
22. Zhu, W., Chen, C., Wang, X., Lin, X.: K-core minimization: an edge manipulation approach. In: Proceedings of the 27th ACM International Conference on Information and Knowledge Management, pp. 1667–1670 (2018)

Adaptive Label Cleaning for Error Detection on Tabular Data

Yaru Zhang[1]($^{(\boxtimes)}$), Jianbin Qin[1], Rui Mao[1], Yan Ji[1], Yaoshu Wang[1], and Muhammad Asif Ali[2]

[1] Shenzhen Institute of Computing Sciences, Shenzhen University, Shenzhen, China
zhangyaru2020@email.szu.edu.cn, {qinjianbin,mao,jy197541}@szu.edu.cn,
yaoshuw@sics.ac.cn
[2] King Abdullah University of Science and Technology, Jeddah, Saudi Arabia
muhammadasif.ali@kaust.edu.sa

Abstract. Existing supervised methods for error detection require access to clean labels to train the classification model. While the majority of error detection algorithms ignore the harm of noisy labels to detection models. In this paper, we design an effective approach for error detection when both data values and labels may be noisy. Nevertheless, we present AdaptiveClean, a method for error detection on tabular data with noisy training labels. We introduce an effective strategy that can choose the most representative instance to clean. For feature extraction, we use the existing four error detection algorithms for handling multiple types of errors. To reduce the negative effect of noisy training labels on the classification model, we use an adaptive label-cleaning method by training any arbitrary ML models iteratively. Our approach can not only prioritize erroneous instances but also clean noisy labels that affect the classifier primarily. Performance evaluation using five different datasets shows that AdaptiveClean excels over the best baseline error detection system by 0.01 to 0.12 in terms of F1 score.

Keywords: Error Detection · Data Quality · Noisy Labels

1 Introduction

Error detection is a prominent task in data-driven applications, which involves identifying incorrect, missing, and duplicate values [9]. Traditionally, data curators manually perform error detection based on prior knowledge and experience [30] which can be time-consuming and requires domain expertise.

Much existing research was proposed to address the issue of erroneous values There are many traditional methods focusing on identifying violations of integrity constraints [7,15,40], patterns [10,17,31], and outlier [13,29,38]. While these methods often require prior knowledge of the data and rely on predefined detection schemes, making it challenging for non-expert users to configure. These methods have limitations in terms of their applicability to different types

X. Song et al. (Eds.): APWeb-WAIM 2023, LNCS 14334, pp. 63–78, 2024.
https://doi.org/10.1007/978-981-97-2421-5_5

of data. Therefore, there is a need for more accessible and flexible approaches that can handle different types of errors. A different approach is learning-based error detection. Such methods can be used to detect erroneous values by learning from labeled samples [15,21,24,28]. They aim to reduce the workload of data scientists through automated or semi-automated error detection methods. However, the accuracy of the learning-based approach depends on the correctness of the training data. Heterogeneous tabular data are collected from multiple sources and often suffer from erroneous attributes and labels. Both noisy features and labels have a negative impact on the performance of the classification model, while noise in labels is more harmful than features [42]. Therefore, it is essential to consider both erroneous attributes and labels when training the classifier.

Noisy labels are a natural consequence of the dataset collection process. Labeling usually involves some degree of automatic labeling or crowdsourcing, which are error-prone. It will be a burden for users to train the model. The simple approach to dealing with the corrupted training labels is correcting or deleting the noisy labels. Relabeling will introduce more noise, and filtering may remove actual correct data [18,26]. Learning with noisy data and effectively improving the model's accuracy is a non-trivial task. On the other hand, most of the studies on noisy labels focus on homogeneous data, such as images, texts, audio, etc. While in practical applications of many fields, the most common data type is tabular data. Research of noisy labels in heterogeneous tabular data is still underexplored, especially when there are both erroneous values and labels in the training dataset.

To cope with the label corruption and support the learning with noisy training data on tabular data, we present a new architecture, called AdaptiveClean. We address the following questions: i. How effectively do we represent the characteristics of the erroneous values? ii. How can we detect errors using a classification model efficiently when there are noisy values and labels in the training data? AdaptiveClean introduces a new structure for error detection on tabular data with noisy training labels. Considering the previous learning-based approach neglecting noisy training labels, we overcome the negative performance impact for error detection based on the feedback of the ML model. Inspired by Raha [24] and TAR [11], we aggregate the results of existing four error detection algorithms to construct feature engineering, which can enhance the representation ability of data values and identify multiple types of actual errors. In addition, we propose an adaptive label-cleaning strategy that selects noisy labels and iteratively improves the quality of training data.

Contribution. The main contribution of this paper is the new holistic framework for error detection tasks, AdaptiveClean, which can effectively reduce the negative impact of noise in the training data and select the noisy labels to clean using Bayesian probability and model performance improvement. Our solution proposes two key technologies: (i) An informative representation for values that aggregate four existing error detection algorithms, such as rule-based violation [7,40], pattern-based violation [17], outlier [29], and external knowledge-base [6], to identify multiple erroneous values accurately. (ii) Considering the negative influence of noisy training labels, we adaptively and iteratively improve the quality of labels and classifiers based on the feedback of the ML model.

2 Error Detection with Noisy Labels

In this paper, we focus on detecting errors on tabular data in a learning-based manner. We define the problem of this paper and discuss how we clean noisy labels using our approach to solve the problem effectively.

2.1 Problem Definition

Manual labeling for tabular data is often expensive and challenging, particularly in real-world scenarios. Furthermore, the labels that we used in the training and testing dataset usually contain a certain proportion of noise. It is essential to address the negativeness of noisy labels for error detection. While this paper only focuses on the training process. We define the problem as detecting errors using a learning-based approach in the presence of noisy training labels.

Definition (Error Detection with Noisy Labels). The goal of error detection is to identify erroneous entries in a dataset. Given a dirty training dataset d with m tuples as $T = \{t_1, t_2, ..., t_m\}$ and n attributes as $A = \{a_1, a_2, ..., a_n\}$. Each cell in the dataset d is associated with a label, denoted by $y = \{l_{d[1,1]}, l_{d[1,2]}, ..., l_{d[i,j]}..., l_{d[m,n]} | l \in \{0 \text{ or } 1\}\}$, where $l_{d[i,j]} = 0$ if $d[i,j]$ is an erroneous value and $l_{d[i,j]} = 1$ is a correct value. What we need to do is to train a classification model M to predict the correctness of each cell $d[i,j]$ effectively. Specifically, the model is trained using the dirty training dataset (X_{train}, Y_{train}), where both the feature X_{train} and labels Y_{train} may be erroneous. The test dataset (X_{test}, Y_{test}) where the features X_{test} might be incorrect but the labels Y_{test} are assumed to be correct. Although the training labels contain noise, the test labels must be clean to ensure that the evaluation is not affected by noisy labels. Our goal is to develop a classification model that can identify erroneous entries with high precision, recall, and F1 score, making it comparable to the model trained on clean data.

2.2 Overall Approach

The main purpose of this paper is to detect erroneous values effectively when there exist noisy training labels. Figure 1 shows the overview of our framework. The system starts by using the external knowledge-base and the existing error detection algorithms to configure the strategies (Step 1). Next, these strategies will identify erroneous values, and use the results to extract features for each cell (Step 2). For each iteration, we use the feature vectors generated by step 2 and the cleaned training labels to train the current optimal error detection model. During iterations, only the labels are cleaned, and the feature vectors remain unchanged. Then we calculate the expected model performance improvement (MPI) for each sample (Step 3). We perform adaptive label cleaning based on

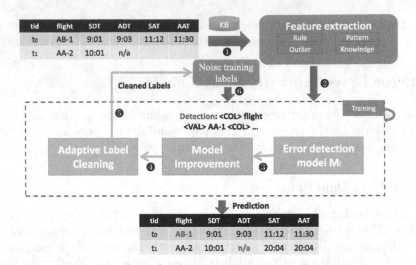

Fig. 1. The framework of AdaptiveClean

feedback from MPI. The examples that have the greatest impact on the model are selected as cleaned (Step 4). Finally, we update the classification model and the training labels to enter the next iteration process (Steps 5 and 6).

2.3 Feature Engineering

To detect multiple types of data errors, we aggregate four existing error detection algorithms for generating feature vectors. As shown in Fig. 2, we use rule-based violation, pattern-based violation, outlier, and knowledge-based violation to form multiple error strategies. Each strategy will mark one cell as either 0 or 1 which means the value is erroneous or correct. All the detection results of strategies as input vectors for the classification model. Vectors constructed in this way can improve the representation ability of data, and detect as many errors as possible.

Strategy Configuration. Real-world data is collected from the web, system, and user input, which is easy to introduce missing values, inconsistency, outliers, etc. Running only one error detection algorithm does well for specific errors, but will result in poor recall when faced with various error types. In order to improve the representation ability of feature vectors, we aggregate the existing four error detection algorithms to form feature engineering. Drawing on the idea of Raha [24] and Metadata-driven [37], we use the four existing error detection algorithms to construct multiple error detection strategies and generate feature vectors. We describe how to aggregate the existing error detection algorithms to form the feature vectors required by the classification model.

- **Rule-based Violation Detection.** It uses integrity constraints to identify inconsistent data [7,40]. This paper mainly focuses on the functional dependency form as $\{X \rightarrow Y$, both X and Y are a single attribute$\}$. In this paper,

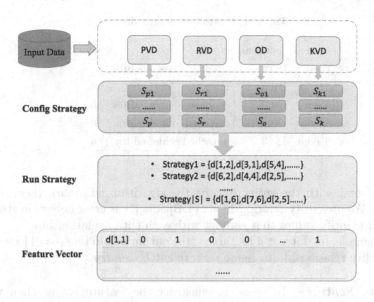

Fig. 2. The generation of feature vector for each cell

we generate all attribute pairs as possible functional dependencies. Regarding each attribute pair $\vee_{a \neq a'} \in A$, the detection strategy $s_{a \to a'}$ marks the value of $d[i, j]$ that violates the functional dependency as erroneous.

For example, on dataset d in Table 1, the output of rule violation detection algorithm would be $s_{r1} = \{d[2,3], d[3,3], d[4,3], d[5,3]\}$ by checking the FD $name \to status$.

- **Pattern-based Violation Detection.** It evaluates data values for correctness based on compatibility with predefined data schemas. In order to avoid an infinite set of possible data patterns, bag-of-character [36] is used to encode all possible data patterns. We generate a set of character-checking strategies s_{ch} to check the validity of each value.

  For example, on dataset d in Table 1, the output of pattern violation detection algorithm would be $s_{p1} = \{\}$ by setting the attribute job to $not\text{-}null$.

- **Outlier Detection.** It detects data values that deviate from the predefined data distribution of values in a specific column. In this paper, the generation of outlier policies will be implemented using statistical and distance-based methods [34]. The detection strategy s_o treats data cells with a frequency less than a threshold as errors using a histogram model. The distance-based approach, named Gaussian model, builds a Gaussian distribution based on the magnitude of data values. It deems data cells whose distances normalized to the mean are greater than a threshold as errors [29].

  For example, on dataset d in Table 1, the output of histogram-based detection algorithm would be $so1 = \{d[2,4]\}$ by setting $\theta_{tf} = \frac{2}{5}$ for attribute job.

- **Knowledge-based Violation Detection.** It matches the columns in data with an external knowledge base [6]. If values in matched two columns are

Table 1. Customer records dataset

	name	company	status	job	salary_level
t_1	Ford	FOX	single	assoc editor	1
t_2	Taylor	IACI	married	journalist	4
t_3	Taylor	IACI	espoused	assoc editor	6
t_4	Taylor	IACI	divorced	chief editor	10
t_5	Taylor	IACI	detached	chief editor	4

inconsistent with the entity library, the algorithm will mark them as erroneous. We use entity-relationships in DBpedia [1] as error detection strategies s_r and identify values that conflict with each entity-relationship r.

For example, on dataset d in Table 1, the output would be $s_{k1} = \{\}$ by setting the entity relationship to *name isPresidentOf country*

Generate Features. In order to construct the feature engineering, we use the results of error detection strategies to map each cell to a feature vector. Inspired by [24,27], we generate multiple types of error detection strategies $S_{col} = \{s_1, s_2, ..., s_{|S|}\}$ for a specific column *col*. Each detection strategy $s \in S_{col}$ will either mark a data cell $d[i,j]$ as erroneous or correct. Specifically expressed as:

$$s\ (d[i,j]) = \begin{cases} False, & \text{s marks } d[i,j] \text{ as error} \\ True, & \text{otherwise} \end{cases} \tag{1}$$

For a specific column, we run each error detection strategy $s \in S_{col}$ on data cell $d[i,j]$ respectively. The output of detection is converted to a logical value into a binary integer. Formally, it can be expressed as:

$$V_{cell}(d[i,j]) = \{int(s(d[i,j])) \mid \forall_s \in S\} \tag{2}$$

Since the feature dimensions of each column are not exactly the same, we need to generate features in a column-based approach. The number of strategies in each column may be different. In order to create an efficient representation, corresponding feature vectors V_{col} are generated for each particular column j.

$$V_{col}(j) = \{v_{cell}(d[i,j]) \mid 1 \le i \le m\} \tag{3}$$

For example, we run error detection strategies $\{s_{r1}, s_{p1}, s_{o1}, s_{k1}\}$ on column *status*. If the value is marked as erroneous by the detection strategy, the corresponding dimension of the feature vector is 0, otherwise, it is marked as 1. Using this, we can obtain $v_{d[1,3]} = \{1,1,1,1\}$, $v_{d[2,3]}, v_{d[3,3]}, v_{d[4,3]}, v_{d[5,3]} = \{0,1,1,1\}$.

2.4 Label Sampling Strategy

Class label noise is a critical issue that affects the quality of the trained model, resulting in decreased accuracy, longer training time, lower interpretability, and increased model complexity potentially [12]. However, manually correcting noisy labels is time-consuming and prone to errors, rendering it infeasible in many situations [42]. Deep learning methods have been successful in handling noisy labels in homogeneous data such as images, audio, and text, but they have shown limited effectiveness in dealing with heterogeneous tabular data. The deep neural network's performance tends to degrade when faced with tabular data.

To address the issue of noisy labels on tabular data, we propose an adaptive label-cleaning method to enhance the quality of label and model. It is commonly acknowledged that the training labels may contain some level of noise, but the specific location of noise is unknown. In order to identify potential noisy samples, we utilize Bayesian probability to calculate the likelihood of samples being noisy and identify samples with high probabilities as candidates for cleaning.

Assuming that M is a model trained on noisy training data d without any modifications, and M_i is trained after correcting the instance label y_i', there are two possible cases for M_i:

Case 1: If the label y_i' of instance i is the same as the ground truth, no cleaning is necessary. The model remains unchanged as $\{y_i = y_i', M_i = M\}$.

Case 2: It requires cleaning if the label y_i' is noisy. The cleaning operation is treated as label flip in the binary classification task where $\{y_i = -y_i' + 1\}$. We then retrain the model using the cleaned labels, and the change in the model's performance is measured as $\{MPI = eval(M_i) - eval(M)\}$, where $eval$ represents the metric of the model.

It should be noted that cleaning instances do not always guarantee an improved model. In some cases, the model's performance may even degrade. We must remain cautious and optimistic about the situation. We cannot know whether the labels are noisy or not until the instances are cleaned. We can estimate the expected improvement of the model based on the probability of each scenario. Pro(Case1) and Pro(Case2) represent the probability of two cases respectively. For an observation label y_i', we can define the MPI resulting from cleaning the label as:

$$MPI_i = \exp^{(Pro(Case1)*0 + Pro(Case2)*(eval(M_i) - eval(M))}$$
$$= \exp^{Pro(Case2)*(eval(M_i) - eval(M))} \qquad (4)$$

Algorithm 1 takes a feature vector V, training noisy label Y', and dirty model M_d trained on noisy data as input. V is obtained by running multiple error detection strategies on data d. The initial value of M_d is trained without any cleaning operation. The system then returns cleaned labels Y_c and model M_c which is training with Y_c. Our cleaning strategy involves computing the expected performance improvement of the classification model, denoted by $\{MPI[i, j] \mid i \in T[t_1, t_2, ..., t_m], j \in A[a_1, a_2, ..., a_n]\}$, resulting from cleaning

---

Algorithm 1. Adaptive label cleaning algorithm

Input: Input: feature vector V of dataset d, noise labels Y', dirty model M_d
Output: Output: clean model M_c
1: Let $M_c = M_d$
2: **for** j in attribute $A[a_1, a_2, ..., a_n]$ **do**
3:     **for** i in tuple $T[t_1, t_2, ..., t_m]$ **do**
4:       $MPI[i,j] = \exp^{Pro(Case2)*(eval(M_i[j])-eval(M_c[j]))}$
5:     **end for**
6:     $i' = \arg\max_i(MPI_i)$
7:     $Y'_j = Y'_j[y_1, y_2, ... - y_i + 1, ..., y_m]$
8:     $M_c[j] = M_{i'}[j]$
9: **end for**
10: **return** M_c

---

each instance $d[i,j]$. We then select and clean the instance labels that have the greatest impact on the model. Formally as:

$$i' = \arg\max_i(MPI_i) \qquad (5)$$

Our goal is to clean the label of instance i that has the largest performance improvement. The sampling strategy should be designed to balance the trade-off between the impact of cleaning a label and the likelihood that it is noisy. Based on the model's feedback, we present an adaptive label-cleaning strategy that aims to clean the noisy labels with the greatest impact on model performance. Inspired by [11], we define the probability that ground truth label $Y_i = -y'_i + 1$ is opposite to the observation label $Y'_i = y'_i$ as Pro(Case2) and calculate the probability using Bayesian conditional probability [3,35]. We formula it as:

$$
\begin{aligned}
Pro(Case2) &= Pro(Y_i = -y'_i + 1 \mid Y'_i = y'_i) \\
&= \frac{Pro(Y'_i = y'_i \mid Y_i = -y'_i + 1) * Pro(Y_i = -y'_i + 1)}{Pro(Y'_i = y'_i)} \\
&\propto Pro(Y'_i = y'_i \mid Y_i = -y'_i + 1) * Pro(Y_i = -y'_i + 1)
\end{aligned}
\qquad (6)
$$

The expression $Pro(Y'_i = y'_i \mid Y_i = -y'_i+1)*Pro(Y_i)$ denotes the probability that the actual label $Y_i = -y'_i + 1$ is different with the observed label $Y'_i = y'_i$, which is equivalent to the label's noise rate $r^{(i)}_{-y'_i+1,y'_i}$. $Y_i = -y'_i + 1$ is equal to the prior $Y = -y'_i + 1$. Therefore, we obtain:

$$Pro(Case2) \propto r^{(i)}_{-y'_i+1,y'_i} * Pro(Y = -y'_i + 1) \qquad (7)$$

Similarly, we can have:

$$Pro(Case2) \propto r^{(i)}_{y'_i,y'_i} * Pro(Y = y'_i) \qquad (8)$$

According to the relationship between the Pro(Case1) and Pro(Case2), we can obtain Pro(Case1) + Pro(Case2) = 1. So $Pro(Case2)$ can be expressed as:

$$Pro(Case1) = \frac{r^i_{-y'_i+1,y'_i} * P(Y = -y'_i + 1)}{r^i_{-y'_i+1,y'_i} * P(Y = -y'_i + 1) + r^i_{y'_i,y'_i} * P(Y = y'_i)} \tag{9}$$

As shown in Algorithm 1, given a model M_d and training data with m tuples containing noisy labels Y', the algorithm is designed to automatically select the most representative noisy labels to be cleaned. For an iterative process, we assign the model M_d trained on the initial dirty data to M_c. For each attribute column as $d[j]$, we traverse each cell $d[i,j]$ and execute the cleaning process separately. As described in line 4, we compute the MPI for all cells in a particular column. Then we select the index value i' with the largest MPI to be cleaned, as described in line 6. We update the model with the cleaned labels and obtain the current best model $M_c[j]$ for column j. We repeat this process for all columns in the dataset to obtain a higher quality model M_c. Using this approach, we can effectively improve the model's accuracy in an adaptive way.

3 Experiments

3.1 Experimental Setup

Datasets. We evaluate the effectiveness of our approach on five datasets that are described in Table 2. (1) Flights is a real-world dataset that we can obtain from previous research [33]. Data errors occur due to inconsistencies in the time of the same flight in the data source. (2) Rayyan is a dataset that describes literature information and covers almost all types of errors. (3) Movies is a dataset taken from the Magellan repository [8]. It is an available dataset that describes specific information about the released movie. (4) Hospital is a real-world dataset that we can obtain from Raha. Errors are artificially introduced by injecting typos. (5) Soccer is a synthetic dataset that describes the information of soccer players and their teams [32]. We randomly sample 20000 tuples for performance evaluation.

Comparison Methods. We evaluate our approach AdaptiveClean against four existing learning-based error detection methods. ActiveClean [21] uses the idea of gradient descent and selects a batch of data with the largest gradient descent for cleaning iteratively. Naive-Mix (NM) samples a batch of candidate data to be cleaned randomly and uses the newly obtained complete dataset for model fitting and prediction. Unlike Naive-Mix, Naive-Sampling (NS) cleans a sampled batch of records, and only uses data that has been cleaned so far for re-training. We also include a configuration-free system Raha [24] that uses multiple error detection algorithms to constitute feature vectors for training. It expands the labeled training data by propagating labels using a clustering-based method.

Table 2. A summary of datasets. The error types are missing value (M), typo (T), data format issue (F), and violated functional dependency (VFD)

Name	Size	Error Rate	Error Types
Flights	2376 * 7	0.30	M,F,VFD
Rayyan	1000 * 11	0.09	M,T,F,VFD
Movies	7390 * 17	0.06	M,F
Hospital	1000 * 20	0.03	T,VFD
Soccer	20000 * 10	0.11	T,F

Table 3. Comparison with error detection baseline

Method	Flights			Rayyan			Movies			Hospital			Soccer		
	P	R	F_1	P	R	F_1	P	R	F_1	P	R	F_1	P	R	F_1
ActiveClean	0.75	0.73	0.74	0.86	0.81	0.83	0.84	0.84	0.84	0.85	0.88	0.86	**0.85**	0.85	0.85
NS	0.78	0.67	0.72	**0.88**	0.74	0.80	0.82	0.81	0.81	**0.87**	0.85	0.86	**0.85**	0.83	0.84
NM	**0.80**	0.73	**0.76**	**0.88**	0.77	0.82	**0.85**	0.83	0.84	**0.87**	0.84	0.85	**0.85**	0.88	0.86
Raha	0.64	0.68	0.66	0.34	0.84	0.48	0.36	0.84	0.50	0.10	0.55	0.17	0.09	0.63	0.16
AdaptiveClean	0.73	**0.74**	0.73	0.85	**0.88**	**0.86**	0.84	**0.86**	**0.85**	0.83	**0.99**	**0.90**	**0.85**	**1.0**	**0.92**

Implementation. In the evaluation, the metric of precision, recall, and F1-score are used to evaluate the performance of error detection. Since the model is trained on each column independently, we report the average score of all columns as the result. We randomly insert 20% of the noisy training labels if not specified in our experiment except for the analysis of noisy labels. We set the test labels to be reliable to ensure the accuracy of model evaluation. There are two parameters $r_{0,1}$ and $r_{1,0}$ we need to configure. For example, given $r_{0,1} = 0.4$ and $r_{1,0} = 0.1$, an instance with ground truth label 1 has a 0.4 probability of being a noisy label, and an instance with ground truth label 0 has a 0.1 probability of being a noisy label. Unless explicitly stated, we use the decision tree model as the default classifier for error detection in all experiments.

3.2   Performance Evaluation

In this experiment, we compare our approach with the baseline on five datasets. As shown in Table 3, AdaptiveClean is superior to the baseline in terms of F1 score and recall. Specifically, AdaptiveClean outperforms the best baseline by 0.01 to 0.06 in terms of F1 score and outperforms the best baseline by 0.01 to 0.12 in terms of recall. Our method cleans the noisy label adaptively based on the feedback of the ML model, which can minimize the negative impact of the noisy training labels. It will improve the performance of the classification model. For precision, AdaptiveClean has a comparable result with the best excellent baseline on Soccer, while it is slightly lower on the other datasets. The superior performance of AdaptiveClean is strongly supported by our high recall and F1 score. The predictive performance of Raha is the worst compared to other methods.

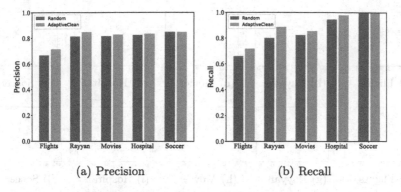

(a) Precision   (b) Recall

Fig. 3. Performance comparison of AdaptiveClean and Random

Raha increases the number of noise labels in the training data by clustering and propagating labels, which seriously affects the performance of the model. The baseline methods shown in Table 3 ignore the impact of noisy training labels, and the sampling strategy does not consider the information delivered by the model. Therefore the average performance is slightly lower than our method.

3.3 Sampling Strategy Analysis

In this experiment, we analyze the effectiveness of the adaptive label-cleaning strategy. We use precision and recall as our metrics to evaluate the implementation of the trained model. We randomly select the labels that need to be cleaned as Random. As shown in Fig. 3, the system that uses the adaptive cleaning method outperforms Random in terms of precision and recall, especially on Flights, Rayyan, and Hospitals. Since we clean the noisy training labels based on the feedback of the classification model which can induce the model to learn corrupted information as much as possible. If we directly predict with the model trained on randomly cleaned labels as Random, the model will be more prone to overfitting to the noisy labels. Therefore, AdaptiveClean gives superior performance compared to using a random way without another strategy.

3.4 System Effectiveness Analysis

In this section, we evaluate the effectiveness of our approach. On the one hand, we analyze the impact of the different proportions of noisy training labels on the model performance. On the other hand, we use a propagation experiment to compare the performance of different α, where α is the ratio of $r_{0,1}$ and $r_{1,0}$.

Influence of Noisy Labels. In this experiment, we study the impact of different proportions of noisy training labels on AdaptiveClean and compare it with Raha, ActiveClean. As shown in Fig. 4, noisy labels in training data can seriously affect the accuracy of the classification model. AdaptiveClean outperforms

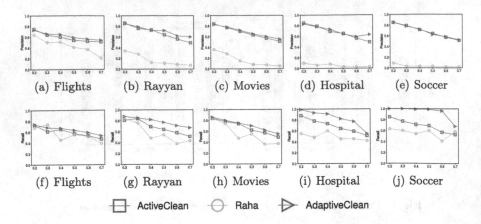

Fig. 4. Different noisy labels rate analysis

Table 4. Performance comparison of different α values

α	Flights			Rayyan			Movies			Hospital			Soccer		
	P	R	F_1	P	R	F_1	P	R	F_1	P	R	F_1	P	R	F_1
$\alpha = 0.25$	**0.73**	**0.74**	**0.73**	0.85	**0.88**	**0.86**	**0.84**	**0.86**	**0.85**	0.83	**0.99**	**0.90**	0.85	1.0	0.92
$\alpha = 0.5$	0.70	0.70	0.70	0.85	0.87	**0.86**	0.83	0.85	0.84	**0.84**	0.96	**0.90**	0.85	1.0	0.92
$\alpha = 0.75$	0.70	0.69	0.69	0.85	0.86	0.85	0.83	0.85	0.84	**0.84**	0.96	0.89	0.85	1.0	0.92
$\alpha = 1.0$	0.69	0.70	0.69	0.85	0.87	**0.86**	0.83	**0.86**	0.84	**0.84**	0.98	**0.90**	0.85	1.0	0.92

Raha and ActiveClean on all five datasets, cause it takes into account the influence of noisy training labels and cleans the noisy labels adaptively based on the feedback of the ML model. The precision and recall of Raha and ActiveClean are always been lower than our method. As the proportion of noise labels in the training data increases, the performance of the three methods decreases, but our method still outperforms others. The purpose of other methods is only to detect erroneous values in the data, but noisy training labels will lead to incorrect information learned by the classification model. In addition, the performance of Raha is slightly worse than ActiveClean, and its accuracy is extremely unstable. The updated strategy of Raha leads to uncertainty in the proportion of noise labels, which will cause fluctuations in the model's performance.

Propagation Experiment. In this experiment, we investigate the performance of different values of α on five datasets, indicating the ratio of different classes. Table 4 shows that changing values of α has a minimal effect on the performance of our approach, especially in Soccer. It is worth noting that the performance of the trained model reported in Table 4 becomes worse with the value of α increasing. In experiments, larger values of α indicate the noise rate is higher, which will lead to negative performance.

4   Related Works

No Learning for Error Detection. Different error detection methods are often tailored for different types of data errors. According to Chu et al. [4], data errors can be divided into two categories: *i.* quantitative errors, which are numeric data that exceed frequency or distance thresholds [16]. DBoost [29] is a comprehensive framework that applies outlier detection algorithms to identify quantitative errors. *ii.* qualitative errors treat violations of predefined rules or patterns as erroneous values [5]. The rule-based systems, such as NADEEF [7], GDR [39], and Holoclean [33], can apply integrity constraints to identify errors. There are also systems that detect values that do not conform to data patterns [20], which help users identify potential quality issues with predefined patterns. While these approaches require users to provide parameters for error detection, only the correct configuration can perform well. This paper proposes a novel approach for error detection that does not require users to participate in the parameter configuration of the system and still achieves good performance.

Learning-Based Error Detection. Naturally, learning-based methods for error detection uses semi-supervised strategies to avoid the need for manually labeled data which rely on pairs of dirty and clean records as training examples [27]. There are some ensembling approaches [15,24,28,37] which can be used to identify the exact location of erroneous values. To enable downstream tasks to work properly, there are several data cleaning systems that serve machine learning have been developed [2,21,23]. The existing methods for error detection do not consider the possibility of noisy labels, which can negatively impact the accuracy of the model. To address this issue, we improve the label quality by cleaning noisy labels iteratively to increase the accuracy of the classification model.

Noisy Labels Processing. Noisy labels can significantly affect the performance of classification models. One obvious solution is to identify and correct suspicious labels to their corresponding ground truth values [41]. Another approach is data pruning, which can be done by using semi-supervised algorithms [22]. By selectively choosing the correct instances, it is possible to guide the network to achieve better performance [19]. Instead of one network supervising another, they collaborate and correct each other's mistakes by leveraging differences [14,25]. The above method works well for homogeneous data, but not for heterogeneous tabular data. While we propose an adaptive method that selects noisy samples based on the feedback of model performance. Prioritizing the most important samples for cleaning can improve prediction accuracy effectively.

5   Conclusion

We present a learning-based approach for error detection which can effectively maintain excellent performance in the case of noise in both attributes and training labels. We introduce an adaptive approach to overcome the challenge of training the error detection model with noisy labels. Our approach iteratively selects the most representative instances, ensuring that the negative impact of noisy labels can be minimized. AdaptiveClean achieves better performance on five experimental datasets when compared to previous state-of-art systems.

Acknowledgements. This research is supported by the National Key R&D Program of China 2021YFB3301500, Shenzhen Continuous Support Grant 20200811104054002, Guangdong Provincial National Science Foundation 2019A1515111047, and Shenzhen Colleges and Universities Continuous Support Grant 20220810142731001.

References

1. Auer, S., Bizer, C., Kobilarov, G., Lehmann, J., Cyganiak, R., Ives, Z.: DBpedia: a nucleus for a web of open data. In: Aberer, K., et al. (eds.) ASWC/ISWC -2007. LNCS, vol. 4825, pp. 722–735. Springer, Heidelberg (2007). https://doi.org/10.1007/978-3-540-76298-0_52
2. Biessmann, F., et al.: DataWig: missing value imputation for tables. J. Mach. Learn. Res. **20**(175), 1–6 (2019)
3. Bolstad, W.M., Curran, J.M.: Introduction to Bayesian Statistics. Wiley, New York (2016)
4. Chu, X., Ilyas, I.F., Krishnan, S., Wang, J.: Data cleaning: overview and emerging challenges. In: Proceedings of the 2016 International Conference on Management of Data (2016)
5. Chu, X., Ilyas, I.F., Papotti, P.: Holistic data cleaning: putting violations into context. In: ICDE. IEEE (2013)
6. Chu, X., et al.: KATARA: a data cleaning system powered by knowledge bases and crowdsourcing. In: 2015 ACM SIGMOD (2015)
7. Dallachiesa, M., et al.: NADEEF: a commodity data cleaning system. In: 2013 ACM SIGMOD (2013)
8. Das, S., Doan, A., Psgc, C.G., Konda, P., Govind, Y., Paulsen, D.: The Magellan data repository (2015)
9. Deng, D., et al.: The data civilizer system. In: CIDR (2017)
10. Dimitriadis, I., Poiitis, M., Faloutsos, C., Vakali, A.: TG-OUT: temporal outlier patterns detection in Twitter attribute induced graphs. World Wide Web **25**(6), 2429–2453 (2022)
11. Dolatshah, M.: Cleaning crowdsourced labels using oracles for statistical classification. Ph.D. thesis, Applied Sciences: School of Computing Science (2018)
12. Frénay, B., Verleysen, M.: Classification in the presence of label noise: a survey. IEEE (2013)
13. Fu, J., Wang, L., Ke, J., Yang, K., Yu, R.: GANAD: a GAN-based method for network anomaly detection. World Wide Web **26**, 2727–2748 (2023)
14. Han, B., et al.: Co-teaching: robust training of deep neural networks with extremely noisy labels. In: Advances in Neural Information Processing Systems, vol. 31 (2018)
15. Heidari, A., McGrath, J., Ilyas, I.F., Rekatsinas, T.: HoloDetect: few-shot learning for error detection. In: Proceedings of the 2019 International Conference on Management of Data (2019)
16. Hellerstein, J.M.: Quantitative data cleaning for large databases. UNECE (2008)
17. Huang, Z., He, Y.: Auto-detect: data-driven error detection in tables. In: Proceedings of the 2018 International Conference on Management of Data (2018)

18. Jeatrakul, P., Wong, K.W., Fung, C.C.: Data cleaning for classification using mis-classification analysis. J. Adv. Comput. Intell. Intell. Inform. **14**(3), 297–302 (2010)
19. Jiang, L., Zhou, Z., Leung, T., Li, L.J., Fei-Fei, L.: MentorNet: learning data-driven curriculum for very deep neural networks on corrupted labels. In: International Conference on Machine Learning. PMLR (2018)
20. Kandel, S., Paepcke, A., Hellerstein, J., Heer, J.: Wrangler: interactive visual specification of data transformation scripts. In: Proceedings of the SIGCHI Conference on Human Factors in Computing Systems (2011)
21. Krishnan, S., Wang, J., Wu, E., Franklin, M.J., Goldberg, K.: ActiveClean: interactive data cleaning for statistical modeling. PVLDB **9**, 948–959 (2016)
22. Li, J., Socher, R., Hoi, S.C.: DivideMix: learning with noisy labels as semi-supervised learning. arXiv preprint arXiv:2002.07394 (2020)
23. Liu, Z., Zhou, Z., Rekatsinas, T.: Picket: self-supervised data diagnostics for ML pipelines. arXiv (2020)
24. Mahdavi, M., et al.: Raha: a configuration-free error detection system. In: SIGMOD (2019)
25. Malach, E., Shalev-Shwartz, S.: Decoupling "when to update" from "how to update". In: Advances in Neural Information Processing Systems (2017)
26. Miranda, A.L.B., Garcia, L.P.F., Carvalho, A.C.P.L.F., Lorena, A.C.: Use of classification algorithms in noise detection and elimination. In: Corchado, E., Wu, X., Oja, E., Herrero, Á., Baruque, B. (eds.) HAIS 2009. LNCS (LNAI), vol. 5572, pp. 417–424. Springer, Heidelberg (2009). https://doi.org/10.1007/978-3-642-02319-4_50
27. Neutatz, F., Chen, B., Abedjan, Z., Wu, E.: From cleaning before ML to cleaning for ML. IEEE (2021)
28. Neutatz, F., Mahdavi, M., Abedjan, Z.: ED2: two-stage active learning for error detection-technical report. arXiv (2019)
29. Pit-Claudel, C., Mariet, Z., Harding, R., Madden, S.: Outlier detection in heterogeneous datasets using automatic tuple expansion (2016)
30. Rahm, E., Do, H.H.: Data cleaning: problems and current approaches. IEEE (2000)
31. Raman, V., Hellerstein, J.M.: Potter's wheel: an interactive data cleaning system. In: VLDB (2001)
32. Rammelaere, J., Geerts, F.: Explaining repaired data with CFDS. VLDB (2018)
33. Rekatsinas, T., Chu, X., Ilyas, I.F., Ré, C.: HoloClean: holistic data repairs with probabilistic inference. arXiv (2017)
34. Ridzuan, F., Zainon, W.M.N.W.: Diagnostic analysis for outlier detection in big data analytics. Procedia Comput. Sci. **197**, 685–692 (2022)
35. van de Schoot, R., et al.: Bayesian statistics and modelling. Nat. Rev. Methods Primers **1**, 1 (2021)
36. Schütze, H., Manning, C.D., Raghavan, P.: Introduction to Information Retrieval. Cambridge University Press, Cambridge (2008)
37. Visengeriyeva, L., Abedjan, Z.: Metadata-driven error detection. In: SSDBM (2018)
38. Xiang, H., Zhang, X.: Edge computing empowered anomaly detection framework with dynamic insertion and deletion schemes on data streams. World Wide Web **25**(5), 2163–2183 (2022)
39. Yakout, M., Elmagarmid, A.K., Neville, J., Ouzzani, M., Ilyas, I.F.: Guided data repair. arXiv (2011)
40. Yan, J.N., Schulte, O., Zhang, M., Wang, J., Cheng, R.: SCODED: statistical constraint oriented data error detection. In: 2020 ACM SIGMOD (2020)

41. Yuan, B., Chen, J., Zhang, W., Tai, H.S., McMains, S.: Iterative cross learning on noisy labels. In: 2018 IEEE Winter Conference on Applications of Computer Vision (WACV). IEEE (2018)
42. Zhu, X., Wu, X.: Class noise vs. attribute noise: a quantitative study. Artif. Intell. Rev. **22**, 177–210 (2004)

A Dual–Population Strategy Based Multi–Objective Yin–Yang–Pair Optimization for Cloud Computing

Hui Xu[✉] and Mingchao Ding

School of Computer Science, Hubei University of Technology, Wuhan 430068, China
xuhui@hbut.edu.cn

Abstract. In order to improve the performance of cloud computing, the multi–objective optimization problems in this field need to be solved efficiently. This paper proposes a novel Dual–Population strategy based Multi–Objective Yin–Yang–Pair Optimization which is termed as DP–MOYYPO. The proposed DP–MOYYPO algorithm makes the following three improvements to Front–based Yin–Yang–Pair Optimization (F–YYPO). First, a population of the same size to explore non–cornered points with large crowding distances is added, so as to enhance the uniformity of the optimized individuals' distribution. Second, the updating method of point P in the splitting stage is modified to reduce the number of function evaluations for the purpose of improving the convergence speed and accuracy. Third, the updating method of P_{2i} in the archive stage is improved. The proposed DP–MOYYPO algorithm has been evaluated using two test suites of the PlatEMO platform. DP–MOYYPO is first compared with F–YYPO and all improved algorithms, and then with five other representative multi–objective optimization algorithms. Experimental results show that, DP–MOYYPO can obtain Pareto fronts with better distribution than F-YYPO, while having faster convergence and higher computational accuracy. And compared with other representative multi-objective optimization algorithms, DP-MOYYPO ranks first in terms of the number of winning test instances in both the convergence metric GD and the integrated performance metric IGD, showing a strong competitive advantage.

Keywords: Cloud computing · Multi-objective optimization · Front-based Yin-Yang-Pair Optimization · Dual-population strategy

1 Introduction

With the progress of the times and technology, data generated by various industries are increasing dramatically, which makes the optimization problems in cloud computing become more and more. A large number of these problems are multi-objective optimization problems. They have multiple optimization objectives, and each of them is interacting and conflicting with each other. Unlike single-objective optimization, the solution of a multi-objective optimization problem is a set of Pareto optimal solutions that represent trade-offs and compromises among the objectives. Intelligent optimization algorithms, also known as meta-heuristic algorithms, with flexibility, gradient-free mechanism and global optimization performance characteristics [1]. Such algorithms not only enable to find multiple solutions in parallel, but also are capable of handling optimization problems such as no-differentiable, discontinuous, and non-convex Pareto front. Therefore, intelligent optimization algorithms can effectively solve multi-objective optimization problems and are widely used in solving problems such as resource allocation [2], task scheduling [3], and virtual machine consolidation [4] in the field of cloud computing.

The Front−based Yin-Yang-Pair Optimization (F-YYPO) [5] is a lightweight multi-objective optimization algorithm with fewer parameters and easy programming implementation, and has been successfully applied to the parameter optimization problem of Stirling engine systems. Punnathanam et al. [6] evaluated the performance of F-YYPO using the UF test suite of CEC2009 [7] and concluded that F-YYPO has the advantages of higher quality of solution results and faster computation compared to other multi-objective optimization algorithms. Nevertheless, there are still many areas for improvement in the algorithmic framework of F-YYPO to further enhance the performance.

Some scholars have proposed improved algorithms for the basic F-YYPO. Li et al. [8] proposed an improved Yin-Yang-Pair multi-objective optimization algorithm based on full-combination strategy (F-ACYYPO). F-ACYYPO made the following three modifications. (1) Complete combination of multiple objective functions was adopted. (2) Adaptive scaling factor was used. (3) Improved the update method of the archive operation. Yang et al. [9] proposed a multi-objective adaptive Yin-Yang-Pair Optimization algorithm (M-AYYPO). M-AYYPO employed adaptive splitting probabilities related to the problem dimension and successfully optimized the fatigue loads problem in complex-terrain wind farms.

In our study, we found that F-YYPO suffers from low convergence and poor distribution of non−dominated solution set. To further improve the performance of F-YYPO, a Dual-Population strategy based Multi-Objective Yin-Yang-Pair Optimization (DP-MOYYPO) is proposed. DP−MOYYPO has adopted the following three improvements. (1) Two sub−populations with separate exploitation and exploration duties are used to search the variable space for improving the distribution of the solutions. (2) The update of point P in the splitting stage is improved to enhance the convergence speed and accuracy of the algorithm by reducing the number of function evaluations. (3) Improve the update of point P_{2i} in the archive stage to reduce the chance of the algorithm falling into a local optimum.

The rest of this paper is organized as follows. Section 2 introduces the F-YYPO algorithm. Section 3 presents the D-MOYYPO algorithm proposed in this paper in detail. Section 4 evaluates the performance of the algorithm through experimental results. Section 5 concludes the work.

2 Front–Based Yin–Yang–Pair Optimization (F–YYPO)

F-YYPO is a simple extension of the single-objective Yin-Yang-Pair Optimization (YYPO) [10] for multi-objective problems. A pair of optimized individuals (P_1 and P_2) and a pair of search radii (δ_1 and δ_2) are assigned to each objective of the problem, and the search of the solution space is performed using the basic mechanism of YYPO. To maintain the low computational complexity of the YYPO algorithm, fast non-dominated sorting and crowding distance calculation from NSGA-II [11] are added to the computational framework.

Similar to YYPO, F-YYPO normalizes all decision variables (between 0 and 1) and scales them appropriately according to the variable boundaries when performing fitness evaluation. I_{min} $I_{max.}$ α are three user-defined parameters, where I_{min} and I_{max} are the minimum value and maximum value of the archive count I, and α is the scaling factor of the radii. SP, SF, and A are three point sets in F-YYPO, in which SP is used to keep the decision variables of the optimized individuals, SF is the Pareto front set, and A is the archive set.

F-YYPO compares the fitness values of each pair of points P_1 and P_2 in the set SP in turn at the beginning of each iteration. This is to ensure that P_1 is the point with the better fitness value, while P_2 is the opposite. If P_1 and P_2 are exchanged, the corresponding radii need to be exchanged as well. Subsequently, the optimized individuals in the SP are saved in an archive A. Next, all points in the SP go through the splitting stage in turn.

2.1 Splitting Stage

Every point (P) in the SP will go through the splitting stage in turn, while entering the radius (δ) corresponding to each point. The splitting stage is designed to produce new points in as many directions as possible in the hypersphere, whereas maintaining randomness. The splitting mode is decided equally by the following two methods.

One–way splitting:

$$\begin{cases} S_j^j = S^j + r \times \delta \\ S_{D+j}^j = S^j - r \times \delta \end{cases}, j = 1, 2, 3, \cdots, D \qquad (1)$$

D–way splitting:

$$\begin{cases} S_k^j = S^j + r \times \left(\delta/\sqrt{2}\right), B_k^j = 1 \\ S_k^j = S^j - r \times \left(\delta/\sqrt{2}\right), B_k^j = 0 \end{cases}, k = 1, 2, 3, \cdots, 2D, j = 1, 2, 3, \cdots, D \qquad (2)$$

In Formula (1) and Formula (2), D is the number of decision variables; S is a matrix consisting of $2D$ identical copies of the point P, which has a size of $2D \times D$; B is a matrix of $2D$ random binary strings of length D (each binary string in B is unique); k denotes the point number and j denotes the decision variable number that will be modified; r is a random number between 0 and 1.

In both splitting methods, random values in the interval [0, 1] are used to correct for out-of-bounds variables. Then, the generated $2D$ new points are evaluated for fitness separately, and the point that undergoes the splitting stage is replaced by the one with the best fitness.

2.2 Archive Stage

The archive stage begins after the required number of archiving updates is completed. In this phase, first, the three sets SP, SF and A are merged and saved into the archive A. Fast non-dominated sorting algorithm is used to compute the points in A and discard the points besides the first front. Second, the remaining points in A are calculated for the crowding distance and sorted in descending order, and the Pareto front set SF is updated with the points in A. Third, M (where M denotes the number of optimization objectives) corner points and the non−corner point with the largest crowding distance are selected from the SF. Update the odd numbered points in SP to the M corner points and all even numbered points to the non-corner points with the largest crowding distance. Then update each pair of search radii using the following equations.

$$\begin{cases} \delta_1 = \delta_1 - (\delta_1/\alpha) \\ \delta_2 = \min(\delta_2 + (\delta_2/\alpha), 0.75) \end{cases} \tag{3}$$

At the end of the archive stage, the archive is cleared, a new value of I for the number of archive updates is randomly generated in the specified range $[I_{\min}, I_{\max}]$, and the archive counter is set to 0.

3 Dual−Population Strategy Based Multi−Objective Yin−Yang−Pair Optimization (DP−MOYYPO)

3.1 Dual−Population Strategy

The algorithmic framework of F-YYPO decomposes the multi−objective optimization problem into several single-objective optimization problems for individual optimization. Each individual only optimizes its own objective function, and does not take into account the diversity of optimization individuals. For example, F-YYPO assigns two pairs of optimizing individuals P_{1i} and P_{2i} to the bi−objective optimization problem. P_{11} and P_{21} always optimizing the solutions with smaller f_1, meanwhile P_{21} and P_{22} always optimizing the solutions with smaller f_2. This optimization mechanism tends to trap the optimizing individuals at both ends of the Pareto front and prevents better global optimization.

Figure 1 shows the solutions obtained by F-YYPO on the multi-objective test instances UF2 and UF8 of CEC2009 with the real Pareto front. UF2 and UF8 are bi-objective and tri-objective optimization problems, respectively. It can be visualized from the figure that on UF2, most of the solutions are gathered at the two ends of the Pareto front; on UF8, most of the solutions are gathered in the $f_1 of_2, f_1 of_3$ planes. This phenomenon indicates that the solutions obtained by F-YYPO are mainly distributed in the vicinity of the true minimum of each objective, and the solutions are poorly distributed.

For more uniform distribution of the Pareto front, DP-MOYYPO uses a dual-population strategy. It considers the original M pairs of optimized individuals in F-YYPO as population I, which is responsible for the optimization of the corner points (points with crowding distance equal to infinity), called the exploitation sub-population. Meanwhile, M pairs of optimized individuals of the same size are added to the algorithm framework

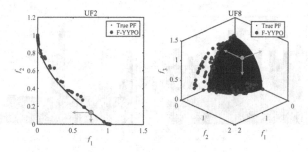

Fig. 1. Pareto fronts of F-YYPO on UF2 and UF8

considered as population II, which is responsible for optimizing the non-corner points with the largest crowding distance, called the exploration sub-population. This calculation not only increases the diversity of optimized individuals, but also improves the distribution of solutions by searching most of the blank areas. For instance, the green points in Fig. 1 are the non-corner points with the largest crowding distance, and iterative computation of them in all directions using the exploration sub-population can produce multiple non-dominated points around them to fill the blank area in the figure.

3.2 Simplified Strategy for the Splitting Stage

In the splitting stage of F-YYPO, $2D$ (where D denotes the number of decision variables) new points are generated for each splitting of point P for fitness evaluation. This allows for a more comprehensive search of the variable space, but can be unnecessarily wasteful when the maximum number of function evaluations is small. In the case of multi-objective optimization problems, selecting only the one with the best fitness for the current objective from a large number of solutions may cause some other non–dominated solutions to be discarded.

Figure 2 shows the number for each index of the selected points as a percentage over the total number of splits when F-YYPO is run once on each of the multi-objective test instances UF2 and UF8. As an example of UF2, each point P splits to generate 20 new points when $D = 10$. The number of times P is replaced by the first new point and the second new point is about 6% and 4.7% of the total number of splits, respectively. It is shown that when new points are evaluated once or twice there is about 10% probability of finding the best point, and all evaluations of the remaining points are discarded. From these two histograms it can also be seen that less than 5% of all points need to be evaluated. This suggests only in fairly few cases does the fitness of all new points need to be evaluated.

In order to improve the search efficiency of F-YYPO and reduce the number of evaluations of the algorithm in the splitting stage, the update of the point P, which has been proven effective in the Reduced Yin-Yang-Pair Optimization [12], is used in DP-MOYYPO. That is, one of the $2D$ new points is evaluated at a time in a random order. When a point with better fitness than P appears, this point is used to replace P and finish the splitting stage. If there is no point with better fitness than P, P is replaced by the point with the best fitness among the new points. The new update method makes the number

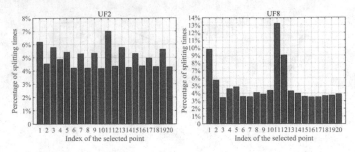

Fig. 2. Percentage of splitting times on UF2 and UF8

of function evaluations for each P in the splitting stage change from a fixed $2D$ to a digit between 1 and $2D$, which greatly reduces the unnecessary computational overhead.

3.3 Enhancement Strategy for the Archive Stage

In the archive stage of F-YYPO, the P_{2i} of each pair of optimized individuals is updated to the same non−corner point with the largest crowding distance. There is a certain probability that this update will result in a transitional exploitation of a region, which will trap the algorithm in a local optimum. Also, other points that may carry potentially important information are ignored, resulting in a less efficient algorithm.

DP-MOYYPO improves this update by selecting non-corner points to update P_{2i} in a random non-repeatable manner when there are sufficient non-corner points, otherwise non-corner points are selected to update P_{2i} in a random repeatable manner. The improved model allows the algorithm to explore around more points. A larger search space not only obtains more high-quality candidate solutions, but also reduces the possibility for the algorithm to fall into a local optimum.

4 Experimental Results and Discussion

To test the performance of DP-MOYYPO, two test suites implemented in PlatEMO [13] platform are used for numerical experiments in this paper. First, the UF test suite is used to compare the performance of four multi-objective algorithms based on YYPO. Then, a total of 19 test instances including UF test suite and WFG test suite are used to compare the performance of DP-MOYYPO with five other representative multi-objective optimization algorithms.

In this study, the effectiveness of the developed DP−MOYYPO against other competing algorithms is verified by two performance metrics: generational distance (GD) and inverted generational distance (IGD). GD is a convergence metric, smaller value means better convergence of the algorithm. IGD is a composite performance metric, whose smaller value indicates better diversity and convergence of the algorithm.

4.1 Comparison Analysis of Algorithms Based on YYPO

In this section, numerical experiments are performed on DP−MOYYPO, F-YYPO, F-ACYYPO and M−AYYPO using the UF test suite of CEC2009. The UF test suite includes seven bi-objective (UF1 ~ UF7) and three tri-objective (UF8 ~ UF10) test instances. Each test instance is run independently 30 times, and the maximum number of function evaluations per run is 300,000. The maximum number of individuals in the non-dominated set generated by the algorithms for the bi-objective and tri-objective problems is 100 and 150, respectively. Referring to the experimental setup in the original paper, the common parameters of the four algorithms are set to $I_{min} = 1$ and $I_{max} = 3$ in the experiment. The scaling factor for DP-MOYYPO, F-YYPO and M-AYYPO is set to $\alpha = 20$. The adaptive tuning parameters used for the scaling factors in F-ACYYPO are set to $k = 0.6$ and $c = 0.4$.

Table 1 shows the mean GD values of each algorithm after running, and it can be seen from the table that DP-MOYYPO outperforms F-YYPO in terms of GD metrics across the board. However, compared to F–ACYYPO, DP–MOYYPO only obtained the optimal GD values on three test instances (UF2 ~ UF4). This shows that F–ACYYPO can be locally closer to the Pareto front with strong convergence. Part of the reason for this phenomenon is that DP–MOYYPO divides half of the computational power to search for the blank areas that exist in the solution process in order to improve the distribution of the solutions. Figure 3 shows the box plots of GD metrics for each algorithm run 30 times for different test instances. As can be seen from this figure, the GD values obtained by DP–MOYYPO are more stable than those of F–YYPO on all test instances. Experimental results show that DP–MOYYPO has faster convergence speed and higher convergence accuracy than F–YYPO.

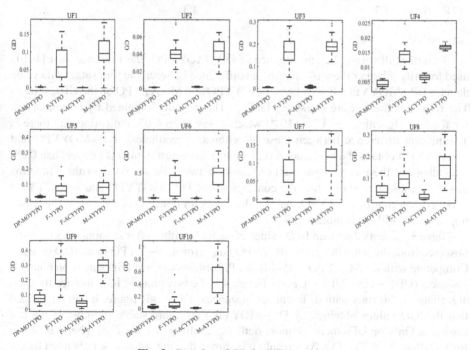

Fig. 3. Boxplot of GD for UF test suite

Table 1. Obtained mean GD values

Problem	M	D	DP−MOYYPO	F−YYPO	F−ACYYPO	M−AYYPO
UF1	2	30	8.27E−04	6.90E−02	**4.33E−04**	9.20E−02
UF2	2	30	**1.38E−03**	4.03E−02	2.07E−03	4.36E−02
UF3	2	30	**4.27E−03**	1.71E−01	7.56E−03	1.96E−01
UF4	2	30	**4.40E−03**	1.44E−02	6.56E−03	1.70E−02
UF5	2	30	1.96E−02	6.75E−02	**1.75E−02**	1.03E−01
UF6	2	30	1.34E−02	2.97E−01	**6.61E−03**	4.83E−01
UF7	2	30	8.70E−04	7.86E−02	**6.64E−04**	1.07E−01
UF8	3	30	4.66E−02	1.02E−01	**1.86E−02**	1.63E−01
UF9	3	30	7.41E−02	2.49E−01	**4.16E−02**	2.87E−01
UF10	3	30	2.81E−02	8.09E−02	**8.57E−03**	4.77E−01

Table 2. Friedman test

p−Value	DP−MOYYPO	F−YYPO	F−ACYYPO	M−AYYPO
3.7E−04	1.7	3	1.3	4
2.1E−01	1.7	-	1.3	-

To further illustrate the convergence of DP−MOYYPO, the Friedman test [14] is used to verify whether there is a significant difference between the four sets of GD value data for DP−MOYYPO, F−YYPO, F−ACYYPO and M−AYYPO on the UF test suite. The results of the test are shown in Table 2. The p−value obtained by Friedman test for the four algorithms is 3.7×10^{-4}, which is less than 0.05, indicating that there is a significant difference between these algorithms. Nevertheless, DP−MOYYPO and F−ACYYPO obtained a p−value of 0.21 by Friedman test, which is greater than 0.05, indicating that there is no significant difference between these two algorithms in terms of GD metrics. This shows that the convergence of DP−MOYYPO and F−ACYYPO is better than that of F−YYPO, but the difference in convergence between these two improved algorithms is quite small.

Table 3 presents the mean IGD values of each algorithm after running, and it can be observed from the table that DP−MOYYPO outperforms F−YYPO in all IGD metrics. Comparing with F−ACYYPO, DP−MOYYPO obtains the best IGD values on nine test instances (UF1 ~ UF5, UF7 ~ UF10). Figure 4 is the box plots of IGD metrics for each algorithm for 30 runs with different test instances. From this figure, it can be noticed that the IGD values obtained by DP−MOYYPO are more stable on most of the test instances. Only on UF6 there are more outliers, resulting in a slightly higher IGD value for UF6 than F−ACYYPO. As a result, it is shown that the solution set obtained by the

DP−MOYYPO is closer to the optimal Pareto front than F−YYPO and other improved algorithms, with better distribution and convergence.

Table 3. Obtained mean IGD values

Problem	M	D	DP−MOYYPO	F−YYPO	F−ACYYPO	M−AYYPO
UF1	2	30	**6.11E−02**	9.62E−02	7.55E−02	1.03E−01
UF2	2	30	**2.85E−02**	7.98E−02	3.67E−02	1.05E−01
UF3	2	30	**1.09E−01**	3.94E−01	1.53E−01	4.78E−01
UF4	2	30	**4.24E−02**	8.82E−02	4.69E−02	1.13E−01
UF5	2	30	**1.92E−01**	1.96E−01	1.97E−01	2.49E−01
UF6	2	30	1.57E−01	1.95E−01	**1.42E−01**	3.05E−01
UF7	2	30	**1.25E−01**	1.99E−01	1.66E−01	1.59E−01
UF8	3	30	**2.37E−01**	3.29E−01	2.94E−01	4.32E−01
UF9	3	30	**2.34E−01**	7.14E−01	2.80E−01	9.91E−01
UF10	3	30	**4.71E−01**	5.88E−01	5.53E−01	1.64E+00

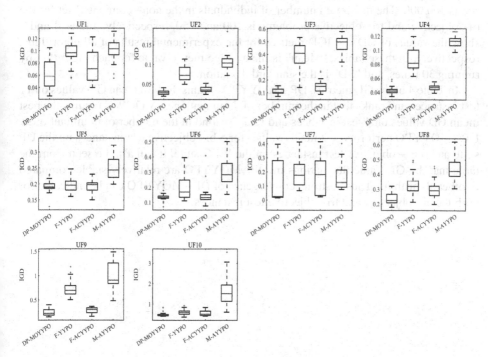

Fig. 4. Boxplot of IGD for UF test suite

The solution obtained for each algorithm on the 10 test instances with the real Pareto front is shown in Fig. 5. It can be visualized from the figure that DP−MOYYPO has better distribution and convergence than F−YYPO on almost all test instances. Combining the data of the above two indicators, DP−MOYYPO is slightly worse than F−ACYYPO in the convergence metric GD, but there is no significant difference. Moreover, there is a greater advantage in the comprehensive metric IGD. It indicates that the improvement strategies adopted by DP−MOYYPO are effective.

4.2 Comparison Analysis with Other Multi−Objective Optimization Algorithms

To further validate the performance of DP−MOYYPO, five representative multi−objective optimization algorithms are selected for comparative analysis. These algorithms include: Multi−Objective Ant Lion Optimizer (MOALO) [15], Multi−Objective Dragonfly Algorithm (MODA) [16], Multi−Objective Grey Wolf Optimizer (MOGWO) [17], Multi−Objective Particle Swarm Optimization (MOPSO) [18], and Multi−objective Salp Swarm Algorithm (MSSA) [1]. The parameters of all algorithms are set according to the original literatures.

Default settings are used for the number of targets (M) and the dimensions of the decision variables (D) for the 19 test instances selected from PlatEMO. Each test instance is run independently 30 times. The maximum number of function evaluations for each run is 100,000. The maximum number of individuals in the non−dominated set for the bi−objective and tri−objective problems is 100 and 150, respectively. Tables 4 and 5 show the values of GD and IGD metrics for the experimental results of all algorithms, respectively. In these tables, "Mean" is the average metric value for each algorithm by running 30 times, and "SD" is the standard deviation.

In Table 4 of the GD metrics, DP−MOYYPO obtains the best mean GD values on 11 test instances and ranks first in the number of dominant ones. MOGWO obtains the best mean GD values on 4 test instances and ranks second in the number of dominant ones. DP−MOYYPO performs best in GD values for 5 bi−objective test instances on the UF suite and 6 tri−objective test instances on the WFG suite. It can also be seen from the table that the GD value fluctuations of DP−MOYYPO are relatively stable compared to other algorithms. Thus, a strong convergence of DP−MOYYPO can be indicated for both the bi−objective and tri−objective test instances.

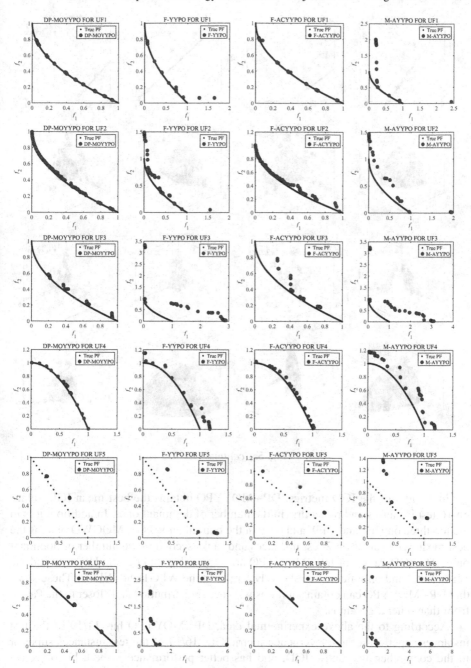

Fig. 5. True and obtained Pareto front for UF test suite

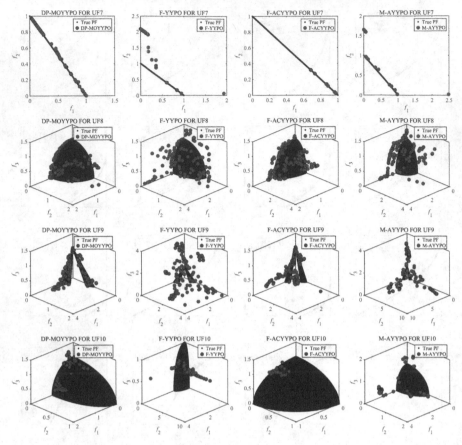

Fig. 5. (*continued*)

In Table 5 of the IGD metrics, DP−MOYYPO obtains the best mean IGD values on 16 test instances and ranks first in the number of dominant ones. In addition, it is in the top three on UF7 and UF9, and only in the bottom on WFG1. MOGWO obtains the best mean IGD values on 2 test instances and ranks second in the number of dominant ones. The IGD values of DP−MOYYPO are very stable on both the UF suite and the WFG suite, and only fluctuate relatively larger on the WFG1 and WFG2. These show that DP−MOYYPO can obtain solutions with better distribution and closer to the Pareto front than other algorithms.

According to the above experimental data, DP−MOYYPO has 11/19 of the test instances superior in the convergence metric GD, 16/19 of the test instances superior in the comprehensive metric IGD, and has better performance in both bi−objective and tri−objective test instances. As a result, it demonstrates that the DP−MOYYPO proposed in this paper has a strong competitive advantage over other multi−objective optimization algorithms in terms of convergence and distribution of solutions.

Table 4. Obtained GD values by DP—MOYYPO and other competitors

Prob—lem		DP—MOYYPO	MOALO	MODA	MOGWO	MOPSO	MSSA
UF1	Mean	**3.43E−03**	1.24E−02	9.87E−02	9.11E−03	3.76E−02	1.27E−02
	SD	**3.91E−03**	4.57E−03	9.97E−02	1.24E−02	2.75E−02	9.27E−03
UF2	Mean	**2.42E−03**	2.14E−02	4.52E−02	7.53E−03	2.90E−02	8.32E−03
	SD	**4.44E−04**	6.62E−03	2.72E−02	2.24E−03	8.87E−03	4.90E−03
UF3	Mean	**2.45E−02**	4.04E−02	1.54E−01	3.64E−02	9.76E−02	6.60E−02
	SD	1.13E−02	**6.09E−03**	1.15E−01	1.74E−02	2.84E−02	2.32E−02
UF4	Mean	9.99E−03	**5.75E−03**	1.35E−02	6.26E−03	8.59E−03	1.20E−02
	SD	1.59E−03	9.11E−04	**3.66E−04**	8.20E−04	6.70E−04	8.92E−04
UF5	Mean	**7.17E−02**	1.79E−01	1.00E+00	3.73E−01	4.71E−01	1.35E−01
	SD	**1.44E−02**	3.99E−02	5.18E−01	2.34E−01	3.43E−01	4.38E−02
UF6	Mean	**5.61E−02**	7.92E−02	9.50E−01	6.36E−02	1.44E−01	6.04E−02
	SD	5.31E−02	**1.81E−02**	6.22E−01	4.22E−02	1.29E−01	2.84E−02
UF7	Mean	4.29E−03	1.54E−02	1.18E−01	**2.77E−03**	1.79E−02	1.36E−02
	SD	3.32E−03	6.12E−03	1.14E−01	**2.07E−03**	1.33E−02	1.00E−02
UF8	Mean	1.17E−01	**4.04E−02**	1.12E−01	9.57E−02	2.61E−01	5.41E−02
	SD	4.81E−02	**1.28E−02**	8.43E−02	7.63E−02	6.48E−02	5.66E−02
UF9	Mean	1.33E−01	4.91E−02	1.59E−01	**2.75E−02**	3.21E−01	1.64E−01
	SD	3.90E−02	2.80E−02	1.06E−01	**2.10E−02**	8.96E−02	8.23E−02
UF10	Mean	2.48E−01	4.02E−01	8.58E−01	4.11E−01	1.39E+00	**1.82E−01**
	SD	1.20E−01	9.47E−02	4.09E−01	2.30E−01	2.84E−01	**4.92E−02**
WFG1	Mean	**1.06E−01**	1.18E−01	1.50E−01	1.19E−01	1.14E−01	1.13E−01
	SD	9.22E−03	4.19E−03	2.25E−02	**4.23E−17**	1.02E−02	5.47E−03
WFG2	Mean	**1.02E−02**	2.47E−02	3.35E−02	1.69E−02	1.97E−02	2.92E−02
	SD	1.53E−03	1.04E−02	7.34E−03	**7.06E−18**	4.68E−03	1.36E−02
WFG3	Mean	6.95E−02	4.74E−02	**4.68E−02**	5.17E−02	7.80E−02	4.88E−02
	SD	**2.30E−03**	8.66E−03	9.79E−03	6.33E−03	3.21E−03	1.52E−02
WFG4	Mean	**6.83E−03**	1.56E−02	1.69E−02	1.06E−02	2.07E−02	1.39E−02
	SD	7.97E−04	2.47E−03	3.67E−03	**6.60E−04**	8.62E−04	2.90E−03
WFG5	Mean	**8.45E−03**	2.39E−02	2.44E−02	1.77E−02	1.45E−02	2.65E−02
	SD	**5.01E−04**	1.24E−02	3.81E−03	4.47E−03	9.70E−04	8.37E−03
WFG6	Mean	8.06E−03	2.44E−02	4.07E−02	**7.28E−03**	2.84E−02	2.42E−02
	SD	1.37E−3	7.98E−03	3.58E−03	3.46E−03	**5.31E−04**	6.75E−03
WFG7	Mean	**1.05E−02**	2.82E−02	3.44E−02	3.79E−02	2.47E−02	2.63E−02
	SD	**8.86E−04**	8.45E−03	9.55E−03	1.91E−03	1.47E−03	7.69E−03
WFG8	Mean	**2.14E−02**	3.56E−02	4.13E−02	2.67E−02	3.42E−02	2.97E−02
	SD	**9.46E−04**	6.71E−03	3.97E−03	1.92E−03	1.18E−03	3.77E−03
WFG9	Mean	9.64E−03	2.53E−02	2.92E−02	**6.75E−03**	2.58E−02	1.98E−02
	SD	9.78E−04	1.38E−02	9.64E−03	2.10E−03	**5.66E−04**	9.23E−03

Table 5. Obtained IGD values by DP−MOYYPO and other competitors

Prob−lem		DP−MOYYPO	MOALO	MODA	MOGWO	MOPSO	MSSA
UF1	Mean	**8.24E−02**	1.71E−01	4.55E−01	1.20E−01	1.72E−01	1.21E−01
	SD	2.15E−02	2.08E−02	1.25E−01	2.27E−02	6.35E−02	**7.12E−03**
UF2	Mean	**3.82E−02**	2.02E−01	2.52E−01	6.26E−02	1.56E−01	1.00E−01
	SD	**5.92E−03**	3.65E−02	5.85E−02	7.40E−03	3.06E−02	2.84E−02
UF3	Mean	**1.53E−01**	3.91E−01	5.32E−01	2.80E−01	5.57E−01	4.71E−01
	SD	**3.31E−02**	4.33E−02	1.04E−01	5.42E−02	7.34E−02	6.45E−02
UF4	Mean	**5.76E−02**	8.93E−02	1.20E−01	6.25E−02	8.07E−02	1.24E−01
	SD	5.70E−03	1.59E−02	**3.36E−03**	5.18E−03	5.04E−03	2.35E−02
UF5	Mean	**2.36E−01**	1.66E+00	3.30E+00	9.52E−01	1.36E+00	1.04E+00
	SD	**4.47E−02**	3.19E−01	3.73E−01	3.60E−01	7.30E−01	2.27E−01
UF6	Mean	**1.84E−01**	7.62E−01	2.42E+00	3.58E−01	6.47E−01	4.54E−01
	SD	7.62E−02	1.51E−01	6.75E−01	**5.24E−02**	3.72E−01	5.53E−02
UF7	Mean	1.60E−01	2.46E−01	5.30E−01	**7.24E−02**	1.75E−01	9.72E−02
	SD	1.45E−01	1.10E−01	1.31E−01	**6.87E−03**	1.51E−01	1.67E−02
UF8	Mean	**3.33E−01**	7.03E−01	7.40E−01	1.03E+00	1.02E+00	4.66E−01
	SD	**4.79E−02**	1.38E−01	1.00E−01	5.76E−01	1.80E−01	7.66E−02
UF9	Mean	4.56E−01	6.58E−01	8.78E−01	**2.81E−01**	1.19E+00	7.04E−01
	SD	**7.37E−02**	1.27E−01	2.23E−01	1.77E−01	2.42E−01	1.54E−01
UF10	Mean	**8.50E−01**	4.09E+00	4.13E+00	2.77E+00	6.22E+00	1.71E+00
	SD	**1.97E−01**	8.64E−01	9.89E−01	1.90E+00	1.27E+00	3.35E−01
WFG1	Mean	1.79E+00	1.71E+00	1.79E+00	1.60E+00	**1.55E+00**	1.57E+00
	SD	7.85E−02	1.17E−01	6.05E−02	**9.03E−16**	1.12E−01	6.10E−02
WFG2	Mean	**1.61E−01**	6.65E−01	4.77E−01	4.59E−01	1.94E−01	7.50E−01
	SD	8.41E−03	1.57E−01	4.42E−02	**1.69E−16**	1.18E−02	1.40E−01
WFG3	Mean	**7.35E−02**	1.26E+00	9.47E−01	8.28E−01	2.40E−01	1.31E+00
	SD	**2.36E−02**	4.39E−01	3.16E−01	4.39E−01	2.43E−02	4.09E−01
WFG4	Mean	**2.19E−01**	1.69E+00	4.77E−01	1.11E+00	3.32E−01	1.93E+00
	SD	**5.64E−03**	2.35E−01	8.34E−02	9.82E−02	1.10E−02	1.74E−01
WFG5	Mean	**2.30E−01**	1.40E+00	4.71E−01	1.49E+00	2.63E−01	1.18E+00
	SD	**4.13E−03**	4.18E−01	8.00E−02	2.80E−01	9.40E−03	3.98E−01
WFG6	Mean	**2.25E−01**	1.85E+00	6.40E−01	1.77E+00	3.93E−01	1.63E+00
	SD	8.55E−03	1.91E−01	6.54E−02	1.33E−01	**6.25E−03**	2.67E−01
WFG7	Mean	**2.49E−01**	1.81E+00	1.15E+00	1.21E+00	3.40E−01	1.78E+00
	SD	**1.08E−02**	2.66E−01	2.00E−01	1.02E−01	1.41E−02	3.08E−01
WFG8	Mean	**3.38E−01**	1.93E+00	7.89E−01	1.91E+00	4.40E−01	1.73E+00
	SD	**1.15E−02**	1.79E−01	6.21E−02	1.20E−01	1.38E−02	2.55E−01
WFG9	Mean	**2.42E−01**	8.57E−01	7.12E−01	1.76E+00	3.72E−01	1.03E+00
	SD	**8.24E−03**	3.06E−01	2.01E−01	3.92E−01	9.33E−03	3.48E−01

5 Conclusions

In this study, the proposed improved algorithm DP—MOYYPO is based on the F—YYPO algorithm framework by using a dual—population strategy to enhance the uniformity of the optimized individuals' distribution. At the same time, the update methods of point P in the splitting stage and the archive stage are improved to enhance the search efficiency of the algorithm. In the experimental section, DP—MOYYPO is first compared with F—YYPO and all improved algorithms, and then with five other representative multi—objective optimization algorithms. Results show that DP—MOYYPO outperforms F—YYPO across the board on all test instances, not only by obtaining Pareto fronts with better distribution, but also by faster and more accurate convergence of the solutions. When compared with other representative multi—objective optimization algorithms, DP—MOYYPO ranks first in all performance metrics, indicating that it has a strong competitive advantage. In the next work, DP—MOYYPO can be applied to multi—objective optimization problems in specific scenarios of cloud computing to further validate the performance of the algorithm.

References

1. Mirjalili, S., Gandomi, A.H., Mirjalili, S.Z., et al.: Salp swarm algorithm: a bio—inspired optimizer for engineering design problems. Adv. Eng. Softw. **114**, 163–191 (2017)
2. Abedi, S., Ghobaei—Arani, M., Khorami, E., et al.: Dynamic resource allocation using improved firefly optimization algorithm in cloud environment. Appl. Artif. Intell. **36**(1) (2022)
3. Mangalampalli, S., Karri, G.R., Kumar, M.: Multi objective task scheduling algorithm in cloud computing using grey wolf optimization. Cluster Comput **26**(6), 3803–3822 (2022). https://doi.org/10.1007/s10586-022-03786-x
4. Li, P., Cao, J.: A virtual machine consolidation algorithm based on dynamic load mean and multi-objective optimization in cloud computing. Sensors **22**(23), 9154 (2022). https://doi.org/10.3390/s22239154
5. Punnathanam, V., Kotecha, P.: Multi—objective optimization of Stirling engine systems using front-based yin-yang-pair optimization. Energy Convers. Manage. **133**, 332–348 (2017)
6. Punnathanam, V., et al.: Front-based Yin-Yang-Pair Optimization and its performance on CEC2009 benchmark problems. In: Proceedings of Smart Innovations in Communication and Computational Sciences 2017, vol. 1, pp. 387–397. Springer, Singapore (2019)
7. Zhang, Q., Zhou, A., Zhao, S., et al.: Multiobjective optimization Test Instances for the CEC 2009 Special Session and Competition. Mechanical engineering, (2008)
8. Li, D., Ai, Z., Wang, Z.: Improved yin-yang-pair multi-objective optimization algorithm based on full-combination strategy. Comput. Eng. Appl. **57**(22), 110–124 (2021)
9. Yang, J., Zheng, S.Y., Song, D.R., et al.: Comprehensive optimization for fatigue loads of wind turbines in complex-terrain wind farms. IEEE Trans. Sustain. Energ. **12**(2), 909–919 (2021)
10. Punnathanam, V., Kotecha, P.: Yin-Yang-pair optimization: a novel lightweight optimization algorithm. Eng. Appl. Artif. Intell. **54**, 62–79 (2016)
11. Deb, K., Pratap, A., Agrawal, S., et al.: A fast and elitist multiobjective genetic algorithm: NSGA—II. IEEE Trans. Evol. Comput. **6**(2), 182–197 (2002)
12. Punnathanam, V., Kotecha, P.: reduced yin—yang—pair optimization and its performance on the CEC 2016 expensive case. In: 2016 IEEE Congress on Evol. Comput. (CEC), pp. 2996–3002 (2016)

13. Tian, Y., Cheng, R., et al.: PlatEMO: A MATLAB platform for evolutionary multi-objective optimization. IEEE Comput. Intell. Mag. **12**(4), 73–87 (2017)
14. Derrac, J., Garcia, S., Molina, D., et al.: A practical tutorial on the use of nonparametric statistical tests as a methodology for comparing evolutionary and swarm intelligence algorithms. Swarm Evol. Comput. **1**(1), 3–18 (2011)
15. Mirjalili, S., Jangir, P., Saremi, S.: Multi-objective ant lion optimizer: a multi-objective optimization algorithm for solving engineering problems. Appl. Intell. **46**(1), 79–95 (2017)
16. Mirjalili, S.: Dragonfly algorithm: a new meta-heuristic optimization technique for solving single-objective, discrete, and multi-objective problems. Neural Comput. Appl. **27**(4), 1053–1073 (2016)
17. Mirjalili, S., Saremi, S., et al.: Multi-objective grey wolf optimizer: a novel algorithm for multi-criterion optimization. Expert Syst. Appl. **47**, 106–119 (2016). https://doi.org/10.1016/j.eswa.2015.10.039
18. Abido, M.: Multiobjective particle swarm optimization for environmental/economic dispatch problem. Electr. Power Syst. Res. **79**(7), 1105–1113 (2009)

Multiview Subspace Clustering of Hyperspectral Images Based on Graph Convolutional Networks

Xianju Li[1,2](✉), Renxiang Guan[1], Zihao Li[1], Hao Liu[1], and Jing Yang[3]

[1] Faculty of Computer Science, China University of Geosciences, Wuhan 430074, China
{ddwhlxj,grx1126,lizihao,liu_hao}@cug.edu.cn
[2] Hubei Key Laboratory of Intelligent Geo-Information Processing, China University of Geosciences, Wuhan 430074, China
[3] Geophysical and Geochemical Exploration Institute of Ningxia Hui Autonomous Region, Yinchuan 750021, China

Abstract. High-dimensional and complex spectral structures make clustering of hyperspectral images (HSI) a challenging task. Subspace clustering has been shown to be an effective approach for addressing this problem. However, current subspace clustering algorithms are mainly designed for a single view and do not fully exploit spatial or texture feature information in HSI. This study proposed a multiview subspace clustering of HSI based on graph convolutional networks. (1) This paper uses the powerful classification ability of graph convolutional network and the learning ability of topological relationships between nodes to analyze and express the spatial relationship of HSI. (2) Pixel texture and pixel neighbor spatial-spectral information were sent to construct two graph convolutional subspaces. (3) An attention-based fusion module was used to adaptively construct a more discriminative feature map. The model was evaluated on three popular HSI datasets, including Indian Pines, Pavia University, and Houston. It achieved overall accuracies of 92.38%, 93.43%, and 83.82%, respectively and significantly outperformed the state-of-the-art clustering methods. In conclusion, the proposed model can effectively improve the clustering accuracy of HSI.

Keywords: Hyperspectral images (HSIs) · multiview clustering · remote sensing · subspace clustering

1 Introduction

The development of spectral imaging technology has enabled hyperspectral image (HSI) to emerge as an effective tool for detection technology, and has promoted the development of various fields including environmental monitoring [1], geological exploration [2], national defense security [3], and mineral identification [4]. In contrast to traditional color images, HSI possess higher resolution and richer spectral information, and provide more accurate ground object information. Accordingly, various HSI processing technologies have emerged to meet the demands of the times.

Supervised HSI classification methods have achieved significant progress in recent decades, including machine learning models such as support vector machines [5] and

X. Song et al. (Eds.): APWeb-WAIM 2023, LNCS 14334, pp. 95–107, 2024.
https://doi.org/10.1007/978-981-97-2421-5_7

deep convolutional neural networks [6, 10]. These methods require artificially labeled data as training samples, but labeling HSI pixels is a time-consuming task that demands professional knowledge [8]. To alleviate this burden, and to overcome the limitation of scarce label information in HSI, unsupervised learning, represented by clustering, has garnered extensive attention [9]. HSI clustering enables automatic data processing and interpretation. However, due to the large spectral variability and complex spatial structure of HSI, clustering remains a challenging task [10].

HSI clustering refers to the process of partitioning pixels into corresponding groups based on their intrinsic similarity to facilitate analysis and interpretation. The objective is to ensure that pixels within the same cluster have high intra-class similarity while those across different clusters have low inter-class similarity [11]. Over the years, numerous clustering algorithms have been proposed and widely used in practice. These algorithms include methods based on cluster centers such as k-means clustering [12], and fuzzy c-means clustering (FCM) [13]. Other clustering techniques rely on feature space density distribution, such as the mean shift algorithm [14], and the clustering algorithm based on integrated density analysis [15]. However, these methods are relatively sensitive to initialization and noise, and rely heavily on similarity measures. To gain a deeper understanding of the underlying structure of HSI data, subspace clustering algorithms have received extensive attention and demonstrated remarkable results [33].

The subspace clustering algorithm is a hybrid of traditional feature selection techniques and clustering algorithms, where feature subsets or feature weights corresponding to each data cluster are obtained during the clustering and division of data samples [16]. Sparse subspace clustering [17] and low-rank subspace clustering [18] are representative examples of this algorithm. The crucial aspect of this algorithm is to identify the sparse representation matrix of the original data, construct a similarity graph based on the corresponding matrix, and then obtain the clustering result using spectral clustering. However, these clustering algorithms only analyze the spectral information of HSI, leading to suboptimal results when used for HSI clustering. To address this issue, Zhang et al. proposed a spectral space sparse subspace clustering method (S_4C) [19] that leverages the rich spatial environment information carried by HSI in the form of a data cube to improve clustering performance. Literature [20] presents the l_2-norm regularized SSC algorithm that integrates adjacent information into the coefficient matrix via l_2-norm regularization constraints.

Subspace clustering methods have proven to be effective in numerous applications, but their performance in complex HSI scenes is often limited due to their lack of robustness. Deep clustering models have been proposed to address this limitation by extracting deep and robust features. Pan et al. [23] proposed a deep subspace clustering model using multi-layer autoencoders to learn self-expression. Moreover, deep clustering models based on graph convolutional neural networks (GCN) have gained popularity due to their ability to capture neighborhood information. Zhang et al. [25] introduced hypergraph convolutional subspace clustering, which fully exploits the high-order relationships and long-range interdependencies of HSI. To extract a deep spectral space representation and robust nonlinear affinity, Cai et al. [26] proposed a graph regularized residual subspace clustering network, which significantly improves the clustering accuracy of HSI.

Despite the success achieved by the algorithms described above in improving clustering performance, they suffer from two significant limitations. Firstly, the direct application of these methods on HSI often produces cluster maps with substantial noise due to the limited discriminative information in the spectral domain, the complexity of ground objects, and the diversity of spectral features in the same class [27]. Secondly, these methods have been tested on a single view, and extensive experiments have demonstrated that incorporating complementary information from multiple views can significantly enhance clustering accuracy [28]. To address these limitations, various multi-view clustering techniques have been proposed. For instance, Tian et al. [29] performed multi-view clustering by using multiple views simultaneously but the method is sensitive to noise. Chen et al. [30] applied multi-view subspace clustering to polarized HSI. Huang et al. [31] combined local and non-local spatial information of views to learn a common intrinsic cluster structure, which led to improved clustering performance. Lu et al. [32] proposed a method that combined spectral and spatial information to build capability regions and employed multi-view kernels for collaborative subspace clustering. However, the information weights of different views have not been considered in these methods when fusing information from multiple views, which can lead to the loss of important information.

To address the challenges, we proposed a novel approach to subspace clustering for HSI. The method combines both texture and space-spectral information and leverages graph convolutional networks. By integrating neighborhood information through GCNs and using the information of two views, we aim to achieve effective representation learning. The contributions of this work are threefold:

1) We provide a novel deep multi-view clustering algorithm for HSI clustering, namely MSCGC, which can simultaneously learn texture information and depth spectral spatial information;

2) We use the powerful feature extraction ability of GCN and the learning ability of topological relationships between nodes to analyze and express the spatial relationship of HSI.

3) An attention-based fusion module is adopted to adaptively utilize the affinity graphs of the two views to build a more discriminative graph.

2 Related Work

2.1 Multi-view Subspace Clustering

Multi-view clustering techniques have demonstrated significant accomplishments in various domains [34, 35]. In HSI analysis, these methods can be leveraged to enhance the clustering performance by integrating multiple sources of information. Multi-view data $\{X^{(p)}\}_{p=1}^{v}$, where $\mathbf{X}^{(p)} \in \mathbf{R}^{d^{(p)} \times n}$ signifies the data from the p th view, with its top dimension $d^{(p)}$. Subspace clustering is premised on the assumption that each data point can be expressed as a linear combination of other points within the same subspace. Based on the above assumptions, the data $X^{(p)}$ of each view itself is used as a dictionary to construct the subspace representation model as:

$$X^{(p)} = X^{(p)}C^{(p)} + E^{(p)}, \tag{1}$$

where $C^{(p)} \in \mathbb{R}^{n \times n}$ is the self-expressive matrix on each view, and $E^{(p)}$ is the representation error. Multi-view subspace clustering methods are usually expressed as follows:

$$\min_{S^{(p)}} \left\| X^{(p)} - X^{(p)} C^{(p)} \right\|_F^2 + \lambda f\left(C^{(p)}\right)$$

$$s.t. \ S^{(p)} \geq 0, \ C^{(p)^T} \mathbf{1} = \mathbf{1}, \tag{2}$$

where $C^{(p)^T} \mathbf{1} = \mathbf{1}$ ensures that each column of $C^{(p)}$ adds up to 1, indicating that each sample point can be reconstructed by a linear combination of other samples. $C_{i,j}^{(p)}$ represents the weight of the edge between the i-th sample and the j-th sample, so $C^{(p)}$ can be regarded as an n × n undirected graph. $f(\cdot)$ is the regularization function, and λ is a parameter to balance regularization and loss. Different $f(\cdot)$ bring self-expression matrix $C^{(p)}$ satisfying different constraints. Examples include imposing sparsity constraints on matrices or finding low-rank representations of data.

After obtaining the self-expression matrix $\left\{C^{(p)}\right\}_{p=1}^{v}$ on each view, they are fused to obtain a unified self-expression matrix $C \in \mathbb{R}^{n \times n}$, in some methods this step is also performed together with the learning phase on each view. Taking the consistent C as the input of spectral clustering, the final clustering result is obtained.

2.2 Hyperspectral Image Clustering

The analysis of HSI data presents a challenging task, primarily due to the high dimensionality, high correlation, and complex distribution of spectral data. Traditional supervised classification methods demand a considerable number of labeled samples, which may be arduous to obtain. As a solution, unsupervised HSI clustering has emerged as a significant research topic in recent years. The primary objective of HSI clustering is to cluster pixels into distinct classes or clusters based on their spectral characteristics, without any prior knowledge of the data. This technique has proven to be effective in overcoming the limitations of traditional supervised classification methods, as it eliminates the need for labeled samples and can reveal previously unknown relationships in the data.

Among existing HSI clustering methods, subspace clustering has shown promising results due to its robustness. In recent years, graph neural network has attracted people's attention due to its robust feature extraction ability, and its combination with subspace clustering is also a research hotspot. Cai et al. [24] effectively combined structure and feature information from the perspective of graph representation learning and proposed a graph convolutional subspace clustering framework. Zhang et al. [25] proposed hypergraph convolutional subspace clustering to fully exploit the high-order relationships and long-range interdependencies of HSI. To extract deep spectral space representation and robust nonlinear affinity, Cai et al. [26] proposed a graph regularized residual subspace clustering network, which greatly improves the HSI clustering accuracy.

3 Method

As shown in Fig. 1, the framework proposed in this paper consists of three important parts, namely the multi-view graph construction module, the dual-branch representation modle and the attention fusion module. We first introduce the multi-view graph building blocks, and then introduce the rest sequentially.

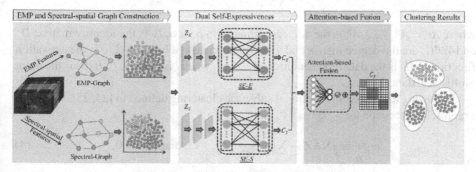

Fig. 1. Illustration of the proposed framework, consisting of three important parts, namely the multi-view graph construction module, the dual-branch representation module and the attention fusion module.

3.1 EMP and Spectral-Spatial Graph Construction

The primary objective of this module is to construct multi-view graphs by extracting texture and spatial spectral features as distinct views. Prior to addressing the primary concern, we must contend with the fact that HSI typically contain a large number of redundant bands. To address this issue, Principal Component Analysis (PCA) is used to reduce the number of bands to d dimensions. Next, the sliding window technique is applied to capture pixel points and their surrounding adjacent pixels, while the patch method is used to represent the data point [24]. Additionally, the EMP algorithm [28] is utilized to corrode the image and retain the texture information. Finally, we obtain two datasets $\{X_1^p, X_2^p\}_{p=1}^n$.

GCN is a neural network that can only operate on data structured as graphs. Therefore, a crucial step in using GCNs is to transform the input data into a graph representation. In this study, we use the K-nearest neighbor (KNN) method to construct a topological graph from the processed data. Specifically, we treat each sample in the data as a node in the graph, and the KNN method is used to determine the connections between nodes. KNN calculates the Euclidean distance between samples within each view and constructs adjacency matrices AE and AS. Finally, by combining the adjacency matrix with the corresponding data, we obtain the EMP map and space spectrum map. This approach facilitates the use of GCNs in multi-view learning tasks by transforming the input data into a graph representation that is amenable to GCN operations.

3.2 Graph Convolution Self-expressive Module

Different from models based on convolutional neural networks, GCN can make full use of the dependencies between nodes and the feature information of each neighbor node through graph convolution operations. The formula for spectrogram convolution is as follows:

$$H^p(l+1) = \sigma\left(\tilde{D}^{-\frac{1}{2}^p}\tilde{A}^p\tilde{D}^{-\frac{1}{2}^p}H^p(l)W^p(l)\right), \tag{3}$$

where $\tilde{A^p} = A^p + I_N$, A^p is the adjacency matrix in p view, I_N is the identity matrix; $\tilde{D}^p$ and $W^p(l)$ are the degree matrix and The weight matrix of layer $l+1$, σ is the activation function; $H(l)$ is the data representation of layer l, when l is equal to 0, $H(0)$ is the input multi-view data.

Following the graph convolutional subspace clustering defined in [24, 25], we define graph self-expression as:

$$\arg\min_{Z} \|X\bar{A}Z - X\|_q + \frac{\lambda}{2}\|Z\|_q, \text{ s.t. diag}(Z) = 0, \tag{4}$$

where q denotes the appropriate matrix norm and λ is the trade-off coefficient. $X\bar{A}Z$ can be viewed as a special linear graph convolution operation parameterized by Z, where Z is the self-expressive coefficient matrix and $\bar{A}$ represents the normalized neighbor matrix.

3.3 Attention Fusion Module

After obtaining affinity matrices Y^p, we need to fuse them together to construct the final affinity graph, and we will apply spectral clustering to the final affinity matrix.. We utilize an attention-based fusion module to learn the importance a^p of each view as follows:

$$a^p = att\left(Y^p\right) \tag{5}$$

where $a^p \in R^{n \times 1}$ are used to measure the importance of each view. The details of the attention module are provided below. In the first step, we concatenate affinity matrices Y^p as $[Y_1 \cdots Y_p] \in R^{n \times pn}$, and introduce a weight matrix $W \in R^{pn \times p}$ to capture the relationships between the self-expression matrices. Next, we apply the tanh function to the product of $[Y_1 \cdots Y_p]$ and W for a nonlinear transformation. Finally, we use softmax and the ℓ_2 function to normalize the attention values, resulting in the final weight matrix:

$$a^p = \ell_2\left(softmax\left(\tanh\left([Y_1 \cdots Y_p] \cdot W\right)\right)\right) \tag{6}$$

Obtaining the weight matrix can realize the fusion operation, and the fused self-expression matrix Y_F is:

$$Y_F = f\left(Y_1 \cdots Y_p\right) = \sum_{i=1}^{N}(a_i1) \odot Y_i \tag{7}$$

where $1 \in R^{1 \times n}$ is a matrix with all elements being 1, and $\odot$ represents the Hadamard product of the matrix.

4 Result

In this section, we show the clustering results of the proposed models on three generic datasets and compare them with several state-of-the-art clustering models.

4.1 Set up

Datasets: We have utilized three authentic HSI datasets, namely, Indian Pines, Pavia University, and Houston2013, which were captured using AVIRIS, ROSIS, and ITRES CASI-1500 sensors, respectively. To ensure computational efficacy, we selected a sub-scene from each dataset for the experimental analysis, and the specifications of these sub-scenes are presented in Table 1. Notably, Houston2013, which is a dataset from the 2013 GRSS competition, exhibits considerable diversity. Given that HSI generally comprise numerous superfluous bands, we employed the Principal Component Analysis (PCA) algorithm to reduce the spectral dimensions to four, which encompass at least 96% of the data's variance. Furthermore, to maintain model accuracy while enhancing computational efficiency, we optimized the hyperparameters of the model, and these adjustments are listed in Table 2.

Baseline: We use six experimental benchmark methods to compare our model. These include two classic clustering algorithms, SC and SSC, as well as excellent clustering algorithms in recent years, including l_2-SSC combined with l_2-norm, robust manifold matrix factorization (RMMF), graph convolutional subspace clustering network (EGCSC) and deep spatial spectral subspace clustering (SCNet). Some experimental results refer to other literature [24, 25], and remaining experiments are implemented using Python 3.9 and run on an Intel i9-12900H CPU.

Table 1. Introduction to the datasets information used in the experiment

Datesets	Indian Pines	Pavia University	Houston-2013
Pixels	85×70	200×100	349×680
Coordinates	30–115, 24–94	150–350, 100–200	0–349, 0–680
Channels	200	103	144
Samples	4391	6445	6048
Clusters	4	8	12

4.2 Quantitative Results

Table 3 illustrates the clustering performance of the proposed MSCGC model and six other models on three benchmark datasets, namely Indian Pines, Pavia University, and Houston2013. The results indicate that the proposed model outperforms the other models in terms of three evaluation metrics: overall accuracy (OA), normalized mutual information (NMI), and Kappa coefficient. Notably, the proposed model exhibits a statistically significant improvement in NMI over the suboptimal method by more than 15% on the Indian Pines and Houston datasets. Furthermore, we can observe the following trends:

Table 2. Experiment important hyperparameters setting information

Datesets	Indian Pines	Pavia University	Houston-2013
Input size	13	11	11
k	30	30	25
λ	100	1000	1000

Table 3. The results of all methods on the experimental data set, the best results of each row are marked in bold

Dataset	Metric	SC	SSC	l_2-SSC	NMFAML	EGCSC	SCNet	MSCGC
InP	OA	0.6841	0.4937	0.6645	0.8508	0.8483	0.8914	**0.9238**
	NMI	0.5339	0.2261	0.3380	0.7264	0.6442	0.7115	**0.8925**
	Kappa	0.5055	0.2913	0.5260	0.7809	0.6422	0.8413	**0.8791**
PaU	OA	0.7691	0.6146	0.5842	0.8967	0.8442	0.9075	**0.9343**
	NMI	0.6784	0.6545	0.4942	0.9216	0.8401	0.9386	**0.9394**
	Kappa	0.8086	0.4886	0.3687	0.8625	0.7968	0.8777	**0.9265**
Hou	OA	0.3661	0.5526	0.4228	0.6346	0.6238	0.7426	**0.8382**
	NMI	0.5067	0.7531	0.5167	0.7959	0.7754	0.7567	**0.8981**
	Kappa	0.2870	0.5022	0.3609	0.5910	0.5812	0.7138	**0.8329**

(1) Introducing deep learning methods and regularization into clustering can better improve accuracy. Methods based on deep learning, such as EGCSC and SCNet, have greatly improved compared with traditional models. In addition, l_2-SSC introduced L2-norm regularization in the traditional model improves the accuracy by 10% on the Indian dataset compared to the SSC method. The introduction of graph regularization by EGCSC has also achieved great improvement.

(2) The introduction of multi-view complementary information is beneficial to improve the clustering accuracy. NMFAML combined with homogeneous information is also used to extract feature information in comparative experiments. Similarly, our model combines texture information and uses GCN to aggregate neighborhood information, which effectively improves the clustering accuracy. It achieved 92.38%, 93.43% and 83.82% accuracy respectively on the three datasets.

(3) The experimental accuracy of MSCGC on the three data sets is better than that of EGCSC. Specifically, compared with EGCSC, our model improves by 7.55%, 9.01% and 21.44% respectively on all data sets. This shows that the introduction of multi-view information and the attention fusion module can improve the clustering accuracy very well. So our model sheds new light on HSI clustering.

4.3 Qualitative Comparison of Different Methods

Figures 2, 3, 4 shows the visualization results of various clustering methods on the Indian, Pavia and Houston2013 datasets. Part (a) of each picture is the true value of the corresponding data set to remove the irrelevant background, and the color of the same class may be different in different methods. Although the SSC algorithm can discover the low-dimensional information of the data from the high-dimensional structure of the data, it does not consider the space constraints. Moreover, traditional subspace clustering methods are difficult to prepare for modeling HSI structures and are weaker than deep learning-based methods to a certain extent. Our model achieves state-of-the-art visualization results on all datasets. Due to the efficient aggregation of texture information, MSCGC has the least salt and pepper noise on the clustering map compared to other methods. At the same time, because the GCN model can aggregate the information of neighbor nodes, it maintains better homogeneity in similar object areas and the boundary of the image remains relatively complete.

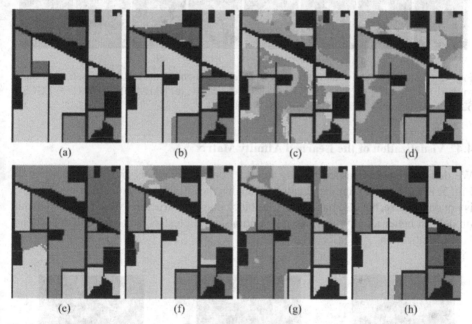

Fig. 2. Clustering results obtained by different methods on the Indian Pines dataset. (a) Ground truth, (b) SC 68.41%, (c) SSC 49.37%, (d) L2-SSC 66.45%, (e) NMFAML 85.08%, (f) EGCSC 84.67%, (g) SCNet 89.14%, and (h) MSCGC 92.38%

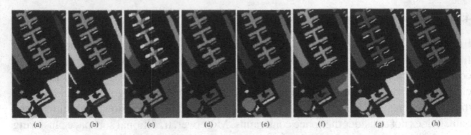

Fig. 3. Clustering results obtained by different methods on the Pavia University dataset. (a) Ground truth, (b) SC 76.91%, (c) SSC 64.46%, (d) L2-SSC 58.42%, (e) NMFAML 89.67%, (f) EGCSC 83.79%, (g) SCNet 90.75%, and (h) MSCGC 93.43%.

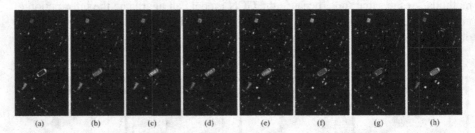

Fig. 4. Clustering results obtained by different methods on the Houston2013 dataset: (a) Ground truth, (b) SC 36:61%, (c) SSC 55:26%, (d) '2-SSC 42:28%, (e) NMFAML 63:46%, (h) EGCSC 62:38%, (i) SCNet 74.26%, and (j) GR-RSCNet 83:82%.

4.4 Visualization of the Learned Affinity Matrix

We demonstrate the learned affinity matrices on three datasets and visualize them in Fig. 5. From the three figures, it can be seen that the affinity matrix learned by MSCGC is not only sparse, but also has an obvious block diagonal structure, which shows that our model can better identify the internal relationship of clusters, so as to realize accurate HSI classification.

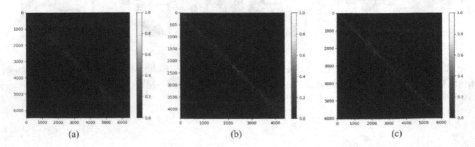

Fig. 5. Visualization of deep affinity matrix learned on (a) Indian Pines, (b) Pavia University, and (c) Houston2013 datasets.

5 Conclusion

We construct a novel multi-view subspace clustering network for HSI clustering, named MSCGC. Specifically, we first extract the texture information and spatial spectral information of the HSI to construct a multi-view map result. Then, a GCN is used to aggregate neighborhood information and the extracted features are fed into a self-expressive network to learn an affinity graph. We then employ an attention-based strategy to fuse the affinity graphs obtained from the two learning branches. We test our proposed model on three general-purpose HSI datasets and compare it with various state-of-the-art models. The experimental results show that MSCGC has achieved the optimal clustering results, and achieved clustering accuracies of 92.38%, 93.43% and 83.82% on the Indian Pines, Pavia University and Houston2013 datasets, respectively.

The self-expression layer determines that our model is difficult to train on large-scale data sets, and minBatch also has an inhibitory effect on the accuracy of the graph neural network. In the future, we will try to explore the potential of MSCGC on large-scale datasets.

Acknowledgments. This work was supported by Natural Science Foundation of China (No. U21A2013 and 42071430), Natural Science Foundation of Ningxia Hui Autonomous Region (2021AAC03453), Opening Fund of Key Laboratory of Geological Survey and Evaluation of Ministry of Education (Grant Number: GLAB2020ZR14 and CUG2022ZR02) and the Fundamental Research Founds for National University, China University of Geosciences (Wuhan) (No. CUGDCJJ202227).

References

1. Camps-Valls, G., Tuia, D., Bruzzone, L., Benediktsson, J.A.: Advances in hyperspectral image classification: Earth monitoring with statistical learning methods. IEEE Signal Process. Mag. **31**, 45–54 (2013)
2. Ghamisi, P., et al.: Advances in hyperspectral image and signal processing: a comprehensive overview of the state of the art. IEEE Geoscience Remote Sens. Mag. **5**, 37–78 (2017)
3. Eismann, M.T., Stocker, A.D., Nasrabadi, N.M.: Automated hyperspectral cueing for civilian search and rescue. Proc. IEEE **97**, 1031–1055 (2009)
4. Guan, R., Li, Z., Li, T., Li, X., Yang, J., Chen, W.: Classification of heterogeneous mining areas based on ResCapsNet and Gaofen-5 imagery. Remote Sens. **14**, 3216 (2022)
5. Tarabalka, Y., Fauvel, M., Chanussot, J., Benediktsson, J.A.: SVM-and MRF-based method for accurate classification of hyperspectral images. IEEE Geosci. Remote Sens. Lett. **7**, 736–740 (2010)
6. Hu, W., Huang, Y., Wei, L., Zhang, F., Li, H.: Deep convolutional neural networks for hyperspectral image classification. J. Sens. **2015**, 1–12 (2015)
7. Chen, Y., Lin, Z., Zhao, X., Wang, G., Gu, Y.: Deep learning-based classification of hyperspectral data. IEEE J. Sel. Top. Appl. Earth Observ. Remote Sens. **7**, 2094–2107 (2014)
8. Ghamisi, P., et al.: New frontiers in spectral-spatial hyperspectral image classification: the latest advances based on mathematical morphology, Markov random fields, segmentation, sparse representation, and deep learning. IEEE Geoscience Remote Sens. Mag. **6**, 10–43 (2018)

9. Zhai, H., Zhang, H., Li, P., Zhang, L.: Hyperspectral image clustering: Current achievements and future lines. IEEE Geoscience Remote Sens. Mag. **9**, 35–67 (2021)
10. Borsoi, R.A., et al.: Spectral variability in hyperspectral data unmixing: a comprehensive review. IEEE Geoscience Remote Sens. Mag. **9**, 223–270 (2021)
11. Chang, J., Meng, G., Wang, L., Xiang, S., Pan, C.: Deep self-evolution clustering. IEEE Trans. Pattern Anal. Mach. Intell. **42**, 809–823 (2018)
12. Kanungo, T., Mount, D.M., Netanyahu, N.S., Piatko, C.D., Silverman, R., Wu, A.Y.: An efficient k-means clustering algorithm: analysis and implementation. IEEE Trans. Pattern Anal. Mach. Intell. **24**, 881–892 (2002)
13. Ghaffarian, S., Ghaffarian, S.: Automatic histogram-based fuzzy C-means clustering for remote sensing imagery. ISPRS J. Photogramm. Remote Sens. **97**, 46–57 (2014)
14. Comaniciu, D., Meer, P.: Mean shift: a robust approach toward feature space analysis. IEEE Trans. Pattern Anal. Mach. Intell. **24**, 603–619 (2002)
15. Chen, Y., Ma, S., Chen, X., Ghamisi, P.: Hyperspectral data clustering based on density analysis ensemble. Remote Sens. Lett. **8**, 194–203 (2017)
16. Li, Q., Liu, W., Li, L.: Affinity learning via a diffusion process for subspace clustering. Pattern Recogn. **84**, 39–50 (2018)
17. Elhamifar, E., Vidal, R.: Sparse subspace clustering: algorithm, theory, and applications. IEEE Trans. Pattern Anal. Mach. Intell. **35**, 2765–2781 (2013)
18. Vidal, R., Favaro, P.: Low rank subspace clustering (LRSC). Pattern Recogn. Lett. **43**, 47–61 (2014)
19. Zhang, H., Zhai, H., Zhang, L., Li, P.: Spectral–spatial sparse subspace clustering for hyperspectral remote sensing images. IEEE Trans. Geosci. Remote Sens. **54**, 3672–3684 (2016)
20. Zhai, H., Zhang, H., Zhang, L., Li, P., Plaza, A.: A new sparse subspace clustering algorithm for hyperspectral remote sensing imagery. IEEE Geosci. Remote Sens. Lett. **14**, 43–47 (2016)
21. Caron, M., Bojanowski, P., Joulin, A., Douze, M.: Deep clustering for unsupervised learning of visual features. In: Ferrari, V., Hebert, M., Sminchisescu, C., Weiss, Y. (eds.) Computer Vision – ECCV 2018. LNCS, vol. 11218, pp. 139–156. Springer, Cham (2018). https://doi.org/10.1007/978-3-030-01264-9_9
22. Min, E., Guo, X., Liu, Q., Zhang, G., Cui, J., Long, J.: A survey of clustering with deep learning: from the perspective of network architecture. IEEE Access. **6**, 39501–39514 (2018)
23. Ji, P., Zhang, T., Li, H., Salzmann, M., Reid, I.: Deep subspace clustering networks. In: Advances in Neural Information Processing Systems, vol. 30 (2017)
24. Cai, Y., Zhang, Z., Cai, Z., Liu, X., Jiang, X., Yan, Q.: Graph convolutional subspace clustering: a robust subspace clustering framework for hyperspectral image. IEEE Trans. Geosci. Remote Sens. **59**, 4191–4202 (2020)
25. Zhang, Z., Cai, Y., Gong, W., Ghamisi, P., Liu, X., Gloaguen, R.: Hypergraph convolutional subspace clustering with multihop aggregation for hyperspectral image. IEEE J. Sel. Top. Appl. Earth Observ. Remote Sens. **15**, 676–686 (2021)
26. Cai, Y., Zeng, M., Cai, Z., Liu, X., Zhang, Z.: Graph regularized residual subspace clustering network for hyperspectral image clustering. Inf. Sci. **578**, 85–101 (2021)
27. Hinojosa, C., Rojas, F., Castillo, S., Arguello, H.: Hyperspectral image segmentation using 3D regularized subspace clustering model. J. Appl. Remote. Sens. **15**, 016508 (2021)
28. Benediktsson, J.A., Palmason, J.A., Sveinsson, J.R.: Classification of hyperspectral data from urban areas based on extended morphological profiles. IEEE Trans. Geosci. Remote Sens. **43**, 480–491 (2005)
29. Tian, L., Du, Q., Kopriva, I., Younan, N.: Kernel spatial-spectral based multi-view low-rank sparse subspace clustering for hyperspectral imagery. In: 2018 9th Workshop on Hyperspectral Image and Signal Processing: Evolution in Remote Sensing (WHISPERS), pp. 1–4. IEEE (2018)

30. Chen, Z., Zhang, C., Mu, T., He, Y.: Tensorial multiview subspace clustering for polarimetric hyperspectral images. IEEE Trans. Geosci. Remote Sens. **60**, 1–13 (2022)
31. Huang, S., Zhang, H., Pižurica, A.: Hybrid-hypergraph regularized multiview subspace clustering for hyperspectral images. IEEE Trans. Geosci. Remote Sens. **60**, 1–16 (2021)
32. Lu, H., Su, H., Hu, J., Du, Q.: Dynamic ensemble learning with multi-view kernel collaborative subspace clustering for hyperspectral image classification. IEEE J. Sel. Top. Appl. Earth Observ. Remote Sens. **15**, 2681–2695 (2022)
33. Peng, Z., Liu, H., Jia, Y., Hou, J.: Adaptive attribute and structure subspace clustering network. IEEE Trans. Image Process. **31**, 3430–3439 (2022)
34. Lu, C., Yan, S., Lin, Z.: Convex sparse spectral clustering: single-view to multi-view. IEEE Trans. Image Process. **25**, 2833–2843 (2016)
35. Wang, Y., Wu, L., Lin, X., Gao, J.: Multiview spectral clustering via structured low-rank matrix factorization. IEEE Trans. Neural Netw. Learn. Syst. **29**, 4833–4843 (2018)

Multi-relational Heterogeneous Graph Attention Networks for Knowledge-Aware Recommendation

Youxuan Wang[1], Shunmei Meng[1(✉)], Qi Yan[2], and Jing Zhang[1,3(✉)]

[1] Nanjing University of Science and Technology, Nanjing, China
{wangyouxuan,mengshunmei,jzhang}@njust.edu.cn
[2] Nanjing Turing Wudao Information Technology Ltd., Nanjing, China
[3] Southeast University, Nanjing, China

Abstract. Knowledge graph (KG) contains rich semantic information and is widely used in recommendation systems. Knowledge graph is a heterogeneous graph, and entities are connected by multiple relations. Users have different preferences for different relations, and the number of different relations varies greatly. Most of the existing recommendation methods may ignore the nodes connected by a small number of relations when sampling, even if it is very important for aggregation. At the same time, the data in the knowledge graph is complex, the relations of various attributes are mixed together, and lack of interpretability. In this paper, we propose a new method named Multi-Relational Heterogeneous Graph Attention Networks for Knowledge-aware Recommendation (MRHGAT). This method splits the knowledge graph into multiple subgraphs through relations and obtains the aggregation information corresponding to the specific relation from the subgraph to prevent important nodes from being ignored. The graph attention mechanism is added to the aggregation process of subgraphs to further optimize the aggregation. We use the representations learned in each subgraph as auxiliary information to supplement the node information that may be lost and improve the effect of aggregation. We also improve the interpretability of recommendations by splitting the knowledge graph. Empirical results on three datasets show that MRHGAT significantly outperforms most methods.

Keywords: Knowledge graph · Graph splitting · Recommendation system · Graph attention networks

1 Introduction

With the development of the Internet, recommendation systems are playing an increasingly important role in our lives. The traditional recommendation technology is collaborative filtering (CF) [5,8,14,22,24], which has been widely used in many fields. However, collaborative filtering has the problems of cold start and sparsity, which can be solved by introducing additional information sources, including social graph [14] and knowledge graph.

X. Song et al. (Eds.): APWeb-WAIM 2023, LNCS 14334, pp. 108–123, 2024.
https://doi.org/10.1007/978-981-97-2421-5_8

In recent years, knowledge graph has received more and more attention in recommendation systems [2,13,17,20,21]. There are many methods for using knowledge graph to assist recommendation systems, which can be roughly divided into three categories: path-based method [6,11,12,21], embedding-based method [2,16,18,23] and graph-based method. However, the path-based and embedding-based methods are not scalable enough to perform end-to-end training. We focus on graph-based recommendation methods. RippleNet [15] stimulates the spread of users' preferences on the set of knowledge entities by iteratively expanding users' potential interests along the links in the knowledge graph. KGCN [19] and KGNN-LS [17] fuse the KG into the recommendation system through the graph method, and obtain the node embedding information through neighborhood aggregation. KGAT [20] introduces the attention mechanism into the graph neural network, which further improves the effect.

However, the recommender system based on the knowledge graph still has obvious problems: 1) Knowledge graph is complex, contains massive information, and the relations of various attributes are mixed together, which makes it impossible to accurately obtain the users' interests for a single relation in knowledge graph, and the interpretability is not strong. 2) In the most recommendation model based on KG, the number of edges connected by a node is unknown, and most sampling method only aggregates a fixed number of neighbors, which may lead to the fact that the edges connected by some kind of relation are not sampled, ignoring their aggregated information, and making the result of aggregation is not satisfactory. In Sect. 5.6 of this paper, we analyze some knowledge graph datasets and find that the number of edges connected by different relations is quite different. For example, there is a film in the knowledge graph, with more than 20 relations connecting actors and only one relation connecting directors. During the routine sampling, because the number of directors connected is small, it is easy to ignore director nodes and lose important director node information. Therefore, it may not be possible to aggregate another work of this director, resulting in poor aggregation effect.

To solve the limitation of knowledge graph in recommendation system, we propose a method named Multi-Relational Heterogeneous Graph Attention Networks (MRHGAT). The model is divided into four steps: Firstly, we split the knowledge graph into knowledge subgraphs according to the relations, and use the subgraphs as the auxiliary information of the original graph to help better aggregate the information of the nodes and edges connected by each relation. Secondly, we aggregate the neighbor of the original graph and each subgraph respectively, extract the user representation of the whole knowledge graph and the user representation of specific relations, so that more information can be considered comprehensively during aggregation. For the original graph, when the neighborhood is aggregated, the weight of the neighborhood depends on the score of the user and the connected relation. For each subgraph, because the relation attributes of the subgraph are the same, it can be regarded as a homogeneous graph. In order to distinguish the importance of each neighbor node in the process of aggregation, attention mechanism is adopted in neighborhood aggregation to enhance the interpretability of the model. Next, we use the information extracted from the subgraph as the auxiliary information of the original graph to merge with the representation of the original graph. Finally, we predict through the prediction layer and get the final recommendation results. Because our method solves the problem of reasonable sampling for each relation, the effect of improvement is more obvious in the datasets with high node degree.

Empirically, we apply the model to three recommendation scenarios. It can be seen from the experiment that compared with many models, the AUC index of MRHGAT in films has increased by 0.6%, 3.1% and 2.0% respectively. And MRHGAT achieves F1 gains of 1.1%, 3.7%, and 1.2% in movie, music and restrautant recommendations.

In summary, the significance of our model is:

- We propose a method MRHGAT. By splitting the knowledge graph, we can accurately obtain users' preferences for a single relation, and enable the model to better capture users' high-level personalized interests. We also compare the effect without knowledge subgraphs information in the experiment, and find that the effect of adding subgraphs is improved, which proved the feasibility of this method.
- We solve the problem that some nodes connected by important relations may be lost during the sampling process, and enhance the aggregation effect of the knowledge graph by aggregating the information of specific edges in the knowledge subgraphs.
- We use the graph attention mechanism to aggregate the neighborhood information of the knowledge subgraphs, further highlight the information of important nodes in the subgraphs, and enhance the effect and interpretability of the model.
- We test three real scenarios, and the experimental results are better than most baselines, which proves the reliability of the model.

2   Related Work

2.1   Graph Neural Network

GCN extends the convolution neural network from Euclidean domains to non-Euclidean domains, so that convolution can be applied to more general scenarios. In the development of graph neural network, Bruna [1] et al. define convolution in Fourier domain, and calculate the eigen decomposition of Laplace matrix of graph. David Hammond [3] et al. propose using Chebyshev multinomial expansion to approximate the convolution filter, and Kipf [7] et al. propose the convolution architecture through first-order approximation. GCN directly operates on the graph and applies "convolution" to neighbors of nodes. However, there is also a potential relationship between items, and additional auxiliary information needs to be introduced to help complete the recommendation.

2.2   Recommendation Methods with Knowledge Graph

Knowledge graph is widely used in recommendation scenarios due to rich connections among items. In entity2rec [9], a method is proposed to split the knowledge graph according to attributes (relations). They believe that different attributes have different semantic values and should have different weights when judging the correlation between two entities, which improves the recommendation effect. RippleNet [15] expands users' potential interests along the links in KG to stimulate the spread of users' preferences on the set of entities. KGCN [19] and KGNN-LS [17] extend this method to the heterogeneous graph field. They improve the recommendation effect by introducing KG information into the interaction graph and using KG neighbors to obtain item embedding. In order to keep the calculation mode fixed and the algorithm efficient, the graph neural network method samples a fixed number of neighbors for each entity when it is embedded. This method makes the edges of some kind of relation connection not sampled, and the information connected by the relation cannot be obtained. Due to its massive information, users' preferences for a certain relation cannot be obtained, leading to poor interpretability.

3   Problem Formulation

We formulate the recommendation problem of MRHGAT. First, for a user-item interaction graph (UIG), we define the users set $U = \{u_1, u_2, ..., u_M\}$, where M is the number of users. And we define the items set $V = \{v_1, v_2, ..., v_N\}$, where N is the number of items. We define user-item interaction matrix $Y \in R^{M \times N}$, where $y_{uv} = 1$ indicates that user u has interacted with item v, otherwise $y_{uv} = 0$. Then, we define knowledge graph G. In datasets, knowledge graph is usually represented as triples. The formula is $G = (h, r, t)$, where h represents the head node, r represents the relation, and t represents the tail node. h, $t \in E, r \in R$. E and R represents entity set and relation set in knowledge graph respectively. We use G_i to represent knowledge subgraph. G_i refers to the knowledge graph split based on relation i. Next, we define the corresponding graph of items and knowledge graph entities, $A = \{(v, e) \mid v \in V, e \in E\}$. (v, e) indicates that item v in the user-item interaction graph is aligned with entity e in the knowledge graph. Finally, we predict the possibility of interaction between user u and item v through Y, G and $\{G_i | i \in R\}$. The prediction function is $y_{uv} = F(u, v| \Theta, G, Y, \{G_i\})$, where y_{uv} denotes the probability of interaction between user u and item v. Θ denotes the model parameters of function F.

4   Methodology

In this section, we introduce the process of MRHGAT. MRHGAT is divided into four parts: 1. Knowledge graph split layer. 2. Embedding propagation layer. 3. Inter graph aggregation layer. 4. Prediction layer. In Fig. 1, we draw the framework of the model.

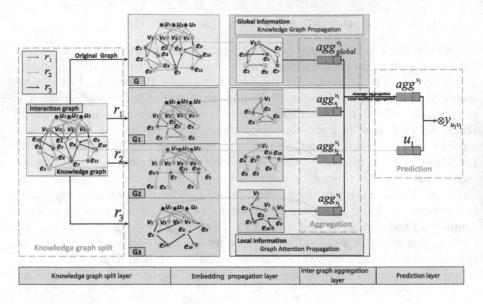

Fig. 1. Overview of MRHGAT.

4.1 Knowledge Graph Split Layer

As shown in the first part of Fig. 1, we split the knowledge graph into subgraphs according to multiple relations, and each subgraph contains the potential factors of users' preferences for the item. we divide the initial knowledge graph G into K subgraphs, where K is the number of relations in knowledge graph.

The knowledge subgraph is expressed as:

$$G_i = \{(h,\ r,\ t) \mid (h,\ r,\ t) \in G,\ r = i\}, i = 1, 2, ..., K. \tag{1}$$

It is worth noting that for knowledge graph G, we split it based on relations. Get the knowledge graph information G_i based on a certain relation i. Here, we still retain the information of the original graph of the knowledge graph as the global information, and the subgraphs after splitting as the local information. Note that after splitting the knowledge graph, item v may not be in the subgraph G_i. When finally aggregating, we do not consider the aggregation in the subgraph, but only return representation v itself. In the code implementation, in order to facilitate the calculation, we add self-loop (v, i, v) in the subgraph to prevent the different number of representations returned by different nodes.

4.2 Embedding Propagation Layer

As shown in the second part of Fig. 1, we use the knowledge graph to obtain global information, and use the knowledge subgraphs to obtain local information. We obtain high-level information in the knowledge graph through neighborhood aggregation. Consider a user-item pair (u, v). We obtain the neighbor

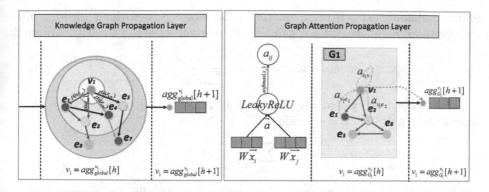

Fig. 2. The left part is Knowledge Graph Propagation Layer, and the right part is Graph Attention Propagation Layer.

of the entity corresponding to the item in the knowledge graph and knowledge subgraphs.

In the graph data in real life, the number of neighbors of nodes may vary greatly. In order to maintain the fixed computing mode and efficient algorithm, we do not take complete neighbors for aggregation, which is why the relations in the recommended algorithm is not fully sampled. We also uniformly sample the same number of neighbors for each entity. The neighbors are represented as:

$$S_v = \{e|e \sim N_v\}, \ |S_v| = N_{neighbor}, \tag{2}$$

$$S_v^i = \{e|e \sim N_v^i\}, \ |S_v^i| = N_{neighbor}, i = 1, 2, ..., K, \tag{3}$$

where S_v represents the global neighbor of the knowledge graph, and S_v^i represents the local neighbor in the subgraph G_i. N_v is the neighbor information of original graph G and N_v^i is the neighbor information of subgraph G_i. $N_{neighbor}$ is a unified neighbor sampling number, which can be configured as a parameter. The sampling here is repeatable. Next, we deal with the aggregation operation of the original graph and knowledge subgraphs in different ways. For the original graph of the knowledge graph, we use the knowledge graph embedding propagation layer to obtain the node neighbor information. For the knowledge subgraphs, we add the graph attention mechanism to further distinguish the importance of neighbor nodes. The method is shown in Fig. 2.

Knowledge Graph Propagation Layer. Next, we build the knowledge graph propagation layer in the original graph. We use the inner dot product operation between vectors to express the importance of users and relations. The importance of users and relations is evaluated by the user-relation score $p(u, r)$:

$$p(u, r) = u \cdot r, \tag{4}$$

where u, r is the embedding of users and relations. Here, we refer to the lightGCN [4] method to delete the two independent parameters of feature transformation matrix and nonlinear activation function, so as to be concise without affecting the recommendation effect. the normalized user-relation score in knowledge graph:

$$\tilde{p}(u, r_{v,e}) = softmax(u \cdot r_{v,e}) = \frac{exp(p(u, r_{v,e}))}{\sum_{e \in S_v} exp(p(u, r_{v,e}))}. \qquad (5)$$

Then we get the global neighborhood aggregation of KG and get the final representation of each corresponding v:

$$v_{S_v}^u = \sum_{e \in S_v} \tilde{p}(u, r_{v,e}) \cdot e, \qquad (6)$$

where $v_{S_v}^u$ is the final representation of graph G. Finally, we aggregate the global neighbor representation. We have implemented Sum aggregation, which can effectively consider neighbor information and its own information:

$$agg_{global}^v = v + v_{S_v}^u, \qquad (7)$$

where agg_{global}^v is the global neighbor representation of G for item v.

Graph Attention Embedding Propagation Layer. For each subgraph, the relation attribute between nodes in the graph is unique. In order to represent the importance of different neighbor nodes in subgraphs, we introduce the graph attention mechanism here. In knowledge subgraph G_i, for any entity e_k, we calculate the similarity coefficient between e_k and its neighbors:

$$e_{kj}^i = a([We_k \| We_j]), j \in S_v^i, i = 1, 2, ..., K, \qquad (8)$$

where $W \in R^{d \times d}$, $a \in R^{d \times 1}$. Next, normalize the similarity coefficient to get the final attention weight a_{kj}^i:

$$a_{kj}^i = \frac{exp(LeakyReLU(e_{kj}^i))}{\sum_{x \in S_v^i} exp(LeakyReLU(e_{kx}^i))}, i = 1, 2, ..., K, \qquad (9)$$

Then the expression after aggregation is:

$$agg_i^v = \sigma(\sum_{x \in S_v^i} a_{kj}^i Wv). \qquad (10)$$

4.3   Inter Graph Aggregation Layer

Through the above two steps, we get the global embedded representation and the local embedded representation. We propagate all the representations between the graphs for aggregation. There are two aggregation methods.

Average aggregation method:

$$agg^v = \frac{(agg_{global}^v + \sum_{i=1}^{K} agg_i^v)}{K + 1}. \qquad (11)$$

Local auxiliary aggregation mode:

$$agg^v = (1 - \sigma)agg^v_{global} + \sigma\frac{\sum_{i=1}^{K} agg^v_i}{K}, \tag{12}$$

where σ is the importance parameter of the global representation.

4.4 Prediction Layer

Through the Inter graph aggregation layer, we get the embedded representation considering the potential factors of relations. Through a single heterogeneous propagation layer and an inter graph aggregation layer, we can get a first-order entity representation. Finally, we use dot product to calculate the score:

$$y_{uv} = f(u, agg^v[H]), \tag{13}$$

where $agg^v[H]$ represents the embedding result obtained after H times heterogeneous propagation. We use the negative sampling method to improve the calculation efficiency of the model and define the loss function:

$$L = \sum_{u\in U}(\sum_{v\in v|y_{uv}=1} J(y_{uv}, \tilde{y}_{uv}) - \sum_{v\in v|y_{uv}=0} J(y_{uv}, \tilde{y}_{uv})) + \lambda||F||_2^2, \tag{14}$$

where J is the cross entropy loss and the items with $y_{uv} = 0$ are obtained through negative sampling. $||F||_2^2$ is created by λ Parametric L_2 regularizer.

5 Experiments

In this section, we evaluate the effect of MRHGAT, and we will verify it in three datasets: movies (MovieLens-20M), music (Last.FM) and restaurant.

We will solve the following research problems:

- **RQ1**: Does MRHGAT perform better than other recommendation methods based on knowledge graph?
- **RQ2**: How does hyper-parameter affect MRHGAT?
- **RQ3**: Is it feasible and necessary to split the knowledge graph?

5.1 Datasets

We test the effect of the model in three real scenarios: movies, music, and restaurant, proving the effectiveness of MRHGAT. Table 1 details the datasets.

Table 1. Basic information of datasets.

	MovieLens-20M	Last.FM	Restaurant
users	138159	1872	2298698
items	16954	3846	1362
interactions	13501622	42346	23416418
entities	102569	9366	28115
relations	32	60	7
triples	499474	15518	160519

5.2 Baselines

- **LibFM** [10] is a classical recommendation algorithm. It is an effective factorization model.
- **CKE** [23] combines CF with structure, text and visual knowledge in a unified recommendation framework.
- **RippleNet** [15] is a method similar to memory network, which spreads user preferences on KG for recommendation.
- **KGNN-LS** [17] takes the relation attribute in the knowledge graph as the user's preference, uses this preference when aggregating neighbors, and uses label smoothing regularization to optimize item embedding.
- **KGCN** [3] extends GCN method to knowledge graph, and can learn structural information and semantic information as well as user personalization and potential interest by selectively gathering neighborhood information.
- **KGAT** [20] combines user-item interaction graph with KG to form a unified CKG. The attention mechanism is used to set weights for each node.

5.3 Experimental Setup

In the experiment, we divide the dataset into train set, validation set and test set. We evaluate our method in two experimental scenarios: 1) In accuracy prediction, we use AUC and F1 indicators to predict the probability of user-item interaction. 2) We use recall@K to predict users select K items with the highest prediction probability.

In the experiment, we compare the current excellent recommendation methods, and the results of these methods have been run on our server. The results of all comparison methods are confirmed by experiments or given according to the original paper. In our model, the number of neighbor samples is $[2, 4, 8, 14, 32, 64]$. The embedding dimension is $[4, 8, 14, 32, 64]$. The number of polymerization layers is $[1, 2, 3]$. Inter graph aggregation methods are average aggregation method and local auxiliary aggregation method. In the auxiliary aggregation method, we adjust the size of the parameter σ to $[0, 0.2, 0.4, 0.6, 0.8, 1]$, where parameter $\sigma = 0$ means that the original graph information is not used, and only the aggregated information in the subgraphs is used for recommendation. If the parameter

Table 2. The results of AUC and F1.

Model	MovieLens-20M		Last.FM		Restaurant	
	AUC	F1	AUC	F1	AUC	F1
RippleNet	0.9757(−0.6%)	0.9290(−1.1%)	0.7888(−7.5%)	0.7153(−7.0%)	0.8330(−3.5%)	0.7551(−4.3%)
libFM	0.9660(−1.6%)	0.9281(−1.2%)	0.7792(−8.6%)	0.7223(−6.1%)	0.8412(−2.5%)	0.7696(−2.5%)
CKE	0.931(−5.1%)	0.8820(−6.1%)	0.7464(−12.4%)	0.6851(−10.9%)	0.8021(−7.0%)	0.7210(−8.7%)
KGNN-LS	0.9750(−0.7%)	0.9295(−1.1%)	0.7881(−7.6%)	0.7019(−8.7%)	0.8424(−2.4%)	0.7754(−1.8%)
KGCN	0.9535(−2.9%)	0.8961(−4.6%)	0.8121(−4.7%)	0.7297(−5.1%)	0.7489(−13.2%)	0.6880(−12.8%)
KGAT	0.9745(−0.7%)	0.9284(−1.2%)	0.8264(−3.1%)	0.7406(−3.7%)	0.8453(−2.0%)	0.7795(−1.2%)
MRHGAT	**0.9815**	**0.9395**	**0.8525**	**0.7692**	**0.8629**	**0.7893**

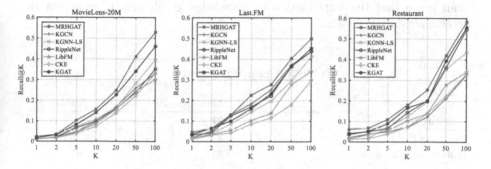

Fig. 3. The result of Recall@K.

$\sigma = 1$, only the original graph information is used. Through this setting, we can also know whether there is valid information in the subgraph and whether the method of splitting the knowledge graph is effective for embedding.

5.4 Performance Comparison (RQ1)

The comparison of experimental results is shown in Table 2 and Fig. 3.

Table 2 evaluates the recommended effect of each model on AUC and F1 of three data sets, and Fig. 3 shows the prediction effect of the model on recall@K. We have the following observations:

- From Table 2, the impact of MRHGAT on the three datasets has been significantly improved, especially in the music and restaurant datasets. The possible reason is that the average degree of nodes in these two datasets is large, and the loss of relation data is more serious during sampling, so splitting the knowledge graph can better improve these two datasets.
- From Fig. 3, Compared with RippleNet and KGNN-LS, our effect in TOPK recommendation evaluation has been improved more significantly. It proves the feasibility of the relation based splitting method.
- In the model aggregation method, our method is similar to KGCN for the aggregation of the original graph. For the aggregation of subgraphs, we introduce the graph attention mechanism. But in the experiment, we got better

results than KGCN, which can also prove the effectiveness of introducing graph attention mechanism and splitting knowledge graph.

- We find that for Last.FM with small dataset, the effect is best when the number of samples is the smallest when sampling in the neighborhood. The possible reason is that in small dataset, the loss of relations is more obvious. Our method better complements the aggregation of lost relations information, leading to better results. In the dataset with high average node degree, our model improvement effect is more obvious, which also confirms this view.
- In the comparison experiment, compared with the recommendation method based on knowledge graph, the effect of libFM is generally the worst, which can prove that the introduction of knowledge graph greatly improves the recommendation system.

Table 3. Optimal parameters.

	MovieLens-20M	Last.FM	Restaurant
Sampling Number	32	4	4
Receptive Field	2	1	1
Embedding Dimension	32	64	16
Global Representation Importance Parameter σ	Average	0.8	0.8

5.5   Hyperparametric Analysis (RQ2)

We conduct hyperparametric analysis of the model to explore the impact of each parameter on the model effect. The parameters we analyzed include the number of neighbor samples, the depth of receptive field, the embedding dimension, and the global representation importance parameter σ. During the analysis of each parameter, the remaining parameters have been adjusted to the best. For each dataset, the best parameters can refer to Table 3.

Table 4. AUC of MRHGAT about the neighbor sampling number K.

K	2	4	8	16	32	64
MovieLens-20M	0.9788	0.9790	0.9793	0.9796	**0.9815**	0.9737
Last.FM	0.8290	**0.8525**	0.8307	0.8216	0.8072	0.8164
Restaurant	0.8603	**0.8629**	0.8545	0.8535	0.8537	0.8537

Influence of Sampling Number K. Different sampling numbers will affect the aggregation effect, so we take the sampling number as a hyperparameter to study its impact on the prediction effect. Table 4 shows that for restaurant and music datasets, the effect is better when the number of samples is small. The possible reason is that in datasets with small node degrees, the relation loss is more obvious when the number of samples is small. Our method complements the aggregation of lost relation information, resulting in better results. It also reflects the feasibility of our method. For the movie dataset, when each node aggregates 32 neighbors, the prediction effect is the best.

Table 5. AUC of MRHGAT about the receptive field depth H.

H	1	2	3
MovieLens-20M	0.9795	**0.9815**	0.9565
Last.FM	**0.8525**	0.8266	0.8065
Restaurant	**0.8629**	0.8519	0.8478

Influence of Receptive Field H. Next, we explore the influence of receptive field depth on the prediction effect. We set the depth of receptive field to 1, 2, and 3. According to the results in Table 5, when H is 1 or 2, the effect is the best. When the receptive field is too large, the effect decreases. The possible reason is that too large H introduces many unnecessary embeddings.

Table 6. AUC of MRHGAT about the dimension d of embedding.

d	4	8	16	32	64
MovieLens-20M	0.9681	0.9776	0.9788	**0.9815**	0.9785
Last.FM	0.8018	0.8211	0.8365	0.8441	**0.8525**
Restaurant	0.8431	0.8454	**0.8629**	0.8538	0.8512

Influence of Embedding Dimension d. Embedding dimension affect the effect of the model. From Table 6, we know that to a certain extent, increasing the dimension will improve the accuracy of the model. However, when the dimension is too large, the model will overfit, resulting in further degradation of performance. The best embedding dimension of movie dataset is 32, music dataset is 64, and restaurant dataset is 16.

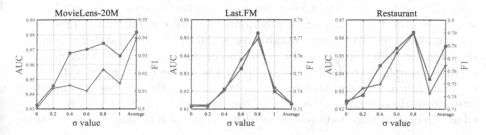

Fig. 4. The influence of importance parameters σ for recommendation.

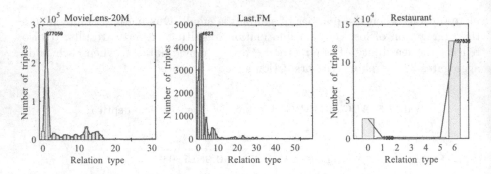

Fig. 5. The number of nodes connected by different relations of the three datasets.

Influence of the Global Representation Importance Parameter σ. We change the aggregation method of inter graph aggregation layer. From Fig. 4, we find that in music and restaurant datasets, local auxiliary aggregation is better. The best effect can be achieved when $\sigma = 0.8$. That is, when the proportion of subgraph information is 0.8, the model is optimal. But in the movie dataset, the average method is better. From the influence of parameter σ on three datasets, we can prove that adding subgraphs information can enhance the effect of the model, and also prove the feasibility of the method.

When $\sigma = 1$, it means that we fully use the information of the original graph. That is, the knowledge graph is not split. From the experimental results, the effect of the model is decreased compared with that of the model containing subgraphs information. It can prove the feasibility and effectiveness of splitting knowledge graph. When $\sigma = 0$, it means that the information of the original graph is not used at all, and only the information of the knowledge subgraphs is used. At this time, the effect is also stronger than some baselines. It can be proved that the knowledge subgraphs contains important information and does not lose information with the splitting of the knowledge graph.

5.6 Datasets Analysis (RQ3)

From Table 1, we can see that each dataset contains rich semantic information. And the correlation coefficient is concentrated within 100, and the number is small. Therefore, the time consumption of splitting knowledge graph is very small. Because of the small size of subgraphs, the time consumption of processing subgraphs is also within an acceptable range. Moreover, the knowledge graph is stored in the form of triple, and the method of splitting the knowledge graph is relatively simple. The feasibility of splitting knowledge graph is verified.

From Fig. 5, we know that there is a large gap in the number of connection nodes of various relations, so that in the knowledge graph, it is only for the edge aggregation information connected by a part of relations, and the smaller number of relations are easy to be ignored in the aggregation process, resulting in the impact of the aggregation effect. From the parameter analysis, we also know that only the original graph is not the best recommendation, and the information in the subgraphs also plays a positive role in the recommendation system. The necessity of splitting knowledge graph is verified.

6 Conclusions and Future Work

The MRHGAT model is proposed in this paper. MRHGAT splits KG according to the relation, learning the global representation of nodes and the local representation based on specific relation. It can accurately capture users' preferences for a single relation and enhance the interpretability of KG. By learning the users' preference representation, the user's potential interest in the items can be obtained. We also solve the problem of poor aggregation effect caused by ignoring certain types of edges in neighbor sampling of KG. By introducing knowledge subgraphs based on relation splitting, user nodes can obtain all kinds of connected relations during aggregation. Prevent loss of important neighbor nodes during aggregation.

As for future work, researchers can: 1. Further analyze and obtain effective information of KG, rather than blindly adding it to the recommendation system. 2. Find a better aggregation method so that the current aggregation problem will not occur. 3. Improve the imbalance of the number of relations in the KG. These are all worth trying to improve the recommended performance.

Acknowledgements. This work was sponsored the National Natural Science Foundation of China under grants 62076130, 91846104, and the Open Research Project of State Key Laboratory of Novel Software Technology (Nanjing University) under grant KFKT2022B28.

References

1. Bruna, J., Zaremba, W., Szlam, A., LeCun, Y.: Spectral networks and locally connected networks on graphs. arXiv preprint arXiv:1312.6203 (2013)
2. Cao, Y., Wang, X., He, X., Hu, Z., Chua, T.S.: Unifying knowledge graph learning and recommendation: towards a better understanding of user preferences. In: The World Wide Web Conference, pp. 151–161 (2019)
3. Defferrard, M., Bresson, X., Vandergheynst, P.: Convolutional neural networks on graphs with fast localized spectral filtering. In: Advances in Neural Information Processing Systems, vol. 29 (2016)
4. He, X., Deng, K., Wang, X., Li, Y., Zhang, Y., Wang, M.: LightGCN: simplifying and powering graph convolution network for recommendation. In: Proceedings of the 43rd International ACM SIGIR Conference on Research and Development in Information Retrieval, pp. 639–648 (2020)
5. He, X., Liao, L., Zhang, H., Nie, L., Hu, X., Chua, T.S.: Neural collaborative filtering. In: Proceedings of the 26th International Conference on World Wide Web. pp. 173–182 (2017)

6. Hu, B., Shi, C., Zhao, W.X., Yu, P.S.: Leveraging meta-path based context for top-n recommendation with a neural co-attention model. In: Proceedings of the 24th ACM SIGKDD International Conference on Knowledge Discovery & Data Mining, pp. 1531–1540 (2018)
7. Kipf, T.N., Welling, M.: Semi-supervised classification with graph convolutional networks. arXiv preprint arXiv:1609.02907 (2016)
8. Koren, Y., Bell, R., Volinsky, C.: Matrix factorization techniques for recommender systems. Computer **42**(8), 30–37 (2009)
9. Palumbo, E., Rizzo, G., Troncy, R.: Entity2rec: learning user-item relatedness from knowledge graphs for top-n item recommendation. In: Proceedings of the Eleventh ACM Conference on Recommender Systems, pp. 32–36 (2017)
10. Rendle, S.: Factorization machines with libFM. ACM Trans. Intell. Syst. Technol. (TIST) **3**(3), 1–22 (2012)
11. Shi, C., Hu, B., Zhao, W.X., Philip, S.Y.: Heterogeneous information network embedding for recommendation. IEEE Trans. Knowl. Data Eng. **31**(2), 357–370 (2018)
12. Sun, Z., Yang, J., Zhang, J., Bozzon, A., Huang, L.K., Xu, C.: Recurrent knowledge graph embedding for effective recommendation. In: Proceedings of the 12th ACM Conference on Recommender Systems, pp. 297–305 (2018)
13. Vijayakumar, V., Vairavasundaram, S., Logesh, R., Sivapathi, A.: Effective knowledge based recommender system for tailored multiple point of interest recommendation. Int. J. Web Portals (IJWP) **11**(1), 1–18 (2019)
14. Wang, H., Wang, J., Zhao, M., Cao, J., Guo, M.: Joint topic-semantic-aware social recommendation for online voting. In: Proceedings of the 2017 ACM on Conference on Information and Knowledge Management, pp. 347–356 (2017)
15. Wang, H., Zhang, F., Wang, J., Zhao, M., Li, W., Xie, X., Guo, M.: RippleNet: propagating user preferences on the knowledge graph for recommender systems. In: Proceedings of the 27th ACM International Conference on Information and Knowledge Management, pp. 417–426 (2018)
16. Wang, H., Zhang, F., Xie, X., Guo, M.: DKN: deep knowledge-aware network for news recommendation. In: Proceedings of the 2018 World Wide Web Conference, pp. 1835–1844 (2018)
17. Wang, H., Zhang, F., Zhang, M., Leskovec, J., Zhao, M., Li, W., Wang, Z.: Knowledge-aware graph neural networks with label smoothness regularization for recommender systems. In: Proceedings of the 25th ACM SIGKDD International Conference on Knowledge Discovery & Data Mining, pp. 968–977 (2019)
18. Wang, H., Zhang, F., Zhao, M., Li, W., Xie, X., Guo, M.: Multi-task feature learning for knowledge graph enhanced recommendation. In: The World Wide Web Conference, pp. 2000–2010 (2019)
19. Wang, H., Zhao, M., Xie, X., Li, W., Guo, M.: Knowledge graph convolutional networks for recommender systems. In: The World Wide Web Conference, pp. 3307–3313 (2019)
20. Wang, X., He, X., Cao, Y., Liu, M., Chua, T.S.: KGAT: knowledge graph attention network for recommendation. In: Proceedings of the 25th ACM SIGKDD International Conference on Knowledge Discovery & Data Mining, pp. 950–958 (2019)
21. Wang, X., Wang, D., Xu, C., He, X., Cao, Y., Chua, T.S.: Explainable reasoning over knowledge graphs for recommendation. In: Proceedings of the AAAI Conference on Artificial Intelligence, vol. 33, pp. 5329–5336 (2019)
22. Weimer, M., Karatzoglou, A., Smola, A.: Adaptive collaborative filtering. In: Proceedings of the 2008 ACM Conference on Recommender Systems, pp. 275–282 (2008)

23. Zhang, F., Yuan, N.J., Lian, D., Xie, X., Ma, W.Y.: Collaborative knowledge base embedding for recommender systems. In: Proceedings of the 22nd ACM SIGKDD International Conference on Knowledge Discovery and Data Mining, pp. 353–362 (2016)

24. Zhang, H., Shen, F., Liu, W., He, X., Luan, H., Chua, T.S.: Discrete collaborative filtering. In: Proceedings of the 39th International ACM SIGIR Conference on Research and Development in Information Retrieval, pp. 325–334 (2016)

Heterogeneous Graph Contrastive Learning with Dual Aggregation Scheme and Adaptive Augmentation

Yingjie Xie[1], Qi Yan[2], Cangqi Zhou[1(✉)], Jing Zhang[1,3(✉)], and Dianming Hu[4]

[1] Nanjing University of Science and Technology, Nanjing, China
{yjx0921,cqzhou,jzhang}@njust.edu.cn
[2] Nanjing Turing Wudao Information Technology Ltd., Nanjing, China
[3] Southeast University, Nanjing, China
[4] SenseDeal Intelligent Technology Co., Ltd., Nanjing, China

Abstract. Heterogeneous graphs are ubiquitous in the real world, such as online shopping networks, academic citation networks, etc. Heterogeneous Graph Neural Networks (HGNNs) have been widely used to capture rich semantic information on graph data, showing strong potential for application in real-world scenarios. However, the semantic information is not fully exploited by existing heterogeneous graph models in the following two aspects: (1) Most HGNNs use only meta-path scheme to model semantic information, which ignores local structure information. (2) The influence of cross-scheme contrast on the model performance is not taken into account. To fill above gaps, we propose a novel Contrastive Learning model on Heterogeneous Graphs (CLHG). Firstly, CLHG encodes local structure and semantic information by a dual aggregation scheme (i.e. network schema and meta-path). Secondly, we perform contrast between views within the same scheme and then comprehensively utilize dual aggregation scheme to collaboratively optimize CLHG. Furthermore, we extend adaptive augmentation to heterogeneous graphs to generate high-quality positive and negative samples, which greatly improves the performance of CLHG. Extensive experiments on three real-world datasets demonstrate that our proposed model achieves competitive results with the state-of-the-art methods.

Keywords: Heterogeneous graph neural networks · Contrastive learning · Dual aggregation scheme · Adaptive augmentation

1 Introduction

Compared with homogeneous graphs, heterogeneous graphs are more realistic to model multiple types of relationships in the real world. For example, the purchase/cancel relationship between consumers and products exist widely in online shopping networks. How to encode complex structure and capture rich semantic information are the key problems of HGNNs [2,7,30]. At the same time, label scarcity problem has long attracted the attention of scholars. Due to various

X. Song et al. (Eds.): APWeb-WAIM 2023, LNCS 14334, pp. 124–138, 2024.
https://doi.org/10.1007/978-981-97-2421-5_9

nodes and edges, this problem is more prominent in the field of heterogeneous networks [22,28].

To capture rich semantic information, a variety of heterogeneous graph representation learning methods have been proposed. These methods can be roughly classified into two categories: (1) The methods are inspired by graph neural networks [4,6,20,24]. For instance, based on graph attention network (GAT) [18], HAN [20] uses hierarchical attention to automatically distinguish importance of meta-paths. (2) The methods are inspired by special structure of heterogeneous graphs [2,3,7,27,30]. For network schema, NSHE [30] focuses to preserve node pairwise and network schema proximity simultaneously. However, these methods ignore various aggregation schemes on heterogeneous networks. Then, contrastive learning methods on heterogeneous graphs [21,32] are proposed recently, which use different aggregation schemes to encode multiple views. They perform cross-scheme contrast between these views to solve the label scarcity problem. However, these methods are still insufficient in their capture of semantic information in two main ways.

- **The lack of using various aggregation schemes.** Most of previous studies do not use multiple aggregation schemes to capture multiple perspective information. For instance, metapath2vec [2] aggregates high-order node information based on meta-path, but ignore local structure information based on network schema. These methods obviously limit the representation ability of heterogeneous graph models. Therefore, modeling multiple perspective information via various aggregation schemes is one of the goals in this paper.
- **The issue of cross-scheme contrast.** The cross-scheme contrast methods [21,32] will most likely force models to focus on differences between aggregation schemes, but ignore differences between nodes. Thus, cross-scheme contrast may undermine the model performance. Moreover, these methods ignore data augmentations that are crucial in graph contrastive learning [23]. And adaptive augmentation is able to help improve embedding quality by preserving important patterns during perturbation [33]. Therefore, another goal of this paper is to avoid the issue of cross-scheme contrast when use various aggregation schemes.

To address these challenges, we propose a novel contrastive representation learning model (CLHG) to model rich semantic information with dual aggregation scheme. Specifically, CLHG first uses an adaptive augmentation algorithm to generate an augmented graph from the original graph. Then, we mine local structure and semantic information from these graphs by a dual aggregation scheme (network schema and meta-path). Finally, we perform contrast between original view and augmented view within scheme (i.e. intra-scheme contrast), which is our proposed novel idea enabling CLHG to focus on finding unique properties of nodes without being disturbed by redundant information from different schemes. In addition, we extend adaptive augmentation to heterogeneous graphs to generate high-quality positive and negative samples, which adaptively perturb possibly unimportant features and relations with different types of nodes.

In summary, our core contributions are three-fold:

- To the best of our knowledge, this is the first time that the concept of intra-scheme contrast has been proposed in heterogeneous graph neural networks.
- We propose a novel contrastive learning model, CLHG, which use a dual aggregation scheme to capture rich semantic information with an extended adaptive augmentation.
- We conduct extensive experiments on three real-world datasets to demonstrate the effectiveness of CLHG and analyze the performance of CLHG by comparing with state-of-the-art methods.

2   Related Work

2.1   Heterogeneous Graph Neural Networks

In the past few years, many heterogeneous graph neural networks (HGNNs) have been developed to learn node representations in heterogeneous graphs. Although the special semantic knowledge of heterogeneous graphs, i.e. meta-path, has been noticed by early scholars [15], the existing HGNNs methods [4,6,20,24] are mostly inspired by graph neural networks (GNNs). For example, based on graph attention network (GAT) [18], HAN [20] uses hierarchical attention to automatically distinguish importance of meta-paths. MAGNN [4] further takes intermediate nodes of meta-paths into account. On this basis, GTN [24] is proposed for automatic identification of useful connections between unconnected nodes. Moreover, HGT [6] is designed for Web-scale heterogeneous graphs via an heterogeneous mini-batch graph sampling algorithm. At the same time, some scholars have noticed the meta-structure of heterogeneous graphs, including network schema [16,26,30] and meta-graph [27,29]. NSHE [30] focuses to preserve both to preserve node pairwise and network schema proximity simultaneously. The meta-graph is proposed on the basis of the meta-path. Mg2vec [27] is proposed to generate multiple meta-graphs, which then guide the learning through both first- and second-order constraints on node embeddings. For unsupervised methods, HetGNN [25] samples a fixed number of nodes and use LSTM to fuse their features. Recently, some works [21,32] leverage contrastive learning to capture high-level information between different schemes.

However, most of the above methods use only one aggregation scheme. Although emerging contrastive learning methods use various aggregation schemes, they ignore the influence of cross-scheme contrast.

2.2   Graph Contrastive Learning

Existing graph contrastive learning models usually learn node or graph representations by training encoders to distinguish positive and negative samples. Specifically, DGI [19] builds local patch representations and corresponding global summaries of graphs as positive pairs, and utilizes Infomax [11] theory to contrast. Then, GCC [13] leverages contrastive learning technique and defines subgraph instance discrimination in and across networks as the pre-training task to learn

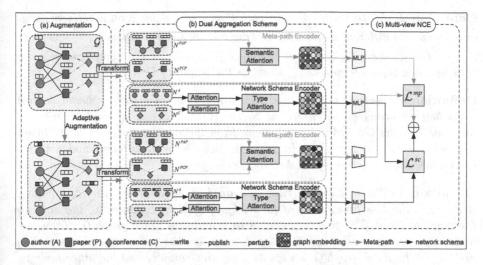

Fig. 1. Illustration of the proposed CLHG.

unique structural representations. Afterwards, data augmentation has been paid attention to recently for improving generalizability of graph contrastive learning models [1,23,31,33]. Zhao et al. [31] study graph data augmentation for GNNs and show that neural edge predictors can effectively encode class-homophilic structure to promote intra-class edges and demote inter-class edges in given graph structure. GraphCL [23] design four types of graph augmentations to generate different views for graph contrastive learning. On this basis, GCA [33] proposes an adaptive augmentation algorithm for graph centrality. For heterogeneous graphs, HeCo [21] employs cross-view (schema view and meta-path view) contrast theory to learn node embeddings.

However, the above methods do not consider extending graph augmentation to heterogeneous graphs, and ignore the influence of cross-scheme contrast with various aggregation schemes.

3 Preliminary

In this section, we formally introduce some concepts we use in the remainder of this paper.

Definition 1. *Heterogeneous Graphs.* *A heterogeneous graph is a graph that preserves diverse node types and relation types, which is formally defined as* $\mathcal{G} = (\mathcal{V}, \mathcal{E}, \mathcal{X}, \mathcal{O}, \mathcal{R})$, *where* $\mathcal{V}$ *is the set of nodes,* $\mathcal{E}$ *is the set of interaction relations,* $\mathcal{O}$ *is the set of node types,* $\mathcal{R}$ *is the set of relation types and* $\mathcal{X} = \{X_1, \ldots, X_{|\mathcal{O}|}\}$ *denotes the set of feature matrices. There is a node type mapping function* $\phi : \mathcal{V} \to \mathcal{O}$ *and a relation type mapping function* $\psi : \mathcal{E} \to \mathcal{R}$ *with* $|\mathcal{R}| + |\mathcal{O}| > 2$.

Definition 2. Meta-path. *A meta-path $\mathcal{P}$ is a path defined on a heterogeneous graph connects different nodes with diverse relations, formally as $O_1 \xrightarrow{R_1} O_2 \xrightarrow{R_2} \dots \xrightarrow{R_l} O_{l+1}$, where $O_i \in \mathcal{O}$, $R = R_1 \cdot R_2 \cdot \dots \cdot R_l$ is the composite relation between node types O_1 and O_{l+1}, and l is the length of $\mathcal{P}$.*

Definition 3. Network Schema. *Network schema is a basic schema defined on a heterogeneous graph, noted as $S_G = (\mathcal{O}, \mathcal{R})$. S_G is a graph defined on the node type set $\mathcal{O}$, and the edges come from the relation type set $\mathcal{R}$. Network schema is used to describe the direct connections between different nodes, which represents local structure of the heterogeneous graph.*

Definition 4. Noise Contrastive Estimation. *In graph contrastive learning, it refers to the loss function used for backpropagation. Specifically, the data on the original graph $\mathcal{G}$ is first augmented to generate a noise graph $\tilde{\mathcal{G}}$. Then one node c_i in the original graph $\mathcal{G}$ is used as an anchor point, the corresponding node x_i in the noise graph $\tilde{\mathcal{G}}$ is used as a positive sample, and the other node x_k is used as a negative sample. Finally, the similarity between positive pair (c_i, x_i) and negative pair (c_i, x_k) is computed. The goal of Noise Contrastive Evaluation (NCE) is to increase the similarity of positive pairs as much as possible while reducing the similarity of negative pairs. The formal definition of the NCE loss function is as follows:*

$$L_c^{NCE} = -\mathbf{E}_V \left[\log \frac{f(c_i, x_i)}{\sum_{x_k \in V, x_k \neq x_i} f(c_i, x_k)} \right], \tag{1}$$

where V is the set of nodes, $f(\cdot, \cdot)$ denotes the similarity of the sample pair.

4   Methodology

We now present our proposed CLHG, which exploits dual aggregation scheme to model rich semantic information. Figure 1 illustrates the framework of CLHG, which consists of three key components: (1) an adaptive augmentation module, which generates augmented graph by adaptive augmentation; (2) a dual aggregation scheme module, which employs meta-path encoder and network schema encoder to learn the representations of nodes; (3) a multi-view NCE module, which performs intra-scheme contrast between original views and augmented views to maximize their agreement.

4.1   Adaptive Augmentation for Heterogeneous Graphs

One of the difficulties in data augmentation on heterogeneous graphs is the handling of various complex structures. For example, nodes of target type are in some sense more important than nodes of other types, since they are connecting entities of most relations. It is necessary to pay attention to these special structure of heterogeneous graphs during the augmentation process. Following GCA [33], adaptive augmentation algorithms in CLHG tend to keep important features and relations unchanged, while perturbing features and structures that may not be important. Note that due to the different types of nodes in heterogeneous graphs, they contribute differently to the final node representation. Therefore, in CLHG, the importance of relations and features is measured between relations and nodes of the same type, respectively.

Node-Level Augmentation. We perturb node features via randomly masking a fraction of dimensions with Gaussian noise. We first formally sample a vector $\tilde{m}^o \in \{0, 1\}^F$, where each of its elements is independently drawn from a Bernoulli distribution with nodes of type o, i.e. $\tilde{m}_k^o \sim Bern\left(1 - p_k^o\right), \forall k$, where p_k^o is the probability of masking the k-th dimension of nodes of type o. Therefore, all types of node features

$$\tilde{\mathcal{X}} = \{\tilde{X}^1, \tilde{X}^2, \ldots, \tilde{X}^o, \ldots, \tilde{X}^{|\mathcal{O}|}\}, \tag{2}$$

where $\tilde{X}^o$ is computed by $\tilde{X}^o = [x_1^o \bullet \tilde{m}^o; x_2^o \bullet \tilde{m}^o; \cdots ; x_N^o \bullet \tilde{m}^o]^\top$. Here $[\cdot; \cdot]$ is the concatenation operator and $\bullet$ is element-wise multiplication.

p_k^o should reflect the importance of the k-th dimension of node features of type o. To preserve the special property that nodes of the target type are more important in heterogeneous graphs, we choose to use degree centrality among degree centrality, eigenvector centrality and PageRank centrality to describe the importance of nodes. So p_k^o should be computed with the following formula:

$$p_k^o = \min\left(\frac{d_{\max}^o - d_k^o}{d_{\max}^o - d_{mean}^o}, p_\tau\right), \tag{3}$$

where d^o is a degree vector of nodes of type o, $d_{\max}$ and d_{mean} are maximum value and average value of d^o respectively, and p_τ is a hyperparameter that controls the overall magnitude of feature augmentation.

Structure-Level Augmentation. Following GCA [33], we extract a subset $\tilde{\mathcal{E}}$ from the original $\mathcal{E}$ with a probability $P\{(u, v) \in \tilde{\mathcal{E}}\} = 1 - p_{uv}^e$, where p_{uv}^e is the probability of removing (u, v). Similar to node-level augmentation, p_{uv}^e should reflect the importance of relation (u, v). We compute p_{uv}^e as

$$p_{uv}^e = \frac{p_u + p_v}{2}, \tag{4}$$

where p_u, p_v are the degree centrality of nodes u and v respectively. Note that if all relations of a node are to be removed, in order to retain the node information, one of the relations will be randomly retained.

Finally, we generate the augmented graph $\tilde{\mathcal{G}}$ by jointly performing node-level and structure-level augmentation. Since we use node degree to measure the importance of nodes and relations, the structure-level adaptive augmentation will be done at the end.

4.2   Dual Aggregation Scheme

After obtaining the augmented graph $\widetilde{\mathcal{G}}$, we adopt two aggregation schemes to aggregate local structure and semantic information of nodes respectively. Since there are different types of nodes in heterogeneous networks, their features are usually located in different spaces [21]. So first, we project features of all types of nodes into a common vector space. Specifically, we design a type-specific mapping matrix W^o to transform the feature x_i of node i of type o into the common space: $h_i = \sigma\left(W^o \cdot x_i + b^o\right)$, where $h_i \in \mathbb{R}^{d \times 1}$ is the projected feature of node i, $\sigma(\cdot)$ is an active function and b^o is a vector bias.

Meta-Path Aggregation Scheme. Our goal now is to learn node embeddings in a meta-path view. Specifically, given a meta-path $\mathcal{P}_n$ starting from node i, we can obtain meta-path-based neighbors $N_i^{\mathcal{P}_n}$. Each meta-path represents a semantic similarity, and we encode this characteristic using a metapath-specific GCN [8]:

$$h_i^{\mathcal{P}_n} = \frac{1}{d_i + 1} h_i + \sum_{j \in N_i^{\mathcal{P}_n}} \frac{1}{\sqrt{(d_i + 1)(d_j + 1)}} h_j, \tag{5}$$

where d_i and d_j are the degrees of nodes i and j, and h_i and h_j are their projected features, respectively. With M meta-paths, we can compute M embeddings $\left\{h_i^{\mathcal{P}_1}, \ldots, h_i^{\mathcal{P}_M}\right\}$ for node i. We then fuse them into a final representation z_i^{mp} on the meta-path view through using semantic-level attention:

$$z_i^{mp} = \sum_{n=1}^{M} \alpha_{\mathcal{P}_n} \cdot h_i^{\mathcal{P}_n}, \tag{6}$$

where $\alpha_{\mathcal{P}_n}$ weights the importance of $\mathcal{P}_n$, which is computed as follows:

$$\alpha_{\mathcal{P}_n} = s\left(\frac{1}{|\mathcal{V}|} \sum_{i \in \mathcal{V}} \mathbf{a}_{mp}^{\top} \cdot \tanh\left(\mathbf{W}_{mp} h_i^{\mathcal{P}_n} + \mathbf{b}_{mp}\right)\right), \tag{7}$$

where $\mathbf{W}_{mp} \in \mathbb{R}^{d \times d}$ and $\mathbf{b}_{mp} \in \mathbb{R}^{d \times 1}$ are learnable parameters, $s(\cdot)$ is a softmax function and $\mathbf{a}_{mp}$ denotes a semantic-level attention vector.

Network Schema Aggregation Scheme. Here, we aim to learn node embeddings from the perspective of network schema. According to the network schema, we define N_i^o as o type neighbors of node i. Different types of neighbors and different neighbors of the same type have different importance to the embedding of node i, respectively. Therefore, we utilize node-level attention layer and type-level attention layer to aggregate neighbor node information to node i. First at the node level, we employ attention mechanism to fuse neighbors of type o:

$$h_i^o = \sigma \left(\sum_{j \in N_i^o} \beta_{i,j}^o \cdot h_j \right),$$ (8)

where $\beta_{i,j}^o$ denotes the attention value of node j with type o to node i. It can be computed as follows:

$$\beta_{i,j}^o = s \left(l \left(\mathbf{b}_o^\top \cdot [h_i; h_j] \right) \right),$$ (9)

where $\mathbf{b}_o \in \mathbb{R}^{2d \times 1}$ is a node-level attention vector for type o and $l(\cdot)$ is LeakyReLU active function. To ensure fairness, each node aggregates information from a fixed number of neighbors, which makes subsequent intra-scheme contrast more challenging.

After we obtain all types of embeddings $\left\{ h_i^1, \ldots, h_i^{|O|} \right\}$, we fuse them through using a semantic-level attention layer to learn the final embedding z_i^{sc} of node i on network schema view. Therefore, we perform a weighted sum over all types of type embeddings to obtain z_i^{sc} :

$$z_i^{sc} = \sum_{o=1}^{|O|} \beta_o \cdot h_i^o,$$ (10)

where β_o is a weight of each node type, computed similarly to $\alpha_{\mathcal{P}_n}$.

In summary, CLHG has learned representions z_i^{mp} and z_i^{sc} of node i on the two views of the original graph $\mathcal{G}$ via dual aggregation scheme. Node representions $\widetilde{z}_i^{mp}$ and $\widetilde{z}_i^{sc}$ are learned on the augmented graph $\widetilde{\mathcal{G}}$ in the same way.

4.3 Multi-view NCE

After obtaining z_i^{mp}, z_i^{sc}, $\widetilde{z}_i^{mp}$ and $\widetilde{z}_i^{sc}$ of node i through the above dual aggregation scheme, we utilize multilayer perceptron (MLP) to project z_i^{mp} into a space for computing node similarity, which is formalized as:

$$r_i^{mp} = W^{(2)} \sigma \left(W^{(1)} z_i^{mp} + b^{(1)} \right) + b^{(2)},$$ (11)

where σ is ELU non-linear function. Other node embeddings r_i^{sc}, $\widetilde{r}_i^{mp}$ and $\widetilde{r}_i^{sc}$ are computed in the same way. Note that $\{W^{(1)}, W^{(2)}, b^{(1)}, b^{(2)}\}$ are shared parameters of the MLP.

Afterwards, we contrast the node embeddings generated by the same aggregation scheme. Our proposed intra-scheme contrast lets the model focus on finding unique properties of nodes without being disturbed by redundant information from different schemes. For example, contrasting between the negative sample pair $(r_i^{mp}, \widetilde{r}_k^{mp})$, the model tends to look for unique properties of node i in a high-order semantic perspective. While HeCo [21] contrasts negative sample pair (r_i^{mp}, r_k^{sc}), which are disturbed by different information from meta-path and network schema. Therefore, we compute the pairwise loss function for each sample pair $(r_i^{mp}, \widetilde{r}_i^{mp})$ in the meta-path scheme as

$$\ell(u_i, v_i) = -\log \frac{\exp(sim(u_i, v_i)/\tau)}{\sum_{v_k \in V, v_k \neq v_i} \exp(sim(u_i, v_k)/\tau)}, \tag{12}$$

$$\mathcal{L}_i^{mp} = \frac{1}{2}[\ell(r_i^{mp}, \widetilde{r}_i^{mp}) + \ell(\widetilde{r}_i^{mp}, r_i^{mp})], \tag{13}$$

where $sim(u, v)$ denotes cosine similarity between u and v, and τ is a temperature parameter. It can be noticed that the meta-path loss of node i is the average of $\ell(r_i^{mp}, \widetilde{r}_i^{mp})$ and $\ell(\widetilde{r}_i^{mp}, r_i^{mp})$. We believe that the augmented graph $\widetilde{\mathcal{G}}$ retains relatively important node features and structural information, so we treat original graph and augmented graph equally.

Also, $\mathcal{L}_i^{sc}$ is computed in the same way. Then the overall loss function is as follows:

$$\mathcal{J} = \frac{1}{|\mathcal{V}|}\sum_{i \in \mathcal{V}}[\lambda \cdot \mathcal{L}_i^{mp} + (1 - \lambda) \cdot \mathcal{L}_i^{sc}], \tag{14}$$

where λ is a hyperparameter used to balance the impact of dual aggregation scheme. Finally, we use $Z^{mp} = [z_1^{mp}; \cdots ; z_N^{mp}]^\top$ for downstream tasks, since Z^{mp} contains more information about nodes of target type.

5   Experiments

In this section, we conduct extensive experiments on three real-world datasets to evaluate the performance of our proposed CLHG by answering the following research questions:

- **RQ1:** How does CLHG perform compared with other baselines?
- **RQ2:** What is the impact of module designs on the improvement of CLHG?

5.1   Experimental Setup

We first introduce the datasets, evaluation metrics, baseline methods, and parameter settings involved in the experiments.

Datasets. To validate our proposed model, we utilize the following three real-world datasets: DBLP[1], ACM[2] and Freebase[3].The basic statistics of the datasets are summarized in Table 1.

- *DBLP Dataset* [4]. It is extracted from the Computer Science Bibliography website. The target type is author, which are divided into four classes. We use three meta-paths, APA, APCPA and APTPA, for the meta-path baselines.

---

[1] https://dblp.org/.
[2] https://dl.acm.org/.
[3] https://www.freebase.com/.

Table 1. Statistics of the datasets.

Dataset	#Node	#Relation	#Meta-path
DBLP	author (A):4057 paper (P):14328 conference (C):20 term (T):7723	P-A:19645 P-C:14328 P-T:85810	APA APCPA APTPA
ACM	paper (P):4019 author (A):7167 subject (S):60	P-A:13407 P-S:4019	PAP PSP
Freebase	movie (M):3492 actor (A):33401 direct (D):2502 writer (W):4459	M-A:65341 M-D:3762 M-W:6414	MAM MDM MWM

- *ACM Dataset* [30]. It is extracted from a subset of ACM datasets. The target type is paper, which are divided into three classes. We use two meta-paths, PAP and PSP, for the meta-path baselines.
- *Freebase Dataset* [10]. It is a movie dataset on the Freebase website. The target type is moive, which are divided into three classes. We use three meta-paths, MAM, MDM and MWM, for the meta-path baselines.

Evaluation Metrics. We use three widely-used metrics *Macro F1-score* (Ma-F1), *Micro F1-score* (Mi-F1), and *Area Under Curve* (AUC) to evaluate the model. F1 score is an indicator used in statistics to measure the accuracy of a binary model, which also considers the accuracy and recalls of the classification model [17].

Baselines. In order to comprehensively verify the performance of CLHG, we consider following baselines:

- **GraphSAGE** [5] is a graph representation learning framework which can be used for inductive applications where the graphs are dynamic.
- **VGAE** [9] is a model that applies the unsupervised learning method of the variational autoencoder to the graph structure using GCN.
- **DGI** [19] is a classical technique used for unsupervised graph contrastive learning, which utilizes Infomax theory to contrast.
- **Mp2vec** [2] is an embedding method for heterogeneous networks that uses meta-paths, it can take only one meta-path as input.
- **HERec** [14] is a heterogeneous network recommendation method based on heterogeneous network node representation learning.
- **HetGNN** [25] is a refined GNN technique that utilizes the heterogeneous information of node and edge types.
- **DMGI** [12] is an unsupervised node embedding method for learning multi-attribute networks.

Table 2. Performance comparisons of all methods in DBLP datasets for the node classification task.

Metric	Ma-F1			Mi-F1			AUC		
Split	20	40	60	20	40	60	20	40	60
GraphSAGE [5]	71.97±8.4	73.69±8.4	73.86±8.1	71.44±8.7	73.61±8.6	74.05±8.3	90.59±4.3	91.42±4.0	91.73±3.8
VGAE [9]	90.90±0.1	9.60±0.3	90.08±0.2	91.55±0.1	90.00±0.3	90.95±0.2	98.15±0.1	97.85±0.1	98.37±0.1
Mp2vec [2]	88.98±0.2	88.68±0.2	90.25±0.1	89.67±0.1	89.14±0.2	91.17±0.1	97.69±0.0	97.08±0.0	98.00±0.0
HERec [14]	89.57±0.4	89.73±0.4	90.18±0.3	90.24±0.4	90.15±0.4	91.01±0.3	98.21±0.2	97.93±0.1	98.49±0.1
HetGNN [25]	89.51±1.1	88.61±0.8	89.56±0.5	90.11±1.0	89.03±0.7	90.43±0.6	97.96±0.4	97.70±0.3	97.97±0.2
HAN [20]	89.31±0.9	88.87±1.0	89.20±0.8	90.16±0.9	89.47±0.9	90.34±0.8	98.07±0.6	97.48±0.6	97.96±0.5
DGI [19]	87.93±2.4	88.62±0.6	89.19±0.9	88.72±2.6	89.22±0.5	90.35±0.8	96.99±1.4	97.12±0.4	97.76±0.5
DMGI [12]	89.94±0.4	89.25±0.4	89.46±0.6	90.78±0.3	89.92±0.4	90.66±0.5	97.75±0.3	97.23±0.2	97.72±0.4
HeCo [21]	91.04±0.3	90.11±0.4	90.61±0.3	91.76±0.3	90.54±0.4	91.59±0.2	98.28±0.2	98.04±0.1	98.65±0.2
CLHG	**92.58±0.3**	**91.08±0.3**	**91.39±0.3**	**93.10±0.3**	**91.42±0.3**	**92.12±0.3**	**98.47±0.1**	**98.16±0.2**	**98.69±0.2**

- **HAN** [20] is a semi-supervised heterogeneous method, which uses hierarchical attention to automatically distinguish the importance of meta-paths.
- **HeCo** [21] is a unsupervised heterogeneous contrastive learning method, which employs cross-view contrast theory to learn node embeddings.

Parameter Settings. For baselines, we adhere to the settings described in their original papers and follow the setup in HeCo [21]. For the proposed CLHG, the optimizer is the Adam algorithm [25] with a learning rate 0.001. We conducted a series of tests to determine the optimal values for p_τ, τ, and λ. For p_τ, we tested values ranging from 0.0 to 1.0 in increments of 0.1. Similarly, for τ, we tested values ranging from 0.1 to 0.9 in increments of 0.1. For λ in the overall loss function, we tested values ranging from 0.0 to 1.0 in increments of 0.1. Based on our tests, we determined that the optimal values were p_τ at 0.7, τ at 0.5 and λ at 0.5. In the splits of the dataset, the test and validation sets of all splits contain 1000 nodes each. And we choose 20, 40 and 60 labeled nodes per class as the training set for different splits.

5.2   Overall Performance (RQ1)

We train a linear classifier for downstream tasks using the node representations obtained above. The performance comparison results are presented in Table 2, Table 3, and Table 4.

Compared to state-of-the-art baseline methods, CLHG achieves the best performance on both of DBLP and ACM datasets. Across all splits of the DBLP dataset, CLHG averages 91.68%, 92.21%, and 98.44% in Ma-F1, Mi-F1, and AUC, which outperform HaCo by 1.21%, 1.00%, and 0.12%, respectively. At the same time, across all splits of the ACM dataset, CLHG averages 89.39%, 89.17%, and 97.63% in Ma-F1, Mi-F1, and AUC, which outperform HaCo by 1.38%, 1.47%, and 1.41%, respectively. On DBLP and ACM datasets, CLHG achieves the best performance among all methods. The above results strongly demonstrate the promise of intra-scheme contrastive learning on heterogeneous graphs.

Table 3. Performance comparisons of all methods in ACM datasets for the node classification task.

Metric	Ma-F1			Mi-F1			AUC		
Split	20	40	60	20	40	60	20	40	60
GraphSAGE [5]	47.13±4.7	55.96±6.8	56.59±5.7	49.72±5.5	60.98±3.5	60.72±4.3	65.88±3.7	71.06±5.2	70.45±6.2
VGAE [9]	62.72±3.1	61.61±3.2	61.67±2.9	68.02±1.9	66.38±1.9	65.71±2.2	79.50±2.4	79.14±2.5	77.90±2.8
Mp2vec [2]	51.91±0.9	62.41±0.6	61.13±0.4	53.13±0.9	64.43±0.6	62.72±0.3	71.66±0.7	80.48±0.4	79.33±0.4
HERec [14]	55.13±1.5	61.21±0.8	64.35±0.8	57.47±1.5	62.62±0.9	65.15±0.9	75.44±1.3	79.84±0.5	81.64±0.7
HetGNN [25]	72.11±0.9	72.02±0.4	74.33±0.6	71.89±1.1	74.46±0.8	76.08±0.7	84.36±1.0	85.01±0.6	87.64±0.7
HAN [20]	85.66±2.1	87.47±1.1	88.41±1.1	85.11±2.2	87.21±1.2	88.10±1.2	93.47±1.5	94.84±0.9	94.68±1.4
DGI [19]	79.27±3.8	80.23±3.3	80.03±3.3	76.63±3.5	80.41±3.0	80.15±3.2	91.47±2.3	91.52±2.3	91.41±1.9
DMGI [12]	87.86±0.2	86.23±0.8	87.97±0.4	87.60±0.8	86.02±0.9	87.82±0.5	96.72±0.3	96.35±0.3	96.79±0.2
HeCo [21]	88.05±0.4	87.38±0.6	89.09±0.4	87.74±0.5	87.09±0.7	88.82±0.5	96.29±0.3	96.12±0.7	96.41±0.4
CLHG	**88.90±0.5**	**90.16±0.3**	**89.11±0.3**	**88.64±0.6**	**89.91±0.4**	**88.97±0.3**	**97.51±0.2**	**97.99±0.2**	**97.40±0.1**

Table 4. Performance comparisons of all methods in Freebase datasets for the node classification task.

Metric	Ma-F1			Mi-F1			AUC		
Split	20	40	60	20	40	60	20	40	60
GraphSAGE [5]	45.14±4.5	44.88±4.1	45.16±3.1	54.83±3.0	57.08±3.2	55.92±3.2	67.63±5.0	66.42±4.7	66.78±3.5
VGAE [9]	53.81±0.6	52.44±2.3	50.65±0.4	55.20±0.7	56.05±2.0	53.85±0.4	73.03±0.7	74.05±0.9	71.75±0.4
Mp2vec [2]	53.96±0.7	57.80±1.1	55.94±0.7	56.23±0.8	61.01±1.3	58.74±0.8	71.78±0.7	75.51±0.8	74.78±0.4
HERec [14]	55.78±0.5	59.28±0.6	56.50±0.4	57.92±0.5	62.71±0.7	58.57±0.5	73.89±0.4	76.08±0.4	74.89±0.4
HetGNN [25]	52.72±1.0	48.57±0.5	52.37±0.8	56.85±0.9	53.96±1.1	56.84±0.7	70.84±0.7	69.48±0.2	71.01±0.5
HAN [20]	53.16±2.8	59.63±2.3	56.77±1.7	57.24±3.2	63.74±2.7	61.06±2.0	73.26±2.1	77.74±1.2	75.69±1.5
DGI [19]	54.90±0.7	53.40±1.4	53.81±1.1	58.16±0.9	57.82±0.8	57.96±0.7	72.80±0.6	72.97±1.1	73.32±0.9
DMGI [12]	55.79±0.9	49.88±1.9	52.10±0.7	58.26±0.9	54.28±1.6	56.69±1.2	73.19±1.2	70.77±1.6	73.17±1.4
HeCo [21]	58.38±0.9	**61.20±0.5**	60.48±0.2	60.91±1.3	**64.11±0.7**	**64.11±0.5**	75.36±1.1	**78.71±0.8**	78.10±1.2
CLHG	**61.91±1.1**	60.63±1.6	**60.91±1.0**	**64.12±0.8**	63.14±1.8	63.83±1.2	**78.27±0.3**	78.14±0.9	**78.32±0.7**

Although CLHG is inferior to HeCo on some splits of the Freebase dataset, it outperforms HaCo by 1.88%, 1.04%, and 1.10% on average in Ma-F1, Mi-F1, and AUC, respectively. We believe this is due to the lack of features for target type nodes in the Freebase dataset. This suggests that adaptive augmentation on heterogeneous graphs depends on important information and further reflects the high performance of our adaptive augmentation algorithm in extracting such information.

5.3   Ablation Study (RQ2)

Impact of Module. To demonstrate the effectiveness of different components in CLHG, we conduct several sets of ablation experiments on node classification and further analyze the contribution of each component in CLHG. We consider five model variants as follows:

- **CLHG_sc:** All nodes in the original and augmented graphs are encoded using only the network schema scheme, ignoring the meta-path scheme.
- **CLHG_mp:** All nodes in the original and augmented graphs are encoded using only the meta-path scheme, ignoring the network schema scheme.
- **CLHG_none:** The augmented graph is copied directly from the original graph without using the adaptive augmentation algorithm.

- **CLHG_rand:** The augmented graph is generated from the original graph by a random augmentation algorithm [23].
- **CLHG_cross:** Cross-scheme contrasts are performed on the original and augmented graphs respectively, and then the model is optimized using the same weights.

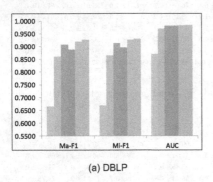

(a) DBLP

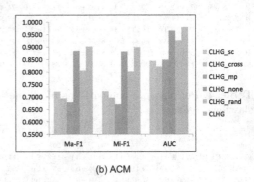

(b) ACM

Fig. 2. Illustration of the Performances of different model variants.

We compare them with CLHG on datasets DBLP and ACM, and report results with 20 and 40 labeled nodes respectively. The results of the ablation study are presented in Fig. 2.

From Fig. 2, we can draw the following conclusions: (1) The dual aggregation scheme is necessary for contrastive learning on heterogeneous graphs. (2) The adaptive augmentation algorithm is effective on heterogeneous graphs, while random augmentation is disadvantageous in some cases. (3) Intra-scheme contrastive learning on heterogeneous graphs holds strong promise.

6   Conclusion and Future Work

In this paper, we propose a concept of intra-scheme contrast on heterogeneous graphs, and further propose a novel heterogeneous contrastive learning model, CLHG. We utilize dual aggregation scheme to aggregate local structure and semantic information of nodes. Then we perform intra-scheme contrast between original views and augmented views to find unique properties of nodes without being disturbed by redundant information from different schemes. Extensive experiments on three real-world datasets demonstrate the superiority of CLHG over other methods. In the future, we plan to utilize more aggregation schemes to capture more perspective information.

Acknowledgements. This work was sponsored the National Natural Science Foundation of China under grants 62076130, 61902186, and 91846104.

References

1. Ding, K., Xu, Z., Tong, H., Liu, H.: Data augmentation for deep graph learning: a survey. ACM SIGKDD Explor. Newsl **24**(2), 61–77 (2022)
2. Dong, Y., Chawla, N.V., Swami, A.: Metapath2vec: scalable representation learning for heterogeneous networks. In: Proceedings of the 23rd ACM SIGKDD International Conference on Knowledge Discovery and Data Mining, pp. 135–144. ACM (2017)
3. Fang, Y., et al.: Metagraph-based learning on heterogeneous graphs. IEEE Trans. Knowl. Data Eng. **33**(1), 154–168 (2021)
4. Fu, X., Zhang, J., Meng, Z., King, I.: MAGNN: metapath aggregated graph neural network for heterogeneous graph embedding. In: Proceedings of The Web Conference 2020, pp. 2331–2341 (2020)
5. Hamilton, W., Ying, Z., Leskovec, J.: Inductive representation learning on large graphs. In: Advances in Neural Information Processing Systems, vol. 30 (2017)
6. Hu, Z., Dong, Y., Wang, K., Sun, Y.: Heterogeneous graph transformer. In: Proceedings of The Web Conference 2020, pp. 2704–2710. ACM (2020)
7. Huang, Z., Zheng, Y., Cheng, R., Sun, Y., Mamoulis, N., Li, X.: Meta structure: computing relevance in large heterogeneous information networks. In: Proceedings of the 22nd ACM SIGKDD International Conference on Knowledge Discovery and Data Mining, pp. 1595–1604. ACM (2016)
8. Kipf, T.N., Welling, M.: Semi-supervised classification with graph convolutional networks (2016). arXiv preprint arXiv:1609.02907
9. Kipf, T.N., Welling, M.: Variational graph auto-encoders (2016). arXiv preprint arXiv:1611.07308
10. Li, X., Ding, D., Kao, B., Sun, Y., Mamoulis, N.: Leveraging Meta-path Contexts for Classification in Heterogeneous Information Networks (2021)
11. Linsker, R.: Self-organization in a perceptual network. Computer **21**(3), 105–117 (1988)
12. Park, C., Kim, D., Han, J., Yu, H.: Unsupervised attributed multiplex network embedding. AAAI **34**(04), 5371–5378 (2020)
13. Qiu, J., et al.: GCC: graph contrastive coding for graph neural network pre-training. In: Proceedings of the 26th ACM SIGKDD International Conference on Knowledge Discovery & Data Mining, pp. 1150–1160 (2020)
14. Shi, C., Hu, B., Zhao, W.X., Yu, P.S.: Heterogeneous information network embedding for recommendation. IEEE Trans. Knowl. Data Eng. **31**(2), 357–370 (2019)
15. Sun, Y., Han, J., Yan, X., Yu, P.S., Wu, T.: PathSim: meta path-based top-k similarity search in heterogeneous information networks. Proc. VLDB Endow. **4**(11), 992–1003 (2011)
16. Sun, Y., Yu, Y., Han, J.: Ranking-based clustering of heterogeneous information networks with star network schema. In: Proceedings of the 15th ACM SIGKDD international conference on Knowledge Discovery and Data Mining, pp. 797–806 (2009)
17. Tong, Z., Chen, X., He, Z., Tong, K., Fang, Z., Wang, X.: Emotion recognition based on photoplethysmogram and electroencephalogram. In: 2018 IEEE 42nd Annual Computer Software and Applications Conference (COMPSAC), vol. 2, pp. 402–407. IEEE (2018)
18. Veličković, P., Cucurull, G., Casanova, A., Romero, A., Lio, P., Bengio, Y.: Graph attention networks (2017). arXiv preprint arXiv:1710.10903

19. Velickovic, P., Fedus, W., Hamilton, W.L., Liò, P., Bengio, Y., Hjelm, R.D.: Deep Graph Infomax p. 46
20. Wang, X., Ji, H., Shi, C., Wang, B., Ye, Y., Cui, P., Yu, P.S.: Heterogeneous graph attention network. In: The World Wide Web Conference, pp. 2022–2032 (2019)
21. Wang, X., Liu, N., Han, H., Shi, C.: Self-supervised Heterogeneous Graph Neural Network with Co-contrastive Learning (2021)
22. Yang, C., Zhang, J., Han, J.: Neural embedding propagation on heterogeneous networks. In: 2019 IEEE International Conference on Data Mining (ICDM), pp. 698–707. IEEE (2019)
23. You, Y., Chen, T., Sui, Y., Chen, T., Wang, Z., Shen, Y.: Graph contrastive learning with augmentations. Adv. Neural. Inf. Process. Syst. **33**, 5812–5823 (2020)
24. Yun, S., Jeong, M., Kim, R., Kang, J., Kim, H.J.: Graph transformer networks. In: Advances in Neural Information Processing Systems, vol. 32 (2019)
25. Zhang, C., Song, D., Huang, C., Swami, A., Chawla, N.V.: Heterogeneous graph neural network. In: Proceedings of the 25th ACM SIGKDD International Conference on Knowledge Discovery & Data Mining, pp. 793–803. ACM (2019)
26. Zhang, M., Hu, H., He, Z., Wang, W.: Top-k similarity search in heterogeneous information networks with x-star network schema. Expert Syst. Appl. **42**(2), 699–712 (2015)
27. Zhang, W., Fang, Y., Liu, Z., Wu, M., Zhang, X.: Mg2vec: learning relationship-preserving heterogeneous graph representations via Metagraph embedding. IEEE Trans. Knowl. Data Eng. **34**(3), 1317–1329 (2022)
28. Zhang, Y., Meng, Y., Huang, J., Xu, F.F., Wang, X., Han, J.: Minimally supervised categorization of text with metadata. In: Proceedings of the 43rd International ACM SIGIR Conference on Research and Development in Information Retrieval, pp. 1231–1240 (2020)
29. Zhao, H., Yao, Q., Li, J., Song, Y., Lee, D.L.: Meta-graph based recommendation fusion over heterogeneous information networks. In: Proceedings of the 23rd ACM SIGKDD International Conference on Knowledge Discovery and Data Mining, pp. 635–644 (2017)
30. Zhao, J., Wang, X., Shi, C., Liu, Z., Ye, Y.: Network schema preserving heterogeneous information network embedding. In: Proceedings of the Twenty-Ninth International Joint Conference on Artificial Intelligence, pp. 1366–1372. International Joint Conferences on Artificial Intelligence Organization (2020)
31. Zhao, T., Liu, Y., Neves, L., Woodford, O., Jiang, M., Shah, N.: Data augmentation for graph neural networks. In: Proceedings of the AAAI Conference on Artificial Intelligence, vol. 35, pp. 11015–11023 (2021)
32. Zhu, Y., Xu, Y., Cui, H., Yang, C., Liu, Q., Wu, S.: Structure-enhanced heterogeneous graph contrastive learning. In: Proceedings of the 2022 SIAM International Conference on Data Mining (SDM), pp. 82–90. SIAM (2022)
33. Zhu, Y., Xu, Y., Yu, F., Liu, Q., Wu, S., Wang, L.: Graph contrastive learning with adaptive augmentation. In: Proceedings of the Web Conference 2021, pp. 2069–2080 (2021)

Ultra-DPC: Ultra-scalable and Index-Free Density Peak Clustering

Luyao Ma[1,2] , Geping Yang[1,2] , Xiang Chen[2] , Yiyang Yang[1(✉)] ,
Zhiguo Gong[3(✉)] , and Zhifeng Hao[4]

[1] Guangdong University of Technology, Guangzhou, Guangdong, China
yyygou@gmail.com
[2] Sun Yat-Sen University, Guangzhou, Guangdong, China
[3] University of Macau, Macau SRA, China
fstzgg@um.edu.mo
[4] Shantou University, Shantou, Guangdong, China

Abstract. Density-based clustering is a fundamental and effective tool for recognizing connectivity structure. The density peak, the data object with the maximum density within a predefined sphere, plays a critical role. However, Density Peak Estimation (DPE), the process of identifying the nearest denser relation for each data object, is extremely expensive. The state-of-the-art accelerating solutions that utilize the index are still resource-consuming for large-scale data. In this work, we propose Ultra-DPC, an ultra-scalable and index-free Density Peak Clustering for Euclidean space, to address the challenges above.

We theoretically study the correlation between two seemly different clustering algorithms: p-means and density-based clustering, and provide a novel p-means density estimator. Based on this, first, p-means is used on a set of samples S to find a set of p Local Density Peaks (LDP), where $p \ll N$, and N is the number of data objects. Second, so as an informative LDP-wise affinity graph is conducted, and then it is enriched by a Random Walk process to incorporate the clues from the non-LDP objects. Third, the importance of LDP is estimated and the most important ones are chosen as the seeds. Finally, the class memberships of the remaining objects are determined according to their relations to the LDP. Ultra-DPC is the fastest DPE method but without reducing the quality of clustering. The evaluation of different medium- and large-scale datasets demonstrates both the efficiency and effectiveness of Ultra-DPC over the state-of-the-art density-based methods.

Keywords: Clustering · Density Peak Estimation · Scalability · Random-Walk

1 Introduction

Clustering, one of the most common unsupervised learning methods, is helpful in automatically assigning the label to data [3, 19, 25] and has been widely used in various downstream tasks. Among them, density-based clustering is popular for its capability of grouping data objects as arbitrary shapes. In the density-based methods, density-connectivity is used to describe the correlation between data objects. According to this, the clustering assignment is determined.

L. Ma and G. Yang—Equal Contribution.

© The Author(s), under exclusive license to Springer Nature Singapore Pte Ltd. 2024
X. Song et al. (Eds.): APWeb-WAIM 2023, LNCS 14334, pp. 139–154, 2024.
https://doi.org/10.1007/978-981-97-2421-5_10

Meanshift [5], a density hill-climbing method, iteratively shifts each data object to the mean vector of data objects within a given radius until converging. Such a policy leads to heavy overhead and is difficult to be applied to large-scale scenarios. To reduce the shifting cost, Quickshift [32] directly shifts each object to its Nearest Denser (ND) object, which dramatically alleviates the search cost. In addition, DBSCAN [8] proposes to continuously and recursively connect the density-reachable objects as clusters. DBSCAN ignores "denser" restrictions in ND, focuses on the density connections between data objects only, and provides an efficient design. However, the methods above need to set a radius parameter τ for density-connectivity computation, which is sensitive and hard to choose, especially for high-dimensional datasets.

To enhance the algorithm robustness, Quickshift++ [16] defines the clustering cores as the groups of densely connected objects, where the mutual k-Nearest Neighbors (mutual k-NN) edge serves as the connection. However, an additional parameter β is used to cut off the graph into clustering cores, which needs extra effort for parameter tuning and diminishes the algorithm's robustness. Further, FastDEC [36] is proposed to partition a given k-NN graph into the dominance components, where k is the only parameter.

On the other hand, splitting the given dataset into exactly K clusters is also a common requirement. DPC [30] supports it by adopting the **Density Peak Estimation (DPE)**, which evaluates the importance of data objects and selects the most K important ones as the density peaks. However, the data object-wise DPE is extremely expensive, and brings in an unacceptable burden for applying DPC to large-scale clustering tasks. To overcome this issue, several works have been proposed. QuickDSC [38] partitions the given k-NN graph into the clustering cores and only executes the clustering-core-wise DPE. Index-DPC [28] builds a query efficient index for accelerating DPE. The methods above require build-in index which dramatically increases the overhead of clustering and becomes the algorithm bottleneck, especially for large-scale datasets.

To face the challenges above, we propose Ultra-DPC, an ultra-scalable and index-free Density Peak Clustering framework. The framework first employs p-means to identify p cluster centers as Local Density Peaks (LDP) and links them as an LDP graph. Random Walk is then used to enhance the graph, followed by Density Peak Estimation (DPE) on the LDP graph. Clustering is determined based on the results of DPE. The contributions of Ultra-DPC include:

- p-**means as Density-based Clustering**. To the best of our knowledge, this is the first work that explores the theoretical correlation between two seemly different clustering methods: p-means and Density-based clustering. Our findings show that p-means is a special density-based clustering method that adopts p-means density.
- **Index-free DPE and Random Walk Enhancement**. Ultra-DPC is index free due to the low resource consumption of p-means and DPE is conducted only on LDP graph, significantly reducing overall workload. Random Walk enhanced graphs are excellent and competitive.
- **Effective and efficient**. Comprehensive experiments on the datasets with various volumes demonstrate the performance of Ultra-DPC is compromised. It outperforms

state-of-the-art density-based methods while being much more efficient. The implementation of Ultra-DPC in Python is publicly available[1].

2 Related Work

To enable the K-way partition, DPC [30] applies two factors to estimate the data objects' importance. Higher importance indicates more possible an object being a seed (e.g., the center of a cluster). A brute force search policy is used to find all possible pairwise Nearest Denser (ND) relations, and results in a computational cost of $O(N^2)$, where N denotes the number of objects. Scalability becomes the primary issue of the existing density-based clustering, it is also the target that this work aims to resolve.

Based on the space index, SNN-DPC [20] retrieves a k-NN graph, and condenses it into a more compact Shared Nearest Neighbors (SNN) graph. The graph edge is reweighed as the count number of the shared neighbors between two objects. A typical DPC is executed on the obtained SNN graph. In [7], Principle Component Analysis (PCA) is applied first to convert the original dataset to low dimensional embedding, and then DPC is applied to the k-NN graph of the resulting embedding. Following a similar idea, QuickDSC [38] first splits the dataset into subgraphs by punning the mutual k-NN graph [16], then DPC is applied to those subgraphs only. On the other hand, Index-DPC [28] constructs several lists to buffer object-wise importance information, and designs an efficient query algorithm to avoid redundant comparisons. Further, FastDEC [36] splits the dataset into several Dominance Components (DCs) according to Nearest Denser (ND) relation for a given k-NN graph. Again, DPC is executed on the resulting DCs only to save computational costs.

To sum up, all existing methods either use brute force computation, or make use of index technology, which cannot fulfill the requirements of applying density peak estimation (DPE) on a large-scale dataset.

3 The Proposed Framework

Preliminaries. Given a set of N data objects $O = \{O_1, O_2, ..., O_N\}$, the dissimilarity between two objects O_i and O_j is measured by the Euclidean distance, and it is denoted as d_{ij}. The density of O_i, indicated by ρ_i, is defined as follows:

$$\rho_i = \sum_{O_j \in O \land \|O_i - O_j\| < d_c} \frac{1}{d_c \sqrt{2\pi}} e^{-\frac{1}{2}(\frac{\|O_i - O_j\|}{d_c})^2} \tag{1}$$

where d_c serves as the pre-defined cut-off distance to eliminate the effects of objects that are located far away. The higher value of ρ_i, the more important O_i.

Besides ρ_i, DPC [30] propose to estimate the importance of an object O_i by an additional factor jointly: δ_i, the **Nearest Denser (ND)** distance, that is the distance from O_i to its Nearest Denser object O_j. For an object O_i, if its ND object O_j exists, δ_i is set as the corresponding pairwise distance d_{ij}. Otherwise, O_i is the global density

---

[1] https://github.com/maluyao17/Ultra-DPC.

Algorithm 1. Ultra-DPC

Require: dataset O; the number of clusters K; the number of samples s; the number of LDP p;
   the number of nearest LDP r; the step size of the Random Walk l
Ensure: cluster labels Y
   ▷ Step 1: Fast LDP generation
1: $S \leftarrow$ Sampling(O, s);   ▷ uniformly sampling without replacement
2: $P \leftarrow p$-means(S, p);   ▷ the obtained p cluster centers are regarded as the LDP
   ▷ Step 2: LDP graph construction
3: $Z \leftarrow$ Construct Samples to LDP similarity graph(P, S, r);   ▷ construct sample to LDP graph
4: $W \leftarrow Z \cdot Z^T$;   ▷ construct LDP-wise similarity graph
   ▷ Step 3: Random Walk on LDP graph
5: $A^l \leftarrow$ Random Walk(W, l);
   ▷ Step 4: DPC on LDP graph
6: $Y_p \leftarrow$ DPE on (A^l);   ▷ DPE on the LDP, and obtain the cluster labels of LDP
7: **for** $O_E \in O$ **do**   ▷ process objects by epoch
8:     $Y_E \leftarrow$ ClusterAssign(O_E, Y_p);   ▷ assign the epoch to LDP
9:     $Y \leftarrow Y \cup Y_E$;
10: **end for**;
11: **return** Y ▷ return the final cluster labels

---

peak, then δ_i is set as the maximum of all possible ND distances. The importance of an object O_i is given by $\gamma_i = \rho_i \cdot \delta_i$. The final step of DPC is to find the most K important objects in terms of γ_i, and according to which, the cluster memberships of remaining objects are determined.

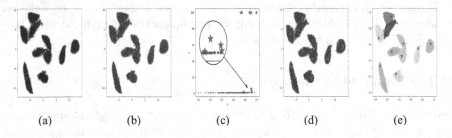

(a)   (b)   (c)   (d)   (e)

Fig. 1. Illustration of a concrete example that is generated by performing Ultra-DPC on the MNIST2D ($N = 70k$, $M = 2$, $K = 10$) with parameter settings $s = 8k$, $p = 200$, $r = 150$, and $l = 50$. Obtained NMI: 0.9173.

3.1   Framework Overview

To solve the computational burden in density-based clustering [5, 8, 16, 32], especially in Density Peak Estimation (DPE) relevant methods [7, 20, 28, 30, 36, 38], Ultra-DPC is proposed in this work.

The primary procedure of Ultra-DPC is briefed as Algorithm 1. (1) a part of the dataset is randomly selected as the sample set S (line 1). Based on S, p-means is used to find Local Density Peaks (LDP) by setting $K = p$, where p is the number of LDP (lines 1–2). (2) a peak-wise similarity graph is built which involves the peak-to-peak affinity only (lines 3–4). (3) a Random Walk with step size l is applied to the obtained graph to iteratively enhance it until converged (line 5). (4) a customized DPC is executed on the converged graph and results in the final clusters (lines 6–10).

A concrete example of Ultra-DPC is provided in Fig. 1. For better visualization, UMAP [23] is used to reduce the feature dimensions of the MNIST dataset from 784 to 2, and export dimension-reduced data objects as a new dataset MNIST2D, and Ultra-DPC is executed on the MNIST2D. In Fig. 1(a), the samples and non-sample objects are respectively plotted as black dots and gray dots, where $s = 8k$ samples are randomly selected from $N = 70k$ objects without replacement. In Fig. 1(b), $p = 200$ Local Density Peaks (LDP), generated by running p-means on the samples, are plotted as blue dots. The LDP graph is created and then enhanced by a Random Walk with step size $l = 50$. Density Peaks Estimation (DPE) is performed on the LDP to select the most important ones, the corresponding Decision Map [30] on LDP is shown in Fig. 1(c). In the Decision Map, two axes correspond to the ρ and δ values of the objects, respectively. The closer to the right upper corner, the more important an object. Based on this, the most 10 important LDP, plotted as red stars, are chosen as the cluster centers. Finally, the less important LDP and objects are assigned to the nearest denser cluster center, the clustering results are shown in Fig. 1(e). The clustering result leads to an NMI value of 0.9173.

Fast Local Density Peak Generation. In large-scale clustering tasks, it is computationally infeasible to execute the Density Peaks Estimation (DPE) for the entire dataset directly. Therefore, the prior step of Ultra-DPC is reducing the scale of the dataset. As described in [30], a density peak denotes the object with the highest density in a certain area, in other words, the density peak is locally "denser" than all objects.

On the other hand, p-means [10,22], a commonly used fast clustering method, is used to pre-process data frequently. Given a set of data objects $O = \{O_1, ..., O_N\}$, the objective of p-means is to partition O into p clusters $C = \{C_1, C_2, ..., C_p\}$ with p cluster centers $c = \{c_1, c_2, ..., c_p\}$, where $c_i = \frac{\sum_{O_j \in C_i} O_j}{|C_i|}$. In this work, rather than the general form of the p-means objective [10], we are more interested in its alternative, which seeks an optimization of the Cluster Internal Variance (CIV). To enable p-means serves as the density-based clustering, a p-means density estimator, which serves as the bridge, is defined. For a given object O_j that is assigned to cluster C_i, its **p-means density** is defined as:

$$\rho_j^{PM} = \frac{|C_i|}{\sum_{O_j,O_k \in C_i \wedge j \neq k} ||O_j - O_k||^2} \approx \frac{1}{\sum_{O_j,O_k \in C_i \wedge j \neq k} ||O_j - O_k||^2} \tag{2}$$

The proposed density estimator is conducted on the p-means, more specifically, the CIV (e.g., the first formula of Eq. 2). For O_j, its p-means density is calculated as the sum of the pairwise distances from O_j to all cluster members e.g., $O_k \in C_i \wedge k \neq j$. Regarding the second form of p-means density, the number of cluster members $|C_i|$ is removed since it is identical for all cluster members within C_i. **Note:** besides p-means density, which is thoroughly used in this work, we might design other forms of density estimators with similar effects, but it is out of the scope of this work.

A theorem describing the correlation between cluster members and the center is defined as follows:

Theorem 1. *Given a cluster assignment of p-means: e.g., object O_j is assigned to the cluster C_i e.g., $O_j \in C_i$. The cluster center c_i is guaranteed as the Nearest Denser (ND) cluster center to O_j if the p-means Density (e.g., ρ_j^{PM}) is used.*

Proof. The theorem consists of two parts:

(1) for O_j, c_i is the nearest cluster center among all cluster centers;
(2) c_i is denser than O_j. s.t. $\rho_{c_i}^{PM} > \rho_j^{PM}$.

The part (1) holds because in each iteration of p-means [10], each object (e.g., O_j) is assigned to its nearest center (e.g., c_i). Regarding part (2), the proof is given as follows. According to the definition of variance in statistics [17,29], the cluster center c_i, known as the expectation of C_i in statistics, is guaranteed as the minimum of the Cluster Internal Variance. It is straightforward to infer that c_i is guaranteed as the maximum in C_i in terms of the p-means density (e.g., Eq. 2). Part (2) holds as well.   □

Based on Theorem 1, we can conclude that $\forall O_j \in C_i, \rho_{c_i}^{PM} > \rho_j^{PM}$, and c_i is the nearest center to O_j over the others. More significantly, **it is unnecessary to perform DPE on all data objects explicitly**: just execute p-means on the input data objects, and the obtained centers serve as the **Local Density Peaks (LDP)** inherently.

LDP differ from the traditional density peaks in several aspects: (1) LDP are the mean vectors of the clusters that are excluded by the dataset, in other words, they are new objects. (2) based on the p-means Density (as Eq. 2), it is guaranteed that the cluster center is denser than all cluster members, therefore once the center is determined, the cluster members are excluded by the DPE explicitly. (3) p-means can be viewed as a good approximation to the object-wise DPE by eliminating the less important objects. If $p \rightarrow N$, p-means tends to compute the p-means density of all objects.

Sample-Based p-Means. p-means might still be too costly to be applied to the large-scale dataset. To make LDP generation even more efficient, a sampling layer is inserted in between the object set O and the p-means. More specially, a set of samples S are randomly selected from O without replacement (line 1 of Algorithm 1, s.t. $|S| = s$ and $s \ll N$. p-means is applied on S rather than the entire O. According to the analysis of p-means [34] and Weighted Kernel p-means [24], such a simple yet effective strategy can abundantly reduce the computational requirements without abating the quality of cluster centers, which serve as the LDP in Ultra-DPC. The visual impacts of Fast LDP generation are demonstrated in Fig. 1(a) and Fig. 1(b).

Sample-based p-means can be viewed as an α-approximation algorithm ($\alpha \geq 1$) to the ideal p-means [35].

Definition 1. *Let $\alpha \geq 1$. $\mathcal{P}$ is an α-approximation algorithm for the p-means problem if the resulting centers, c, return by $\mathcal{P}$ satisfies,*

$$\mathbb{W}(O, c) \leq \alpha \mathbb{W}(O, c_{opt})$$

where $\mathbb{W}(O, c)$ denotes the sum of squared error of every data point $\{O_1, O_2, ..., O_N\}$ to its nearest center in c, and c_{opt} is the ideal optimal centers. The corresponding error bound is given as follow,

Theorem 2. *Let $\alpha \geq 1$, $0 < \beta < 1$, $0 < \gamma < 1$ and $\epsilon > 0$ be approximation parameters. Let $\mathcal{P}$ be an α-approximated algorithm of the p-means problem. Given the data object set $O = \{O_1, O_2, ...O_N\}$, suppose a subset $S \subset O$ of size s is uniformly sampled at random without replacement. s.t.*

$$s \geq \ln(\frac{1}{\gamma})(1 + \frac{1}{n})/(\frac{2\beta^2\epsilon^2}{\Delta^2\alpha^2} + \frac{1}{n}\ln(\frac{1}{\gamma}))$$

where $\Delta = \max_{ij} \|O_i - O_j\|^2$. If p-means is run on S and obtain the LDP P^S, with probability at least $1 - \gamma$,

$$\mathbb{W}(S, P^S) \leq 4(\alpha + \beta)\mathbb{W}(O, P_{opt}) + \epsilon \qquad (3)$$

where P_{opt} is the set of optimal LDP.

According to Theorem 2, if the number of samples s is sufficient, the obtained LDP P are as excellent as the ones that obtained by applying p-means on overall dataset O. In other words, s is a significant parameter that can be used to balance the trade-off between LDP quality and demanded resources. Both the theoretical analysis and experiment demonstrate that a small value of s can lead to an excellent cluster result.

LDP Graph Construction. After obtaining p local density peaks $P = \{P_1, P_2, ..., P_p\}$, it is possible to perform the DPE on P. However, the conducted graph is rather limited if only the information from LDP is involved, thus more clues are essential. Inspired by spare representation methods [4, 13, 33, 34, 37], the clusters are involved in the LDP graph construction.

First, a graph $Z \in \mathbb{R}^{p \times s}$ that describes the similarities between the LDP P and the samples S is defined as:

$$Z_{ij} = e^{-\frac{\|P_i - S_j\|^2}{2\sigma^2}} \qquad (4)$$

where P_i denotes the ith LDP, S_j denotes the jth sample, and the bandwidth parameter σ is defined as the distance average between LDP and samples. We only retain the largest r element of each row to make sure the sparsity of the graph. In other words, for each sample, its r nearest LDP is retained. Based on Z, the LDP graph is conducted:

$$W = ZZ^T \qquad (5)$$

Through the matrix multiplication, $W \in \mathbb{R}^{p \times p}$ contains the information that is propagated from S.

Compared with index-based DPE methods, which require an established data structure on all objects, Ultra-DPC only needs to invoke p-means on a small number of samples, and results in p LDP. The resource consumption of Ultra-DPC is greatly reduced in both the time and space aspects.

Random Walk on LDP Graph. The LDP graph conducted on the samples and LDP might still be incomplete and biased, inspired by the works [1, 2, 12, 27], a Random Walk is used to comprehend and strengthen it. In the Random Walk, it is necessary to define the transitional probabilities between LDP. In Ultra-DPC, the normalized LDP graph A is used as the transition matrix and is defined as follows:

$$
A_{ij} = \begin{cases} \dfrac{W_{ij}}{\sum_{P_k \neq P_i} W_{ik}}, & \text{if } i \neq j \\ 0, & \text{otherwise} \end{cases} \tag{6}
$$

where A_{ij} indicates the probability from the ith LDP to the jth LDP. The normalization step ensures the sum of the probabilities from an LDP to others is 1.

After constructing the transition matrix, based on the Random Walk of Markov process, we can define the multi-step transition probability matrix $A^l = A^{l-1} \cdot A$, where $A^1 = A$. where l denotes the number of steps in the Random Walk.

DPC on LDP Graph. A customized DPC is proposed in this subsection to process the "Random Walked" LDP graph A^l. First, a spectrum-like [31] LDP density is defined:

$$
\rho_i^{LDP} = \sum_{j=1 \wedge i \neq j}^{p} A_{ij}^l \tag{7}
$$

where A_{ij}^l denotes the similarity between the ith and the jth LDP.

To obtain the "distance" from an P_i to its nearest denser LDP, we just find the most similar and meanwhile denser LDP (e.g., P_j) accordingly, and return the corresponding peak-wise similarity. The LDP distance from an LDP P_i to its "nearest" and denser LDP is defined as follows:

$$
\delta_i^{LDP} = \begin{cases} \min_{j: \rho_j^{LDP} > \rho_i^{LDP}} (\dfrac{1}{A_{ij}^l}), & \text{if } P_j \text{ is the most similar \& denser LDP} \\ \max_j (\delta_j^{LDP}), & \text{otherwise} \end{cases} \tag{8}
$$

where only the non-zero entries in A^l are estimated. Therefore, the importance of an LDP is estimated as $\gamma_i^{LDP} = \rho_i^{LDP} \cdot \delta_i^{LDP}$.

Then it follows a typical procedure of DPC: (1) select the most K important LDP as the cluster centers; (2) assign the less important LDP to the nearest center; (3) assign the object to the nearest LDP. To process the extremely large dataset that can not be stored by the main memory, we might split the data into epochs and determine cluster labels epoch by epoch (as the lines 7–10 of Algorithm 1).

Implementation and Complexity Analysis. Ultra-DPC is implemented in Python, and the p-means is conducted on Numpy. Currently, only the CPU-based Ultra-DPC is developed. The time complexity and space complexity analysis of Ultra-DPC is provided as follows. Just recall that, in general $t, l, r < p < s \ll N$, where N denotes the number of objects, s denotes the number of samples, r denotes the number of nearest LDP in constructing sparse representation Z, l denotes the number of steps in a random walk, t is the number of iterations in p-means, and M denotes the number of features.

Suppose the input dataset is given as N vectors with M-dimensional features. In the Fast LDP generation step, the time complexity of performing p-means on s samples is $O(spMt)$, while the space complexity is $O(sp)$. Then, constructing an LDP graph requires $O(sp + p^2)$ space and $O(spM + s^2p)$ time. In the Random Walk step, matrix multiplication is used, which costs $O(lp^3)$ in time and $O(p^2)$ in space. Regarding the third step, it requires $O(p^2)$ in computing the nearest denser LDP, and $O(pM)$ in space. To determine the cluster membership, it requires $O(Np)$ computational time, and the space requirement is $O(|O_E|)$, where $|O_E|$ is the size of the epoch.

According to the analysis above, the time complexity of Ultra-DPC is dominated by $O(spMt + s^2p + lp^3 + Np)$, and the space complexity is dominated by $O(sp + pM + |O_E|)$, such that $p < s \ll N$. On the other hand, the brute-force DPC requires $O(N^2)$ computational time, and $O(N^2)$ space. The Space Index based methods require $O(NlogN)$ computational time, and $O(NM)$ space. Apparently, the efficiency gain and the saved storage of Ultra-DPC are outstanding.

4 Evaluation

In this section, multiple experiments are conducted on the datasets with various volumes to evaluate the capability of Ultra-DPC.

Baselines. Ultra-DPC is compared with state-of-the-art clustering methods, especially the density-based methods. The baselines are summarized as follows: **K-means** [22]: the most classical clustering method. **DBSCAN** [8]: a typical, classical, and efficient density-based method. **DPC** [30]: the method that partitions the object according to the density peak estimation (DPE). **SNN-DPC** [20]: Shared-nearest-neighbor (SNDD) based DPC. **MeanShift** [5]: a robust mode-seeking-based method. **QuickShift** [32]: an accelerated version of Meanshift. **QuickShift++** [16]: an enhanced Quickshift, that is conducted by a mutual k-NN based modes-seeking. **QuickDSC** [38]: a density-based method that combines the advantages of DPC and Quickshift++. **FastDEC** [36]: a clustering by fast dominance estimation.

Evaluation Metric and Pre-processing and Experiment Setup. In this work, three popularly used metrics are chosen to evaluate the experimental results from different perspectives: adjusted mutual information (AMI) [26], adjusted rand index (ARI) [14], and normalized mutual information (NMI) [26]. The audience may refer to [14,26] for the detail. For all metrics, a higher value indicates a better result.

The "min-max normalization" is used to pre-process the data objects, each dimension of data is linearly mapped to the range [0,1]. All experiments are performed on a WIN-10 64-bit machine with an Intel (R) Core I5-10300F CPU (2.50 GHz) with 16 GB of main memory. For all methods, we use the implementations on the Python 3.6 platform. In the numerical evaluation, each experiment is executed 10 times and the average performance is reported. To comprehend the evaluation, the datasets with different orders of magnitude are selected. The employed datasets are described in Table 1.

Table 1. Datasets in Evaluation

Datasets	# Object	# Dimension	# Category
phoneme [11]	4,509	258	5
S2 [9]	5,000	2	15
PenDigits	10,992	16	10
USPS [15]	11,000	256	10
banknote	1372	4	2
MNIST [18]	70,000	3472	10
MNIST-2D	70,000	2	10
emnist-letter [6]	145,600	784	26
emnist-digits [6]	280,000	784	10
emnist-byclass [6]	814,255	784	62
yahoo	1,400,000	100	10
MNIST8m [21]	8,000,000	784	10

Table 2. Performance comparison on medium size datasets.

Algorithm	ARI	AMI	NMI	ARI	AMI	NMI
MNIST				PenDigits		
K-means	0.3666	0.4998	0.4999	0.5986	0.6915	0.6920
QuickShift	N/A	N/A	N/A	0.3225	0.5261	0.5264
QuickShift++	0.0231	0.2598	0.5333	0.5750	0.6697	0.7689
DPC	N/A	N/A	N/A	N/A	N/A	N/A
SNN-DPC	N/A	N/A	N/A	N/A	N/A	N/A
DBSCAN	0.1835	0.0973	0.4059	0.4041	0.6434	0.6533
MeanShift	N/A	N/A	N/A	0.6088	0.7117	0.7165
QuickDSC	0.3341	0.5406	0.5408	0.4305	0.6465	0.6469
FastDEC	0.4526	0.6609	0.6610	0.6445	0.7835	0.7838
Ultra-DPC	**0.7186**	**0.7932**	**0.7931**	**0.7056**	**0.7981**	**0.7984**
MNIST-2D				yahoo		
K-means	0.9020	0.8979	0.8477	0.4855	0.5551	0.5551
QuickShift	N/A	N/A	N/A	N/A	N/A	N/A
QuickShift++	0.1142	0.4208	0.6252	N/A	N/A	N/A
DPC	N/A	N/A	N/A	N/A	N/A	N/A
SNN-DPC	N/A	N/A	N/A	N/A	N/A	N/A
DBSCAN	0.7784	0.7784	0.5496	0	0	0.2799
MeanShift	0.8081	0.8595	0.8595	N/A	N/A	N/A
QuickDSC	0.8576	0.8845	0.8845	N/A	N/A	N/A
FastDEC	0.7718	0.8467	0.8468	N/A	N/A	N/A
Ultra-DPC	**0.9298**	**0.9172**	**0.9173**	**0.4995**	**0.5777**	**0.5777**

Table 3. Performance comparison on large size datasets.

Algorithm	ARI	AMI	NMI	ARI	AMI	NMI
emnist-byclass				emnist-digits		
K-means	0.2063	0.4736	0.4739	0.6061	0.4936	0.4936
QuickShift	N/A	N/A	N/A	N/A	N/A	N/A
QuickShift++	N/A	N/A	N/A	N/A	N/A	N/A
DPC	N/A	N/A	N/A	N/A	N/A	N/A
SNN-DPC	N/A	N/A	N/A	N/A	N/A	N/A
DBSCAN	0	0	0.4259	0	0	0.3102
MeanShift	N/A	N/A	N/A	N/A	N/A	N/A
QuickDSC	N/A	N/A	N/A	N/A	N/A	N/A
FastDEC	N/A	N/A	N/A	N/A	N/A	N/A
Ultra-DPC	**0.3349**	**0.5180**	**0.5183**	**0.6625**	**0.6535**	**0.6535**
emnist-letter				MNIST8M		
K-means	0.2227	0.4030	0.4034	N/A	N/A	N/A
QuickShift	N/A	N/A	N/A	N/A	N/A	N/A
QuickShift++	N/A	N/A	N/A	N/A	N/A	N/A
DPC	N/A	N/A	N/A	N/A	N/A	N/A
SNN-DPC	N/A	N/A	N/A	N/A	N/A	N/A
DBSCAN	0	0	0.4302	N/A	N/A	N/A
MeanShift	N/A	N/A	N/A	N/A	N/A	N/A
QuickDSC	N/A	N/A	N/A	N/A	N/A	N/A
FastDEC	N/A	N/A	N/A	N/A	N/A	N/A
Ultra-DPC	**0.2095**	**0.4132**	**0.4135**	**0.5112**	**0.6471**	**0.6471**

Table 4. Running time comparison.

	KM	QS	QS++	DPC	SDPC	FDEC	QDSC	DBS	MS	UDPC
S2	0.21	2.58	0.64	10.00	215.62	_0.17_	0.95	**0.036**	18.89	6.37
phoneme	_0.35_	2.49	9.11	4.43	171.51	6.95	14.48	**0.33**	22.06	1.42
USPS	_1.79_	12.44	53.10	29.73	1142	52.53	24.93	**1.47**	1550	17.42
banknote	_0.06_	0.22	0.09	0.44	23.79	0.09	_0.06_	**0.01**	3.45	0.43
PenDigits	12.41	**0.41**	2.75	N/A	N/A	4.58	60.44	_1.33_	187.92	15.03
MNIST	**42.78**	N/A	5467	N/A	N/A	3629	32058	_79.13_	971.66	285.93
MNIST-2D	**1.04**	N/A	10.01	N/A	N/A	7.77	83.26	_4.16_	2031.35	5.37
yahoo	_84.68_	N/A	N/A	N/A	N/A	N/A	N/A	27704	N/A	**8.19**
emnist-byclass	_2356_	N/A	N/A	N/A	N/A	N/A	N/A	27090	N/A	**341.03**
emnist-digits	_149.56_	N/A	N/A	N/A	N/A	N/A	N/A	2260	N/A	**48.69**
emnist-letter	_358.24_	N/A	N/A	N/A	N/A	N/A	N/A	649.32	N/A	**35.41**
MNIST8m	N/A	N/A	N/A	N/A	N/A	N/A	N/A	N/A	N/A	**521.86**

Overall Comparison. The performance of distinctive methods is reported. Because of the space limitation, only the results of medium/large datasets (Table 2/Table 3) are reported The bold represents the best result, and the second-best result is highlighted by underline, "N/A" denotes that we cannot obtain the clustering result on the experimental host, e.g. out-of-memory or execution time exceeds 24 h. For each method, we tune the primary parameters, and the best clustering result in terms of the NMI, and the corresponding execution time are reported (Table 4).

Based on the results the Tables above, several findings are drawn as follows. DPC and SNN-DPC perform well on small-scale and medium-scale datasets but fail in large-scale datasets. The reason is that both methods execute the Density Peak Estimation (DPE) in a brute-force manner, that is estimate the importance of every object. This policy is computationally infeasible if the number of objects N is large. QuickShift++, QuickDSC, and FastDEC can handle this case by taking advantage of the space index. However, the space index might still fail, if the dataset is too large to be held by the experimental host. In that case, the index building and the k-NN graph retrieval will be extremely expensive. In contrast, Ultra-DPC can effectively reduce the required resources. In addition, the clustering quality in terms of three metrics is excellent: Ultra-DPC can obtain the best or the second-best result on most datasets discarding their volumes, which demonstrates the effectiveness of Ultra-DPC.

Ultra-DPC is the only method that is able to process the million-scale dataset in a host with 16 GB of main memory. The underlying reasons are two folders: (1) Ultra-DPC does not perform the object-wise DPE or conduct the object-wise k-NN graph; (2) the random-walked LDP graph is excellent and informative enough so that based on it, Ultra-DPC can determine the cluster label of the objects epoch by epoch. Thus the overall execution time and memory consumption are largely restricted. It shows the efficiency of our framework is compromised.

Regarding non-DPE method, e.g., DBSCAN, it is one of the fastest methods, its speed is even competitive with that of the K-means. The former normally is conducted on the space index, and requires $O(NlogN)$ to retrieve neighborhood information. The latter iteratively assigns each object to the nearest cluster center. The complexity of both methods is object-scale: as N increases, the execution time increases sharply.

Among all, Ultra-DPC provides the best result (e.g., NMI = 0.9478). The performance of Ultra-DPC on small to medium-sized datasets is compromised and may not provide the shortest runtime. Its benefits are more apparent on large-scale datasets. For the yahoo dataset, where $N = 1,400k$, only three methods K-means, DBSCAN, and Ultra-DPC can work, and Ultra-DPC performs the best. If N is even larger, e.g. $N = 8,000k$, only the Ultra-DPC can provide output for MNIST8m. Even the index-free method K-means cannot handle such an extremely large dataset, it also demonstrates the necessity of adding a sampling-layer.

Robustness Testing on Parameters. In this subsection, we demonstrate the Robustness of the parameters that are critical to Ultra-DPC. Several experiments are conducted on the proposed framework, and several datasets in different scales are chosen: phoneme ($N = 4,509$), S2 ($N = 5k$), MNIST ($N = 70k$), and yahoo ($N = 1,400k$). Four parameters are involved: the number of samples s, the number of LDP p, the number of nearest

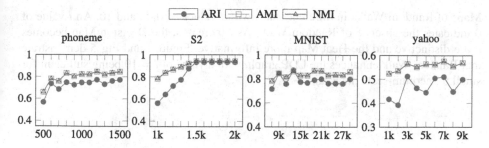

Fig. 2. Robust Testing of parameter s (x-axis) on phoneme, S2, MNIST and yahoo datasets. Three metrics ARI, AMI, and NMI are used.

samples for each LDP r, and the number of steps in Random Walk l. In the individual experiment, only the target parameter varies, while the remaining are fixed as default. The default settings of parameters are: $s = 1k$, $p = 100$, $l = 10$, $r = 100$ for phoneme, $s = 1400$, $p = 300$, $l = 5$, $r = 100$ for S2, $s = 7k$, $p = 300$, $l = 50$, $r = 100$ for MNIST and $s = 7k$, $p = 100$, $l = 50$, $r = 100$ for yahoo.

Effect of the Parameter s. Based on Fig. 2, a clear trend can be found: the NMI, ARI, and AMI of Ultra-DPC increase as the increase of s, and become stable as it reaches a relatively large value. The identified trend is consistent with our aforementioned claim: if s is chosen appropriately, Ultra-DPC is able to find a set of high-quality LDP. Besides, s is an insensitive parameter, and based on this, it is easy to trade off between the algorithm efficiency and clustering accuracy. Further, in a large-scale dataset, such as yahoo, only $3k$ samples can provide results that are both stable and excellent, which shows the excellence of the sampling schema of Ultra-DPC.

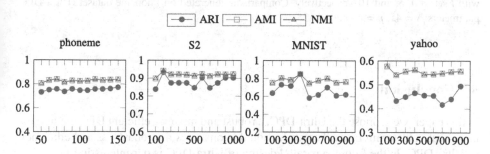

Fig. 3. Robust Testing of parameters p (x-axis) on phoneme, S2, MNIST and yahoo datasets. Three metrics ARI, AMI, and NMI are used.

Effect of the Parameter p and l. As shown in Fig. 3, the NMI, ARI, and AMI of Ultra-DPC keep waving as p varies, but the waving magnitude is constrained in a small range. It demonstrates that p is a robust parameter, and a limited number of LDP can result in good clustering results. Figure 4 and Fig. 5 show the Decision Maps and Heat

Maps of Random Walks in different states with l values of 0, 5, and 10. An l value of 0 indicates the absence of Random Walk. As l increases, the Decision Map becomes more distinctive and the Heat Map more informative. Figure 4 and Fig. 5 demonstrate that Random Walk enhances the LDP graph, with a value of $l = 10$ being sufficient for small to medium-scale datasets.

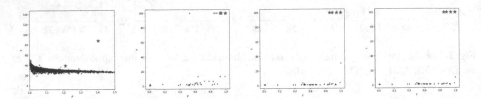

Fig. 4. Decision Map Comparison: (a) Decision Map of DPC. (d_c=0.05); (b)-(d) Decision Map of Ultra-DPC with l set to 0, 5, and 10, respectively. Comparison generated on Phoneme dataset (Ultra-DPC parameters: $s = 4k$, $p = 45$ and $r = 25$).

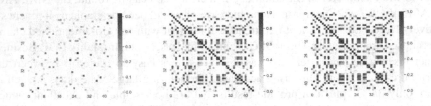

Fig. 5. Heat Map (matrix form of LDP Graph) Comparison. (a)-(c) Heat Map of Ultral-DPC with l set to 0, 5, and 10, respectively. Comparison generated on Phoneme dataset (Ultra-DPC parameters: $s = 4k$, $p = 45$ and $r = 25$).

5   Conclusion

In this paper, we propose the Ultra-DPC, a robust and memory efficient DPC method for large-scale clustering tasks. The evaluation demonstrates the robustness and efficiency of Ultra-DPC. In the future, a parallel design of Ultra-DPC is our interesting.

Acknowledgments. This work was supported by National Key D&R Program of China (2019YFB1600704, 2021ZD0111501), NSFC (61603101, 61876043, 61976052, 71702065), NSF of Guangdong Province (2021A1515011941), State's Key Project of Research and Development Plan (2019YFE0196400), NSF for Excellent Young Scholars (62122022), Guangzhou STIC (EF005/FST-GZG/2019/GSTIC), NSFC & Guangdong Joint Fund (U1501254), the Science and Technology Development Fund, Macau SAR (0068/2020/AGJ, SKL-IOTSC(UM)-2021–2023, GDST (2020B1212030003), MYRG2022-00192-FST, Guangdong Provincial Key Laboratory of Big Data Computing, The Chinese University of Hong Kong, Shenzhen (B10120210117-OF09).

References

1. Barnes, G., Feige, U.: Short random walks on graphs. In: STOC, pp. 728–737. ACM (1993)
2. Brakensiek, J., Guruswami, V.: Bridging between 0/1 and linear programming via random walks. In: STOC, pp. 568–577. ACM (2019)
3. Chan, T.H., Guerquin, A., Hu, S., Sozio, M.: Fully dynamic k-center clustering with improved memory efficiency. IEEE Trans. Knowl. Data Eng. **34**(7), 3255–3266 (2022)
4. Chen, X., Cai, D.: Large scale spectral clustering with landmark-based representation. In: AAAI. AAAI Press (2011)
5. Cheng, Y.: Mean shift, mode seeking, and clustering. IEEE Trans. Pattern Anal. Mach. Intell. **17**(8), 790–799 (1995)
6. Cohen, G., Afshar, S., Tapson, J., van Schaik, A.: EMNIST: extending MNIST to handwritten letters. In: IJCNN, pp. 2921–2926. IEEE (2017)
7. Du, M., Ding, S., Jia, H.: Study on density peaks clustering based on k-nearest neighbors and principal component analysis. Knowl. Based Syst. **99**, 135–145 (2016)
8. Ester, M., Kriegel, H.P., Sander, J., Xu, X.: A density-based algorithm for discovering clusters in large spatial databases with noise. In: KDD, pp. 226–231 (1996)
9. Fränti, P., Virmajoki, O.: Iterative shrinking method for clustering problems. Pattern Recognit. **39**(5), 761–775 (2006)
10. Hartigan, J.A., Wong, M.A.: Algorithm as 136: a k-means clustering algorithm. J. Roy. Stat. Soc. Ser. C (Appl. Stat.) **28**(1), 100–108 (1979)
11. Hastie, T., Friedman, J.H., Tibshirani, R.: The Elements of Statistical Learning: Data Mining, Inference, and Prediction. Springer Series in Statistics. Springer, Cham (2001). https://doi.org/10.1007/978-0-387-84858-7
12. Huang, D., Wang, C., Peng, H., Lai, J., Kwoh, C.: Enhanced ensemble clustering via fast propagation of cluster-wise similarities. IEEE TSMC. Syst. **51**(1), 508–520 (2021)
13. Huang, D., Wang, C., Wu, J., Lai, J., Kwoh, C.: Ultra-scalable spectral clustering and ensemble clustering. TKDE **32**(6), 1212–1226 (2020)
14. Hubert, L., Arabie, P.: Comparing partitions. J. Classif. **2**(1), 193–218 (1985)
15. Hull, J.J.: A database for handwritten text recognition research. IEEE Trans. Pattern Anal. Mach. Intell. **16**(5), 550–554 (1994)
16. Jiang, H., Jang, J., Kpotufe, S.: Quickshift++: Provably good initializations for sample-based mean shift. In: ICML, vol. 80, pp. 2299–2308. PMLR (2018)
17. Kriegel, H., Schubert, E., Zimek, A.: The (black) art of runtime evaluation: Are we comparing algorithms or implementations? Knowl. Inf. Syst. **52**(2), 341–378 (2017)
18. LeCun, Y., Bottou, L., Bengio, Y., Haffner, P.: Gradient-based learning applied to document recognition. Proc. IEEE **86**(11), 2278–2324 (1998)
19. Liu, B., Bai, B., Xie, W., Guo, Y., Chen, H.: Task-optimized user clustering based on mobile app usage for cold-start recommendations. In: KDD, pp. 3347–3356. ACM (2022)
20. Liu, R., Wang, H., Yu, X.: Shared-nearest-neighbor-based clustering by fast search and find of density peaks. Inf. Sci. **450**, 200–226 (2018)
21. Loosli, G., Canu, S., Bottou, L.: Training invariant support vector machines using selective sampling. In: Large Scale Kernel Machines, vol. 2 (2007)
22. MacQueen, J.: Classification and analysis of multivariate observations. In: 5th Berkeley Symposium on Mathematical Statistics and Probability, pp. 281–297 (1967)
23. McInnes, L., Healy, J.: UMAP: uniform manifold approximation and projection for dimension reduction. CoRR abs/1802.03426 (2018)
24. Mohan, M., Monteleoni, C.: Beyond the nyström approximation: speeding up spectral clustering using uniform sampling and weighted kernel k-means. In: IJCAI (2017)

25. Najafi, M., He, L., Yu, P.S.: Outlier-robust multi-view subspace clustering with prior constraints. In: ICDM, pp. 439–448. IEEE (2021)
26. Nguyen, X.V., Epps, J., Bailey, J.: Information theoretic measures for clusterings comparison: Variants, properties, normalization and correction for chance. J. Mach. Learn. Res. **11**, 2837–2854 (2010)
27. Paudel, B., Bernstein, A.: Random walks with erasure: diversifying personalized recommendations on social and information networks. In: WWW, pp. 2046–2057. ACM (2021)
28. Rasool, Z., Zhou, R., Chen, L., Liu, C., Xu, J.: Index-based solutions for efficient density peak clustering. IEEE Trans. Knowl. Data Eng. **34**(5), 2212–2226 (2022)
29. Rice, J.A.: Mathematical statistics and data analysis. Cengage Learning (2006)
30. Rodriguez, A., Laio, A.: Clustering by fast search and find of density peaks. Science **344**(6191), 1492–1496 (2014)
31. Shi, J., Malik, J.: Normalized cuts and image segmentation. IEEE Trans. Pattern Anal. Mach. Intell. **22**(8), 888–905 (2000)
32. Vedaldi, A., Soatto, S.: Quick shift and kernel methods for mode seeking. In: Forsyth, D., Torr, P., Zisserman, A. (eds.) ECCV 2008. LNCS, vol. 5305, pp. 705–718. Springer, Heidelberg (2008). https://doi.org/10.1007/978-3-540-88693-8_52
33. Yang, G., et al.: Litewsec: a lightweight framework for web-scale spectral ensemble clustering. In: TKDE, pp. 1–12 (2023)
34. Yang, G., Deng, S., Yang, Y., Gong, Z., Chen, X., Hao, Z.: LiteWSC: a lightweight framework for web-scale spectral clustering. In: Bhattacharya, A., et al. (eds.) DASFAA. LNCS, vol. 13246, pp. 556–573. Springer, Cham (2022). https://doi.org/10.1007/978-3-031-00126-0_40
35. Yang, G., et al.: RESKM: a general framework to accelerate large-scale spectral clustering. Pattern Recogn. **137**, 109275 (2022)
36. Yang, G., Lv, H., Yang, Y., Gong, Z., Chen, X., Hao, Z.: FastDEC: clustering by fast dominance estimation. In: Amini, M.R., Canu, S., Fischer, A., Guns, T., Kralj Novak, P., Tsoumakas, G. (eds.) ECML-PKDD. LNCS, vol. 13713. Springer, Cham (2022)
37. Yang, Y., et al.: Graphlshc: towards large scale spectral hypergraph clustering. Inf. Sci. **544**, 117–134 (2021)
38. Zheng, X., Ren, C., Yang, Y., Gong, Z., Chen, X., Hao, Z.: QuickDSC: clustering by quick density subgraph estimation. Inf. Sci. **581**, 403–427 (2021)

Lifelong Hierarchical Topic Modeling via Non-negative Matrix Factorization

Zhicheng Lin, Jiaxing Yan, Zhiqi Lei, and Yanghui Rao[✉]

School of Data and Computer Science, Sun Yat-sen University, Guangzhou, China
{linzhch26,yanjx6,leizhq5}@mail2.sysu.edu.cn,
raoyangh@mail.sysu.edu.cn

Abstract. Hierarchical topic modeling has been widely used in mining the latent topic hierarchy of documents. However, most of such models are limited to a one-shot scenario since they do not use the identified topic information to guide the subsequent mining of topics. By storing and exploiting the previous knowledge, we propose a lifelong hierarchical topic model based on Non-negative Matrix Factorization (NMF) for boosting the topic quality over a text stream. In particular, we construct a knowledge graph by the accumulated topic hierarchy information and use the knowledge graph to guide the training of our model on future documents. Moreover, the structure information in the knowledge graph is completed by supervised learning. Experiments on real-world corpora validate the effectiveness of our approach on lifelong learning paradigms.

Keywords: Hierarchical topic model · Semantic knowledge graph · Non-negative matrix factorization

1 Introduction

Lifelong learning [32,38] is a paradigm of exploiting knowledge gained from past learning to assist in the learning of new data. Besides being equipped with classification and clustering, this paradigm has also been applied to topic modeling [8,19,35], information extraction [6] and text classification [11]. A so-called topic model is dedicated to mining the potential semantic structure of topics in a corpus. However, many traditional topic models based on Latent Dirichlet Allocation (LDA) [4] are one-shot models, which mine the topics from a fixed corpus in one single run. These models cannot make full use of the previously mined thematic information over a text stream, which motivates the research of lifelong topic modeling [19,35]. The key to the success of lifelong topic models is that the accumulated topical knowledge can effectively guide the topic mining of the corpus at future moments. Unfortunately, the existing lifelong topic models, such as NMF-LTM [8], LDA-LTM [10], and LCM [35], assume the topics to be independent. These models construct knowledge graphs that only contain similarities between words or topics, without considering any information about the hierarchical structure among topics.

X. Song et al. (Eds.): APWeb-WAIM 2023, LNCS 14334, pp. 155–170, 2024.
https://doi.org/10.1007/978-981-97-2421-5_11

The application of a rational hierarchical topic structure has been demonstrated by many areas, including the hierarchical classification of Web pages [31] and the discovery of hierarchical topics in academic repositories [33]. Most of the existing hierarchical topic models are based on probabilistic graphical models. For instance, hLDA [18] captured the hierarchical structure of topics through a nested Chinese Restaurant Process (nCRP). In the hierarchical PAM [30], the topic relationships were constructed as Directed Acyclic Graphs (DAGs). Recently, a Tree-Structured Neural Topic Model (TSNTM) [20] was proposed to learn the hierarchical structure of topics by parameterizing the hierarchical topic distribution. The other model named nTSNTM [12] captured the relational dependency between topics based on a non-parametric prior. However, both TSNTM and nTSNTM generated a topic structure with only one root node, which brings unrealistic limitations and thus reduces the reasonableness of the topic structure. More importantly, most of the existing hierarchical topic models are limited to a one-shot scenario since they do not use the information of the mined topic structure to guide the topic mining on future documents.

NMF [26] introduces a non-negative constraint on the traditional matrix factorization method to ensure the non-negativity of matrix elements, which guarantees the model performance with powerful interpretability [8,35]. Considering the aforementioned advantages of NMF, we propose a **L**ifelong **H**ierarchical topic model based on **NMF** (LHNMF). We utilize the hierarchical topic structure information to construct a knowledge graph. Moreover, we fully exploit this knowledge graph to mine the potential information which guides the model for hierarchical topic mining on new documents. Among the existing lifelong topic models, knowledge graphs are often set as undirected for graph Laplacian. In contrast, the knowledge graph in our LHNMF is directed because it carries the information of hierarchical topic structures. Inspired by [36], we develop a graph Laplacian for our directed knowledge graph in order to integrate those knowledge into the objective function of our model. The main contributions of this paper can be summarized as follows:

1. We propose an NMF-based lifelong hierarchical topic model. To the best of our knowledge, this is the first time that hierarchical topic models are developed and evaluated on the lifelong learning paradigm.
2. We construct the knowledge graph using the information of hierarchical topic structure and introduce the graph Laplacian for directed graphs.
3. We mine the knowledge graph for potential hierarchical relationships to improve the completeness and the quality of knowledge.

2   Related Work

In a preliminary study, hLDA [18] was proposed to capture the hierarchical structure information among topics by introducing the nCRP, and the construction of topic trees was implemented by Gibbs sampling [3]. However, hLDA could only obtain the topic distribution of a single-path. In order to overcome this drawback, the nested Hierarchical Dirichlet Process (nHDP) [33] extended HDP [40] by using a borrowing mechanism between topics when several different topic trees share the same topic. In addition, the nested Chinese Restaurant Franchise Process (nCRF) [1] that combines the advantages of HDP and nCRP was proposed. However, these nCRP-based models required Gibbs sampling to calculate posterior probabilities, which were computationally expensive. Based on Neural Variational Inference (NVI) [23], TSNTM [20] utilized Doubly Recurrent Neural Networks [2] to parameterize the topic distribution. In addition, nTSNTM [12] introduced the Stick Breaking Process (SBP) [37] to achieve non-parametric parameterization of topic distributions, but the lack of constraints on the over-concentration of topics led to the degradation of topic quality. To address this issue, NMF-based HSOC [28] introduced three optimization constraints, including global independence, local independence, and information consistency to ensure the soundness of the topic structure. CluHTM [42], on the other hand, combined pre-trained word embeddings and NMF to improve the quality of topics. Unfortunately, all of them did not use the previously mined information of topic structures to assist the subsequent topic mining.

The accumulation and maintenance of knowledge over continuous tasks are the most critical features of lifelong learning [10]. In lifelong topic modeling, both NMF-LTM [8] and LDA-LTM [10] constructed knowledge graphs to assist model training. However, they were limited to mining flat topics and ignored the hierarchical structure among topics. This led to a relatively unreasonable quality of the topic structure for real-world corpora.

3   Methodology

In this section, we describe the proposed hierarchical NMF in a lifelong learning process, i.e., LHNMF. The main components of our model include: (i) NMF for topic modeling; (ii) the construction, update, and integration of our semantic knowledge graph in the lifelong learning paradigm.

3.1   NMF for Topic Modeling

NMF has been widely used in the field of topic models [8] due to its success in clustering high-dimensional data [13]. Given a lifelong learning paradigm, it is assumed that a document stream $\{D_t\}_{t=1}^{T}(T = +\infty)$ is to be processed, where $D_t \in \mathbb{R}_+^{M \times N}$ denotes the word-document matrix at time t. According to the factorization of non-negative matrices, we decompose D_t into two low-rank matrices $U_t \in \mathbb{R}_+^{M \times K}$ and $V_t \in \mathbb{R}_+^{K \times N}$, i.e., $D_t \approx U_t V_t$, where $K \ll min(M, N)$ is the number of latent factors (i.e., topics). For convenience of describing the proposed method, we summarize the frequently used notations in Table 1. Generally, the objective of NMF-based topic models can be formulated as follows:

$$\min_{U_t, V_t \geq 0} \| D_t - U_t V_t \|_F^2 . \tag{1}$$

By decomposing D_t according to Eq. (1), we can obtain the topic words of the corpus through U_t, and discover the implicit topic structure of the corpus

Table 1. Frequently-used notations.

Notation	Description
M	The number of words
$N^{(l)}$	The number of documents at the l-th layer
$K^{(l)}$	The number of topics at the l-th layer
$D_t^{(l)} \in R^{M \times N^{(l)}}$	Word-document matrix of the l-th layer at the t-th moment
$U_t^{(l)} \in R^{M \times K^{(l)}}$	Word-topic matrix of the l-th layer at the t-th moment
$V_t^{(l)} \in R^{K^{(l)} \times N^{(l)}}$	Topic-document matrix of the l-th layer at the t-th moment
$K_t \in R^{M \times M}$	Semantic knowledge graph at the t-th moment

through V_t. For a hierarchical topic model, topics with a hierarchical structure need to be mined. By exploiting the characteristics of NMF, we can decompose the word-document matrix layer by layer. Figure 1(a) shows an illustration of such processes, and the steps are formulated as follows:

1. Decompose D_t to get U_t and V_t.
2. Through the topic-document matrix V_t, the documents at time t are grouped into blocks according to their topics, which generate k document blocks $D_{t1}, \cdots, D_{tk}$.
3. Perform NMF on $D_{t1}, \cdots, D_{tk}$ respectively to obtain topic-document matrices and word-topic matrices which contain more fine-grained topics.

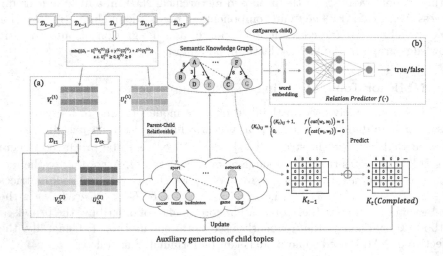

Fig. 1. The architecture of LHNMF.

Accordingly, Eq. (1) can be rewritten as:

$$\min_{U_t^{(l)}, V_t^{(l)} \geq 0} \| D_t^{(l)} - U_t^{(l)} V_t^{(l)} \|_F^2,$$ (2)

where l represents the index of layers. In this case, the k-th topic at the l-th layer and the topics at the $(l+1)$-th layer obtained by factorizing $D_{tk}^{(l+1)}$ naturally contain the parent-children relationships.

3.2   Semantic Knowledge Graph

In the fields of data mining and natural language processing, knowledge graphs are often constructed to assist a model for lifelong learning [19,35]. For the hierarchical topic model proposed in this paper, we develop a semantic knowledge graph with contextual relationships between topic words to assist the training of our model parameters.

Structure Formalization. As mentioned above, we perform a layer-by-layer matrix decomposition of the document blocks at each moment. By decomposing the same document at different layers, the resulting topic words have a parent-child relationship. We utilize a matrix $K_t \in \mathbb{R}^{M \times M}$ to store the parent-child relationship between topic words, which is the adjacency matrix of the knowledge graph at the t-th moment. Based on each adjacency matrix, the knowledge graph also contains the similarity information among topic words in the form of edge weights. The construction of K_t is formulated as follows:

$$(K_t)_{ij} = \begin{cases} (K_t)_{ij} + 1, & isParentChild(w_i, w_j); \\ 0, & i = j, \end{cases}$$ (3)

where w_i and w_j are topic words. The function $isParentChild(w_i, w_j)$ indicates that word w_i is in the parent topic and word w_j is in the child topic.

Supervised Learning for Knowledge Graph Completion. Since our model achieves hierarchical clustering based on data division, it makes the relationship between the superior and subordinate nodes that are divided into different clusters missing. Therefore, the aforementioned knowledge graph constructed by the existing topic information only is imperfect, and there are potential parent-child relationships between certain words remain to be discovered. To address this issue, we propose a supervised learning method of completing our knowledge graph, as shown in Fig. 1(b). Firstly, we use GloVe [34] to obtain word embeddings of topic words, and then count the topic word pairs with parent-child relationships and the topic word pairs with non parent-child relationships. The topic word pairs with parent-child relationships are concatenated and given a real label of true. Similarly, the topic word pairs with non parent-child relationships are concatenated and given a real label of false. Then, a *relation predictor* $f(\cdot)$

is constructed and the above word pairs are used as input for training. Finally, the potential as well as undetermined relationships of the knowledge graph (i.e., edges) are predicted to determine whether they are real. If the prediction is true, the weights of the edges between the word nodes are set to +1; otherwise, the weights of the corresponding edges are set to 0. By doing so we can improve the hierarchical information carried by the knowledge graph and help our model achieve better performances in future moments. The above process is formulated as follows:

$$(K_t)_{ij} = \begin{cases} (K_t)_{ij} + 1, & f(cat(w_i, w_j)) = 1; \\ 0, & f(cat(w_i, w_j)) = 0, \end{cases} \tag{4}$$

where $cat(\cdot)$ means the concatenation of two vectors. As shown in Fig. 1(b), we use a two-layer Multilayer Perceptron (MLP) as the *relation predictor* to complete the knowledge graph for simplicity.

Graph Laplacian Regularization of Directed Graph. We add the knowledge graph to our objective function using a graph Laplacian regularization [14]. According to the construction strategy of our knowledge graph, it is directed due to the presence of hierarchical information. However, the existing graph Laplacian is designed for undirected graphs. Inspired by [36], we develop a novel graph Laplacian regularization for our directed knowledge graph. Given the adjacency matrix K_t, the graph Laplacian is estimated as follows:

$$(H_t)_{ij} = \frac{(K_t)_{ij}}{\sqrt{(O_t)_{ij}^\tau (P_t)_{ij}^\tau}} = [((O_t)^\tau)^{-\frac{1}{2}} (K_t)((P_t)^\tau)^{-\frac{1}{2}}]_{ij}, \tag{5}$$

where H_t denotes the graph Laplacian of the knowledge graph at the t-th moment, $(O_t)^\tau$ and $(P_t)^\tau$ are diagonal matrices with $(O_t)_{ii}^\tau = \sum_k (K_t)_{ik} + \tau$ and $(P_t)_{jj}^\tau = \sum_k (K_t)_{kj} + \tau$. The regularization parameter $\tau > 0$ is set to the average out-degree, i.e., $\tau = \sum_{i,k} (K_t)_{ik}/M$. In this situation, $(O_t)^\tau$ and $(P_t)^\tau$ represent the out-degree and in-degree of the knowledge graph, respectively. For $l > 1$, we intend to get more topic-specific information, which is related to the child nodes in the semantic knowledge graph. Considering that the in-degree of the knowledge graph should play a more important role for the aforementioned case, we weaken the out-degree's effect by adding a decay coefficient associated with the level l. Consequently, the general formula of calculating $(O_t)^\tau$ is $(O_t)_{ii}^\tau = \sum_k (K_t)_{ik} + 10^{-l}\tau$.

When compared with the graph Laplacian of undirected graphs written as $L = D - W$, in which W is a symmetric adjacency matrix and D is a diagonal matrix whose entries are the degrees of the graph, i.e., $D_{jj} = \sum_l W_{jl}$, we can see that the graph Laplacian of undirected graphs never distinguishes in-degree or out-degree information. On the contrary, in our proposed method, the adjacency matrix K is not a symmetric matrix, and in the process of performing the graph Laplacian, we separate the in-degree from out-degree information, which enables our graph Laplacian matrix to imply hierarchical semantic information.

Therefore, to exploit such hierarchical semantic information for LHNMF, our graph Laplacian regularization is given below:

$$\Psi(U_p, U_c, K_t) = Tr(U_c^T H_t U_p), \qquad (6)$$

where $Tr(\cdot)$ denotes the trace of a matrix. Besides, U_p and U_c are the parent-topic matrix and the child-topic matrix, respectively. Such a regular term can be used to guide the generation of high-quality child topics using the parent topic.

3.3 Objective Function

As shown in Fig. 1, our LHNMF generates topic words with parent-child relations through a semantic knowledge graph. Formally, the objective function of discovering the first level's topics is given below:

$$\mathcal{L}_1 = ||D_t^{(1)} - U_t^{(1)} V_t^{(1)}||_F^2 + \gamma^{(1)} U_t^{(1)} + \lambda^{(1)} V_t^{(1)} \quad s.\,t. \quad U_t^{(1)} \geq 0, V_t^{(1)} \geq 0. \;\; (7)$$

In the first term at the first level, the word-document matrix $D_t^{(1)}$ is decomposed into the word-topic matrix $U_t^{(1)}$ and the topic-document matrix $V_t^{(1)}$. Given a hyperparameter $\gamma^{(1)}$, the sparsity constraint on $U_t^{(1)}$ is introduced as the second term $\gamma^{(1)} U_t^{(1)}$, which can help the model generate more sparse generic topic words [7,44]. Similarly, given a hyperparameter $\lambda^{(1)}$, the sparsity constraint on $V_t^{(1)}$ is introduced as the third term $\lambda^{(1)} V_t^{(1)}$. This ensures that each document covers limited numbers of topics [8], and such a sparsity constraint also encourages the model to generate interpretable topics [5], which corresponds with the intuition that a document usually focuses on several salient topics instead of covering a wide variety of topics [27].

For the second level, the objective function is derived as follows:

$$\mathcal{L}_2 = \sum_{k=1}^{K^{(1)}} \left\{ ||D_{tk}^{(2)} - U_{tk}^{(2)} V_{tk}^{(2)}||_F^2 + \Psi(U_t^{(1)}, U_{tk}^{(2)}, K_{t-1}) + \Theta(U_{tk}^{(2)}) \right.$$
$$\left. + \gamma^{(2)}(U_{tk}^{(2)}) + \lambda^{(2)}(V_{tk}^{(2)}) \right\} \quad s.t. \quad U_{tk}^{(2)} \geq 0, V_{tk}^{(2)} \geq 0, \qquad (8)$$

where $K^{(1)}$ is the topic number at the first layer.

At the second layer, we need to generate more specific topics to satisfy the characteristics of the child-topics. According to Eq. (6), we can obtain:

$$\Psi(U_t^{(1)}, U_{tk}^{(2)}, K_{t-1}) = \alpha \cdot Tr((U_{tk}^{(2)})^T H_{t-1} U_t^{(1)}), \qquad (9)$$

where H_{t-1} is calculated by Eq. (5) and α is the hyperparameter of this term. This regularization term can effectively use the parent topic $U_t^{(1)}$ to guide the generation of the child topic $U_{tk}^{(2)}$. Also, in order to make the topic words at the sub-topic layer more diverse, we add the orthogonal regularization in the third term [9]. The last two terms are used to encourage sparse topics and topic words so that the generated topics and topic words will not be too redundant.

Although more layers of hierarchical topics can be applied in our model, we adopt $\mathcal{L}_1$ and $\mathcal{L}_2$ for topic discovery in this study to be consistent with the existing NMF-based hierarchical topic model, i.e., CluHTM[1] [42].

Finally, we use binary cross entropy (BCE) as the objective function of the *relation predictor*, which is formulated as follows:

$$\mathcal{L}_{rp} = -\sum_{i=1}^{N} \left[y^{(i)} log(\hat{y}^{(i)}) + (1 - y^{(i)}) log(1 - \hat{y}^{(i)}) \right], \tag{10}$$

where $y^{(i)}$ denotes the real label of the i-th input $x^{(i)}$. The input $x^{(i)}$ is a combination of word embeddings of parent-child topic words (i.e., $y^{(i)}$ is 1) or non parent-child topic words (i.e., $y^{(i)}$ is 0). Besides, $\hat{y}^{(i)}$ is the predicted value of the i-th output, which is formulated as follows:

$$\hat{y}^{(i)} = W_3\sigma(W_2\sigma(W_1 x^{(i)} + b_1) + b_2) + b_3, \tag{11}$$

where W and b represent the weight and bias of the neural network, respectively. σ is a rectified linear unit (ReLU) [16].

3.4   Iterative Updates for Parameters

For parameters in $\mathcal{L}_1$ and $\mathcal{L}_2$, we adopt the iterative algorithm of updating the factor matrices in NMF. Based on the multiplicative update rule [26], we obtain the update formulas for each parameter in $\mathcal{L}_1$ and $\mathcal{L}_2$, as follows:

$$U_t^{(1)} \leftarrow U_t^{(1)} \odot \frac{D_t^{(1)}(V_t^{(1)})^T}{U_t^{(1)}V_t^{(1)}(V_t^{(1)})^T + \frac{\gamma^{(1)}}{2}}. \tag{12}$$

$$V_t^{(1)} \leftarrow V_t^{(1)} \odot \frac{(U_t^{(1)})^T D_t^{(1)}}{(U_t^{(1)})^T U_t^{(1)} V_t^{(1)} + \frac{\lambda^{(1)}}{2}}. \tag{13}$$

$$U_{tk}^{(2)} \leftarrow U_{tk}^{(2)} \odot \frac{D_{tk}^{(2)}(V_{tk}^{(2)})^T + 2\beta \cdot U_{tk}^{(2)}}{U_{tk}^{(2)}V_{tk}^{(2)}(V_{tk}^{(2)})^T + \frac{\alpha}{2} \cdot H_{t-1}U_t^{(1)} + 2\beta \cdot U_{tk}^{(2)}(U_{tk}^{(2)})^T U_{tk}^{(2)} + \frac{\gamma^{(2)}}{2}}. \tag{14}$$

$$V_{tk}^{(2)} \leftarrow V_{tk}^{(2)} \odot \frac{(U_{tk}^{(2)})^T D_{tk}^{(2)}}{(U_{tk}^{(2)})^T U_{tk}^{(2)} V_{tk}^{(2)} + \frac{\lambda^{(2)}}{2}}. \tag{15}$$

In addition, the parameters in $\mathcal{L}_{rp}$ are updated by Adam [22].

---

[1] Note that several dummy root topics were used in the original CluHTM. However, we experimentally observed that such a model quite concentrated on a few topics, especially for a relatively small corpus. To achieve a reasonable topic structure in lifelong learning paradigms, we discard such dummy root topics.

3.5 Computational Complexity

In this section, we analyze the time complexity of LHNMF. The matrices to be updated in the model include $U_t^{(1)}$, $V_t^{(1)}$, $U_{tk}^{(2)}$, $V_{tk}^{(2)}$, and H_t. The time complexity of one iteration is $O(3MN^{(1)}K^{(1)} + 3MK^{(1)})$, $O(MN^{(1)}K^{(1)} + (K^{(1)})^2(M + N^{(1)}) + 3K^{(1)}N^{(1)})$, $O(3MN^{(2)}K^{(2)} + 9MK^{(2)} + 2M^2K^{(2)})$, $O(MN^{(2)}K^{(2)} + (K^{(2)})^2(M + N^{(2)}) + 3K^{(2)}N^{(2)})$, and $O(5M^2 + 2M^3)$. Note that the number of $U_{tk}^{(2)}$ and $V_{tk}^{(2)}$ at the second layer is determined by $K^{(1)}$. Therefore, the time complexity of each iteration is $O(4MN^{(1)}K^{(1)} + (K^{(1)})^2(M + N^{(1)}) + 3MK^{(1)} + 3N^{(1)}K^{(1)} + 4K^{(1)}MN^{(2)}K^{(2)} + 9K^{(1)}MK^{(2)} + 2K^{(1)}M^2K^{(2)} + K^{(1)}(K^{(2)})^2(M + N^{(2)}) + 3K^{(1)}N^{(2)}K^{(2)} + 5M^2 + 2M^3)$.

4 Experiments

In this section, we validate the performance of our LHNMF on hierarchical topic modeling using three widely used corpora. We employ metrics of evaluating the topic quality and hierarchical topic affinity to compare LHNMF with several representative baselines. Also, all lifelong topic models are evaluated in terms of knowledge transfer and catastrophic forgetting.

Table 2. The statistics of corpora.

Datasets	#Documents	#Words	#Topics	#Chunks
20News	18,208	1,992	55	5
Reuters	10,763	2,000	30	4
NIPS	1,499	3,531	30	3

4.1 Experimental Setting

We run our experiments on three benchmark corpora with potential hierarchical topic structures, including 20News [29], Reuters [43], and NIPS [39]. These datasets are tokenized with non UTF-8 characters and stop words eliminated. Furthermore, words with low frequency are removed by following [12,39]. Table 2 summarizes the statistics of the three corpora.

In our experiments, we use three hierarchical topic models, two flat topic models, and two lifelong topic models as baselines for a comprehensive comparison. The hierarchical topic models are SawETM [15], nTSNTM [12], and CluHTM [42]. The flat topic models are GSM [29] and NSTM [45]. Finally, the lifelong topic models are LDA-LTM [10] and NMF-LTM [8]. The models of GSM and NMF-LTM are implemented according to their original papers by ourselves. For the other models, we use their open source codes.

The values of all hyperparameters mentioned in Sect. 3.3 are set by grid searching. Accordingly, a set of hyperparameter values with better comprehensive performance is derived, where $\alpha = 5$, $\beta = 0.3$, $\gamma^{(1)} = 0.01$, $\lambda^{(1)} = 0.06$, $\gamma^{(2)} = 0.003$ and $\lambda^{(2)} = 0.03$. For the number of topics per dataset, we refer to the stability measure [17] to give a range for the number of topics in the dataset. This measure automatically determines the appropriate number of topics for datasets using the average Jaccard coefficient. For the other models, the parameters are set as the same values given in the original paper.

In addition, since our model uses a lifelong learning paradigm, the metrics of our model are the average of the metrics obtained from the training of all document chunks by following [10].

4.2 Topic Evaluation

Topic coherence refers to the degree of relevance between topic words within the same topic. This metric can be used to measure the quality of topics. We use the Normalized Pointwise Mutual Information (NPMI) [25] to evaluate topic quality. Since this metric has been shown to be close to human judgment, it also represents the interpretability of topics. This metric is calculated as follows:

$$\text{NPMI}(w_i, w_j) = \frac{log \frac{P(w_i, w_j)}{P(w_i)P(w_j)}}{-log(P(w_i, w_j))}. \tag{16}$$

Table 3 presents the performance of our model and all baselines in terms of topic coherence, which indicates that our model achieves the best results on both Reuters and NIPS datasets, and performs better than most baselines on 20News. Although NSTM works well on 20News, it performs much worse on the other two datasets, i.e., the performance of NSTM is quite unstable. LDA-LTM uses the generated prior knowledge to guide the mining of topics, but its prior knowledge may contain information from future moments, which brings unfair comparison under the single-pass stream processing scenario. Therefore, we can conclude that our model can mine more highly relevant topics when compared with the existing flat, lifelong, and hierarchical topic models generally.

Table 3. The average NPMI scores using top 5, 10, and 15 words for each topic.

Model		20News	Reuters	NIPS
Flat	GSM	0.206	0.192	0.115
	NSTM	**0.294**	0.173	0.064
Lifelong	NMF-LTM	0.205	0.170	0.011
	LDA-LTM	0.291	0.255	0.191
Hierarchical	nTSNTM	0.256	0.262	0.180
	SawETM	0.254	0.230	0.161
	CluHTM	0.142	0.134	0.137
Ours (Lifelong & Hierarchical)		0.274	**0.276**	**0.204**

4.3 Topic Hierarchy Analysis

For a reasonable topic hierarchy, a considerable semantic similarity between parent-child topics should be ensured. In contrast, a certain degree of independence should be maintained between non parent-child topics. To evaluate the quality of the hierarchical topics mined by the hierarchical topic model, we use the metrics of Cross-Layer Pointwise Mutual Information (CLNPMI) [12] and hierarchical affinity [21] for evaluation.

Hierarchical Topic Quality. To evaluate the soundness of topic hierarchy, we measure the coherence between parent and child topics by $CLNPMI(W_p, W_c) = \sum_{w_i \in W_p'} \sum_{w_j \in W_c'} \frac{NPMI(w_i, w_j)}{|W_p'||W_c'|}$, where $W_p' = W_p - W_c$ and $W_c' = W_c - W_p$. In the above, W_p and W_c represent the top N words of the parent topic and its child topics, respectively.

Table 4. CLNPMI and topic affinity ratio (S_{ratio}) of hierarchical topic models.

Model	20News		Reuters		NIPS	
	CLNPMI	S_{ratio}	CLNPMI	S_{ratio}	CLNPMI	S_{ratio}
SawETM	0.127	8.882	0.035	4.448	0.018	**19.090**
nTSNTM	0.137	4.630	**0.091**	3.024	0.036	1.850
CluHTM	0.047	4.280	0.031	2.463	0.026	0.153
Ours	**0.157**	**9.025**	0.060	**6.602**	**0.052**	11.651

As shown in Table 4, our model performs well on these datasets overall in terms of CLNPMI. This illustrates that the prior knowledge of hierarchical topic relationships carried by the semantic knowledge graph is helpful for mining high-quality hierarchical topics.

Topic Hierarchical Affinity. Topics that form parent-child relationships should show greater similarity in their word distributions than topics without parent-child relationships. According to [21], we use the hierarchical affinity to measure the similarity between parent-child topics. We assume that $S_\lambda(k)$ represents parent-child topic affinity, and $S_{\bar\lambda}(k)$ represents non parent-child topic affinity. Then, we use the ratio $S_{ratio} = \frac{S_\lambda(k)}{S_{\bar\lambda}(k)}$ to measure whether the parent-child topic relationships mined by the model exhibit better similarity than non parent-child topic relationships. The formulas are given below:

$$S_\lambda(k) = \frac{1}{|\lambda(k)|} \sum_{i \in \lambda(k)} \frac{\phi_k \cdot \phi_i}{|\phi_k||\phi_i|}; \quad S_{\bar\lambda}(k) = \frac{1}{|\bar\lambda(k)|} \sum_{j \in \bar\lambda(k)} \frac{\phi_k \cdot \phi_j}{|\phi_k||\phi_j|}, \quad (17)$$

where $\Phi(k)$ is the topics at layer k, $\lambda(k)$ is the set of child topics of $\Phi(k)$, and $\bar\lambda(k)$ is the set of non-child topics of $\Phi(k)$. A higher value of S_{ratio} means that parent-child topics are more similar than non parent-child topics.

The results are shown in Table 4, from which we can observe that our model achieves good performance on the aforementioned metric. This fully indicates that the parent-child topics mined by our model show better similarity in distribution when compared to non parent-child topics. Although SawETM achieves an impressive value of S_{ratio} on NIPS, it has the lowest CLNPMI and its hierarchy is not reasonable when evaluated comprehensively. The overall results also indicate that the hierarchical topic structure generated by our model is reasonable for each moment.

4.4 Analysis on Lifelong Learning

For lifelong learning paradigms, catastrophic forgetting [24] is a non-negligible problem. To assess the ability of our model in avoiding catastrophic forgetting, the 20News and Reuters datasets whose chunk numbers are relatively large are used. For both datasets, we re-learn the first document chunk after our model has learned all the document chunks. The results of CLNPMI and NPMI under the above setting are shown in Fig. 2(a), where a blue bar represents the metric's value obtained by training our model on the first document chunk, and an orange bar represents the metric's value obtained by retraining our model on the first document chunk after training on all document chunks. From the results we can observe that our model does not experience excessive forgetting of prior knowledge, thus effectively avoiding the significant performance degradation associated with catastrophic forgetting.

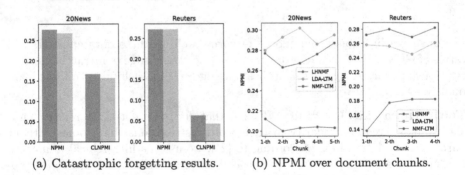

(a) Catastrophic forgetting results.   (b) NPMI over document chunks.

Fig. 2. Evaluation of lifelong topic models on 20News and Reuters.

In addition, we compare the topic coherence of our model with LDA-LTM and NMF-LTM for each document chunk on 20News and Reuters to test whether positive knowledge transfer happened. The results are shown in Fig. 2(b). It can be concluded that there is an overall upward trend of our model when compare with LDA-LTM and NMF-LTM. This indicates that our model not only makes good use of the auxiliary function of the knowledge graph, but also does not suffer from the catastrophic forgetting of the emergent prior knowledge.

4.5   Ablation Study

Our LHNMF contains a semantic knowledge graph and mines the knowledge graph for latent parent-child relationship information. Therefore, ablation experiments need to be conducted on these two modules to verify the effective enhancement for our LHNMF. We conduct ablation experiments on two datasets, i.e., 20News and NIPS. The results are shown in Table 5, where "Ours w/o KG" means the knowledge graph module is removed, and "Ours w/o Com" means the completion of our knowledge graph is not conducted.

From the experimental results we can conclude that there is a good improvement on the topic quality of our model for both modules. Moreover, the potential information mining of the knowledge graph can be supplemented to enable the model to make better use of the auxiliary information brought by the knowledge graph. We have also used different classifiers (e.g., MLP and Logistic Regression) as relational predictors, and the model is stable in terms of the above metrics.

Table 5. Ablation results on 20News and NIPS.

Model	20News		NIPS	
	CLNPMI	NPMI	CLNPMI	NPMI
Ours w/o KG	0.138	0.265	0.027	0.189
Ours w/o Com	0.153	0.268	0.040	0.180
Ours	**0.157**	**0.274**	**0.052**	**0.204**

4.6   Running Time Comparison

As an illustration, we run models on a server equipped with Intel(R) Xeon(R) Silver 4214R CPU @ 2.40 GHz, 48 cores and 128G memory, and 2 × NVIDIA GTX 1080Ti with 2 × 12G memory. On the 20News dataset with the same number of topics and topic structure, it costs about 5 min to train our LHNMF until the model converged. The time cost is nearly the same for the existing NMF-based lifelong flat topic model, i.e., NMF-LTM [8]. However, CluHTM [42], which is one of the latest NMF-based one-shot hierarchical topic model, costs about 1 h for convergence.

Different from CluHTM, our model maintains a knowledge graph and incorporates it into parameter updating. Besides, our knowledge graph stores the semantic relationships of topic words, thus its adjacency matrix is sparser and we can reduce the running speed of the model very well. In contrast, each topic generation of CluHTM involves the computation of the Cluwords-based [41] TFIDF matrix, which greatly increases the running time of CluHTM.

5   Conclusion

In this paper, we propose an NMF-based lifelong hierarchical topic model named LHNMF. A knowledge graph with topic words' hierarchical semantic information is constructed to guide the training of LHNMF, and the potential parent-child relationships between words are completed by supervised learning. Our model ensures the reuse of mined topic information, which is critical for lifelong learning in real-world applications. Extensive experiments validate the effectiveness of our LHNMF on generating a more reasonable hierarchical topic structure than baselines. Besides, our model can be trained fast in terms of the time complexity.

Acknowledgements. This work has been supported by the National Natural Science Foundation of China (61972426).

References

1. Ahmed, A., Hong, L., Smola, A.: Nested Chinese restaurant franchise process: applications to user tracking and document modeling. In: ICML, pp. 1426–1434 (2013)
2. Alvarez-Melis, D., Jaakkola, T.S.: Tree-structured decoding with doubly-recurrent neural networks. In: ICLR (2017)
3. Andrieu, C., De Freitas, N., Doucet, A., Jordan, M.I.: An introduction to MCMC for machine learning. Mach. Learn. **50**(1), 5–43 (2003)
4. Blei, D.M., Ng, A.Y., Jordan, M.I.: Latent dirichlet allocation. J. Mach. Learn. Res. **3**, 993–1022 (2003)
5. Card, D., Tan, C., Smith, N.A.: Neural models for documents with metadata. In: ACL, pp. 2031–2040 (2018)
6. Carlson, A., Betteridge, J., Wang, R.C., Hruschka Jr, E.R., Mitchell, T.M.: Coupled semi-supervised learning for information extraction. In: WSDM, pp. 101–110 (2010)
7. Chen, X.: Learning with sparsity: Structures, optimization and applications. Ph.D. thesis, Carnegie Mellon University (2013)
8. Chen, Y., Wu, J., Lin, J., Liu, R., Zhang, H., Ye, Z.: Affinity regularized nonnegative matrix factorization for lifelong topic modeling. IEEE Trans. Knowl. Data Eng. **32**(7), 1249–1262 (2020)
9. Chen, Y., Zhang, H., Wu, J., Wang, X., Liu, R., Lin, M.: Modeling emerging, evolving and fading topics using dynamic soft orthogonal NMF with sparse representation. In: ICDM, pp. 61–70 (2015)
10. Chen, Z., Liu, B.: Topic modeling using topics from many domains, lifelong learning and big data. In: ICML. vol. 32, pp. 703–711 (2014)
11. Chen, Z., Ma, N., Liu, B.: Lifelong learning for sentiment classification. arXiv preprint arXiv:1801.02808 (2018)
12. Chen, Z., Ding, C., Zhang, Z., Rao, Y., Xie, H.: Tree-structured topic modeling with nonparametric neural variational inference. In: ACL/IJCNLP, pp. 2343–2353 (2021)
13. Choo, J., Lee, C., Reddy, C.K., Park, H.: Weakly supervised nonnegative matrix factorization for user-driven clustering. Data Min. Knowl. Discov. **29**(6), 1598–1621 (2015)

14. Dai, L., Zhu, R., Wang, J.: Joint nonnegative matrix factorization based on sparse and graph laplacian regularization for clustering and co-differential expression genes analysis. Complex. **2020**, 3917812:1–3917812:10 (2020)
15. Duan, Z., et al.: Sawtooth factorial topic embeddings guided gamma belief network. In: ICML, pp. 2903–2913 (2021)
16. Glorot, X., Bordes, A., Bengio, Y.: Deep sparse rectifier neural networks. In: AIS-TATS, vol. 15, pp. 315–323 (2011)
17. Greene, D., O'Callaghan, D., Cunningham, P.: How many topics? stability analysis for topic models. In: ECML/PKDD, vol. 8724, pp. 498–513 (2014)
18. Griffiths, T., Jordan, M., Tenenbaum, J., Blei, D.: Hierarchical topic models and the nested Chinese restaurant process. In: NIPS, vol. 16, pp. 17–24 (2003)
19. Gupta, P., Chaudhary, Y., Runkler, T.A., Schütze, H.: Neural topic modeling with continual lifelong learning. In: ICML, vol. 119, pp. 3907–3917 (2020)
20. Isonuma, M., Mori, J., Bollegala, D., Sakata, I.: Tree-structured neural topic model. In: ACL, pp. 800–806 (2020)
21. Kim, J.H., Kim, D., Kim, S., Oh, A.: Modeling topic hierarchies with the recursive Chinese restaurant process. In: CIKM, pp. 783–792 (2012)
22. Kingma, D.P., Ba, J.: Adam: a method for stochastic optimization. In: ICLR (2015)
23. Kingma, D.P., Welling, M.: Auto-encoding variational bayes. In: ICLR (2014)
24. Kirkpatrick, J., et al.: Overcoming catastrophic forgetting in neural networks. CoRR abs/1612.00796 (2016)
25. Lau, J.H., Newman, D., Baldwin, T.: Machine reading tea leaves: automatically evaluating topic coherence and topic model quality. In: EACL, pp. 530–539 (2014)
26. Lee, D., Seung, H.S.: Algorithms for non-negative matrix factorization. In: NIPS. vol. 13, pp. 556–562 (2000)
27. Lin, T., Hu, Z., Guo, X.: Sparsemax and relaxed wasserstein for topic sparsity. In: WSDM, pp. 141–149 (2019)
28. Liu, R., Wang, X., Wang, D., Zuo, Y., Zhang, H., Zheng, X.: Topic splitting: a hierarchical topic model based on non-negative matrix factorization. J. Syst. Sci. Syst. Eng. **27**(4), 479–496 (2018)
29. Miao, Y., Grefenstette, E., Blunsom, P.: Discovering discrete latent topics with neural variational inference. In: ICML, pp. 2410–2419 (2017)
30. Mimno, D., Li, W., McCallum, A.: Mixtures of hierarchical topics with pachinko allocation. In: ICML, pp. 633–640 (2007)
31. Ming, Z.Y., Wang, K., Chua, T.S.: Prototype hierarchy based clustering for the categorization and navigation of web collections. In: SIGIR, pp. 2–9 (2010)
32. Mitchell, T., et al.: Never-ending learning. Commun. ACM **61**(5), 103–115 (2018)
33. Paisley, J.W., Wang, C., Blei, D.M., Jordan, M.I.: Nested hierarchical dirichlet processes. IEEE Trans. Pattern Anal. Mach. Intell. **37**(2), 256–270 (2015)
34. Pennington, J., Socher, R., Manning, C.D.: Glove: Global vectors for word representation. In: EMNLP, pp. 1532–1543 (2014)
35. Qin, X., Lu, Y., Chen, Y., Rao, Y.: Lifelong learning of topics and domain-specific word embeddings. In: ACL/IJCNLP (Findings), pp. 2294–2309 (2021)
36. Rohe, K., Qin, T., Yu, B.: Co-clustering directed graphs to discover asymmetries and directional communities. PNAS **113**(45), 12679–12684 (2016)
37. Sethuraman, J.: A constructive definition of dirichlet priors. Statistica Sinica 639–650 (1994)
38. Silver, D.L.: Machine lifelong learning: challenges and benefits for artificial general intelligence. In: AGI, vol. 6830, pp. 370–375 (2011)
39. Tan, C., Card, D., Smith, N.A.: Friendships, rivalries, and trysts: characterizing relations between ideas in texts. In: ACL, pp. 773–783 (2017)

40. Teh, Y., Jordan, M., Beal, M., Blei, D.: Sharing clusters among related groups: hierarchical dirichlet processes. In: NIPS, vol. 17, pp. 1385–1392 (2004)
41. Viegas, F., et al.: Cluwords: exploiting semantic word clustering representation for enhanced topic modeling. In: WSDM, pp. 753–761 (2019)
42. Viegas, F., Cunha, W., Gomes, C., Pereira, A., Rocha, L., Goncalves, M.: Cluhtm-semantic hierarchical topic modeling based on cluwords. In: ACL, pp. 8138–8150 (2020)
43. Wu, J., et al.: Neural mixed counting models for dispersed topic discovery. In: ACL, pp. 6159–6169 (2020)
44. Xu, Z., Chang, X., Xu, F., Zhang, H.: $L_{1/2}$ regularization: A thresholding representation theory and a fast solver. IEEE Trans. Neural Networks Learn. Syst. **23**(7), 1013–1027 (2012)
45. Zhao, H., Phung, D., Huynh, V., Le, T., Buntine, W.L.: Neural topic model via optimal transport. In: ICLR (2021)

Time-Aware Preference Recommendation Based on Behavior Sequence

Jiaqi Wu, Yi Liu, Yidan Xu, Yalei Zang, Wenlong Wu, Wei Zhou, Shidong Xu, and Bohan Li$^{(\boxtimes)}$

Nanjing University of Aeronautics and Astronautics, Nanjing, China
bhli@nuaa.edu.cn

Abstract. Sequential recommendation (SR) has become an important schema to assist people in rapidly finding their interest in the progressively growing data. Especially, long and short-term based methods capture user preferences and provide more precise recommendations. However, they rarely consider the effect of time intervals and limit the short-term preferences' weight in predicting the next items. In this paper, we propose a novel model called TPR-BS (**T**ime-aware **P**reference **R**ecommendation based on **B**ehavior **S**equence) to address these issues. We model the user's long and short-term behavioral sequence separately and fuse sequence features to obtain the user's comprehensive preferences' representation. Specifically, we first use the sparse attention layer to filter the effect of irrelevant information on long-term preferences. Then we modify the Gated Recurrent Unit (GRU) based on time intervals and encode the user's short-term behavior sequence into the hidden states for the corresponding moment. Besides, we construct a target attention network layer to highlight the last-moment interaction behavior. TPR-BS aims to dynamically capture user preferences' changes which can reflect the user's general preferences and the latest intentions. The experimental results indicate that our model outperforms state-of-the-art methods on three public datasets.

Keywords: Sequential recommendation · User preferences · Time intervals · Behavior sequence

1 Introduction

Recommender Systems (RS) aim to explore the vast amount of available information, filter it according to user preferences and assist users make decisions. However, user preferences and requirements change over time in the real world, while the explosive growth of data can lead to information overload, making it difficult to fulfill the tasks of making recommendations. User-based behavioral sequence modeling is an efficient way to solve that issue. Traditional sequential recommendation methods learn the relationships in historical behavior with Markov Chain (MC), which tends to examine static preferences and have difficulty modeling higher-order dependency [15]. Recently, deep learning approaches have been

X. Song et al. (Eds.): APWeb-WAIM 2023, LNCS 14334, pp. 171–185, 2024.
https://doi.org/10.1007/978-981-97-2421-5_12

employed in recommender systems to process the sequences, such as Recurrent Neural Network (RNN) [4,6], Convolutional Neural Network (CNN) [9,16,21], Graph Neural Network (GNN) [1,18,19], and Attention Mechanism [7,10,20].

The user behavior information in the sequence affects the recommended next item differently. Most existing recommender systems model the user's interaction sequences as an integrated whole and treat user preferences as individual vectors. User preferences are difficult to track in practice, while the time intervals serve an important role. As a result, the key in sequential recommendation research is how to model sequential user behaviors, which find user preferences and achieve accurate matching of users with information. As shown in Fig. 1, a user who enjoys sports may always browse fitness equipment (i.e., long-term preferences). In cases where a bonsai suddenly appears, we consider that it is the result of an accidental click by the users while browsing and needs to be filtered out. However, the user has paid more attention to computer and related accessories products in the last two weeks (i.e., short-term preferences), and time intervals can guide the next item recommended.

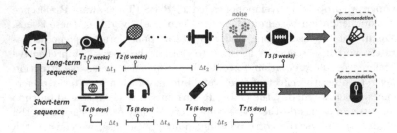

Fig. 1. The example of sequential recommendation.

In this paper, we propose a new recommendation model TPR-BS which decomposes interactive user behavior into long-term preferences' representation and short-term interests evolution to solve the recommendation problem for the next item. Compared with the previous work, we consider the decay issue in the long-term sequence and replace the activation function in the attention network. Moreover, an optimized gate recurrent unit model with time interval vectors and a target attention mechanism block are progressively used to process the short sequence. With the proposed model, we can obtain separate long and short-term preferences, which are then adaptively combined to obtain a more accurate representation of user preferences and complete the recommendation task. The main contributions of this paper are summarized as follows:

- We utilize the sparse attention network to process long-term sequence. The long-term module adds a decay function to ensure the effect of previous states fades over time. In addition, a sparse function is introduced to reduce the computational weight of the irrelevant items in the sequence.
- We modify the gating structure of the GRU and add a time gate to learn short-term preferences changing. A target attention network layer is designed to

enhance the preferences' representation. Finally, an adaptive fusion attention module is introduced to process the long and short-term preferences, which improves the model's performance.

- Extensive experiments are conducted on three real-world datasets to demonstrate the effectiveness of our proposed model in modeling user behavior data. The comparison results indicate TPR-BS achieves significant performance on various evaluation metrics.

2   Related Work

2.1   Sequential Recommendation

Sequential Recommendation focuses on capturing sequential information in users' history interactions to recommend relevant items to them. In early work, the typical learning model is the MC [5]. For instance, FPMC [15] employs MCs to learn the sequences, but it fails to analyze dynamic feature changes and limits the learning capability of the model. With the rising of deep learning research, some methods develop their scalability and high-intensity computing and show great effectiveness in the sequential recommendation. Specifically, Caser [16] assumes that the user's behavioral sequence is the image and uses CNN for feature extraction. GRU4Rec [6] utilizes the gating structure of GRU to model user sequences and alleviate the gradient disappearance problem. In addition, the attention mechanism and transformer are also widely employed in sequence modeling tasks. For example, SASRec [7] employs a multi-layer self-attention block to achieve better recommendation results. BST [2] uses a transformer structure to uncover the user's sequence information and improve the overall recommendation effectiveness of the model.

2.2   Long and Short-Term Sequence Modeling for Recommendation

In recommendation scenarios, user preferences are difficult to track since the stable long-term preferences and dynamic short-term preferences. Both play significant roles in the accuracy of recommendation results. However, most models treat both as an entirety, assigning the same weights, which leads to undesirable recommendations. Therefore, it has been studied in depth by many scholars. For example, Zhao et al. [24] use an adversarial framework to train users' long-term and short-term sequences separately and obtain top-k recommendation results. Wu et al. [17] applied attention networks to the long-term sequence and Long Short-Term Memory (LSTM) to short-term sequences. Lv et al. [11] proposed to divide long and short-term sequences and employ multi-head self-attention to process short-term sequences. They extracted users' current preferences from multiple dimensions. Zheng et al. [25] extracted long and short-term preferences using two attention encoders separately, then devise a decoupling preference scheme using contrastive learning. Inspired by the existing work, our work captures long-term and short-term user preferences based on their behavioral characteristics. Then, we use adaptive fusion for dynamic preferences.

3   Problem Formulation

In the sequential recommendation, let $\mathcal{U} = \{u_1, u_2, \ldots, u_{|U|}\}$ denote the user set and $\mathcal{V} = \{v_1, v_2, \ldots, v_{|V|}\}$ denote the item set, where $|U|$ and $|V|$ are the number of users and items respectively. Based on historical data, we can obtain the user's behavior sequence, long and short-term sequence, which are defined as follows.

For each user $u \in \mathcal{U}$, $\mathcal{X}_u = \{X_u^1, X_u^2, \ldots, X_u^{L-1}, X_u^L\}$ denotes a sequence of user's behavior history. Each session $X_u^i = \{(v_u^1, t_u^1), (v_u^2, t_u^2), \ldots, (v_u^t, t_u^t)\}$ consists of items and timestamps, where (v_u^t, t_u^t) means user u interacted with the item v_u^t at time t_u^t. Given the raw sequence, we transform long-term input sequence into fixed-length m and long-term sequence turns into $\mathcal{L}_u = \{l_u^1, l_u^2, \ldots, l_u^m\}$. We define $L - m$ as n, $\mathcal{S}_u = \{s_u^1, s_u^2, \ldots, s_u^n\}$ denotes the short-term sequence, where n is the maximum length of the sequence.

Following the previous denotation, we define the fused preferences as $\mathcal{H}_u = \{h_u^1, h_u^2, \ldots, h_u^L\}$. The aim of sequential recommendation is to recommed top-K items for each user.

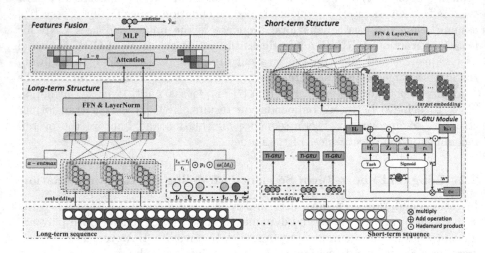

Fig. 2. The general architecture of the proposed TPR-BS.

4   Methodology

In this section, we will present a detailed introduction to the proposed model TPR-BS. It is shown in Fig. 2. The long-term structure and short-term structure model user behavior sequence separately to get user preferences. Then, the feature fusion structure is used to integrate the long and short-term preferences adaptively. Finally, we obtain the hybrid feature representation and predict the items for users.

4.1 Long-Term Preferences Model Structure

Position Embedding with Time intervals. In the long-term sequence, users' behaviors change over time and overlap in nature. However, fewer weights and memories should be assigned to apply to the next calculation with the longer time elapsed since the previous interaction. Hence, we utilize the d-dimensional relative temporal embedding to transform the temporal information in the sequence.

In practice, there might be situations with the same timestamp. Thus, we introduce the position embedding and apply it to simulate the temporal decay of the historical sequence. More specifically, giving the learnable position embedding $p_i \in \mathbb{R}^d$, time interval $\Delta t_l = t_i - t_j$ between item i and item j and decay function $\omega(\Delta t_l)$. We denote the difference between timestamps divided by the absolute value of the current interaction timestamp t_i as the time position value. As Matzner $et\ al.$ [13] suggested, we employ the effective decay function $\omega(\Delta t_l) = 1/log(e + \Delta t_l)$. The personalized time position PT_i is computed as follows:

$$PT_i = p_i \odot \left| \frac{t_n - t_i}{t_i} \right| \odot \frac{1}{log(e + (t_{i+1} - t_i))}, \tag{1}$$

where $\odot$ is an element-wise multiplication, and t_n represents the timestamp of the last interaction in the session. The temporal position embedding PT_i represents the position of the action in the user's behavior sequence.

Sparse Self-Attention Network. Since the attention network can assign weights to feedback data for dynamically capturing focused information. We design a self-attentive network and replace the activation function for synthetically learning the user's long-term sequence. In this way, we can obtain a potential vector representation of users' long-term preferences and minimize the influence of irrelevant information.

Motivated by the work [14], a sparse transform function α-entmax is introduced to replace softmax, where the former assigns a tendency to zero probability to data with no value in the sequence, such as error clicking items. The specific formulation of the activation function is defined as follows:

$$\alpha\text{-entmax}(\mathbf{z}) = Relu((\alpha - 1)\mathbf{z} - \tau\mathbf{1})^{\frac{1}{\alpha - 1}}, \tag{2}$$

where τ is the Lagrange multiplier, and $\mathbf{1}$ is the all-one vector. α is the parameter that controls the sparsity and smoothness of the α-entmax($\cdot$) function. When $\alpha = 1$, α-entmax is equivalent to the softmax function. When $1 < \alpha < 2$, α-entmax tends to generate sparse mappings and $\alpha = 2$ recovers sparsemax [12]. The higher value of α has the higher weight value that the attention network gives to historical data with high relevance scores.

We choose the general value $\alpha = 1.5$, which has a dedicated fast calculation and enhances the interpretation and certainty of the results [14]. For a given user u, we first formulate the attention score for the i-th interactive item in the sparse attention layer.

$$\alpha_{ij}^l = \alpha\text{-entmax}(\frac{Q_l K_l^\top}{\sqrt{d}}), \tag{3}$$

where $\sqrt{d}$ is used to scale the dot product.

Then we project the positional embedding into the sequence embedding matrix. The input matrix function of *Query*, *Key* and *Value* in the self-attention block is: $Q_l = e_{li}W_l^q + PT_i$, where e_{li} is the current input vector. The similar operation to get K_l and V_l. Through the above process, the output is defined as:

$$\ell_i = \sum_{j=1}^m \alpha_{ij}^l(e_{lj}W_l^v + PT_j). \tag{4}$$

where $W_l^q, W_l^k, W_l^v \in \mathbb{R}^{d\times d}$ is the linear projection matrix, which in the specific operation according to Q_l, K_l and V_l respectively.

Finally, we apply feed-forward neural networks to introduce non-linearity for the model. Also, we insert the normalization layer, dropout layer and residual connection into the model to alleviate the overfitting problem of the network.

$$FFN(\ell_i) = Relu(\ell_i W_{l1} + b_{l1})W_{l2} + b_{l2}, \tag{5}$$
$$F_{\mathcal{L}i} = LayerNorm(\ell_i + Dropout(FFN(\ell_i))), \tag{6}$$

where $W_{l1}, W_{l2} \in \mathbb{R}^{d\times d}$ are weight matrices and $b_{l1}, b_{l2} \in \mathbb{R}^d$ are bias vectors. Now we capture the user's long-term preferences, which will integrate short-term preferences features in the later stage to predict the users' comprehensive preferences.

4.2   Short-Term Preferences Model Structure

Time interval-aware GRU Module. The gate structures of GRU could learn remote relations and selectively forgets certain information. It has been shown to be effective in many temporal modeling tasks. Meanwhile, it also plays a key role in the recommendation scene. The structure of the GRU module is described in the following definition.

$$z_t = \sigma(g_t W_z^1 + h_{t-1} W_z^2 + b_z), \tag{7}$$
$$r_t = \sigma(g_t W_r^1 + h_{t-1} W_r^2 + b_r), \tag{8}$$
$$h_t' = tanh(g_t W_h^1 + (r_t \odot h_{t-1})W_h^2 + b_h), \tag{9}$$
$$h_t = z_t \odot h_t' + (1 - z_t) \odot h_{t-1}, \tag{10}$$

where g_t is the input to the GRU. Specifically, it is the embedding representation matrix of the user interaction behavior. W_* and b_* are parameters. $\sigma(\cdot)$ is the sigmoid activation function.

Users' short-term preferences frequently change over short periods. Embedded learning time intervals is helpful for mining user preferences from the sequential sequence. Thus, in this section, we allow the time interval to be used as a filter to capture potential preferences in the sequence and process changes in user preferences within a short period. To capture temporal correlation in a short-term sequence, we design a time gate d_t to upgrade the architecture [3]. It is identified by the current input e_{st}, the previous message h_{t-1}, and the interaction behavior time interval $\Delta t_s = t_t - t_{t-1}$, that is:

$$d_t = \sigma(e_{st}W_d^1 + h_{t-1}W_d^2 + \Delta t_s W_d^3 + b_d). \tag{11}$$

Based on the original equations, the time interval Δt_s is further integrated into the update gate z_t as:

$$Z_t = \sigma(e_{st}W_z^1 + h_{t-1}W_z^2 + \Delta t_s W_z^3 + b_z). \tag{12}$$

For the candidate state h'_t, d_t and r_t jointly control the extent of preserved historical information. For the current state h_t, d_t and Z_t control the update direction. We update both as:

$$H'_t = tanh(e_{st}W_h^1 + (r_t \odot h_{t-1} \odot d_t)W_h^2 + b_h), \tag{13}$$

$$H_t = Z_t \odot d_t \odot H'_t + (1 - Z_t) \odot h_{t-1}. \tag{14}$$

The update gate Z_t, reset gate r_t and time gate d_t control the updating of information in each hidden layer. The structure of the modified GRU can be seen in Fig. 2, and we name it Ti-GRU for short.

Ti-GRU encodes user behavior sequence into the hidden state of the corresponding instant. It retains the relevant information from the previous moment with time and transfers them to the next unit. After that, we formulate F_S as users' current preferences.

Current Target Attention Mechanism. Ti-GRU can be considered as a features extractor, in which each item embedding contains information about the time interval and each other. However, final interaction behavior tends to contain intense user preference expectations in the short-term sequence. Inspired by the work [23], we add an attention network in this layer to learn further short-term preferences' representation with the learnable target embedding and the sequence outputs from the last module.

Our attention mechanism reinforces the output of the Ti-GRU at the last session as the target embedding. Then we introduce biases b_{ij}^k, $b_{ij}^v \in \mathbb{R}^d$ to encode the dependence of other session embeddings and the target embedding. In this way, the model allows focusing more on considering the effect of recent interactions on users' short-term preferences.

The final attention weights α_{ij}^s in this mechanism as follows:

$$\alpha_{ij}^s = \frac{exp(m_{ij}^s)}{\sum_{k=1}^n exp(m_{ik}^s)}, \tag{15}$$

$$m_{ij}^s = \frac{(W_s^q f_i^s)(W_s^k f_j^s + b_{ij}^k)^\top}{\sqrt{d}}, \tag{16}$$

After obtaining the attention scores, we use them to adjust the update status of network nodes, which are defined below:

$$sa_i = \sum_{j=1}^n \alpha_{ij}^s (W_s^v f_j^s + b_{ij}^v), \tag{17}$$

where $W_s^q, W_s^k, W_s^v \in \mathbb{R}^{d \times d}$ are the input projection of *Query*, *Key*, *Value* respectively. b_{ij}^k and b_{ij}^v enable the attention to model global dependencies, it follows the traditional query-key-value operation when we set $b_{ij}^k, b_{ij}^v = 0$. As before (c.f., Sect. 4.1), we add the feed-forward network layer and layer normalization.

Under this mechanism, the attention weight is based on known information about the short-term and the recent interaction sequence. It can be used to enhance the dynamic modeling of user's current preferences, which helps in modeling user behavior.

4.3   Adaptive Fusion and Prediction

Different users show different reliance on long and short-term preferences in real-life situations. To combine them for integrated modeling, we design an attention-based fusion method in TPR-BS. Specifically, we assign different weights to the embedding of long-term preferences $F_\mathcal{L}$ and short-term preferences $F_\mathcal{S}$ to achieve adaptive fusion.

$$\eta = \sigma([F_\mathcal{L}; F_\mathcal{S}] W_{ls} + b_{ls}), \tag{18}$$

$$F_{ls} = \eta \cdot F_\mathcal{L} + (1 - \eta) \cdot F_\mathcal{S}, \tag{19}$$

where $[\cdot; \cdot]$ is concatenate operation. $W_{ls} \in \mathbb{R}^{d \times d}$ can attribute weights to $F_\mathcal{L}$ and $F_\mathcal{S}$ based on their potential features.

Besides, instead of fusing only two features like other models, we append the current target attention output $\mathcal{S}_A$ to the fusion vector F_{ls} of them with the Multi-Layer Perceptron (MLP) layer (i.e., $MLP[F_{ls}; \mathcal{S}_A]$). In this way, the enhanced short-term preferences can be applied as the auxiliary information for the integrated preferences.

Finally, we calculate the predicted scores $p_{i,t}$ of candidate items by dot product operations and select the top-K items with higher scores to recommend to the user.

$$p_{i,t} = \mathcal{H}_u^t \cdot \mathcal{M}_i^\top, \tag{20}$$

where $\mathcal{H}_u^t$ is the final hybrid representation of user preferences. $\mathcal{M}$ indicates an item embedding matrix. For the loss function, we apply the cross-entropy between the predicted matching $\hat{y}_{ui}$ and the real feedback y_{ui} as follows:

$$\mathcal{L}_{loss} = -\frac{1}{N} \sum_{(u,i)\in\mathcal{D}} y_{ui}log(\hat{y}_{ui}) + (1 - y_{ui})log(1 - \hat{y}_{ui}) + \lambda_* \left\| \Theta \right\|^2, \qquad (21)$$

where N is the number of training samples. $\mathcal{D}$ denotes sample sets respectively. $\hat{y}_{ui}$ represents the recommended probability of items' output by the softmax layer. Θ is the model's set of hyperparameters, and λ_* is the l2-loss weight.

5   Experiments

In this section, we first introduce our experimental settings, including experimental data, evaluation metrics, and baseline comparison methods. And then we conduct experiments to answer the following research questions (RQs):

- RQ1: How does TPR-BS perform compared with other state-of-the-art sequential recommendation models?
- RQ2: Can the time intervals be helpful in the TPR-BS?
- RQ3: How does each TPR-BS component contribute to the recommendation task's performance?
- RQ4: What effect can different hyper-parameter settings lead the performance of TPR-BS?

5.1   Datasets

We evaluate our model on three different sizes of public benchmark datasets: MovieLens-1M (ML-1M), Tmall and Steam. In the experiments, we delete sequence and items with less than 5 actions for each dataset. Then the sequence is divided into ordered sessions and items are sorted based on the timestamp of the user's action. Finally, for all datasets, 70% of users are randomly chosen as the training set, 20% as the test set, and 10% as the validation set. Table 1 shows the statistics of the datasets.

Table 1. Statistics of the datasets.

Dataset	Users	Items	Avg.action/user	Ratings	Sparsity
ML-1M	6040	3623	165.47	999416	0.9543
Tmall	52957	47095	34.99	1852968	0.9992
Steam	84130	13536	15.04	1265315	0.9989

5.2   Evaluation Metrics

The matching phase of the recommendation is the main focus. For each user, we adopt three top-K evaluation metrics to evaluate the performance of the recommendation model. These metrics are: Mean Reciprocal Rank (MRR@K), Hit Rate (HR@K) and Normalized Discounted Cumulative Gain (NDCG@K). In our experiments, we set K={5, 10, 20} while repeating each experiment 5 times and reporting the average results.

5.3   Baselines

In order to show the effectiveness of our proposed model, we compare various recommendation methods.

- POP: A simple and effective recommendation method, which ranks items and is often used as the baseline in experiments.
- FPMC [15]: A hybrid model, which combines matrix factorization and Markov chains for the recommendation.
- GRU4Rec [6]: An RNN-based approach, which emulates the user's session sequence based on the GRU layer.
- Caser [16]: A CNN recommendation system, which uses networks to model the user behavior sequence and treats the embedding matrices as pictures.
- SHAN [22]: A hierarchical attention sequential recommendation network, which learns users' long and short-term preferences separately.
- TiSASRec [8]: A self-attention model, which sets up the self-attention blocks and adds time intervals to learn user preferences.
- SURGE [1]: A GNN recommendation method, which constructs item-item graphs of interest from sequences by metric learning.
- TPR-BS: Our proposed model.

5.4   Parameters Setting Detail

For the TPR-BS model structure, we take *Adam* as the optimizer. The dimensionality of the potential vector is fixed at 64. Furthermore, we set the learning rate as 0.001 and the batch size as 256. The dropout rate is adjusted between {0.1, 0.2, 0.3, 0.5} according to the sparsity of the datasets. We also perform a grid search for other hyperparameters to achieve optimization.

5.5   Overall Comparison (RQ1)

Table 2 shows the recommended performance of all models on datasets. The best performance is marked in bold, and the second best is underlined.

In general, deep learning methods outperform traditional sequential models (i.e., POP and FPMC). GRU4Rec and Caser model the dependencies between users and sequence to provide more appropriate recommended items for users. Compared with traditional models, they contain more information from user behaviors and enrich the learning vectors of the model.

Table 2. Performance comparisons of all models on datasets.

Dataset	K	Metric	POP	FPMC	GRU4Rec	Caser	SHAN	TiSASRec	SURGE	TPR-BS	Improvement
ML-1M	5	MRR	0.0336	0.0728	0.0903	0.0973	0.1046	0.1287	0.1201	**0.1326**	3.03%
	10	HR	0.0789	0.1403	0.1542	0.1635	0.2112	0.2459	0.2358	**0.2537**	3.17%
		NDCG	0.0671	0.1085	0.1173	0.1415	0.1709	0.1815	0.1884	**0.1923**	2.07%
	20	HR	0.0843	0.1519	0.1679	0.2092	0.2760	0.2973	0.2839	**0.3104**	4.41%
		NDCG	0.0708	0.1311	0.1526	0.1894	0.2192	0.2201	0.2325	**0.2362**	1.59%
Tmall	5	MRR	0.0263	0.0751	0.0805	0.0864	0.1034	0.1073	0.1168	**0.1207**	3.34%
	10	HR	0.0695	0.1283	0.1458	0.1594	0.1824	0.2039	0.2256	**0.2364**	4.79%
		NDCG	0.0413	0.0735	0.1025	0.1172	0.1325	0.1587	0.1722	**0.1794**	4.18%
	20	HR	0.0556	0.1279	0.1526	0.2058	0.2464	0.2682	0.2758	**0.2915**	5.69%
		NDCG	0.0591	0.1075	0.1162	0.1393	0.1579	0.1813	0.1902	**0.2036**	7.05%
Steam	5	MRR	0.0105	0.0491	0.0526	0.0675	0.0874	0.1095	0.1112	**0.1160**	4.32%
	10	HR	0.0426	0.0773	0.1154	0.1399	0.1568	0.2014	0.2158	**0.2279**	5.61%
		NDCG	0.0221	0.0736	0.0974	0.1083	0.1349	0.1462	0.1673	**0.1732**	3.53%
	20	HR	0.0382	0.1024	0.1421	0.1835	0.2365	0.2659	0.2568	**0.2831**	6.47%
		NDCG	0.0479	0.1028	0.0965	0.1171	0.1475	0.1642	0.1801	**0.1894**	5.16%

SHAN considers the long and short-term semantic relationships in the sequence. TiASARec adds time intervals and models different time intervals as relationships between interactions in the sequence. SURGE dynamically captures the diverse preferences of users at different times. They perform better than GRU4Rec and Caser. That proves long and short-term behaviors are essential for recommendations. Meanwhile, it also illustrates the benefit of time intervals as auxiliary information to improve the performance of recommendations.

Our proposed TPR-BS has improved significantly over baselines on all datasets and the evaluation metrics. The reason is that TPR-BS fully captures the user's behavioral sequence and considers time intervals in the model. The experiments results prove the TPR-BS model's effectiveness in adapting to dynamic user preferences changes.

5.6 Ablation Analysis

Impact of Time Intervals (RQ2). Time intervals are the critical technique proposed in our model. In this section, we explore different methods of time intervals and conduct experiments. They are: (1) Ignoring the time decay in the long-term module. (2) Removing the time gate of Ti-GRU. (3) Using the standard timestamps. (4) Using personalized time intervals (i.e., TPR-BS).

Table 3 shows that the last method achieves the best performance regardless of the datasets. Compared with other methods, we observe that it is improved mainly because our proposed personalized time intervals consider both the long-term decay and short-term variations in practical problems. It includes personalized position embedding with time intervals and an improved GRU. This indicates time intervals are valuable and meaningful for sequential recommendation tasks.

Table 3. Comparison of timestamp processing methods (NDCG@10).

Method	ML-1M	Tmall	Steam
w/o Time Decay	0.1793	0.1664	0.1649
w/o Time Gate	0.1855	0.1502	0.1486
w/ Timestamps	0.1674	0.1523	0.1391
w/ Time Intervals	**0.1923**	**0.1794**	**0.1732**
Improvement	3.67%	7.81%	5.03%

Key Components Influence (RQ3). In TPR-BS, we further conduct several ablation experiments to verify the effectiveness of different designed components. The model's components are split and combined into new models as follows.

- **TPR-BS_NS**: removes the long-term module and retains the short-term preferences network.
- **TPR-BS_NL**: removes the short-term module and only uses the long-term network to process user sequence.
- **TPR-BS_NA**: ignores short-term target attention networks and no longer emphasizes the vital effect of short-term user preferences.
- **TPR-BS_NH**: processes the user sequence using the long and short-term networks separately, leaving out the final fusion operation.

Table 4. Comparison of different ablation settings.

Method	Tmall			Steam		
	MRR@5	HR@20	NDCG@20	MRR@5	HR@20	NDCG@20
TPR-BS	**0.1207**	**0.2915**	**0.2036**	**0.1160**	**0.2831**	**0.1894**
_NS	0.1065	0.2564	0.1877	0.0913	0.2713	0.1703
_NL	0.1168	0.2736	0.1658	0.1081	0.2539	0.1786
_NA	0.0927	0.2390	0.1601	0.0823	0.2374	0.1568
_NH	0.0835	0.2287	0.1563	0.0794	0.2265	0.1456
Improvement	3.34%	6.54%	8.47%	7.31%	4.35%	6.05%

Table 4 shows the results of different experiment settings on Tmall and Steam. We can observe that TPR-BS_NL has mostly better-assessed values than TPR-BS_NS, which demonstrates the effectiveness of modeling users' short-term preferences. Similarly, TPR-BS_NA does not perform as well as TPR-BS_NL. The reason is our target layer emphasizes the weight of the user's current interaction with the attention network. Fusion structure helps the model capture and adaptively learn user behavior sequence. However, TPR-BS_NH deletes this and leading to its poor performance. Finally, TPR-BS achieves the best performance on both datasets. In a word, the experimental results show that both long-term and short-term user preferences positively affect the recommendation tasks. The fusion of the two is better than considering only long-term or short-term features.

5.7   Hyper-parameter Analysis (RQ4)

Influence of Latent Dimensionality. Fig. 3 shows how changing the latent dimensionality affects the model's performance on the datasets. We observe that the performance of TPR-BS is consistently better than two highest baselines (i.e., TiSASRec and SURGE). Furthermore, excessive dimensionality weakens performance, which may be caused by overfitting. The experimental results show that TPR-BS performs best when $d = 64$.

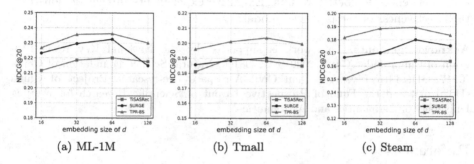

Fig. 3. Comparison model performance with different d settings (NDCG@20).

Influence of Sequence Length. We set multiple random combinations of $\{m, n\}$ for three datasets and select the best effect point in each interval to draw the fold line. As shown in Fig. 4, NDCG@20 keeps increasing as m increases in most cases. The reason is that the information on users' long-term preferences enriches. Similarly, the value of n influences the learning outcome of users' short-term preferences. As the data sparsity varies, the best combination obtained varies. In comparative experiments, we select the appropriate combination values of $\{m, n\}$ to achieve better performance.

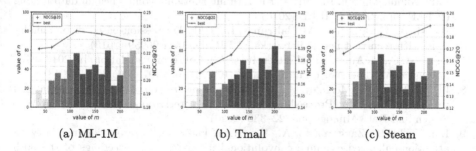

Fig. 4. Comparison performance of TPR-BS with different $\{m, n\}$ (NDCG@20).

6   Conclusion

In this work, we investigate the user preferences problem in sequential recommendation tasks and design an effective model named TPR-BS. We use time interval information as a filter to construct long and short-term preferences learning mechanisms separately. The model embeds the user's sequential behavior into the latent preferences' space and incorporates preferences' features adaptively. In this way, TPR-BS provides personalized recommendations based on the target user preferences for better results. Extensive experiments further demonstrate the effectiveness of the proposed model.

Acknowledgements. This work is supported by the National Key R&D Program of China under Grant 2020YFB1708100, Natural Science Foundation of China (62172351), 14th Five-Year Plan Civil Aerospace Pre-research Project of China (D020101), and the Fund of Prospective Layout of Scientific Research for Nanjing University of Aeronautics and Astronautics.

References

1. Chang, J., et al.: Sequential recommendation with graph neural networks. In: Proceedings of the 44th International ACM SIGIR Conference on Research and Development in Information Retrieval, pp. 378–387 (2021)
2. Chen, Q., Zhao, H., Li, W., Huang, P., Ou, W.: Behavior sequence transformer for e-commerce recommendation in Alibaba. In: Proceedings of the 1st International Workshop on Deep Learning Practice for High-Dimensional Sparse Data, pp. 1–4 (2019)
3. Chen, X., Zhang, Y., Qin, Z.: Dynamic explainable recommendation based on neural attentive models. In: Proceedings of the AAAI Conference on Artificial Intelligence. vol. 33, pp. 53–60 (2019)
4. Dai, S., Yu, Y., Fan, H., Dong, J.: Spatio-temporal representation learning with social tie for personalized poi recommendation. Data Sci. Eng. **7**(1), 44–56 (2022)
5. He, R., McAuley, J.: Fusing similarity models with Markov chains for sparse sequential recommendation. In: 2016 IEEE 16th international conference on data mining (ICDM), pp. 191–200. IEEE (2016)
6. Hidasi, B., Karatzoglou, A., Baltrunas, L., Tikk, D.: Session-based recommendations with recurrent neural networks. CoRR abs/1511.06939 (2015)
7. Kang, W.C., McAuley, J.: Self-attentive sequential recommendation. In: 2018 IEEE international conference on data mining (ICDM), pp. 197–206. IEEE (2018)
8. Li, J., Wang, Y., McAuley, J.: Time interval aware self-attention for sequential recommendation. In: Proceedings of the 13th International Conference on Web Search and Data Mining, pp. 322–330 (2020)
9. Liu, Y., Li, B., Zang, Y., Li, A., Yin, H.: A knowledge-aware recommender with attention-enhanced dynamic convolutional network. In: Proceedings of the 30th ACM International Conference on Information & Knowledge Management, pp. 1079–1088 (2021)
10. Liu, Y., Xuan, H., Li, B., Wang, M., Chen, T., Yin, H.: Self-supervised dynamic hypergraph recommendation based on hyper-relational knowledge graph. arXiv preprint arXiv:2308.07752 (2023)

11. Lv, F., et al.: SDM: sequential deep matching model for online large-scale recommender system. In: Proceedings of the 28th ACM International Conference on Information and Knowledge Management, pp. 2635–2643 (2019)
12. Martins, A., Astudillo, R.: From softmax to sparsemax: a sparse model of attention and multi-label classification. In: International Conference on Machine Learning, pp. 1614–1623. PMLR (2016)
13. Nguyen, A., Chatterjee, S., Weinzierl, S., Schwinn, L., Matzner, M., Eskofier, B.: Time matters: time-aware LSTMs for predictive business process monitoring. In: Leemans, S., Leopold, H. (eds.) ICPM 2020. LNBIP, vol. 406, pp. 112–123. Springer, Cham (2021). https://doi.org/10.1007/978-3-030-72693-5_9
14. Peters, B., Niculae, V., Martins, A.F.: Sparse sequence-to-sequence models. In: Proceedings of the 57th Annual Meeting of the Association for Computational Linguistics, pp. 1504–1519 (2019)
15. Rendle, S., Freudenthaler, C., Schmidt-Thieme, L.: Factorizing personalized markov chains for next-basket recommendation. In: Proceedings of the 19th International Conference on World Wide Web, pp. 811–820 (2010)
16. Tang, J., Wang, K.: Personalized top-n sequential recommendation via convolutional sequence embedding. In: Proceedings of the Eleventh ACM International Conference on Web Search and Data Mining, pp. 565–573 (2018)
17. Wu, Y., Li, K., Zhao, G., Qian, X.: Personalized long-and short-term preference learning for next poi recommendation. IEEE Trans. Knowl. Data Eng. **34**(4), 1944–1957 (2020)
18. Xuan, H., Li, B.: Temporal-aware multi-behavior contrastive recommendation. In: Wang, X., et al. (eds.) DASFAA 2023, Part II. LNCS, vol. 13944, pp. 269–285. Springer, Cham (2023). https://doi.org/10.1007/978-3-031-30672-3_18
19. Xuan, H., Liu, Y., Li, B., Yin, H.: Knowledge enhancement for contrastive multi-behavior recommendation. In: Proceedings of the Sixteenth ACM International Conference on Web Search and Data Mining, pp. 195–203 (2023)
20. Yang, Y., Ye, Z., Zhao, H., Meng, L.: A novel link prediction framework based on gravitational field. Data Sci. Eng. **8**(1), 47–60 (2023)
21. Yin, H., Yang, S., Song, X., Liu, W., Li, J.: Deep fusion of multimodal features for social media retweet time prediction. World Wide Web **24**, 1027–1044 (2021)
22. Ying, H., et al.: Sequential recommender system based on hierarchical attention network. In: IJCAI International Joint Conference on Artificial Intelligence, pp. 3926–3932 (2018)
23. Yuan, J., Song, Z., Sun, M., Wang, X., Zhao, W.X.: Dual sparse attention network for session-based recommendation. In: Proceedings of the AAAI Conference on Artificial Intelligence, vol. 35, pp. 4635–4643 (2021)
24. Zhao, W., Wang, B., Ye, J., Gao, Y., Yang, M., Chen, X.: Plastic: prioritize long and short-term information in top-n recommendation using adversarial training. In: IJCAI, pp. 3676–3682 (2018)
25. Zheng, Y., et al.: Disentangling long and short-term interests for recommendation. In: Proceedings of the ACM Web Conference 2022, pp. 2256–2267 (2022)

Efficient Multi-object Detection for Complexity Spatio-Temporal Scenes

Kai Wang[1], Xiangyu Song[2]([✉]), Shijie Sun[1], Juan Zhao[3], Cai Xu[4],
and Huansheng Song[1]

[1] School of Information Engineering, Chang'an University, Xi'an, China
{kwang,shijieSun,hshsong}@chd.edu.cn
[2] School of Software and Electrical Engineering, Swinburne University of Technology,
Melbourne, Australia
x.song@deakin.edu.au
[3] New Network Research Division, Peng Cheng Laboratory, Shenzhen, China
zhaoj09@pcl.ac.cn
[4] School of Computer Science and Technology, Xidian University, Xi'an, China
cxu@xidian.edu.cn

Abstract. Multi-Object detection in traffic scenarios plays a crucial role in ensuring the safety of people and property, as well as facilitating the smooth flow of traffic on roads. However, the existing algorithms are inefficient in detecting real scenarios due to the following drawbacks: (1) a scarcity of traffic scene datasets; (2) a lack of tailoring for specific scenarios; and (3) high computational complexity, which hinders practical use. In this paper, we propose a solution to eliminate these drawbacks. Specifically, we introduce a Full-Scene Traffic Dataset (FSTD) with Spatio-temporal features that includes multiple views, multiple scenes, and multiple objectives. Additionally, we propose the improved YOLOv7 model with redesigned **B**iFusion, **N**WD and SPP**F**CSPC modules (**BNF-YOLOv7**), which is a lightweight and efficient approach that addresses the intricacies of multi-object detection in traffic scenarios. BNF-YOLOv7 is achieved through several improvements over YOLOv7, including the use of the BiFusion feature fusion module, the NWD approach, and the redesign of the loss function. First, we improve the SPPCSPC structure to obtain SPPFCSPC, which maintains the same receptive field while achieving speedup. Second, we use the BiFusion feature fusion module to enhance feature representation capability and improve positional information of objects. Additionally, we introduce NWD and redesign the loss function to address the detection of tiny objects in traffic scenarios. Experiments on the FSTD and UA-DETRAC dataset show that BNF-YOLOv7 outperforms other algorithms with a 3.3% increase in mAP on FSTD and a 2.4% increase on UA-DETRAC. Additionally, BNF-YOLOv7 maintains significantly better real-time performance, increasing the FPS by 10% in real scenarios.

Keywords: Computer vision · Object detection · Intelligent transportation · Spatio-temporal database · YOLOv7

X. Song et al. (Eds.): APWeb-WAIM 2023, LNCS 14334, pp. 186–200, 2024.
https://doi.org/10.1007/978-981-97-2421-5_13

1   Introduction

Multi-Object detection is a critical task across various industrial scenarios. For example, in the Internet of Things (IoT) [34], multi-object detection is important for intelligent security [39] and intelligent transportation [21], while in social media [36], it is used for face detection [20]. In recent years, with the widespread availability of traffic data analytic, multi-object detection in traffic scenarios has become increasingly significant. It has a profound impact on the safety of individuals and property and plays a crucial role in ensuring the smooth flow of traffic on roads. Therefore, developing effective multi-object detection systems for traffic big data scenarios is of utmost importance.

Multi-Object detection of traffic data analytic scenarios is a challenging task due to occlusion and scene complexity [30]. Currently, the methods based on Convolutional Neural Network (CNN) are predominantly used to tackle this problem [32]. CNN-based multi-object detection tasks have witnessed significant progress in recent years [26]. Some notable benchmark datasets like MS COCO [17] and UA-DETRAC [33] greatly promote the development of multi-object detection application. One-stage object detection algorithm prevails in practical applications due to its optimal balance between speed and accuracy. SSD [19] is the pioneer of the one-stage object detection algorithm, YOLO [22–24] series provides new ideas for one-stage detection algorithm, YOLOv4 [1] designed the detection algorithm with three parts: backbone, neck and detection head. YOLOv5 [12], YOLOv6 [15] and YOLOv7 [28] introduced more bag-of-freebies and bag-of-specials to enhance the detection effect.

However, Most detectors aim for common object detection without specific scenario improvements, and existing traffic scene datasets are relatively simple. As a result, there are three main challenges in directly applying current algorithm models to solve multi-object detection in traffic scenarios. These challenges are intuitively illustrated in Fig. 1. Firstly, variations in object size can occur due to the distance of the traffic object from the camera. Secondly, images captured by roadside cameras in traffic scenarios often include many confounding factors due to their large coverage area. These factors may cause adverse effects, such as the occlusion of traffic objects, and reduce detection performance. Thirdly, different traffic scenarios exhibit considerable variation and lack diversity in realistic traffic scene datasets.

In this paper, we present an enhanced algorithm BNF-YOLOv7 based on YOLOv7 [28] and propose a Full-Scene Traffic Dataset (FSTD) with Spatio-temporal features for the above three challenges. The FSTD contains 23,887 images, 243,814 annotations, and 4 categories common to traffic scenes. Unlike UA-DETRAC [33], our proposed dataset exhibits greater balance and encompasses additional non-motorized categories. BNF-YOLOv7 is accomplished by the following improvments over YOLOv7. Firstly, We enhance the SPPCSPC structure to improve the efficiency of the algorithm. Moreover, we designed the BiFusion module to replace the upsample module in the neck of BNF-YOLOv7 to achieve the reduction of parameters and more precise location information. Finally, we use the combination of NWD [31] and IoU [38] for the algorithm loss

(a)Size variation   (b) Vehicle obscuration   (c) Scenario variation

Fig. 1. The challenges of multi-object detection in traffic scenarios. The cases respectively shows the object size variation, vehicle obscuration and scenario variation of objects on roadside cameras-captured images.

function computation to improve the detection precision of tiny objects. Compared with YOLOv7, our enhanced algorithm can better perform traffic object detection.

The main contributions of this paper are listed as follows:

- We propose a lightweight and efficient model for multi-object detection in traffic scenarios, BNF-YOLOv7. The model first improves the SPPCSPC module and obtains a speed increase while keeping the original receptive field unchanged. In addition, the BiFusion module is designed to enhance the location information. Besides, NWD is introduced to improve the detection effect of tiny objects to a new level.
- We propose a Full-Scene Traffic Dataset (FSTD) with Spatio-temporal features that contains 23,887 images of different highway scenes, 243,814 annotations of traffic objects, and 4 categories common to traffic scenes, including vehicle and person.
- Experiments on both specific and commonly benchmark datasets validate the effectiveness of the proposed models. BNF-YOLOv7 achieved a performance gain of about 3.3% mAP@0.5 compared to the state-of-the-art method on FSTD. In addition, the gain of mAP become 2.4% on UA-DETRAC.

2   Related Work

We review the related literature along Multi-Object Detection with YOLO Series.

2.1   Multi-object Detection

Multi-object detection has been a prominent research area in intelligent transportation problems. In the initial stage of research, researchers often employ multiple sensor fusion such as LIDAR to detect traffic objects on the road. Liang et al. [16] proposes an end-to-end learnable architecture that reasons about 2D and 3D object detection by fusing information at various levels. YOdar [14] proposes an uncertainty-based approach for sensor fusion with camera and radar data, combining the results of the camera data and the results of the radar data in an uncertainty-aware manner and post-processing the two outputs by a gradient boosting method to generate a joint prediction of the two networks. However, the multi-object detection algorithm using multi-sensor fusion suffers from the drawbacks of sluggish detection speed, low detection precision and high cost, among others. Current multi-object detection algorithms are mainly enhanced by inheriting the merits of generic object detection algorithms, such as R-CNN [8], Fast R-CNN [7], Faster R-CNN [25], SSD [19], etc. Cascade R-CNN [2]consists of a series of detectors, each of which further improves on the previous detector to significantly improve detection performance at low signal-to-noise ratios, large objects, and high occlusions. CenterNet [4] treats each object as a triplet and designs cascading corner sets and center sets which enrich the information collected in the upper left and lower right corners and provide more identifiable information from the center region, which improves precision and recall.

The vast majority of existing datasets of traffic scene are captured from the perspective of roadside cameras or autonomous driving. KITTI [6] dataset contains multi-sensor data for road driving, which can be used for autonomous driving, object detection, road segmentation, etc. Cityscapes [3] is a data set for autonomous driving research in urban scenes, which contains a large number of high-resolution images and semantic segmentation annotations. BDD100K [37] encompasses scenarios of diverse times, weathers, and locations, and is conducive for examining issues such as the stability and robustness of autonomous driving. UA-DETRAC [33] is a data set for vehicle detection, which comprises vehicle objects in various scenarios, and can be employed to study the detection and tracking of vehicle objects. The dataset also furnishes vehicle labeling information, including bounding boxes, categories, and ID of the object.

2.2   YOLO Series

The one-stage algorithm directly generates the class probability and position coordinates of the object, so it has a faster detection speed. The pioneer of the one-stage algorithms, YOLOv1 [22], draws inspiration from the GoogleNet [27] for image classification, which employs a cascaded module of a smaller convolutional network. YOLOv2 [23] pioneered the joint use of classification and detection training methods, which significantly improved the prediction accuracy while maintaining the advantage of fast inference. YOLOv3 [24] utilizes the residual structure of ResNet [10] to deepen the network structure while avoiding the problem of challenging network convergence caused by network gradient explosion. YOLOv4 [1] introduced more bag-of-freebies and bag-of-specials to enhance the detection effect. It also makes a significant contribution to the real-time performance of multi-object detection, enabling better models to be obtained by single GPU as well. YOLOv5 [12] constructs models of varying sizes according to the scale reduction of different channels. YOLOv6 [15] supports the deployment of various platforms simultaneously, greatly simplifying the adaptation work during project deployment. YOLOv7 [28] optimizes the model

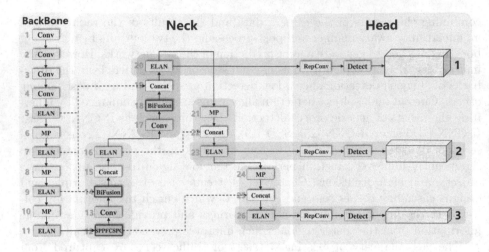

Fig. 2. The architecture of the BNF-YOLOv7. Compared to original version, we mainly improve the neck by applying SPPCSPC. We also add BiFusion to further improve the feature representation capability. In addition, we redesign loss function to make BNF-YOLOv7 stronger.

structure and reparameterizes the model to further improve detection accuracy and speed.

3 Methodology

3.1 Overview of YOLOv7

YOLOv7 [28] is a one-stage object detection algorithm proposed by Wang et al. based on the YOLO [1,12] series. The algorithm is further enhanced on the basis of YOLOv5 [12], and its detection speed and accuracy in the range of 5 FPS to 160 FPS surpass those of currently known multi-object detection algorithms.

YOLOv7 network structure adheres to the previous YOLO series algorithm design principles. The network comprises four components: Input, BackBone, Neck and Head. The algorithm first resizes the input image to a uniform size. Then the input image is fed into the backbone network. The backbone network initially passes through four layers of Convolution-BatchNormalization-SiLU(CBS) module, which consists mainly of convolutional layer, BatchNormalization [11] layer and SiLU [5] activation function and can extract features of varying sizes. ELAN [28] enhances the deeper network's effectiveness by regulating the shortest and longest gradient path. Finally, after three MP [28] and ELAN combinations, features of different sizes are output to the Neck layer. The Neck layer adopts PAFPN [18] structure, and for the final output of BackBone, the 32 times C5 sampling feature map, after SPPCSPC reduces the number of channels and fuses with the output C4 and C3 in BackBone to obtain P3, P4, and P5. The RepConv [28] structure in the Head layer adjusts the number of channels and uses 1×1 convolution to predict the object.

3.2 BNF-YOLOv7

The framework of BNF-YOLOv7 is illustrated in Fig. 2. We modify the original YOLOv7 to make it more suitable for traffic object detection. Next we will concentrate on three enhanced modules: (1) SPPFCSPC. (2) BiFusion. (3) Normalized Gaussian Wasserstein Distance (NWD).

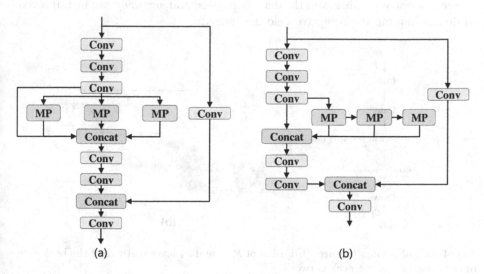

Fig. 3. SPP structure in the neck network. (a) SPPCSPC structure in the original YOLOv7.(b) Improved SPPFCSPC structure.

SPPFCSPC. The SPPCSPC [28] structure in YOLOv7 is a module obtained by applying the SPP [9] and the CSP [29] structure, as illustrated in Fig. 3. SPP can increase the receptive field, which enables the algorithm to adapt to different resolution images and can effectively avoid the problems of image distortion caused by cropping and scaling operations on image regions, and it can solve the problem of redundant feature extraction related to the graph by CNN, which significantly improves the speed of generating candidate frames and saves computational costs. The CSP structure first splits the features into two parts, one part is processed conventionally, the other part is processed by the SPP structure, and finally the results of the two parts are merged together, which can reduce the amount of computation, accelerate the processing speed and enhance the detection accuracy.

In the SPPCSPC structure, there are four branches passing through the MP structure, four different scales of MP have four receptive fields, which can be used to distinguish between large and small objects. In the traffic scene, different traffic objects are at varying distances from the camera and their scales are different. However, the structure obtains fewer types of features and it results in the decline of detection precision.

We improve the SPPCSPC to obtain the SPPFCSPC structure, as depicted in Fig. 3. Adding connections between multiple MP structures, so that the features obtained after max pooling can be partly concat operation, while the other part can continue pooling, and the features at different levels can be acquired by different times of pooling, and the detection speed and accuracy can be improved while maintaining the receptive field unchanged.

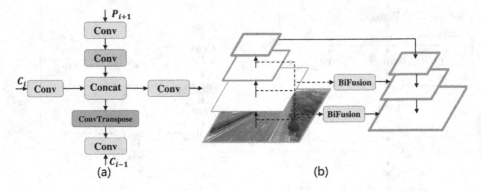

(a)   (b)

Fig. 4. (a) BiFusion structure. (b)Fusion of P_{i-1} and P_i layer features of the backbone network into P_i of the neck network.

Feature Fusion Module. We designed the BiFusion module for feature fusion. We incorporates BiFusion neck fusion module to augment the location information without excessively increasing the computational effort. The feature maps of both backbone and neck are integrated, C_i and the C_{i-1} underlying features in the backbone network are merged into the P_i of the neck network. This can acquire more precise location information and play a vital role in object localization.

BiFusion neck fusion involves the following steps: Firstly, the same scale feature map is downsampled using 1×1 convolution. Secondly, the large scale feature map is firstly downsampled using 1×1 convolution and then further downsampled using 3×3 convolution with a stride size of 2. Then the small scale feature map is upsampled using 2×2 transposed convolution, and Finally the feature map concat obtained from these three parts is stitched together and downsampled again using 1×1 convolution, as illustrated in Fig. 4.

NWD for Tiny Objects. The loss function is a criterion of the performance of a neural network, and selecting the appropriate loss function can yield better results when training the model, and the training can converge faster. YOLOv7 employs CIoU [40] to measure the discrepancy between the detection bounding box predicted by the model and the ground-truth. However, since the traditional IoU Loss employs IoU to measure the similarity between two bounding boxes, the displacement of the bounding box causes a significant error in the IoU of the two boxes when the object to be detected is small, as illustrated in Fig. 5.

(a)   (b)

Fig. 5. The orange box in the figure is the ground-truth, and the blue and green boxes in the a and b figures represent the object detection results. The IoU between the detection result and the ground-truth in figure a is 0.53, and the IoU between the detection result and the ground-truth in figure b is 0.14. It can be seen that for tiny objects, the position offset of the detection bounding box has a greater impact on the IoU. (Color figure online)

In this paper, we introduce NWD [31] to measure the similarity of two boxes based on the original loss function. We use a two-dimensional Gaussian distribution to model arbitrarily oriented bounding boxes for multi-object detection and approximate the IoU-induced loss between the two bounding boxes by calculating the NWD of the two detection bounding boxes.

For the Gaussian distributions $\mathcal{N}_a$ and $\mathcal{N}_b$ modeled according to the bounding boxes $A = (cx_a, cy_a, w_a, h_a)$ and $B = (cx_b, cy_b, w_b, h_b)$, The equation of GWD [35] can be represented as

$$W_2^2(\mathcal{N}_a, \mathcal{N}_b) = \left\| \left(\left[cx_a, cy_a, \frac{w_a}{2}, \frac{h_a}{2} \right]^T, \left[cx_b, cy_b, \frac{w_b}{2}, \frac{h_b}{2} \right] \right) \right\|_2^2 \quad (1)$$

where (cx, cy) denotes the center point of the bounding box, w and h denote the width and height of the bounding box, respectively. The GWD is normalized to obtain the NWD, which can be defined as

$$NWD(\mathcal{N}_a, \mathcal{N}_b) = exp\left(-\frac{\sqrt{W_2^2(\mathcal{N}_a, \mathcal{N}_b)}}{C}\right) \qquad (2)$$

Normalizing NWD yields $Loss_{NWD}$, which can be defined as

$$Loss_{NWD} = 1 - NWD(\mathcal{N}_a, \mathcal{N}_b) \qquad (3)$$

Based on NWD and CIoU, we design a compound loss function. Therefore, the optimized regression loss can be expressed as shown in Eq. 4.

$$Loss_{box} = \alpha Loss_{NWD} + \beta Loss_{CIoU} \qquad (4)$$

Datasets. Since existing publicly traffic datasets such as UA-DETRAC [33] are all urban roads and have a limited diversity of scenarios. We collected and constructed FSTD by employing highway roadside cameras to capture images. An exemplar dataset can be observed in Fig. 6.

(a) sunny scenario   (b) cloudy scenario   (c) low view scenario   (d) night scenario

Fig. 6. Examples of different scenarios in FSTD.

Highway video records the road conditions, traffic flow over a period of time. It can also record the vehicles in different lanes, the distance between the vehicles and the road, and the speed of the vehicles to provide information about the vehicle density, speed and lane occupancy at various locations. Thus highway video contains rich spatio-temporal information. FSTD is a snapshot with temporal characteristics drawn from the highway video. Based on the collected images, we propose four common categories in traffic scenarios. They are car, truck, bus and person. FSTD comprises 23,887 images of diverse traffic scenarios, camera angles, and weather conditions. We annotate the FSTD following the UA-DETRAC object detection data format, and randomly sample 20% of FSTD as the test set and the rest as the training set. We show the statistics of the FSTD in Fig. 7(a).

UA-DETRAC is a widely used benchmark for object detection of traffic scenarios, but since it contains only 100 traffic scenes, the scene variety is limited, so most of the images are redundant data. We sampled one image every five frames from the UA-DETRAC training and testing sets, resulting in 14,242 training images and 10,722 testing images. The quantity of traffic objects of various categories in the UA-DETRAC is displayed in Fig. 7(b).

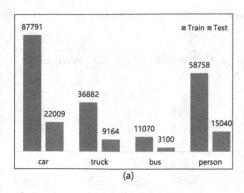 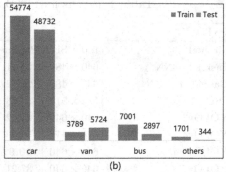

Fig. 7. The number of labels of each category. (a) FSTD. (b) UA-DETRAC.

4   Experiments

We evaluate our model using a portion of the UA-DETRAC [33] dataset in conjunction with the FSTD.

4.1   Implementation Details and Evaluation Metrics

We implemented BNF-YOLOv7 on Pytorch 1.11. All of our models were trained and evaluated using NVIDIA Tesla V100 GPU. In the training phase, we combined both FSTD and UA-DETRAC training sets for 500 epochs. We employed the Adam [13] optimizer for training and set 0.01 as the initial learning rate, which was updated using the OneCycleLR strategy. The size of the input image of our model are 640×640 and the batch size is 32. Several experiments show that the algorithm is optimal when the α is 0.2 and β is 0.8 in $Loss_{box}$. In addition,This paper employs Precision, Recall, mAP@0.5, FPS and parameters as evaluation metrics.

4.2   Comparisons with the State-of-the-Art

We have conducted extensive comparative experiments on FSTD and UA-DETRAC. The experimental results of FSTD and UA-DETRAC are shown in Table 1. From the results in the table, we draw the following conclusions.

BNF-YOLOv7 achieves the highest mAP@0.5 on both FSTD and UA-DETRAC. On FSTD, the mAP@0.5 of BNF-YOLOv7 is 0.83. Likewise, BNF-YOLOv7 attains the highest mAP@0.5 on UA-DETRAC, reaching 0.7. Both outperform the original YOLOv7 algorithm. In terms of FPS, BNF-YOLOv7 still has the advantage. This result fully illustrates the superiority of our network structure.

Table 1. Results for FSTD and UA-DETRAC datasets.

FSTD

Method	Input Size	Parameters(M)	Precison	Recall	mAP@0.5	FPS
Faster R-CNN	640 × 480	125.41	0.78	0.695	0.76	10.3
Cascade R-CNN	640 × 480	126.72	0.8	0.69	0.77	11.2
SSD	512 × 512	<u>26.53</u>	0.828	0.69	0.776	85.8
YOLOv3	640 × 640	61.62	0.812	0.72	0.79	70.2
YOLOv4	416 × 416	52.66	0.823	0.723	0.793	92.5
YOLOv5m	640 × 640	**20.9**	<u>0.832</u>	0.712	0.793	94.5
YOLOv7	640 × 640	37.21	0.811	<u>0.728</u>	<u>0.797</u>	<u>120.4</u>
BNF-YOLOv7(Ours)	640 × 640	37.11	**0.839**	**0.755**	**0.83**	**132.5**

UA-DETRAC

Method	Input Size	Parameters(M)	Precison	Recall	mAP@0.5	FPS
Faster R-CNN	640 × 640	125.43	0.66	0.61	0.642	10.5
Cascade R-CNN	640 × 640	126.7	0.638	0.623	0.644	11.4
SSD	512 × 512	<u>26.56</u>	0.695	0.616	0.652	85.4
YOLOv3	640 × 640	61.52	0.71	0.6	0.6	70.3
YOLOv4	416 × 416	52.64	0.659	0.629	0.65	92.3
YOLOv5m	640 × 640	**20.82**	<u>0.712</u>	0.586	0.645	94.7
YOLOv7	640 × 640	37.23	0.711	<u>0.639</u>	<u>0.676</u>	<u>120.3</u>
BNF-YOLOv7(Ours)	640 × 640	37.12	**0.742**	**0.658**	**0.7**	**132.6**

Compared with the two-stage methods, our model has significant advantages in various aspects such as parameters, mAP@0.5, and FPS. In contrast to the one-stage methods, all methods lag far behind us except YOLOv5m and YOLOv7 algorithms. YOLOv7 is close to our method in mAP@0.5. YOLOv5m exceeds us in parameters, but has a large gap in mAP@0.5. Our method achieves a better balance between parameters and detection precision.

Table 2. Comparisons of the performance of the BNF-YOLOv7 model for each category on the FSTD and UA-DETRAC testsets.

Classes	FSTD					UA-DETRAC				
	all	car	truck	bus	person	all	car	van	bus	others
Precision	0.825	**0.885**	<u>0.871</u>	0.74	0.805	0.679	**0.73**	0.606	<u>0.711</u>	0.67
Recall	0.764	<u>0.894</u>	**0.896**	0.674	0.591	0.736	0.806	<u>0.863</u>	**0.899**	0.375
mAP@0.5	0.83	<u>0.933</u>	**0.935**	0.736	0.717	0.7	<u>0.819</u>	0:655	**0.863**	0.464

We list the Precision, Recall, mAP@0.5 of our final results for each category on FSTD and UA-DETRAC in Table 2. It can be seen that BNF-YOLOv7 achieves promising results in most categories on both FSTD and UA-DETRAC. Unfortunately, there are still some hard categories, such as person on FSTD and others on UA-DETRAC, in which BNF-YOLOv7 performs less than ideal. It may be caused by the imbalance in the number of samples between different categories. Therefore, these problems are still the focus of future research.

We have selected some representative images as a demonstration of the test results. Figure 8 shows the results under different scenarios of FSTD and UA-DETRAC. columns 1–3 are from FSTD, and columns 4–5 are from UA-DETRAC.

(a)FSTD   (b)UA-DETRAC

Fig. 8. Some Detection results from our BNF-YOLOv7 on FSTD and UA-DETRAC, different category use bounding boxes with different color.

It can be seen that BNF-YOLOv7 can effectively recognize different sizes of objects in different traffic scenarios. The excellent performance on FSTD and UA-DETRAC demonstrates the generalization ability of our proposed model, which can be applied to various traffic scenes.

4.3   Ablation Experiments

In this section, we ablated the different components of the proposed BNF-YOLOv7 using FSTD. The original YOLOv7 parameters are used as the benchmark, '↓' means the parameters are reduced and '−' means the parameters are unchanged. The experimental results are shown in Table 3. In comparison to the original YOLOv7, the BiFusion module substantially reduces the model parameters and has a slight increase in mAP. Furthermore, the SPPFCSPC module and NWD loss function also effectively enhanced the mAP of the model while not causing an increase in parameters.

In summary, our proposed BNF-YOLOv7 achieves significant improvements in both mAP (↑ 3.3%) and FPS(120.4 versus 132.5) compared to the original YOLOv7.

Table 3. Ablation Study on FSTD testset.

YOLOv7	SPPFSCPC	BiFusion	NWD	Parameters	Precision	Recall	mAP@0.5	FPS
✓				37212738	0.828	0.719	0.797	120.4
✓	✓			37212738(−)	0.825	0.729	0.803	121.4
✓		✓		37118146(↓)	0.833	0.721	0.801	121.5
✓			✓	37212738(−)	0.803	0.738	0.805	122.3
✓	✓	✓		37118146(↓)	0.829	0.734	0.818	127.4
✓	✓		✓	37212738(−)	0.822	0.749	0.82	123.3
✓		✓	✓	37118146(↓)	0.831	0.733	0.817	128.6
✓	✓	✓	✓	**37118146(↓)**	**0.839**	**0.755**	**0.83**	**132.5**

5   Conclusion

We present BNF-YOLOv7 and FSTD in this work. BNF-YOLOv7 addresses the issues of existing detection algorithms regarding insufficient positional information, low accuracy of tiny object detection, and high computational complexity. Three techniques, SPPFCSPC, BiFusion and NWD, are employed to minimize parameters and achieve a better trade-off between speed and accuracy. Moreover, we constructed a dataset comprising various real traffic scenarios. Numerous experiments validate the performance and efficiency of BNF-YOLOv7. BNF-YOLOv7 performs well on both FSTD and UA-DETRAC datasets with low computational cost. Using mAP@0.5, FPS, and parameters as evaluation metrics, our model performs best. This result fully demonstrates the superior performance of our model.

As mentioned in the article, although BNF-YOLOv7 performs well in most scenarios and categories, there are still shortcomings. In the future, we will try to address these issues to provide a more accurate model.

Acknowledgments. This work was supported in part by the Major Key Project of PCL under Grant PCL2023A09 and PCL2022A03, Guangdong Major Project of Basic and Applied Basic Research under Grant 2019B030302002.

References

1. Bochkovskiy, A., Wang, C.Y., Liao, H.Y.M.: YOLOv4: optimal speed and accuracy of object detection. arXiv:2004.10934 [cs, eess] (2020)
2. Cai, Z., Vasconcelos, N.: Cascade R-CNN: high quality object detection and instance segmentation. IEEE Trans. Pattern Anal. Mach. Intell. **43**(5), 1483–1498 (2019)
3. Cordts, M., et al.: The cityscapes dataset for semantic urban scene understanding. In: Proceedings of the IEEE Conference on Computer Vision and Pattern Recognition (CVPR) (2016)
4. Duan, K., Bai, S., Xie, L., Qi, H., Huang, Q., Tian, Q.: Centernet: keypoint triplets for object detection. In: Proceedings of the IEEE/CVF International Conference on Computer Vision, pp. 6569–6578 (2019)

5. Elfwing, S., Uchibe, E., Doya, K.: Sigmoid-weighted linear units for neural network function approximation in reinforcement learning. Neural Netw. **107**, 3–11 (2018)
6. Geiger, A., Lenz, P., Urtasun, R.: Are we ready for autonomous driving? The kitti vision benchmark suite. In: Conference on Computer Vision and Pattern Recognition (CVPR) (2012)
7. Girshick, R.: Fast R-CNN. In: Proceedings of the IEEE International Conference on Computer Vision, pp. 1440–1448 (2015)
8. Girshick, R., Donahue, J., Darrell, T., Malik, J.: Rich feature hierarchies for accurate object detection and semantic segmentation. In: Proceedings of the IEEE Conference on Computer Vision and Pattern Recognition, pp. 580–587 (2014)
9. He, K., Zhang, X., Ren, S., Sun, J.: Spatial pyramid pooling in deep convolutional networks for visual recognition. IEEE Trans. Pattern Anal. Mach. Intell. **37**(9), 1904–1916 (2015)
10. He, K., Zhang, X., Ren, S., Sun, J.: Deep residual learning for image recognition. In: Proceedings of the IEEE Conference on Computer Vision and Pattern Recognition, pp. 770–778 (2016)
11. Ioffe, S., Szegedy, C.: Batch normalization: accelerating deep network training by reducing internal covariate shift. In: International Conference on Machine Learning, pp. 448–456. PMLR (2015)
12. Jocher, G.: YOLOv5 by Ultralytics (2020). https://doi.org/10.5281/zenodo.3908559, https://github.com/ultralytics/yolov5
13. Kingma, D.P., Ba, J.: Adam: a method for stochastic optimization. arXiv preprint arXiv:1412.6980 (2014)
14. Kowol, K., Rottmann, M., Bracke, S., Gottschalk, H.: YOdar: uncertainty-based sensor fusion for vehicle detection with camera and radar sensors (2020). https://doi.org/10.48550/arXiv.2010.03320
15. Li, C., et al.: YOLOv6 v3.0: a full-scale reloading (2023).https://doi.org/10.48550/arXiv.2301.05586
16. Liang, M., Yang, B., Chen, Y., Hu, R., Urtasun, R.: Multi-task multi-sensor fusion for 3D object detection. In: Proceedings of the IEEE/CVF Conference on Computer Vision and Pattern Recognition, pp. 7345–7353 (2019)
17. Lin, T.-Y., et al.: Microsoft COCO: common objects in context. In: Fleet, D., Pajdla, T., Schiele, B., Tuytelaars, T. (eds.) ECCV 2014. LNCS, vol. 8693, pp. 740–755. Springer, Cham (2014). https://doi.org/10.1007/978-3-319-10602-1_48
18. Liu, S., Qi, L., Qin, H., Shi, J., Jia, J.: Path aggregation network for instance segmentation. In: Proceedings of the IEEE Conference on Computer Vision and Pattern Recognition, pp. 8759–8768 (2018)
19. Liu, W., et al.: SSD: single shot multibox detector. In: Leibe, B., Matas, J., Sebe, N., Welling, M. (eds.) ECCV 2016. LNCS, vol. 9905, pp. 21–37. Springer, Cham (2016). https://doi.org/10.1007/978-3-319-46448-0_2
20. Marriott, R.T., Romdhani, S., Chen, L.: A 3D GAN for improved large-pose facial recognition. In: Proceedings of the IEEE/CVF Conference on Computer Vision and Pattern Recognition, pp. 13445–13455 (2021)
21. Qin, L., et al.: Id-yolo: real-time salient object detection based on the driver's fixation region. IEEE Trans. Intell. Transp. Syst. **23**(9), 15898–15908 (2022)
22. Redmon, J., Divvala, S., Girshick, R., Farhadi, A.: You only look once: unified, real-time object detection. In: Proceedings of the IEEE Conference on Computer Vision and Pattern Recognition, pp. 779–788 (2016)
23. Redmon, J., Farhadi, A.: Yolo9000: better, faster, stronger. In: Proceedings of the IEEE Conference on Computer Vision and Pattern Recognition, pp. 7263–7271 (2017)

24. Redmon, J., Farhadi, A.: Yolov3: an incremental improvement. arXiv preprint arXiv:1804.02767 (2018)
25. Ren, S., He, K., Girshick, R., Sun, J.: Faster R-CNN: towards real-time object detection with region proposal networks. Adv. Neural Inf. Process. syst. **28** (2015)
26. Song, X., et al.: A survey on deep learning based knowledge tracing. Knowl.-Based Syst. **258**, 110036 (2022)
27. Szegedy, C., et al.: Going deeper with convolutions. In: Proceedings of the IEEE Conference on Computer Vision and Pattern Recognition, pp. 1–9 (2015)
28. Wang, C.Y., Bochkovskiy, A., Liao, H.Y.M.: Yolov7: trainable bag-of-freebies sets new state-of-the-art for real-time object detectors. arXiv preprint arXiv:2207.02696 (2022)
29. Wang, C.Y., Liao, H.Y.M., Wu, Y.H., Chen, P.Y., Hsieh, J.W., Yeh, I.H.: CSPNet: a new backbone that can enhance learning capability of CNN. In: Proceedings of the IEEE/CVF Conference on Computer Vision and Pattern Recognition Workshops, pp. 390–391 (2020)
30. Wang, F., Xu, J., Liu, C., Zhou, R., Zhao, P.: On prediction of traffic flows in smart cities: a multitask deep learning based approach. World Wide Web **24**, 805–823 (2021)
31. Wang, J., Xu, C., Yang, W., Yu, L.: A normalized gaussian wasserstein distance for tiny object detection. arXiv preprint arXiv:2110.13389 (2021)
32. Wang, L., et al.: Model: motif-based deep feature learning for link prediction. IEEE Trans. Comput. Soc. Syst. **7**(2), 503–516 (2020)
33. Wen, L., et al.: UA-DETRAC: a new benchmark and protocol for multi-object detection and tracking. Comput. Vis. Image Underst, **193**, 102907 (2020)
34. Xu, C., et al.: Uncertainty-aware multi-view deep learning for internet of things applications. IEEE Trans. Industr. Inf. **19**(2), 1456–1466 (2022)
35. Yang, X., Yan, J., Ming, Q., Wang, W., Zhang, X., Tian, Q.: Rethinking rotated object detection with gaussian wasserstein distance loss. In: International Conference on Machine Learning, pp. 11830–11841. PMLR (2021)
36. Yin, H., Yang, S., Song, X., Liu, W., Li, J.: Deep fusion of multimodal features for social media retweet time prediction. World Wide Web **24**, 1027–1044 (2021)
37. Yu, F., et al.: Bdd100k: a diverse driving dataset for heterogeneous multitask learning. In: Proceedings of the IEEE/CVF Conference on Computer Vision and Pattern Recognition, pp. 2636–2645 (2020)
38. Yu, J., Jiang, Y., Wang, Z., Cao, Z., Huang, T.: Unitbox: an advanced object detection network. In: Proceedings of the 24th ACM International Conference on Multimedia, pp. 516–520 (2016)
39. Zhang, Wei, Gao, Xian-zhong, Yang, Chi-fu, Jiang, Feng, Chen, Zhi-yuan: A object detection and tracking method for security in intelligence of unmanned surface vehicles. J. Ambient Intell. Humanized Comput. **13**(3), 1279–1291 (2020). https://doi.org/10.1007/s12652-020-02573-z
40. Zheng, Z., et al.: Enhancing geometric factors in model learning and inference for object detection and instance segmentation. IEEE Trans. Cybern. **52**(8), 8574–8586 (2021)

PERTAD: Towards Pseudo Verification for Anomaly Detection in Partially Labeled Graphs

Wenjing Chang[1,2], Jianjun Yu[1(✉)], and Xiaojun Zhou[1]

[1] Computer Network Information Center, Chinese Academy of Sciences,
Beijing, China
[2] University of Chinese Academy of Sciences, Beijing, China
{changwenjing,xjzhou}@cnic.cn, yujj@cnic.ac.cn

Abstract. The graph-based anomaly detection task aims to identify nodes with patterns that deviate from those of the majority nodes in a large graph, where only a limited subset of nodes is annotated. However, inadequate supervised knowledge and uncertainty of anomalous structure restrict the performance of detection. In this paper, we propose PERTAD (**P**seudo V**ER**ifica**T**ion for **A**nomaly **D**etection), a novel semi-supervised learning method for detecting anomalies in partially labeled graphs. Specifically, we first propose a self-verification framework that comprises a target network (T-model) and a verification network (V-model). The framework employs pseudo-labeled graphs to transfer the knowledge learned by T-model to V-model, and then corrects T-model with the performance error between the two networks on the labeled set. Furthermore, we introduce a layer-level aggregation mechanism for node representation in deep GNNs to address the uncertainty of anomalous structures. The proposed mechanism re-aggregates neighborhood information across layers, aiming to preserve low-order neighborhood characteristics and alleviate the over-smoothing effect. Extensive experiments on real-world graph-based anomaly detection tasks demonstrate that PERTAD significantly outperforms state-of-the-art baselines.

Keywords: Anomaly Detection · Pseudo-Labeling · Semi-supervised learning

1 Introduction

Recently, graph-structured data has gained popularity as a powerful method for representing complex relations and attributes in various fields. Detecting anomalous nodes in graphs, which can propagate malicious information to their neighboring nodes, has attracted sufficient attention in academic. With the developments of graph neural networks (GNNs) [5,8,15], numerous GNN-based anomaly detection methods have been widely proposed [3,4,21]. Despite their success, two key challenges remain in graph-based anomaly detection.

Anomalous nodes in graphs may not solely be attributed to specific node features, but also to particular neighborhood structures. The diversity and uncertainty of anomalous structures make shallow models perform poorly in detecting

X. Song et al. (Eds.): APWeb-WAIM 2023, LNCS 14334, pp. 201–216, 2024.
https://doi.org/10.1007/978-981-97-2421-5_14

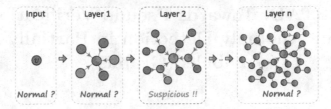

Fig. 1. An illustration of information aggregation at different layers for an anomalous node in a graph. Anomalies can exhibit normal features, and their suspicious characteristics may be either exposed or disguised during the information aggregation process.

anomalies. Deep architectures appear to be a better option for extracting complex information from high-order neighbors. Nevertheless, deepening GNN models may lead to over-smoothing [12], which results in indistinguishable representations across different categories and enables anomalies to disguise themselves. As shown in Fig. 1, an anomalous node v appears normal before aggregating its neighbors' information. At the second layer, it has a higher anomalous score, indicating a higher likelihood of being an anomaly. However, at deeper layers, the anomalous score decreases again, potentially causing the node to evade detection if only the final output is considered. Hence, maintaining neighbor information at multiple layers is crucial to capture unique node characteristics.

Besides, in many real-world applications, nodes are usually unlabeled due to huge data volume and high annotation cost. The lack of annotated data poses a significant challenge for deep learning models. To address this issue, pseudo-labeling [11,20] has been proposed as a semi-supervised learning framework. It involves training a teacher model on a well-labeled dataset and subsequently refining a student model by fine-tuning it on pseudo-labeled sets. A natural problem arises that pseudo-labeled sets can easily lead to confirmation bias if the labels are inaccurate or the distribution is skewed. Several studies [1,17] have been explored to construct high-quality pseudo-labeled sets that contain more beneficial information for the target model, i.e., the student model. However, these methods overlook the characteristics of data distribution in anomaly detection tasks, where the pseudo-labels with high confidence are mostly normal samples, which will aggravate the class imbalance problem.

In this work, we introduce a novel semi-supervised deep model for anomaly detection in partially labeled graphs, as **P**seudo VE**R**ifica**T**ion for **A**nomaly **D**etection (PERTAD). PERTAD comprises two neural networks, namely the target network (T-model) and the verification network (V-model), which are trained on the truly labeled data and pseudo-graphs, respectively. The pseudo-graphs are constructed based on a balanced anomalous confidence measurement. After being trained on pseudo-graphs, the V-model can update the training status of the T-model by leveraging the differences between their predictions on labeled nodes. To address the uncertainty of anomalous structure, PERTAD employs deep GNNs to capture information from distant neighbors, while also using a layer-level aggregation mechanism to preserve the characteristics in short-distance. To the best of our knowledge, PERTAD is among the first works that have explored the use of pseudo-labeling in graph anomaly detection tasks.

The key contribution of the present work are summarized as follows:

- We propose a novel semi-supervised model PERTAD based on pseudo verification for graph anomaly detection. Furthermore, we propose a balanced anomalous confidence measurement to construct high-quality pseudo-graphs.
- To alleviate the uncertainty of anomalous structures, we devise a layer-level aggregation mechanism, which maintains neighborhood information in multiple distances.
- Extensive experiments on three real-world datasets demonstrate the effectiveness of the proposed framework and its superiority compared to a variety of state-of-the-art methods.

2   Related Work

2.1   Anomaly Detection on Graphs

The expressive capacity of graph neural networks (GNNs) has resulted in their widespread application for graph anomaly detection. GEM [16] is the first heterogeneous graph neural network approach for detecting malicious accounts. SemiGNN [23] is a semi-supervised model that detect fraudsters from features and relations with hierarchical attention mechanism. GraphConsis [14] investigates the inconsistency problems from context, feature and relation. CAREGNN [4] employs a label-aware similarity measure and a similarity-aware neighbor selector to enhance the GNN aggregation process against camouflages. DCI [24] is a new graph SSL scheme that injects a clustering step to reduce data inconsistency. PC-GNN [13] is a pick and choose graph neural network to alleviate the imbalance and inconsistency problem in anomaly detection. BWGNN [21] observes anomalies from the lens of the graph spectrum and develops its architecture invoking graph wavelet theory. Different from the methods mentioned above, we focus on detecting anomalies in graphs with fewer annotations with the help of pseudo-labeling.

2.2   Pseudo Labeling

Pseudo-labeling has been successfully applied in many applications, such as image classification [6], object detection [2,19], and object re-identification [27]. These methods often contain a pair of networks: a pre-trained teacher model generates pseudo labels on the unlabeled set, and a student model trains on both the labeled and pseudo-labeled sets. In order to construct high-quality pseudo-labeled sets, Pseudo-Label [11], UDA [25], and FixMatch [20] choose unlabeled samples by confidence-based thresholds. However, these methods keep a fixed teacher model during the student's learning, resulting in a confirmation bias when the pseudo-labels are inaccurate. Meta Pseudo Labels [18] proposes to address this problem by continuing to adapt the pre-trained teacher by learning

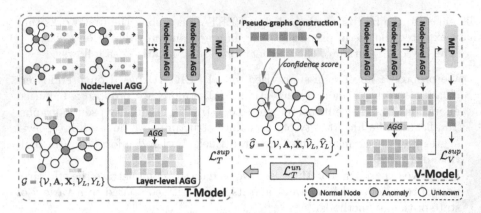

Fig. 2. The architecture of the proposed PERTAD for anomaly detection in partially labeled graphs. The framework is composed of three modules: V-model, pseudo-graphs construction, and T-model. The T-model provides pseudo-graphs to the V-model, while the V-model feedback its training status on the truly labeled set to the T-model.

from the student's feedback to improve the student's performance on the labeled set. Unlike the methods mentioned above, graph-based anomaly detection tasks are typically developed on non-independent and identically distributed data and exhibit poor performance in cases where the amount of available annotation is limited. Constructing a high-quality pseudo-labeled set and alleviating the issue of confirmation bias are the primary challenges for implementing the pseudo-labeling strategy for anomaly detection in graphs.

3   Problem Definition

In this work, we focus on the problem of anomaly detection in partially labeled graphs, which can be formulated as a semi-supervised imbalanced binary node classification problem in graphs.

We define a partially labeled graph as $\mathcal{G} = \{\mathcal{V}, \mathbf{A}, \mathbf{X}, \mathcal{V}_L, Y_L\}$, where $\mathcal{V} = \{v_1, ..., v_n\}$ is a set of n nodes and $\mathcal{V}_L$ is the set of labeled nodes. The unlabeled nodes can be represented as $\mathcal{V}_U = \mathcal{V} \setminus \mathcal{V}_L$. $\mathbf{A} \in \mathbb{R}^{n \times n}$ is the adjacency matrix, and $\mathbf{X} \in \mathbb{R}^{n \times d}$ denotes the attribute matrix. Y_L contains the class information for the node subset $\mathcal{V}_L$, where the distribution of classes are imbalanced. Given $\mathcal{G}$, we aim to learn an anomaly detector f that can well infer anomalous scores for each node in $\mathcal{V}_U$, according to the graph structure, attribute information, and label information of $\mathcal{V}_L$.

4   Proposed Model

In this section, we introduce the details of the proposed PERTAD framework, as shown in Fig. 2. Firstly, we give an overview of the whole framework. Secondly, we introduce the generation of node embeddings from node-level and layer-level in Sect. 4.2. Then we introduce the construction of pseudo-graphs and the optimization of PERTAD in detail in Sects. 4.3 and 4.4, respectively.

4.1   Overview

As illustrate in Fig. 2, our framework comprises three key modules, namely a target network (T-model), a verification network (V-model) and a knowledge transfer bridge (pseudo-graphs). In the proposed self-verification framework, the T-model serves as a teacher while the V-model acts as a student, and both of them can mutually learn from each other. To obtain node representations of the input graph, we feed the raw graph into the T-model, which consists of two steps. Firstly, we aggregate the locality neighborhood information for all nodes at each layer with a node-level aggregator. Secondly, we adopt a layer-level aggregation in message passing along with convolution layers to preserve the shallow information. Afterward, we construct high-quality pseudo-graphs by measuring the certainty of pseudo-labels and the proportion of anomalies, which contain more abundant information in the T-model and facilitate the V-model training. Subsequently, the V-model learns from the constructed pseudo-graphs in a supervised manner and provides feedback to the T-model. As a result, the T-model optimizes two objectives, one of which is the supervised loss of detection performance, while the other is the unsupervised loss of knowledge transfer.

4.2   Embedding Generation

The main idea of GNN-based anomaly detection methods is to generate distinguish node representations by aggregating neighborhood information. However, shallow GNNs have limited ability to capture information in long distance, whereas deep GNNs are unable to retain raw features in short distance. To mitigate the problem of over-smoothing and reinforce the effect of original characteristics, we propose to fusion low-order and high-order node representations during the training process. There are two steps in node embedding generation.

Node-Level Aggregation. First, we calculate node representations by aggregating first-order neighbors information at each layer. We adopt the GIN [26] as an integral part of the encoder to get ideal embeddings for its effectiveness. The node representations are calculated as follows:

$$\mathbf{h}_v^{(l)} = \mathrm{MLP}^{(l)} \left(\left(1 + \epsilon^{(l)}\right) \cdot \mathbf{h}_v^{(l-1)} + \sum_{j \in \mathcal{N}(v)} \mathbf{h}_j^{(l-1)} \right) \qquad (1)$$

where $\mathrm{MLP}^{(l)}$ is the multi-layer perceptron at the l-th layer, $\mathbf{h}_v^{(l)}$ is the embedding of node v at the l-th layer, $\mathcal{N}(v) = \{u \in \mathcal{V} | A(v, u) > 0\}$ is neighbors of node v, and ϵ can be set as a learnable parameter or a fixed scalar.

Layer-Level Aggregation. During the propagation of information in the graph, the succeeding layer in the graph neural network aggregates the information from the preceding order of neighbors. Each node may be influenced by the information of all the nodes in the graph, thus causing them to lose their individual or low-order neighborhood information as the number of layers increases. In light of the successful implementation of DenseNet [7] in retaining low-dimensional features in deep convolutional neural networks, we devise a skip connection for message passing through layer-level aggregation. To maintain information across multiple distances, we combine node representations at each layer as illustrated in Eq. (2) where $W \in \mathbb{R}$ is the weight matrix.

$$\mathbf{h}_v = \sigma \left(W \left(\mathbf{h}_v^{(l)} \oplus \text{AGG} \left\{ \mathbf{h}_v^{(0)}, \mathbf{h}_v^{(1)}, ..., \mathbf{h}_v^{(l-1)} \right\} \right) \right) \tag{2}$$

where σ is a nonlinear activation function and AGG represents an aggregator function. The purpose of the aggregator function is to preserve the inherent properties of embeddings across various layers. In this study, we evaluate three potential aggregator functions, namely, the mean aggregator, max aggregator, and min aggregator.

4.3 Pseudo-graphs Construction

Upon obtaining the final node representations, we proceed to compute the hard pseudo-labels for the unlabeled set $\mathcal{V}_U$, which are utilized to construct pseudo-graphs.

$$y'_v = \varphi(p_v, \tau) \tag{3}$$

where $p_v = \text{MLP}(\mathbf{h}_v)$ is the anomalous score of the node v, τ denotes the discrimination threshold for anomalous class, and $\varphi(p, q) \in \{0, 1\}$ is a symbolic function, where 1 indicates that p is greater than q.

A balanced and accurately annotated input graph is beneficial for anomaly detection tasks. Nonetheless, the generated labels may contain errors, which could bring misleading information during verification model training. Furthermore, in anomaly detection tasks, the initial distribution of normal and anomalous nodes may cause the generated labels to be biased. To address the aforementioned issues, we propose to evaluate the quality of a graph for anomaly detection from two perspectives: the certainty of pseudo-labels and the proportions of anomalies. The former ensures the precision of labels, while the latter mitigates the class imbalance problem in the anomaly detector. Following this two evaluations, we construct pseudo-labels as follows.

First, to increase the likelihood of selecting anomalous nodes and ensure the selection of relatively certain nodes, we calculate the anomalous confidence for each node in $\mathcal{V}_U$ with a discrimination threshold of τ.

$$c_v = |p_v - \tau| \tag{4}$$

To address the class imbalance problem in pseudo-labeled node selection, we apply a down-sampling strategy by selecting nodes from $\mathcal{V}_U$ with high anomalous confidence scores and then re-sampling them to ensure class balance. Specially, truly labeled nodes in the original graph $\mathcal{G}$ are not involved in constructing pseudo-labeled sets, which serves as a measure of knowledge transfer during the verification process. The pseudo-labeled set $\widehat{\mathcal{V}}_L$ in the pseudo graph $\widehat{\mathcal{G}}$ can be defined as follows:

$$\widehat{\mathcal{V}}_L = \left\{ v \in \mathcal{V}_U \mid c_v > \rho \text{ and SUM} \left(\widehat{Y}_L \right) = \frac{|\widehat{\mathcal{V}}_L|}{2} \right\} \tag{5}$$

where ρ is determined by the top-50% confidence scores in each iteration, and $\widehat{Y}_L = \left\{ y'_v \mid v \in \widehat{\mathcal{V}}_L \right\}$ is the set of pseudo labels. The condition SUM $\left(\widehat{Y}_L \right) = \frac{|\widehat{\mathcal{V}}_L|}{2}$ ensures that the number of normal and anomalous nodes in $\widehat{\mathcal{V}}_L$ are equal.

Finally, we describe the pseudo-graph as $\widehat{\mathcal{G}} = \left\{ \mathcal{V}, \mathbf{A}, \mathbf{X}, \widehat{\mathcal{V}}_L, \widehat{Y}_L \right\}$. Note that $\widehat{\mathcal{G}}$ can be regarded as the transfer bridge between T-model and V-model. A high-quality $\widehat{\mathcal{G}}$ can mitigate the loss in knowledge transfer process, which may be beneficial to evaluate the performance of T-model.

4.4   Training and Verification Framework

Effectively utilizing unlabeled nodes is a significant challenge for anomaly detection in partially labeled graphs. Inspired by generative adversarial networks (GANs) and pseudo labeling, we propose to build a pair of networks, including a target network (T-model) and a verification network (V-model). The T-model is supervised by the few ground-truth and feedback from the V-model. The V-model, on the other hand, learns from the pseudo-graphs generated by the T-model. By doing so, the V-model can evaluate whether the T-model is capable of generating accurate pseudo labels that contain enough information to infer the true labels. The full optimization process is described as follows:

Firstly, the T-model learns from the labeled nodes in a supervised manner, aiming to minimize the cross-entropy loss between its predictions and the ground-truth labels:

$$\underset{\theta_T}{argmin} \left[\text{CE} \left(Y_L, T(\mathcal{V}_L; \theta_T) \right) \right] \tag{6}$$

where $T(\mathcal{V}_L; \theta_T)$ denotes the predictions of the T-model on the labeled set $\mathcal{V}_L$ under the parameters θ_T and CE (p, q) denotes the cross-entropy loss between two distributions p and q.

Next, we initiate the V-model to monitor whether the T-model has gained sufficient knowledge from the labeled nodes. The primary principle of the V-model is to accurately fit the pseudo-graphs generated by the T-model. Consequently, the V-model is optimized via the following loss function:

$$\underset{\theta_V}{argmin} \left[\text{CE} \left(\widehat{Y}_L, V(\widehat{\mathcal{V}}_L; \theta_V) \right) \right] \tag{7}$$

where $V(\widehat{\mathcal{V}}_L; \theta_V)$ denotes the predictions of the V-model on the pseudo-labeled set $\widehat{\mathcal{V}}_L$ under the parameters θ_V.

Algorithm 1: The training process of PERTAD

Input: A graph: $\mathcal{G} = (\mathcal{V}, \mathbf{A}, \mathbf{X}, \mathcal{V}_L, Y_L)$; Number of layers, epochs, verification
interval: L, N, μ;

Output: Labels $\{\widehat{y}_u : u \in \mathcal{V}_U\}$ for unlabeled nodes.

1 **for** $e = 1, \ldots, N$ **do**
2   **for** $l = 1, \ldots, L$ **do**
3         $\mathbf{h}_v^{(l)} \leftarrow$ Eq. (1), $\forall v \in \mathcal{V}_L$;   ▷ Node-level Aggregation
4      $\mathbf{h}_v \leftarrow$ Eq. (2), $\forall v \in \mathcal{V}_L$;   ▷ Layer-level Aggregation
5      $\mathcal{L}_T^{sup} \leftarrow$ CE $(Y_L, T(\mathcal{V}_L; \theta_T))$;   ▷ Supervised Loss on T-model
6   **if** $e \bmod \mu == 0$ **then**
7         $\widehat{\mathcal{G}} = \left\{ \mathcal{V}, \mathbf{A}, \mathbf{X}, \widehat{\mathcal{V}}_L, \widehat{Y}_L \right\}$;   ▷ Construct pseudo-graphs $\widehat{\mathcal{G}}$
8         $\theta_V \leftarrow$ train(V-model, $\widehat{\mathcal{G}}$) ;   ▷ V-model Training
9         $\mathcal{L}_T^{un} \leftarrow$ CE $(T(\mathcal{V}_L; \theta_T), V(\mathcal{V}_L; \theta_V))$;   ▷ Unsupervised Loss on T-model
10   Update T-model using $\mathcal{L}_T^{sup}$ and $\mathcal{L}_T^{un}$;

Once the V-model has been thoroughly trained, we can evaluate its performance on the labeled set $\mathcal{V}_L$ and then optimize the T-model. First, we calculate the verification loss between the predictions generated by the T-model and V-model. Next, the T-model minimizes the following unsupervised loss:

$$\underset{\theta_T}{argmin} \left[\text{CE} \left(T(\mathcal{V}_L; \theta_T), V(\mathcal{V}_L; \theta_V) \right) \right] \tag{8}$$

where θ_V is the fixed parameter, $T(\mathcal{V}_L; \theta_T)$ represents the learned knowledge of the T-model from the labeled set and $V(\mathcal{V}_L; \theta_V)$ represents the transferred knowledge from the T-model to the V-model.

Note that the difference between predictions of the T-model and V-model mostly reflects the quality of the T-model, under the assumption that the loss in pseudo-graphs construction is sufficiently small. Specifically, the T-model only needs to consider the loss incurred by labeled nodes if the V-model produces similar predictions, which means the transferred knowledge is adequate to accurately infer the true labels. If the T-model achieves correct predictions while the V-model fails, it implies that pseudo-graphs might contain erroneous information. Conversely, if the T-model produces incorrect predictions while the V-model gets it right, then the loss can be perceived as a complement to the supervised loss, which could potentially enhance the training of T-model. Therefore, we posit that integrating the verification loss is effectiveness to T-model. The loss functions of the T-model and V-model can be formulated as follows:

$$\begin{aligned} \mathcal{L}_T &= \mathcal{L}_T^{sup} + Q(e, \mu) \mathcal{L}_T^{un} \\ &= \text{CE} \left(Y_L, T(\mathcal{V}_L; \theta_T) \right) + Q(e, \mu) \text{CE} \left(T(\mathcal{V}_L; \theta_T), V(\mathcal{V}_L; \theta_V) \right) \end{aligned} \tag{9}$$

$$\mathcal{L}_V = \mathcal{L}_V^{sup} = \text{CE} \left(T(\widehat{\mathcal{V}}_L; \theta_T), V(\widehat{\mathcal{V}}_L; \theta_V) \right) \tag{10}$$

where μ is the verification interval and $Q(e, \mu) \in \{0, 1\}$ is a symbolic function that performs pseudo verification if the current epoch e is an integer multiple of

μ. The verification interval decides the times of self-supervision of the T-model. The overall training algorithm is summarized in Algorithm 1.

5   Experiments

In this section, we conduct experiments to evaluate the effectiveness of PER-TAD for the node anomaly detection task when the label information is lacking. Particularly, we mainly answer the following questions:

- **RQ1:** Does the proposed PERTAD outperform state-of-the-art methods?
- **RQ2:** How exactly do the key components contribute to the predictions?
- **RQ3:** How do the key hyperparameters affect PERTAD, such as the verification interval, the threshold in graph construction and graph layers?

5.1   Datasets

We evaluate our work on three real-world datasets: (1) Wiki [9] is an editor-page graph consisting of user edit records made in a single month on Wikipedia pages. (2) Reddit [9] is a user-subreddit graph collected from Reddit. It displays an edge if a user posts to a subreddit. (3) Alpha [10] is a user-user trust graph of Bitcoin users trading on the Bitcoin Alpha platform.

The ground truth of these three datasets is established by their administrators. Specifically, Wiki and Reddit are completely labeled, whereas Alpha is partially labeled. To study the robustness of PERTAD, we construct partially labeled graphs by randomly sampling different proportions of instances from the Wiki and Reddit datasets. Detailed statistics of datasets are shown in Table 1.

Table 1. Statistics of datasets.

Dataset	Relation	#Users (% normal, abnormal, unknown)	#Objects	#Edges
Wiki	Editor-Page	8,227 (97.36%, 2.64%, 0%)	1,000	18,227
Reddit	User-Subreddit	10,000 (96.34%, 3.66%, 0%)	984	78,516
Alpha	User-User	3,286 (3.99%, 2.53%, 93.49%)	3,754	24,186

5.2   Baselines

To demonstrate the superiority of PERTAD, we compare it with two categories of baselines: (1) general GNN models, including GCN [8], GraphSAGE [5], GAT [22], GeniePath [15] and GIN [26]. (2) GNN-based anomaly detection models, including GEM [16], CAREGNN [4], DCI [24], PC-GNN [13] and BWGNN [21].

- GCN is a graph convolution network which obtains node embeddings by a localized first-order approximation of spectral graph convolutions.
- GraphSAGE is an inductive GNN model which aggregates embeddings from a fixed sample number of local neighbors.

- GAT is a graph attention network that employs attention mechanism for neighbor aggregation.
- GeniePath designs a novel aggregate method of GNNs to filter graph signals from neighbors of different hops away.
- GIN proposes a sum-like aggregation function to effectively preserve the structure homophily.
- GEM adaptively learns discriminative embeddings from heterogeneous graphs, including an attention mechanism to learn the importance of different types of nodes and a sum operator to aggregate embeddings in each type.
- CAREGNN employs a label-aware similarity measure and a similarity-aware neighbor selector to enhance the GNN aggregation process.
- DCI adopts graph SSL scheme and injects a clustering step to reduce the inconsistency between the structural patterns and the label semantics.
- PC-GNN combines label distribution information into sampling process and chooses part of labeled nodes for training to remedy the imbalance problem.
- BWGNN leverages Beta graph wavelet to generate band-pass filters to better capture anomaly information in spectral and spatial domains on the graph.

5.3   Experimental Settings

We unify the learning rate (0.01), embedding dimension (128), number of training epochs (300) with the early stopping strategy and optimizer (Adam) for all model. For PERTAD and its variants, we set the number of layers L as 10, the verification interval μ as 50, and the threshold in graph construction τ as 0.4. PERTAD, PERTAD-Min, PERTAD-Max adopts mean aggregator, min aggregator and max aggregator in layer-level aggregation respectively.

We employ two widely adopted metrics to comprehensively measure the performance of all methods: AUC and Recall. To improve the training efficiency, we employ under-sampling and threshold-moving techniques to train PERTAD and other baselines. We use 10-fold cross-validation to evaluate all models and report the average performance for all methods.

All experiments were conducted on Linux server with 10 Intel Silver 4120 Processors and 2 NVIDIA GeForce RTX 3090 GPUs. GCN, GraphSAGE, GAT, GeniePath and GIN are implemented based on DGL[1]. GEM is implemented based on DGFraud-TF2[2]. For CAREGNN, DCI, PC-GNN and BWGNN, we use the source code provided by their authors.

5.4   Performance Comparison (RQ1)

We compare the performance of PERTAD along with the baselines on aforementioned three datasets. The average results are reported in Table 2 and Fig. 3, and we can make the following observations.

---

[1] https://github.com/dmlc/dgl.
[2] https://github.com/safe-graph/DGFraud-TF2.

Table 2. Performance comparison in anomaly detection tasks. We conduct 10-fold cross-validation on each model and then report the average experimental results. The best and second-best overall results are highlighted with bold and underline text.

DataSet	Wiki				Reddit				Alpha	
Ratio	10%		60%		10%		60%		\	
Metric	AUC	Recall	AUC	Recall	AUC	Recall	AUC	Recall	AUC	Recall
GCN	58.19	54.59	70.73	59.63	65.41	57.37	70.46	59.34	83.18	55.82
GraphSAGE	58.78	53.79	68.29	53.62	64.91	59.06	72.20	60.94	79.24	56.39
GAT	62.43	52.26	62.83	50.48	65.59	54.93	66.87	54.60	77.25	53.08
GeniePath	62.61	52.78	72.95	62.92	66.17	57.05	71.55	59.65	81.20	66.67
GIN	60.97	53.63	70.96	63.82	66.96	57.34	71.33	62.44	82.62	66.85
GEM	61.72	54.57	68.64	62.36	60.48	54.96	67.90	61.29	82.71	66.71
CAREGNN	58.16	55.82	69.13	63.30	60.84	58.66	68.21	62.57	83.88	67.11
DCI	64.12	53.16	73.37	53.40	**68.05**	52.91	72.00	51.68	<u>89.37</u>	77.91
PC-GNN	55.41	56.96	63.58	59.11	52.87	55.86	59.59	58.93	76.39	76.84
BWGNN	61.27	53.50	70.11	64.03	65.75	55.99	71.21	62.87	81.50	70.30
PERTAD-Max	64.99	54.94	<u>73.49</u>	**69.07**	66.63	<u>59.63</u>	**73.99**	**65.50**	89.34	75.09
PERTAD-Min	<u>66.77</u>	**60.67**	71.79	62.59	65.47	58.17	72.50	62.00	87.78	**81.40**
PERTAD	**67.53**	<u>58.03</u>	**74.64**	<u>66.88</u>	<u>67.63</u>	**61.29**	<u>73.94</u>	<u>63.22</u>	**92.12**	<u>78.98</u>

Firstly, we can observe from Table 2 that PERTAD outperforms all baseline methods, which indicates the superiority of its innovative mechanism that effectively leverages vast amounts of unlabeled data. Unlike other baselines that focus less on model evaluation during training, PERTAD makes use of self-supervision by training a plain model on pseudo-labeled sets, while preserving neighborhood information from multiple distances to discover anomalous structures.

Secondly, GNN-based anomaly detection methods demonstrate significant superiority over general GNN models in Recall score. General GNN models suffer from the imbalance problem, where the anomalous class can not be sufficiently represented. The situation is even worse in GAT due to the huge attention parameters that cannot be well-trained without sufficient knowledge of the anomalous class. Furthermore, CAREGNN and PC-GNN, two state-of-the-art methods for graph-based fraud detection, perform worse than other anomaly detection methods on the Wiki and Reddit datasets. This could be attributed to their complex neighbor aggregator focus more on relation-level characteristics in homogeneous graphs, while Wiki and Reddit are bipartite graphs. Hence, the consistent node attributes are crucial to downstream tasks.

Finally, Fig. 3 illustrates the performance of PERTAD and three superior models on the Wiki dataset for varying labeling ratios from 10% to 60%. We observe that PERTAD consistently achieves the best performance under each labeling ratio in terms of AUC and Recall. Moreover, PERTAD has a smaller improvement than others as the labeling ratios increase, indicating its robustness to labeling ratios. Therefore, we conclude that PERTAD consistently outperforms GeniePath, DCI, and BWGNN and is robust to labeling ratios.

5.5 Ablation Study (RQ2)

We disable one key component at a time to explore its effect on PERTAD. We evaluate PERTAD: (1) without the V-model (NoV); (2) without the layer-level aggregation (NoL). The results are shown in Tables 3 and 4.

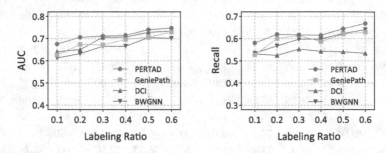

Fig. 3. A comparison of PERTAD to several state-of-the-art models with varying labeling ratios. Each point represents the average score of 10-fold evaluation.

Table 3. Ablation study of two key components on three datasets under different labeling ratios. The best scores of 10-fold evaluation are highlighted with bold text.

DataSet(Ratio)	Wiki(10%)		Wiki(60%)		Reddit(10%)		Reddit(60%)		Alpha	
Metric	AUC	Recall	AUC	Recall	AUC	Recall	AUC	Recall	AUC	Recall
NoV	66.94	57.83	73.73	64.19	67.38	59.40	**74.10**	60.95	91.25	73.28
NoL	**68.15**	**60.32**	70.27	59.68	66.17	57.86	69.33	57.53	86.29	66.82
PERTAD	67.53	58.03	**74.74**	**66.88**	**67.63**	**61.29**	73.94	**63.22**	**92.12**	**78.98**

PERTAD significantly outperforms NoV, which demonstrates that using an auxiliary network to provide feedback for model training status is effective. PERTAD also has a better performance than NoL comprehensively. It indicates that aggregating neighbor information at multiple layers is practical for generating precision node embeddings. Note that the improvement of NoL is slight when the labeling ratio increases, which means the proposed self-verification framework is robustness to labeling ratios and thus more suitable to less labeling tasks.

However, PERTAD gets lower scores than NoL on the Wiki (10%) dataset, especially in Recall score. Intuitively, we infer that layer-level aggregation can increase the dimension of node representations, which brings more parameters to learn, and leads to worse performance with limited labeled knowledge. To explore the impact of each component in benign and fraud classes, we further exhibit the Recall score of each class in Table 4.

As Table 4 shows, on Wiki (10%) dataset, NoL gets a 19.12% decrease in the Recall-Fraud than the full model, which indicates the layer-level aggregation is more focused on maintaining the unique features of anomalies. In contrast, NoV gets a 4.09% decrease in the Recall-Benign, which means the self-verification framework is more helpful in learning the representativeness embeddings of normal nodes. Therefore, the poor performance in the Recall score on Wiki (10%) dataset is because the gain of the V-model in benign class is less than that of layer-level aggregation in fraud class. We argue that the reason is that the graph size of Wiki is relatively tiny, and when the labeling ratio is low, pseudo-graphs cannot provide enough knowledge for V-model. We conclude that PERTAD is more effective on large-scale graphs.

Table 4. Recall scores of each class in the ablation study for anomaly detection. "-F" and "-B" denote "Fraud nodes as Positive" and "Benign nodes as Positive".

DataSet(Ratio)	Wiki(10%)		Wiki(60%)		Reddit(10%)		Reddit(60%)		Alpha	
Metric	R-F	R-B	R-F	R-B	R-F	R-B	R-F	R-B	R-F	R-B
NoV	**92.47**	18.51	**83.71**	44.20	**83.47**	34.38	65.14	64.08	**80.16**	61.67
NoL	64.89	**55.00**	63.51	**60.58**	49.64	**70.14**	58.70	**65.54**	61.04	**73.06**
PERTAD	84.01	22.60	77.18	56.69	78.27	39.92	**78.07**	53.51	80.16	69.31

5.6   Parameter Sensitivity (RQ3)

We investigate the performance of PERTAD with respect to three key hyperparameters on the Wiki (10%) and Reddit (10%) datasets. The results are shown in Figs. 4 and 5.

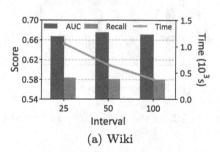

(a) Wiki

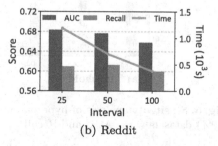

(b) Reddit

Fig. 4. Sensitivity analysis of hyperparameter μ on the Wiki (10%) and Reddit (10%) datasets w.r.t. AUC, Recall and Time.

Verification Interval. We investigate the influence of verification intervals on the performance and computation cost of PERTAD in Fig. 4. We observe that decreasing the verification interval improves the performance of PERTAD, but significantly increases the computation cost. A lower verification interval indicates a higher frequency of validation, which in turn provides more feedback from the V-model to the T-model, but it also demands more training time for the V-model. Thus, the selection of the verification interval is crucial to achieve a balance between performance and computational cost.

Threshold of Labels in Graph Construction. Figure 5(a) illustrates the effects of the threshold for pseudo labels in graph construction. PERTAD achieves peak performance when the threshold is set to 0.4. It indicates that both higher and lower thresholds result in less accurate pseudo-labels, which can hinder the training of the V-model. Hence, it can be inferred that the construction of pseudo-graphs plays a pivotal role in the pseudo verification process that enhance the training of PERTAD.

Number of Graph Layers. Figure 5(b) demonstrates that the number of layers affects the performance of PERTAD. It has been observed that shallow networks usually have worse representation ability, while deep networks are prone to over-smoothing. Both of them may lead to a decrease in the overall performance of the model. Therefore, we deduce that a decent number of layers can better develop the ability of layer-level aggregation.

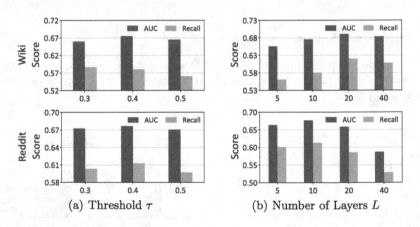

Fig. 5. Sensitivity analysis of hyperparameter τ and L on the Wiki (10%) and Reddit (10%) datasets w.r.t. AUC and Recall.

6   Conclusion

In this paper, we design a novel semi-supervised anomaly detection framework on partially labeled graphs, named PERTAD. PERTAD leverages an auxiliary network trained on the constructed pseudo-graphs to provide feedback on the training status of the target network. A layer-level aggregation mechanism is proposed to preserve the neighborhood information of various distance while also mitigating the over-smoothing problem inherent in deep GNNs. Extensive experiments demonstrate the superior performance of PERTAD from various aspects. In the future, further investigation into aggregation strategies in layer-level node representation may be a fruitful area of exploration.

Acknowledgment. This research is supported by a grant from the National Key Research and Development Program of China (No. 2022YFF0711801), the National Natural Science Foundation of China (Grant No. J2224012) and the CAS 145 Informatization Project CAS-WX2022GC-0301. Jianjun Yu is the corresponding author.

References

1. Cascante-Bonilla, P., Tan, F., Qi, Y., Ordonez, V.: Curriculum labeling: revisiting pseudo-labeling for semi-supervised learning. In: AAAI, pp. 6912–6920 (2021)
2. Deng, J., Li, W., Chen, Y., Duan, L.: Unbiased mean teacher for cross-domain object detection. In: CVPR, pp. 4091–4101 (2021)
3. Ding, K., Li, J., Bhanushali, R., Liu, H.: Deep anomaly detection on attributed networks. In: SIAM, pp. 594–602 (2019)
4. Dou, Y., Liu, Z., Sun, L., Deng, Y., Peng, H., Yu, P.S.: Enhancing graph neural network-based fraud detectors against camouflaged fraudsters. In: CIKM (2020)
5. Hamilton, W.L., Ying, Z., Leskovec, J.: Inductive representation learning on large graphs. In: NeurIPS, pp. 1024–1034 (2017)
6. Han, K., Rebuffi, S.A., Ehrhardt, S., Vedaldi, A., Zisserman, A.: Automatically discovering and learning new visual categories with ranking statistics. In: ICLR (2019)
7. Huang, G., Liu, Z., Van Der Maaten, L., Weinberger, K.Q.: Densely connected convolutional networks. In: CVPR, pp. 4700–4708 (2017)
8. Kipf, T.N., Welling, M.: Semi-supervised classification with graph convolutional networks. In: ICLR (2017)
9. Kumar, S., Hooi, B., Makhija, D., Kumar, M., Faloutsos, C., Subrahmanian, V.: Rev2: fraudulent user prediction in rating platforms. In: WSDM (2018)
10. Kumar, S., Zhang, X., Leskovec, J.: Predicting dynamic embedding trajectory in temporal interaction networks. In: SIGKDD, pp. 1269–1278 (2019)
11. Lee, D.H., et al.: Pseudo-label: the simple and efficient semi-supervised learning method for deep neural networks. In: ICML, p. 896 (2013)
12. Liu, M., Gao, H., Ji, S.: Towards deeper graph neural networks. In: SIGKDD, pp. 338–348 (2020)
13. Liu, Y., et al.: Pick and choose: a GNN-based imbalanced learning approach for fraud detection. In: WWW 2021: Proceedings of the Web Conference 2021, pp. 3168–3177 (2021)
14. Liu, Z., Dou, Y., Yu, P.S., Deng, Y., Peng, H.: Alleviating the inconsistency problem of applying graph neural network to fraud detection. In: SIGIR (2020)
15. Liu, Z., et al.: GeniePath: graph neural networks with adaptive receptive paths. In: AAAI, pp. 4424–4431 (2019)
16. Liu, Z., Chen, C., Yang, X., Zhou, J., Li, X., Song, L.: Heterogeneous graph neural networks for malicious account detection. In: CIKM, pp. 2077–2085 (2018)

17. Lokhande, V.S., Tasneeyapant, S., Venkatesh, A., Ravi, S.N., Singh, V.: Generating accurate pseudo-labels in semi-supervised learning and avoiding overconfident predictions via hermite polynomial activations. In: CVPR, pp. 11435–11443 (2020)
18. Pham, H., Dai, Z., Xie, Q., Le, Q.V.: Meta pseudo labels. In: CVPR (2021)
19. RoyChowdhury, A., et al.: Automatic adaptation of object detectors to new domains using self-training. In: CVPR, pp. 780–790 (2019)
20. Sohn, K., et al.: FixMatch: simplifying semi-supervised learning with consistency and confidence. In: NeurIPS, pp. 596–608 (2020)
21. Tang, J., Li, J., Gao, Z., Li, J.: Rethinking graph neural networks for anomaly detection. In: ICML, pp. 21076–21089 (2022)
22. Velickovic, P., Cucurull, G., Casanova, A., Romero, A., Liò, P., Bengio, Y.: Graph attention networks. In: ICLR (2018)
23. Wang, D., et al.: A semi-supervised graph attentive network for financial fraud detection. In: ICDM, pp. 598–607 (2019)
24. Wang, Y., Zhang, J., Guo, S., Yin, H., Li, C., Chen, H.: Decoupling representation learning and classification for GNN-based anomaly detection. In: SIGIR (2021)
25. Xie, Q., Dai, Z., Hovy, E., Luong, T., Le, Q.: Unsupervised data augmentation for consistency training. In: NeurIPS, pp. 6256–6268 (2020)
26. Xu, K., Hu, W., Leskovec, J., Jegelka, S.: How powerful are graph neural networks. In: ICLR (2019)
27. Zhang, X., Ge, Y., Qiao, Y., Li, H.: Refining pseudo labels with clustering consensus over generations for unsupervised object re-identification. In: CVPR (2021)

STTR-3D: Stereo Transformer 3D Network for Video-Based Disparity Change Estimation

Qitong Yang[1], Lionel Rakai[1], Shijie Sun[1], Huansheng Song[1]([envelope]), Xiangyu Song[2], and Naveed Akhtar[3]

[1] Chang'an University, Xi'an 710000, ShaanXi, China
hshsong@chd.edu.cn
[2] Swinburne University of Technology, Hawthorn, VIC 3122, Australia
[3] The University of Western Australia, Crawley, WA 6009, Australia

Abstract. In the field of computer vision and stereo depth estimation, there has been little research in obtaining high-accuracy disparity change maps from two-dimensional images. This map offers information that fills the gap between optical flow and depth which is desirable for numerous academic research problems and industrial applications, such as navigation systems, driving assistance, and autonomous systems. We introduce STTR3D, a 3D extension of the STereo TRansformer (STTR) which leverages transformers and an attention mechanism to handle stereo depth estimation. We further make use of the Scene Flow FlyingThings3D dataset which openly includes data for disparity change and apply 1) refinements through the use of MLP over relative position encoding and 2) regression head with an entropy-regularized optimal transport to obtain a disparity change map. This model consistently demonstrates superior performance for depth estimation tasks as compared to the original model. Compared to the existing supervised learning methods for estimating stereo depth, our technique simultaneously handles disparity estimation and the disparity change problem with an end-to-end network, also establishing that the addition of our transformer yields improved performance that achieves high precision for both issues.

Keywords: Stereo Estimation · Disparity Change · Scene Flow · Optimal Transport

1 Introduction

Scene flow estimation is one of the fundamental computer vision tasks that computes the depth and 3D motion vectors in a dynamic scene. It can benefit high-level applications, e.g. navigation systems [6,31], driving assistance [10,20] and autonomous systems [29,44], which are highly relevant to dynamic road scenes. Recently, deep learning-based methods have shown significant improvements over manually engineered feature-based techniques in the scene flow domain [23,35] alongside addressing challenges with multi-view clustering [9,37,38]. To estimate occlusion areas with optical flow or disparities, Ilg

X. Song et al. (Eds.): APWeb-WAIM 2023, LNCS 14334, pp. 217–231, 2024.
https://doi.org/10.1007/978-981-97-2421-5_15

et al. [17] presented a learning-based approach that uses a generic neural model for the task. A deep neural network architecture is also proposed in [14] for large-scale 3D point clouds. Similarly, Jiang et al. [19] introduced a compact network for holistic scene estimation, which shares common encoder features among four closely related tasks of optical flow estimation, disparity information from stereo, occlusion estimation, and semantic segmentation. The goal of other recent works has also been achieving competitive accuracy and real-time performance [16,34] particularly in image clustering and feature fusion [8,42].

Traditional scene flow estimation has three critical tasks: optical flow estimation, disparity estimation, and disparity change estimation. 3D points and 3D motion vectors can be deduced from them if camera parameters are known. Mayer et al. [24] showed that disparity change estimation can be handled well by deep learning. Recent research in this direction is mainly based on 3D point cloud scene flow estimation. The techniques need point cloud data to accomplish 3D reconstruction. Beh et al. [3] attempted to overcome the problem of 3D motion estimation changes by exploiting recognition while [4] proposed to estimate 3D motion from unstructured point clouds using a deep neural network. However, point cloud data is often voluminous [28] and sometimes has limited access. On the other hand, it is easier to obtain camera data and disparity maps.

Nowadays, a large number of stereo matching algorithms show remarkable performance, such as [5,40]. Vegeshna et al. [33] devised a stereo matching algorithm with careful handling of disparity, discontinuity, and occlusion, while [36] proposed a fully event-based stereo depth estimation algorithm that relies on message passing. These techniques are limited to small and single-scale receptive fields, and use traditional methods for cost aggregation or altogether ignore it. Yee et al. [41] proposed a network architecture that focuses on improving computational efficiency and robustness with a dense sub-pixel disparity estimation algorithm while Dong et al. [7] proposed an approach to computing dense disparity maps that takes the characteristics of man-made environments into account.

Mukherjee et al. [26] combined the known block-based and region-based stereo matching in their technique. However, the fact that their distribution is learned indirectly through a regression loss, causes the problem of ambiguous regions around object boundaries. This limitation is addressed by [11] using a neural network that uses a new loss function that is derived from the Wasserstein distance [32] between the ground truth and the predicted distributions. Most of the existing deep learning methods in scene flow domain are based on CNN, and disparity change estimation is rarely addressed. Like stereo depth estimation, disparity change estimation tracks the pixel point changes between two images. In stereo depth estimation, this problem has been recently considered by STereo TRansformer (STTR) [21]. The authors propose to use Transformers for the problem, which greatly improves the matching accuracy.

Inspired by STTR, we also rethink this problem from a sequence-to-sequence perspective. Since both disparity estimation and disparity change estimation can be viewed as a sequence-to-sequence problem, we make the following contributions to learn an effective neural model for the problem.

- We propose a new network, named STTR3D, which takes motivation from STTR, and deals with the task of disparity change. Our model offers a new way to handle the disparity change problem while maintaining high performance for disparity estimation.
- We take advantage of transformers and attention mechanisms by introducing a new transformer (STTR3D-TR) for disparity change estimation. Feature matching is also taken into account by adapting optimal transport from [30] in our disparity change estimation part.
- Our technique achieves the SOTA performance for disparity change task on FlyingThings3D [24] while achieving highly competitive results on both Scene Flow Driving and Monkaa dataset [24] without model fine-tuning.

2 Proposed STTR3D

The following section provides details of the proposed STereo TRansformer 3D (STTR3D). As shown in Fig. 1, the network consists of three sub-networks; namely, feature extractor, STTR3D-TR sub-network for disparity change estimation, and STTR-TR sub-network. The blocks additionally take help from Multi-head scaled dot-product attention, Multilayer perceptron (MLP) for position encoding, entropy-regularized Optimal Transport (OT), Regression, Context Adjustment Layer (CAL), and a specialized loss function. Below we discuss the components of the proposed STTR3D.

2.1 Feature Extractor

The original feature extractor of STTR is adopted in our model as it takes advantage of multiple resolutions. The original CNN is applied for extracting feature maps from four input frames: left and right images at different timestamps (t_1 and t_2). The Encoder-Decoder (hour-glass shaped) extractors first down samples via residual blocks and up samples by using transpose convolution alongside residual blocks and dense blocks. The two frames' stereo pairs t_1 and t_2 share the same weight. Our feature extractor gives a full-resolution feature map where each pixel will be denoted as a feature vector $p_f \in R^c$.

2.2 STTR3D-TR

In our Transformer, an alternating attention mechanism is leveraged to calculate self-attention and cross-attention. Self-attention computes the attention between pixels in the same image, while cross-attention computes the attention between pixels in the image pairs. We will discuss attention modules further in Sect. 2.2. In the feature matching problem, the task of handling ambiguous pixels is a key issue. To solve this problem, we use Multilayer Perceptron (MLP) for position encoding, and we use Optimal Transport (OT) to handle this problem with improved results, we explain this in detail in Sects. 2.2 and 2.2.

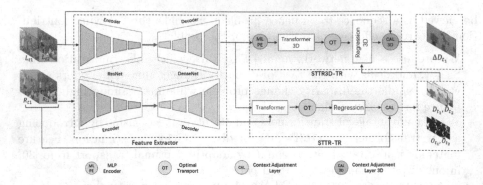

Fig. 1. Schematics of STTR3D. The network extracts features of left and right frames in t_1, t_2 by using a shared Encoder and Decoder. The extracted features are processed by STTR3D-TR and STTR-TR subnets, where "TR" denote Transformer and Regression. For STTR3D-TR, two left image feature maps are processed by Tansformer3D after adding positional encoding from the MLP encoder. The two maps are sent for optimal transport to generate the matching score map. Regression3D takes the disparity map from two depth frames and matching score maps to gain the raw disparity change map. The raw disparity change map is then sent to Context Adjustment Layer 3D with two left frames to get the final disparity change map. For STTR-TR, it takes 4 feature maps. It also uses a Transformer to update the feature and uses optimal transport with regression to gain a raw disparity map. The context adjustment layer 3D leverages all the origin stereo images to adjust the disparity maps.

Multi-head Scaled Dot-Product Attention Similar to the original STTR, we apply the same attention mechanism of alternating self-attention and cross-attention within the transformer. The changes to the implementation of attention within the transformer module as compared to STTR are as follows: The original STTR model implements a Multi-head attention model [21] which involves alternating self and cross-attention in the transformer module. It applies relative position encoding to all the self-attention and cross-attention layers.

Inspired by the SuperGlue network [30], the key point encoder allows the graph network generated to be motivated by both the appearance and position mutually when the vector is combined with a Multi-Layer Perceptron (MLP). Our model removes the position encoding from each attention module and applies Multi-Layer Perceptron encoding on the features which allow embedding and augmentation by positional information before implementing the transformer function as displayed in Fig. 2. This method offers the advantage of building more dynamic graphs between the essential key points and reducing computational complexity.

MLP for Position Encoding. For handling the problem of ambiguous pixels, we encode the pixels with both positional information and visual information. There are many instances of encoders being used in NLP [27]. In STTR, researchers choose the relative positional encoding to handle large textureless

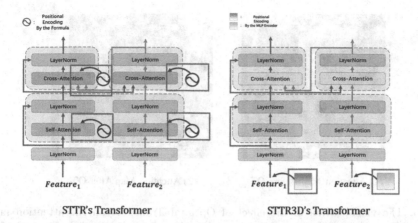

Fig. 2. A comparison of our transformer with the original STTR transformer illustrates the difference in placement of the encoding mechanism for both visual and location information. STTR applies relative position encoding to the self-attention and cross-attention layers while STTR3D feeds the already MLP-encoded feature descriptors into the transformer module before the attention module.

areas. In contrast, our STTR3D applies Multilayer Perceptron (MLP) to improve the objective of encoding the position information. We build the MLP by using several 1×1 convolutions without bias. The input feature map is updated as

$$p_f = p_f + \mathrm{MLP}_{en}(p_f). \tag{1}$$

This encoder allows the model to match pixels using both appearance information and position information so MLP performs well with our Transformer.

Entropy-Regularized Optimal Transport and Regression. In the STTR3D subnet, following the attention layer, we get the latest matching score as a cost matrix $\mathcal{M}$ of two marginal distributions a and b, representing the pixel points from two moments. The optimal coupling matrix [21] is found by iteratively normalizing along rows and columns by using the entropy regularization [21]. Figure 3 visualizes the effect of the Optimal Transport (OT) on the attention map patch.

For the application of Regression 3D after OT, the regression head specifically leverages the attention updated by the OT multiplied with the disparity map to obtain a new disparity map. This is subtracted from the disparity map of a different time stamp after re-aggregation. We get a coupling matrix S and create a unique index of each pixel in the two frames, as shown in Fig. 4. Then, we choose the highest responded pixel index to create a new disparity map D_{new} as given in Eq. (2).

$$D_{new}[i] = D_{t2}[x] \arg\max_{x \in r_1} S(x) \tag{2}$$

where $D_{t2}[x]$ is the selected pixel, r_1 is the index of frame 1, and $S(x)$ is the matching score.

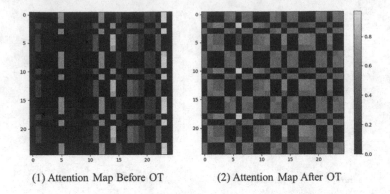

(1) Attention Map Before OT   (2) Attention Map After OT

Fig. 3. The visualization of the effect of Optimal Transport on the attention map patch. The left and right images represent the attention map patch before and after the Optimal Transport respectively. After Optimal Transport, ambiguous matching points(like the pixels with high light) in the attention map have been eliminated.

However, this method prevents the gradient flow and makes it hard to train the Transformer. To this end, we regard attention as a matching score. Given a cost matrix M of two marginal distributions a and b of length I_{hw}, the entropy-regularized optimal transport attempts to find the optimal coupling matrix T by solving:

$$\mathcal{T} = \underset{T \in R_+^{I_{hw}*I_{hw}}}{\arg\min} \sum_{i,j=1}^{I_{hw},I_{hw}} T_{ij}M_{ij} - \gamma E(T) \tag{3}$$

$$s.t. \quad T1_{I_{hw}} = a, T^T 1_{I_{hw}} = b$$

Let's denote the attention matrix after OT as $\mathcal{T}$. Let's denoting the matching probability in $\mathcal{T}$ as t, we re-weigh the assignment matrix by building a 3px-window

$$\widetilde{t_{\mathcal{L}}} = \frac{t_{\mathcal{L}}}{\sum_{\mathcal{L} \in \mathcal{N}_3(k)}}, for \mathcal{L} \in \mathcal{N}_3(k) \tag{4}$$

We use the disparity map from t_2 calculated by STTR-TR to build the new disparity map for computing a disparity change map. We first compute

$$D_{new} = \sum_{\mathcal{L} \in \mathcal{N}_3(k)} d_{t2}\widetilde{t_{\mathcal{L}}} \tag{5}$$

where d_{t2} represents the disparity value of each pixel in t_2. After getting the new disparity map, we can get the disparity change map by

$$\Delta D = D_{new} - D_{t1}. \tag{6}$$

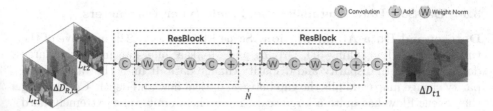

Fig. 4. Overview of CAL3D. L_{t1}, L_{t2} are left images from t_1 and t_2. $\Delta D_{R,t1}$ is the output of the "Regression 3D". The CAL3D consists of N ResBlock.

2.3 Context Adjustment Layer 3D (CAL3D)

Similar to stereo estimation, disparity change estimation also has its adjustment layers, named **C**ontext **A**djustment **L**ayer **3D** (**CAL3D**), as shown in Fig. 4. The CAL3D consists of N ResBlock. We concatenate the t_1 and t_2 left image with a disparity change map as the input of the CAL3D. The residual block expands the channels of the concatenated data and merges the information from different channels.

2.4 Loss

Based on the loss of STTR [21], we design the loss for both the disparity and disparity change tasks. The final loss is a linear combination of the sub-losses, computed as the following

$$L = w_1 L_{t1} + w_2 L_{t2} + w_3 L_{\Delta d}, \qquad (7)$$

where w_1, w_2, w_3 are the weights, L_{t1} and L_{t2} are the STTR loss from t_1, t_2 respectively. $L_{\Delta d}$ is the L1 loss [12] for disparity change task.

The sub-losses L_{t1}, L_{t2} are denoted as L_{ti} where $i \in \{1,2\}$.

$$L_{ti} = w_{i_1} L_{rr} + w_{i_2} L_{d1,r} + w_{i_3} L_{d1,f} + w_{i_4} L_{be,f}, \qquad (8)$$

where w_{i1}, w_{i2}, w_{i3} and w_{i4} are the loss weights, which are shown in the Sect. 3. L_{rr} is the Relative Response loss proposed in [22], $L_{d1,r}$ is the smooth L1 function for raw disparity, $L_{d1,f}$ represents the smooth L1 loss for the final disparities, and $L_{be,f}$ represents the binary-entropy loss for the final occlusion map.

3 Experiments

In this section, we first explain the dataset employed for training and evaluation followed by the introduction of the hyperparameters used for the transformer and attention layers. We describe the experimental details, including the application of the MLP encoder, attention stride for memory feasible implementation, and the disparity change task. Cross-domain tests are also included to affirm the generalization of the method. Finally, quantitative and qualitative results are reported and compared with state-of-the-art methods along with the ablation study and limitations.

3.1   Dataset, Data Augumentation, and Hyper Parameters

Dataset and Data Augumentation. Scene Flow dataset [24], inclusive of the three subsets FlyingThings3D, Driving and Monkaa, offers an innovative large dataset with both disparity and disparity change data. It finds the value of disparity change maps and fills the gap between optical flow and depth. Considering that Scene Flow has quite a large scale, researchers omit some extremely hard samples from the FlyingThings3D dataset. We use the FlyingThings3D, Driving and Monkaa subsets to train STTR3D. For data augmentation, we use random crops to reduce the usage of GPU memory. Additionally, we also add Gaussian noise and random RGB shift to the data pre-processing procedure to simulate realistic situations.

Hyper-parameters and Training. In our experiments, the output dimension of the feature is $C_f = 128$. The number of attention layers for both Transformers is set to 6. The attention strides are set to $s = 3$. The layer channel of MLP are set to $16, 32, 64, 128$. The number of iterations for OT is set to 10. For the learning rate, we set the initial learning rate to 1e-4, and the learning rate decay to 0.99. We use AdamW optimizer with a weight decay of 1e-4. All loss function weights of STTR are set to 1, and the STTR3D's loss function weight is also 1. We load the weights of STTR and train the STTR separately. After training the STTR, we train the whole model. We use the whole FlyingThings3D dataset to train the STTR3D for 10 epochs. The experiments are conducted on one NVIDIA RTX TITAN GPU. For the evaluation of our model, we use 3px error and EPE (end-point error) as the evaluation metrics for disparity estimation and disparity change estimation respectively. Similar to STTR, we only evaluate the non-occlusion area. We conduct several ablation experiments to explore the impact of different improvements on the model.

3.2   Ablation Experiments

To test if our Transformer, MLP encoder, and regression head affect the original STTR performance, the results for disparity estimation are tabulated in Table 1 for comparison. We train the whole model and apply the disparity evaluation metrics 3px error and EPE for the performance evaluation. STTR3D is then compared to the original STTR and other prior works which cover the primary learning-based stereo depth standards. They include the deformable convolution and correlation-based AANet [40], a spatial pyramid pooling based PSMNet [5], 3D convolution based GANet-11 [43], a combination approach of 3D convolution and group-wise correlation GwcNet-g [15], and the binary classification based Bi3D [2]. The results in Table 1 show the introduction of the new model can cooperate with the original STTR model by displaying matching results. This confirms the superior performance of the original model is still reflected with the extensions in STTR3D. For the disparity change estimation task, we elaborate on the MLP encoder, attention stride, and context adjustment layers. Consistent with STTR, we validate using the test split directly since Scene Flow is only used for pre-training.

Table 1. Ablation experiments to test disparity estimation. All the models are trained and tested on the Scene Flow dataset with the same data augmentation. ↓ indicates lower values are more desirable, ↑ shows higher values are better. Since we use the module from STTR to predict the disparity change, the results confirm the disparity estimation task specifically does not degrade.

	STTR3D	STTR	AANet	GwcNet	PSMNet	GANet	Bi3D
3px error ↓	**1.13**	1.13	1.86	1.57	2.87	1.6	1.7
epe↓	**0.42**	0.42	0.49	0.48	0.95	0.48	0.54
IoU↑	**0.92**	0.92	N/A	N/A	N/A	N/A	N/A

MLP Encoder. Ambiguous matching problems exist both in disparity estimation and disparity change estimation. STTR used relative positional encoding to let the Transformer perceive the tiny differences between two ambiguous pixels. Since the disparity change task matches the relevant pixels between two frames, the MLP encoder is applied in place of Relative Positional Encoding used in the original STTR framework. It encodes the keypoint jointly with its visual features and positional information, similarly applied in [30]. The difference in the information carried by the MLP encoder improves the two metrics of disparity change estimation. The ablation study results are shown in Table 2.

Table 2. Ablation experiments about MLP. We train the model on the Scene Flow dataset. The introduction of MLP improved the performance of STTR3D

	STTR3D	STTR3D (without MLP)
epe↓	**0.55**	0.67

Memory-Feasible Implementation About STTR3D. Although STTR3D is effective in solving the disparity change problem, the introduction of the Transformer results in a significant increase in computation and GPU memory usage similar to STTR. The matching step flattens the pixel's index, so the memory usage in the Transformer will be

$$\text{bits} = 32H^2W^2N_hN, \tag{9}$$

where H and W represent the height and width of the image. N_h represents the number of attention heads, N represents the number of attention layer, the number 32 represents the unit data size during float32 precision training. Such a huge consumption is generally not feasible. To handle this problem, we also use gradient checkpoint [13]when computing the attention. Gradient checkpoint sacrifices part of the computing efficiency in exchange for the utilization of GPU memory where the memory usage will only be relevant to the single attention layer memory. Furthermore, attention stride s is also adopted to reduce memory usage [1,39]. We tried different kinds of attention stride values s to find the balance between accuracy and computational speed. Mixed precision training [25] is a method applied during the training process to speed up the training and reduce the consumption of GPU memory [18].

The introduction of the attention stride can greatly decrease the usage of GPU memory and the time for computation. We test several values of attention strides to explore their impact. While certain models have applied a stride value of 2 [1], our training finds optimal results with a stride value of 3 similar to [21,39] based on the limitation of a single GPU (experiments performed on the NVIDIA RTX TITAN GPU with 24GB GPU memory) where reducing the attention stride further would squeeze the crop size of the data and further harm the performance. The results of the attention stride about disparity change task are shown in Table 3. We also show in Table 3 that the attention stride mainly influenced the performance of the attention layer, but it has only a slight effect on the final result. This is mainly because the context adjustment layers can learn how to upsample a low-resolution disparity change map to a high-resolution disparity change map.

Table 3. Ablation experiments for the attention stride s in disparity change task. We train the model using different attention strides on the Scene Flow dataset and validate on the test split. OOM means **Out Of Memory**.

	STTR3D (s=2)	**STTR3D (s=3)**	STTR3D (s=4, 5)	STTR3D (s=6)
epe↓	OOM	**0.55**	0.57	0.60

3.3   Disparity Change Task

We test our method on the Scene Flow dataset and compare the results for the disparity change task. The results affirm that our method achieves considerable improvements over FlowNet, see Table 5. In Fig. 5, we illustrate the disparity change ground truths alongside the corresponding RGB left and right images with the predicted maps generated by our model and FlowNet as a comparison.

The disparity change predictions in Fig. 5 are analyzed as follows:

1. The appearance of objects in the disparity change map for STTR3D shows a significantly less blurred effect.
2. Our model shows sharper and clearer boundaries and edges compared to FlowNet.
3. The improved clarity in the disparity change map generated by STTR3D demonstrates a better representation ability of smaller structures.

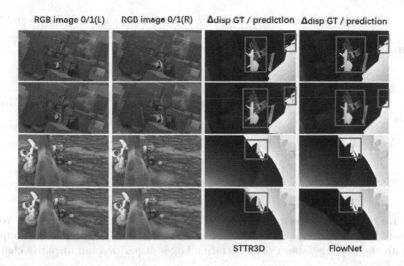

Fig. 5. Representative examples of the RGB left-right images, the disparity change ground truths (rows 1 and 3), and the comparison of the disparity change predictions (rows 2 and 4) by both STTR3D and FlowNet. It demonstrates that STTR3D is superior to FlowNet in the disparity change task. STTR3D can predict the detailed disparity change due to the optimal transport matching.

3.4   Cross Domain Test

Since STTR has a strong generalization ability, the generalization of STTR3D is also re-tested for the disparity estimation task to prove that this new task does not interfere with the generalization of the original STTR model. We apply the same approach as STTR, which means we train only on the Scene Flow dataset and avoid further fine-tuning the model on the target dataset. This approach also ensures that prior models are trained using the same augmentation technique. In Table 4, results confirm the generalization of our model is consistent with STTR, proving that the new task does not bring negative side effects on the disparity estimation task.

Correspondingly, since the key innovation of our model is based on disparity change, we also test the generalization of our method for the disparity change task. We test our model on the Scene Flow driving dataset without fine-tuning. Scene Flow Driving is closer to the actual environment compared to the synthetic data of FlyingThings3D and Monkaa. Our model displays improved results with all three subsets of Scene Flow and the greatest improvement on the disparity change task with the Driving dataset, proving superior generalization. The results comparing our model with FlowNet [24] are summarized in Table 5.

228   Q. Yang et al.

Table 4. Cross-domain test for the disparity estimation task on KITTI2015 and MPI Sintel dataset. Results ascertain that STTR3D can retain the high performance in disparity estimation since it inherits the parts from STTR.

	KITTI2015			MPI Sintel		
	3px error↓	epe↓	IoU↑	3px error ↓	epe↓	IoU↑
PSMNet	7.43	1.39	N/A	7.93	3.7	N/A
GwcNet-g	6.75	1.59	N/A	5.83	1.32	N/A
STTR	6.74	1.5	0.98	5.75	3.01	0.86
STTR3D	6.74	1.5	0.98	5.75	3.01	0.86

Table 5. Results of the epe for disparity change task using Scene Flow dataset. We do not fine-tune STTR3D on the Driving dataset. All measurements represent endpoint error with the Driving dataset demonstrating larger disparities and disparity change.

	STTR3D	FlowNet
FlyingThings	**0.55**	0.8
Driving	**5.76**	16.34
Monkaa	**0.46**	0.78

3.5 Limitations

While STTR3D performed well for the tasks of both disparity and disparity change estimation, it is worth mentioning that the availability of data for the disparity change task is currently limited in terms of public availability. Currently, the Scene Flow dataset offers the full data for the disparity, optical flow and disparity change, which limits the generalization test for the disparity change task. Since 3D motion vectors can be decomposed into optical flow components and disparity change components, we hope there will be more datasets for disparity changes in the future.

4 Conclusion

We presented a new model called STTR3D to estimate and represent both disparity maps and disparity change maps in a video sequence. STTR3D takes advantage of combining a CNN and Transformer of the original STTR setup and includes a new transformer for the task of disparity change. It considerably improves the accuracy of disparity change estimation as compared to traditional CNNs. Experimental results confirm that the introduction of the new Transformer also does not degrade STTR's performance as the new STTR3D architecture accomplished two tasks simultaneously. STTR3D further performs well in the new environment with a small amount of fine-tuning. Future work will take advantage of disparity change maps to complete 3D reconstruction.

Acknowledgment. This work was supported in part by the National Natural Science Foundation of China under Grant No. 62072053 and the National Natural Fund Joint Fund Project under Grant No. U21B2041.

References

1. Aich, S., Vianney, J.M.U., Islam, M.A., Liu, M.K.B.: Bidirectional attention network for monocular depth estimation. In: 2021 IEEE International Conference on Robotics and Automation (ICRA), pp. 11746–11752. IEEE (2021)
2. Badki, A., Troccoli, A., Kim, K., Kautz, J., Sen, P., Gallo, O.: Bi3D: stereo depth estimation via binary classifications. In: Proceedings of the IEEE/CVF Conference on Computer Vision and Pattern Recognition, pp. 1600–1608 (2020)
3. Behl, A., Hosseini Jafari, O., Karthik Mustikovela, S., Abu Alhaija, H., Rother, C., Geiger, A.: Bounding boxes, segmentations and object coordinates: How important is recognition for 3D scene flow estimation in autonomous driving scenarios? In: Proceedings of the IEEE International Conference on Computer Vision, pp. 2574–2583 (2017)
4. Behl, A., Paschalidou, D., Donné, S., Geiger, A.: PointFlowNet: learning representations for rigid motion estimation from point clouds. In: Proceedings of the IEEE/CVF Conference on Computer Vision and Pattern Recognition, pp. 7962–7971 (2019)
5. Chang, J.R., Chen, Y.S.: Pyramid stereo matching network. In: Proceedings of the IEEE Conference on Computer Vision and Pattern Recognition, pp. 5410–5418 (2018)
6. Diamantas, S.C., Oikonomidis, A., Crowder, R.M.: Depth estimation for autonomous robot navigation: a comparative approach. In: 2010 IEEE International Conference on Imaging Systems and Techniques, pp. 426–430. IEEE (2010)
7. Dong, Q., Feng, J.: Outlier detection and disparity refinement in stereo matching. J. Vis. Commun. Image Represent. **60**, 380–390 (2019)
8. Fang, U., Li, J., Lu, X., Mian, A., Gu, Z.: Robust image clustering via context-aware contrastive graph learning. Pattern Recognit. **138**, 109340 (2023)
9. Fang, U., Li, M., Li, J., Gao, L., Jia, T., Zhang, Y.: A comprehensive survey on multi-view clustering. IEEE Trans. Knowl. Data Eng. **35**, 12350–12368 (2023)
10. Fletcher, L., Loy, G., Barnes, N., Zelinsky, A.: Correlating driver gaze with the road scene for driver assistance systems. Robot. Auton. Syst. **52**(1), 71–84 (2005)
11. Garg, D., Wang, Y., Hariharan, B., Campbell, M., Weinberger, K.Q., Chao, W.L.: Wasserstein distances for stereo disparity estimation. Adv. Neural. Inf. Process. Syst. **33**, 22517–22529 (2020)
12. Girshick, R.: Fast r-CNN. In: Proceedings of the IEEE International Conference on Computer Vision, pp. 1440–1448 (2015)
13. Griewank, A., Walther, A.: Algorithm 799: revolve: an implementation of check-pointing for the reverse or adjoint mode of computational differentiation. ACM Trans. Math. Softw. **26**(1), 19–45 (2000)
14. Gu, X., Wang, Y., Wu, C., Lee, Y.J., Wang, P.: HPLFlowNet: hierarchical permutohedral lattice FlowNet for scene flow estimation on large-scale point clouds. In: Proceedings of the IEEE/CVF Conference on Computer Vision and Pattern Recognition, pp. 3254–3263 (2019)
15. Guo, X., Yang, K., Yang, W., Wang, X., Li, H.: Group-wise correlation stereo network. In: Proceedings of the IEEE/CVF Conference on Computer Vision and Pattern Recognition, pp. 3273–3282 (2019)

16. Hur, J., Roth, S.: Self-supervised monocular scene flow estimation. In: Proceedings of the IEEE/CVF Conference on Computer Vision and Pattern Recognition, pp. 7396–7405 (2020)
17. Ilg, E., Saikia, T., Keuper, M., Brox, T.: Occlusions, motion and depth boundaries with a generic network for disparity, optical flow or scene flow estimation. In: Proceedings of the European Conference on Computer Vision (ECCV), pp. 614–630 (2018)
18. Jia, X., et al.: Highly scalable deep learning training system with mixed-precision: training imagenet in four minutes. arXiv preprint arXiv:1807.11205 (2018)
19. Jiang, H., Sun, D., Jampani, V., Lv, Z., Learned-Miller, E., Kautz, J.: Sense: A shared encoder network for scene-flow estimation. In: Proceedings of the IEEE/CVF International Conference on Computer Vision, pp. 3195–3204 (2019)
20. Kukkala, V.K., Tunnell, J., Pasricha, S., Bradley, T.: Advanced driver-assistance systems: a path toward autonomous vehicles. IEEE Consum. Electron. Mag. 7(5), 18–25 (2018)
21. Li, Z., et al.: Revisiting stereo depth estimation from a sequence-to-sequence perspective with transformers. In: Proceedings of the IEEE/CVF International Conference on Computer Vision, pp. 6197–6206 (2021)
22. Liu, X., et al.: Extremely dense point correspondences using a learned feature descriptor. In: Proceedings of the IEEE/CVF Conference on Computer Vision and Pattern Recognition, pp. 4847–4856 (2020)
23. Ma, W.C., Wang, S., Hu, R., Xiong, Y., Urtasun, R.: Deep rigid instance scene flow. In: Proceedings of the IEEE/CVF Conference on Computer Vision and Pattern Recognition, pp. 3614–3622 (2019)
24. Mayer, N., et al.: A large dataset to train convolutional networks for disparity, optical flow, and scene flow estimation. In: Proceedings of the IEEE Conference on Computer Vision and Pattern Recognition, pp. 4040–4048 (2016)
25. Micikevicius, P., et al.: Mixed precision training. arXiv preprint arXiv:1710.03740 (2017)
26. Mukherjee, S., Guddeti, R.M.R.: A hybrid algorithm for disparity calculation from sparse disparity estimates based on stereo vision. In: 2014 International Conference on Signal Processing and Communications (SPCOM), pp. 1–6. IEEE (2014)
27. Özçift, A., Akarsu, K., Yumuk, F., Söylemez, C.: Advancing natural language processing (NLP) applications of morphologically rich languages with bidirectional encoder representations from transformers (BERT): an empirical case study for Turkish. Automatika: časopis za automatiku, mjerenje, elektroniku, računarstvo i komunikacije 62(2), 226–238 (2021)
28. Pajić, V., Govedarica, M., Amović, M.: Model of point cloud data management system in big data paradigm. ISPRS Int. J. Geo Inf. 7(7), 265 (2018)
29. de Queiroz Mendes, R., Ribeiro, E.G., dos Santos Rosa, N., Grassi, V., Jr.: On deep learning techniques to boost monocular depth estimation for autonomous navigation. Robot. Auton. Syst. 136, 103701 (2021)
30. Sarlin, P.E., DeTone, D., Malisiewicz, T., Rabinovich, A.: Superglue: learning feature matching with graph neural networks. In: Proceedings of the IEEE/CVF Conference on Computer Vision and Pattern Recognition, pp. 4938–4947 (2020)
31. Shen, M., Gu, Y., Liu, N., Yang, G.Z.: Context-aware depth and pose estimation for bronchoscopic navigation. IEEE Robot. Autom. Lett. 4(2), 732–739 (2019)
32. Vallender, S.: Calculation of the wasserstein distance between probability distributions on the line. Theory Probab. Appl. 18(4), 784–786 (1974)
33. Vegeshna, V.P.K.V.: Stereo matching with color-weighted correlation, hierarchical belief propagation and occlusion handling. arXiv preprint arXiv:1708.07987 (2017)

34. Wang, L., Ren, J., Xu, B., Li, J., Luo, W., Xia, F.: Model: motif-based deep feature learning for link prediction. IEEE Trans. Comput. Soc. Syst. **7**(2), 503–516 (2020)
35. Wang, Z., Li, S., Howard-Jenkins, H., Prisacariu, V., Chen, M.: FlowNet3D++: geometric losses for deep scene flow estimation. In: Proceedings of the IEEE/CVF Winter Conference on Applications of Computer Vision, pp. 91–98 (2020)
36. Xie, Z., Chen, S., Orchard, G.: Event-based stereo depth estimation using belief propagation. Front. Neurosci. **11**, 535 (2017)
37. Xu, C., Guan, Z., Zhao, W., Wu, H., Niu, Y., Ling, B.: Adversarial incomplete multi-view clustering. In: IJCAI, vol. 7, pp. 3933–3939 (2019)
38. Xu, C., Zhao, W., Zhao, J., Guan, Z., Song, X., Li, J.: Uncertainty-aware multiview deep learning for internet of things applications. IEEE Trans. Industr. Inf. **19**(2), 1456–1466 (2022)
39. Xu, D., Wang, W., Tang, H., Liu, H., Sebe, N., Ricci, E.: Structured attention guided convolutional neural fields for monocular depth estimation. In: Proceedings of the IEEE Conference on Computer Vision and Pattern Recognition, pp. 3917–3925 (2018)
40. Xu, H., Zhang, J.: AANet: adaptive aggregation network for efficient stereo matching. In: Proceedings of the IEEE/CVF Conference on Computer Vision and Pattern Recognition, pp. 1959–1968 (2020)
41. Yee, K., Chakrabarti, A.: Fast deep stereo with 2D convolutional processing of cost signatures. In: Proceedings of the IEEE/CVF Winter Conference on Applications of Computer Vision, pp. 183–191 (2020)
42. Yin, H., Yang, S., Song, X., Liu, W., Li, J.: Deep fusion of multimodal features for social media retweet time prediction. World Wide Web **24**, 1027–1044 (2021)
43. Zhang, F., Prisacariu, V., Yang, R., Torr, P.H.: GA-Net: guided aggregation net for end-to-end stereo matching. In: Proceedings of the IEEE/CVF Conference on Computer Vision and Pattern Recognition, pp. 185–194 (2019)
44. Zhou, C., Yan, Q., Shi, Y., Sun, L.: DoubleStar: long-range attack towards depth estimation based obstacle avoidance in autonomous systems. arXiv preprint arXiv:2110.03154 (2021)

HM-Transformer: Hierarchical Multi-modal Transformer for Long Document Image Understanding

Xi Deng, Shasha Li$^{(\boxtimes)}$, Jie Yu, and Jun Ma

College of Computer, National University of Defense Technology,
Changsha 410000, China
{shashali,yj,majun}@nudt.edu.cn

Abstract. Transformer plays a massive role in document image understanding. However, it has difficulty handling text in long document images due to the increasing quadratic complexity along the text length. To solve this problem, we propose the hierarchical multi-modal transformer (HM-Transformer) for long document image understanding. HM-Transformer hierarchically models document images. It learns the block representation first and then the document image representation. Specifically, we first use the pre-trained model LayoutLMv3 to generate a block representation containing image, layout, and text information. Then we utilize a document multimodal transformer to model the global multimodal document text representations and image representations. We then reconstruct the in-block representation in the block multimodal Transformer with the help of global multimodal document text and image representation. Finally we conduct different operations depending on the downstream tasks. Experimental results show that HM-Transformer achieves new state-of-the-art performance on two downstream document image understanding tasks, including FUNSD and CORD. The code and models are publicly available at https://github.com/dx233333/picture_project.

Keywords: Hierarchical · Transformer · Document image understanding

1 Introduction

Document image understanding aims to analyze and extract the required information from diverse document images. These document images include business forms, receipts, academic papers, etc. Figure 1 shows samples from two benchmark datasets. Automatically processing document images using artificial intelligence techniques is essential for industry and academia.

On the one hand, document images contain textual and equally significant visual and layout information. Multimodal models assist in fully exploiting information from different modalities simultaneously [19,32]. On the other hand,

X. Song et al. (Eds.): APWeb-WAIM 2023, LNCS 14334, pp. 232–245, 2024.
https://doi.org/10.1007/978-981-97-2421-5_16

self-supervised pre-training models [15,20] of the transformer [27] family have emerged in recent years. Pre-trained models trained on large-scale unlabelled data help to achieve better performance for downstream tasks. Therefore, using pre-trained multimodal models [1,23,30,31] on document image understanding is an excellent option; in particular, LayoutLMv3 [13] has obtained state-of-the-art results. However, in practice, the computational complexity of the model is proportional to the square of the text length. So it is often needed to truncate the document into several parts for input into the model, which is a great challenge for long document image understanding [4].

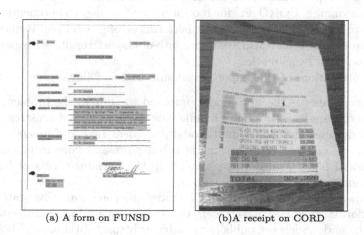

(a) A form on FUNSD   (b)A receipt on CORD

Fig. 1. Samples from two benchmark datasets.

There are several ways to handle long documents in the text modality. One direction is the emergence of pre-trained models for longer texts [2,5,6,26,34]. But it isn't easy to directly transfer to the multimodal downstream task of document image understanding. Another direction is in variations on frameworks such as hierarchical structures [33,35] or sparse self-attention mechanisms [3]. The former sentence representation requires more context, and the latter might lose some information. Recently, the Hi-Transformer [29] uses global document representations to enhance sentence representations for efficient and effective long document modeling. However, there is little concern about long documents in the downstream task of a multimodal pre-trained model for document image understanding.

To address the problem of truncating long documents in the downstream task of multimodal pre-trained models, we propose a hierarchical multimodal transformer (HM-Transformer) for long document image understanding. As shown in Fig. 2, inspired by Hi-transformer, HM-transformer learns the block representation first and then the document image representation. And it always focuses on layout and image information throughout the process. Specifically, we employ the current most advanced multimodal pre-trained model LayoutLMv3 to generate the text block representation. The block representation contains the entire image information, the text within the block, and the corresponding layout information for the text. We then use a document transformer that enables all text block and image representations to interact, resulting in a global document image representation. Next, we use the global document image representation to enhance the text block and image representation in a block transformer. Finally, we use two pooling methods to obtain the document image representation for the downstream task of document image understanding.

We select two publicly available benchmark datasets as downstream tasks to evaluate the performance of the HM-Transformer model: FUNSD [14] for form understanding, CORD [22] for receipt understanding. Experimental results demonstrate that HM-Transformer outperforms strong baselines, including the vanilla LayoutLMv3, and achieves new state-of-the-art results with parameter efficiency on both benchmarks.

The contributions of this paper are summarized as follows:

- HM-Transformer is the first model proposed to address the problem of how multimodal pre-trained models handle long documents in downstream tasks. The model uses the global multimodal document text and image representation to to enhance the modeling of token representations in blocks in the multimodal block Transformer, alleviating the harmful effects of long text truncation.
- Experimental results show that HM-Transformer achieves new state-of-the-art performance in two downstream tasks of document image understanding. The code and models are publicly available at https://github.com/dx233333/picture_project.

2   Related Works

The **multimodal pre-trained models** perform excellently on various downstream tasks of document image understanding. LayoutLM [31] uses the BERT [15] architecture as the backbone to model text and layout information jointly. LayoutLMv2 [30] follows the Transformer [27] framework to integrate text, layout, and visual information in the pre-training phase. ERNIE-Layout [23] applies a multimodal Transformer to learn better representations that combine text, layout, and image features. LayoutLMv3 [13] utilizes a unified multimodal Transformer with inputs of text and images to learn cross-modal representations. These pre-trained models have powerful representation learning ability, with LayoutLMv3 currently achieving state-of-the-art performance. However, the model complexity of the Transformer is proportional to the square of the text sequence length. Due to the limitation of memory and computational resources, the maximum sequence length of all multimodal pre-trained models for document image understanding is only 512. In practice, text exceeding this length is truncated, resulting in partial loss of contextual information. Our model is designed to alleviate the information loss due to truncation on long documents on document image understanding tasks.

Handling long documents by document segmentation to adapt pre-trained models has attracted academic interest recently. It splits long text into short parts according to specific rules and generates separate part representations. These representations are then aggregated by pooling, attention mechanisms, MLP, LSTM, or transformer methods. Yang et al. [33] use hierarchical attention networks at word and sentence levels. Pappagari et al. [21] use Bert to obtain smaller block representations and then aggregate these representations by LSTM or transformer. Yongli Hu [12] employed scibert to generate section representations and then used transformer for information interaction. Hibert [35] proposed a pre-training model for text summarization exploiting sentence representations. These models have achieved some success, but they need more context on smaller part representations. Wu et al. [29] use global document representations to augment sentence representations to allow sentence representations to see the full text. Inspired by Wu, we sliced the document into chunks in the text modality of the multimodal model, used the global document representation to augment the chunk representation, and carefully considered image information. Also, our work is the first to handle long documents in document image understanding for the downstream task of pre-training multimodal models.

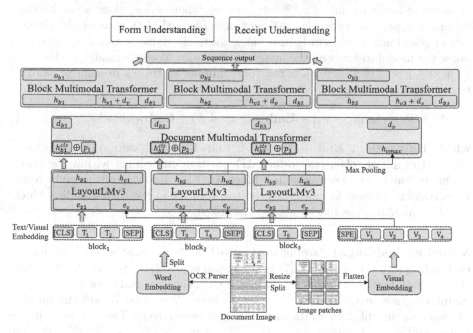

Fig. 2. The architecture of HM-Transformer.

3   Approach

We build the multimodal transformer architecture as the backbone of the HM-Transformer. The transformer adopts a multi-layer architecture, where each layer consists mainly of multi-head attention and point-wise, fully connected layers. Figure 2 presents an overview of the HM-Transformer. Detailed descriptions of the model are as follows.

3.1 Input Representation

Blocked Text Embedding. The text embedding consists of word embedding, 1D position embedding, and 2D layout position embedding. We obtain the text and the corresponding 2D layout information from the document image in the pre-processing stage with officially provided OCR annotations. Following LayoutLMv3, we initialize our word embedding with RoBERTa's embedding matrix. The difference is that we split the tokenized token into multiple blocks by the maximum length L. And we add [CLS] at the beginning of each block and [SEP] at the end. The 1D position refers to the index of tokens within the corresponding block. And the 2D layout position (x_1, y_1, x_2, y_2) indicates the coordinates of the bounding box of text blocks, where (x_1, y_1) and (x_2, y_2) correspond to the bounding box's upper left and lower right positions, respectively. Like layoutlmv3, We normalize all coordinates by the image size and use position embedding layers to embed x_1, y_1, x_2, and y_2, where x_1, x_2 share an embedding table and y_1, y_2 share another. Different from LayoutLMv3, we adopt block-level layout positions, where words in a block share the same layout information to allow the model to focus on the image of the entire block. Formally, we have the j-th ($0 \leq j <$ len, where len is the total number of tokens in a single sample) text token embedding:

$$\mathbf{T}_j = \mathbf{Emb}_{tok}(w_j) + \mathbf{Emb}_{1D}(j \bmod L) + \mathbf{Emb}_{2D}(block_{\lfloor j/L \rfloor}) \qquad (1)$$

where $\mathbf{Emb}_{tok}(w_j)$, $\mathbf{Emb}_{1D}(j \bmod L)$, and $\mathbf{Emb}_{2D}(block_{\lfloor j/L \rfloor})$ respectively denote the initial token embedding, 1D position embedding within the corresponding block, and corresponding block-level 2D layout position embedding. In particular, $(j \bmod L)$ calculates the token index in the corresponding block, $(\lfloor j/L \rfloor)$ calculates the block index of the token.

Visual Embedding. Following the LayoutLMv3, we take the patch projection embedding introduced by ViT [7] as the visual embedding. Specifically, we resize the original document image for an image with the specified resolution. We can define the new image as $I \in \mathbb{R}^{C \times H \times W}$, where C, H, and W are the number of channels, height, and width of the image, respectively. Then we split image I into uniform 2D patches $I_p \in \mathbb{R}^{C \times P \times P}$ of number $N = HW/P^2$. We map patches to D dimensions with a trainable linear projection and flatten them into a sequence of visual embeddings. And we add [SPE] at the beginning of the sequence as the token for the whole image representation. We also add a learnable position embedding for each patch to mark their order.

3.2 LayoutLMv3

LayoutLMv3 is a generic multimodal pre-training model for document image understanding and has achieved state-of-the-art results on several downstream tasks. However, due to its maximum input text sequence length being 512, long texts in document images need to be truncated. Our model is an improved version of LayoutLMv3 to alleviate the information loss due to truncation. HM-Transformer$_{\text{BASE}}$ adopts LayoutLMv3$_{\text{BASE}}$ with a 12-layer standard Transformer encoder to process the input, while HM-Transformer$_{\text{LARGE}}$ utilizes LayoutLMv3$_{\text{LARGE}}$ with a 24-layer standard Transformer encoder. (For convenience, we define the range of values of that appears in the full text. $0 \leq i <$ toknum, where toknum is the number of blocks in a single sample.)

The pre-trained model LayoutLMv3 is at the bottom of the HM-transformer framework. We concatenate the same visual token sequence e_v with each block's text token sequence e_{bi} to form a new input sequence. Then we feed each of these information-rich input sequences into the parameter-sharing LayoutLMv3. After sufficient interaction within the LayoutLMv3 model, the output embeddings incorporate the textual information of the whole block, the layout information, and the information of the entire image on top of itself. The front part of the output is new embeddings h_{bi} for the sequence of text tokens in the block. Among them, the token embedding h_{bi}^{cls} of [CLS] is a blocked textual representation containing the context within the block, layout, and image information. Similarly, the later part of the output is the new visual embedding sequence h_{vi} for patch tokens of the image. In particular, we consider the embedding corresponding to [SPE] as the image representation of the whole image after interacting with the block's text with layout information.

In addition, our neat model is not only suitable for LayoutLMv3. The LayoutLMv3 module in our model can be replaced with other multimodal pre-trained models, no or slightly modified.

3.3 Document Multimodal Transformer

The input of the document multimodal transformer consists of two parts. One part is text embedding. The text embedding e_{bi} with layout information becomes contextual multimodal representations h_{bi} of intra-block tokens after interacting with contextual text tokens and image patches in the previous module LayoutLMv3. The embedding h_{bi}^{cls} of the "[CLS]" token in h_{bi} is considered the block text representation. We add a block position embedding (denoted as p_i for the i-th block) to the representation h_{bi} to identify block orders. This block representation with positional information is the text embedding input for the document multimodal transformer. The other part of the input is image embedding. In the LayoutLMv3 module, all embeddings e_v of image patches interact with different block texts, focusing on distinct features, to obtain various new image embeddings h_{vi}. We perform max pooling in each dimension of these embeddings h_{vi} to obtain the most important features while reducing the model parameters and speeding up the training. The pooled result h_{vmax} is the image embedding input of the multimodal document transformer, concatenated behind the text embedding input. We apply the multimodal document transformer to the concatenated inputs to capture the document context and crucial image information. We further learn the document context-aware multimodal block representation d_{bi} and image representation d_v.

3.4 Block Multimodal Transformer

This module serves to reconstruct block representations with the aid of document context-aware block representations and image representations. We divide the input of the module into three parts to illustrate it. Specifically, part one is the intra-block context-aware multimodal text representation h_{bi} produced by LayoutLMv3. The second part concatenates the image representation h_{vi} produced by LayoutLMv3 with the d_v output by the multimodal document transformer. h_{vi} focuses on the image information corresponding to each block, while d_v concerns the core image information of all blocks. The combination of both allows the block representation to fuse image features better. Next, the third part of the input is d_{bi}, generated by the multimodal document transformer. Document context-aware block representations d_{bi} propagate the global document context with fused layout and image information into each block to improve the block representation modeling. We apply a multimodal block transformer to each block's three-part input to fully model the interactions. It outputs a document context-aware token representation sequence for each block, denoted as o_{bi}. Finally, we do different operations on the sequence depending on the downstream task. The downstream tasks in this paper are form understanding and receipt understanding. We provide the implementation details of tasks in the experimental section.

4 Experiments

4.1 Model Configurations

HM-Transformer has a multimodal block transformer and a multimodal document transformer. The multimodal block transformer utilizes a 2-layer Transformer encoder with 12-head self-attention, hidden size of $D = 768$, and 2048 intermediate size of position-wise feed-forward networks. And the multimodal document transformer uses a 1-layer Transformer with the same settings as the multimodal block Transformer. In the text preprocessing stage, we tokenize the text sequence and set the maximum sequence length $L = 512$ in each block. We add a [CLS] token at the beginning of each block and a [SEP] token at the end. When the text sequence length of the last block is shorter than L, we add extra [PAD] tokens to the end of it. In the image preprocessing stage, the parameters are $C = 3$, $H = 224$, $W = 224$, $P = 16$, and $N = 196$, respectively.

4.2   Compared Methods

Since we are the first to address the long document truncation problem for the downstream task of multimodal pre-trained models in document image understanding, there is no suitable baseline to compare. We use popular pre-trained models as less reasonable baselines in Table 1. Our model can work on these pre-trained models with minor modifications, but we have not yet completed it and have only chosen the best one, LayoutLMv3, for our experiments. We classify these compared methods into the following categories according to their modalities, determining how to combine them with HM-transformer in the future.

[T] Text Modality. BERT [15] and RoBERTa [20] are classical pre-trained language models with the input of text information only. We directly report the FUNSD and CORD results of BERT, RoBERTa from LayoutLMv3 [13]. After removing image-related and layout-related embeddings and substituting the LayoutLMv3 module with BERT or Roberta, our model can facilitate the reconstruction of in-block text embeddings with global text information in future work.

[T+L] Text and Layout Modalities. LayoutLM$_{LARGE}$ [31] combines word-level layout information with word embedding. StructuralLM [17] leverages cell-level positional embeddings. Similarly, BROS [11], LiLT [28], TILT$_{LARGE}$ [24], and FormNet [16] combine text and spatial layout information from document images. Our model could likewise use global text embeddings with spatial position relations to assist in reconstructing text embeddings within blocks.

[T+L+I] Text, Layout and Image Modalities. Under this category, apart from utilizing textual and layout information, LayoutLM$_{BASE}$ [31], TILT$_{BASE}$ [24], SelfDoc [18], and UDoc [8] rely on Faster R-CNN [25] to extract visual features. Whereas LayoutLMv2 [30], DocFormer [1], and XYLayoutLM [9] generate image embeddings from CNN models. LayoutLMv3 [13] utilizes linear embeddings to encode image patches instead. All these pre-trained models strive to align image modalities and text, helping our models to allow truncated document blocks to interact with each other in multiple modalities on downstream tasks. We selected LayoutLMv3, which currently works best, as the component of HM-Transformer. Nevertheless, our model equally applies to other multimodal pre-trained models, and we leave this for future work.

4.3   Main Results

We evaluate our model on two publicly available benchmarks for multimodal tasks. Table 1 shows the model performance on dataset FUNSD, for the form understanding task and dataset CORD, for the receipt understanding task. Details are as follows.

Task I: Form Understanding. FUNSD [14], a subset of RVL-CDIP [10], is a noisy scanned document dataset for form understanding. It comprises 199 real, scanned forms with 9,707 fully annotated semantic entities. Each semantic entity is described by one of the four labels: "question", "answer", "header", or "other". We concentrate on the semantic entity labeling task, assigning the correct label to each entity. The training and test sets contain 149 and 50 samples, respectively. We use the official OCR annotations on the images to obtain text and layout information. And we fine-tune HM-Transformer on the FUNSD dataset with a learning rate of $1e{-}5$ and a batch size of 4 for 300 epochs.

Table 1. Comparison with existing popular models on the CORD and FUNSD datasets. "T/L/I" denotes "text/layout/image" modality. In the UDoc[†] model, CORD is split into 626/247 training/test receipts rather than the official 800/100 training/test receipts adopted by other works. Therefore, the score[†] is not comparable for other scores.

Model	Par	Modality	FUNSD F1↑	CORD F1↑
BERT$_{BASE}$ [15]	110M	T	60.26	89.68
RoBERTa$_{BASE}$ [20]	125M	T	66.48	93.54
BROS$_{BASE}$ [11]	110M	T+L	83.05	95.73
LiLT$_{BASE}$ [28]	–	T+L	88.41	96.07
LayoutLM$_{BASE}$ [31]	160M	T+L+I	79.27	–
SelfDoc [18]	–	T+L+I	83.36	–
UDoc [8]	272M	T+L+I	87.93	98.94[†]
TILT$_{BASE}$ [24]	230M	T+L+I	–	95.11
XYLayoutLM$_{BASE}$ [9]	–	T+L+I	83.35	–
LayoutLMv2$_{BASE}$ [30]	200M	T+L+I	82.76	94.95
DocFormer$_{BASE}$ [1]	183M	T+L+I	83.34	96.33
LayoutLMv3$_{BASE}$ [13]	133M	T+L+I	90.29	96.56
HM-Transformer$_{BASE}$ (Ours)	134M	T+L+I	**91.79**	**96.86**
BERT$_{LARGE}$ [15]	340M	T	65.63	90.25
RoBERTa$_{LARGE}$ [20]	355M	T	70.72	93.80
LayoutLM$_{LARGE}$ [31]	343M	T+L	77.89	–
BROS$_{LARGE}$ [11]	340M	T+L	84.52	97.40
StructuralLM$_{LARGE}$ [17]	355M	T+L	85.14	–
FormNet [16]	217M	T+L	84.69	–
FormNet [16]	345M	T+L	–	97.28
TILT$_{LARGE}$ [24]	780M	T+L	–	96.33
LayoutLMv2$_{LARGE}$ [30]	426M	T+L+I	84.20	96.01
DocFormer$_{LARGE}$ [1]	536M	T+L+I	84.55	96.99
LayoutLMv3$_{LARGE}$ [13]	368M	T+L+I	92.08	97.46
HM-Transformer$_{LARGE}$ (Ours)	381M	T+L+I	**92.75**	**97.64**

We report entity-level F1 scores for this task. In Table 1, from top to bottom, we find that the pre-trained model using only text information performs the worst. The model that utilizes both text and layout information achieves suboptimal performance. And the model that additionally exploits image features has the best result. The above indicates that the effective use of more modal information can boost the model's performance in document image understanding. Among them, LayoutLMv3 is the most outstanding model in the baseline and is an essential component of our model. Our model achieves new state-of-the-art results on the FUNSD dataset. We lead the performance with 91.79 and 92.75 F1 scores on the base and large model, superior to LayoutLMv3 by 1.50 and 0.67 points, respectively. The relatively small gain of HM-Transformer$_{\text{LARGE}}$ may be an adverse effect of too many model parameters with a limited amount of data for downstream tasks. Experimental results show that HM-Transformer can significantly improve the performance of multimodal pre-trained models on downstream tasks of long document image understanding with a tiny increase in parameter number.

Task II: Receipt Understanding. CORD [22] is a receipt key information extraction dataset. It defines 30 fields that fall under 4 categories. This receipt understanding task intends to label each word to the correct field. This dataset has 1000 receipts, containing 800/100/100 receipts for training/validation/testing, respectively. We use the official images and OCR annotations as inputs to our model. And we fine-tune HM-Transformer on the CORD dataset with a learning rate of $2e$–5 and a batch size of 12 for 50 epochs.

We report entity-level F1 scores for this task. The results on the CORD dataset in Table 1 likewise demonstrate that text, layout, and image information are all essential in the downstream task of document image understanding. For the base model size, the HM-Transformer achieves an F1 score of 96.86, which outperforms the state-of-the-art result of 96.56 presented by LayoutLMv3. The HM-Transformer$_{\text{LARGE}}$ also shows superior performance on the evaluation metric and achieves a state-of-the-art F1 score of 97.64. HM-Transformer can benefit both the form understanding and the receipt understanding task. Experimental results again illustrate that our model can effectively help multimodal pre-trained models improve performance by reconstructing in-block representations with the aid of global multimodal document representations and enhancing interactions using image information.

4.4 Ablation Study

Table 2. Ablation study of each module

Model#	Block-Transformer			Doc-transformer		FUNSD	CORD
	OT&OI	text	image	text	image	F1↑	F1↑
1	✗	✗	✗	✗	✗	90.29	96.56
2	✓	✗	✗	✗	✗	90.52	96.38
3	✓	✓	✗	✓	✗	91.61	96.75
4	✓	✓	✗	✓	✓	91.24	96.68
5	✓	✓	✓	✓	✓	**91.79**	**96.86**

The Effect of Each Module. We added each module individually to vanilla LayoutLMv3 to study the effect of each module. We first take vanilla LayoutLMv3 as model #1. Then we use the original text embedding h_{bi} and image embedding h_{bi} (OT&OI) generated by model #1 as the only input to the multimodal block Transformer, denoted as model #2. Model #3 is obtained by adding the multimodal document Transformer without image input h_{vi} to Model #2. Model #4 is the same as Model #5, except that no new image embedding input d_v exists in the multimodal block. Model #5 refers to the full version of the HM-Transformer.

The results are shown in Table 2. Model #2 has a higher F1 score on FUNSD than model #1 but a lower F1 score on CORD. Model #2 is equivalent to adding another Transformer layer behind Model #1. This unstable effect shows that our model does not rely on the Transformer layer to be effective. Model 3 works better than model #2 on both datasets, which proves that global document text embedding benefits the reconstruction of in-block embedding and improves the model's effectiveness. Surprisingly, the performance of model #4 decreases concerning model #3. This indicates that in multimodal block Transformer, discarding image embedding suffers the model effect even with global multimodal text embedding. Model #5 performs best on all datasets, illustrating that the multimodal document Transformer output's global multimodal text embeddings db_i and image embeddings d_v can only work together to benefit from each other. To summarize, the hierarchical structure of using global document information to assist in-block embedding reconstruction and the design of utilizing image information to enhance interactions are both significant.

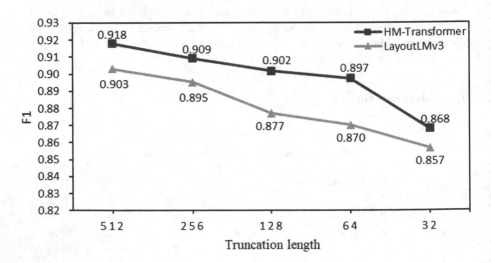

Fig. 3. Variation curves of F1 scores with truncation length on the FUNSD dataset.

The Effect of Truncation Length. Long samples might be truncated on the downstream task of a multimodal pre-training model for document image understanding, leading to a loss of contextual information and affecting the model's effectiveness. Our model uses global multimodal document text embedding and image embedding to alleviate the harmful effects of truncation. We study the impact of truncation length on the HM-Transformer and LayoutLMv3 models, where the truncation length varies between 32 and 512. We take the truncation length as the maximum length of the text block and then split the text in the long document image sample into several blocks. Figure 3 shows that the effect of both models decreases as the truncation length decreases. This result is reasonable because when the truncation length is shorter, a long sample is fragmented more, and thus more contextual information is lost. However, we can note that our model's decline (slope of the curve) is significantly smaller than that of LayoutLMv3. This proves our model can alleviate the negative effect of long document truncation for multimodal pre-training models.

5   Conclusion and Future Work

To address the problem of truncating long documents in the downstream task of multimodal pre-trained models, we propose a hierarchical multimodal transformer (HM-Transformer) for long document image understanding. Experimental results again illustrate that our model can effectively help multimodal pre-trained models improve performance by reconstructing in-block representations with the aid of global multimodal document representations and enhancing interactions using image information. In future work we will use our model for more multimodal pre-training models to further test its effectiveness and generality. In addition we will also evaluate our model on longer and more complex datasets.

Acknowledgements. This work was supported by Hunan Provincial Natural Science Foundation Project (No. 2022JJ30668) and (No. 2022JJ30046).

References

1. Appalaraju, S., Jasani, B., Kota, B.U., Xie, Y., Manmatha, R.: Docformer: end-to-end transformer for document understanding. In: Proceedings of the IEEE/CVF International Conference on Computer Vision, pp. 993–1003 (2021)
2. Beltagy, I., Peters, M.E., Cohan, A.: Longformer: the long-document transformer. arXiv preprint arXiv:2004.05150 (2020)
3. Child, R., Gray, S., Radford, A., Sutskever, I.: Generating long sequences with sparse transformers. arXiv preprint arXiv:1904.10509 (2019)
4. Cui, L., Xu, Y., Lv, T., Wei, F.: Document ai: benchmarks, models and applications. arXiv preprint arXiv:2111.08609 (2021)
5. Dai, Z., Yang, Z., Yang, Y., Carbonell, J.G., Le, Q., Salakhutdinov, R.: Transformer-xl: attentive language models beyond a fixed-length context. In: Proceedings of the 57th Annual Meeting of the Association for Computational Linguistics, pp. 2978–2988 (2019)

6. Ding, S., et al.: Ernie-doc: a retrospective long-document modeling transformer. In: ACL-IJCNLP, pp. 2914–2927 (2021)
7. Dosovitskiy, A., et al.: An image is worth 16×16 words: transformers for image recognition at scale. In: ICLR (2021)
8. Gu, J., et al.: Unidoc: unified pretraining framework for document understanding. Adv. Neural. Inf. Process. Syst. **34**, 39–50 (2021)
9. Gu, Z., et al.: Xylayoutlm: towards layout-aware multimodal networks for visually-rich document understanding. In: CVPR, pp. 4583–4592 (2022)
10. Harley, A.W., Ufkes, A., Derpanis, K.G.: Evaluation of deep convolutional nets for document image classification and retrieval. In: ICDAR, pp. 991–995. IEEE (2015)
11. Hong, T., Kim, D., Ji, M., Hwang, W., Nam, D., Park, S.: Bros: a pre-trained language model focusing on text and layout for better key information extraction from documents. In: AAAI, vol. 36, pp. 10767–10775 (2022)
12. Hu, Y., Ding, W., Liu, T., Gao, J., Sun, Y., Yin, B.: Hierarchical multiple granularity attention network for long document classification. In: IJCNN, pp. 1–7. IEEE (2022)
13. Huang, Y., Lv, T., Cui, L., Lu, Y., Wei, F.: Layoutlmv3: pre-training for document AI with unified text and image masking. In: Proceedings of the 30th ACM International Conference on Multimedia, pp. 4083–4091 (2022)
14. Jaume, G., Ekenel, H.K., Thiran, J.P.: FUNSD: a dataset for form understanding in noisy scanned documents. In: ICDARW, vol. 2, pp. 1–6. IEEE (2019)
15. Kenton, J.D.M.W.C., Toutanova, L.K.: Bert: pre-training of deep bidirectional transformers for language understanding. In: NAACL-HLT, pp. 4171–4186 (2019)
16. Lee, C.Y., et al.: Formnet: structural encoding beyond sequential modeling in form document information extraction. In: ACL, pp. 3735–3754 (2022)
17. Li, C., et al.: Structurallm: structural pre-training for form understanding. In: ACL, vol. pp. 6309–6318 (2021)
18. Li, P., et al.: Selfdoc: self-supervised document representation learning. In: CVPR, pp. 5652–5660 (2021)
19. Liu, X., Gao, F., Zhang, Q., Zhao, H.: Graph convolution for multimodal information extraction from visually rich documents. In: NAACL-HLT, pp. 32–39 (2019)
20. Liu, Y., et al.: Roberta: a robustly optimized bert pretraining approach. arXiv preprint arXiv:1907.11692 (2019)
21. Pappagari, R., Zelasko, P., Villalba, J., Carmiel, Y., Dehak, N.: Hierarchical transformers for long document classification. In: ASRU, pp. 838–844. IEEE (2019)
22. Park, S., et al.: Cord: a consolidated receipt dataset for post-ocr parsing. In: Workshop on Document Intelligence at NeurIPS 2019 (2019)
23. Peng, Q., et al.: Ernie-layout: layout knowledge enhanced pre-training for visually-rich document understanding. arXiv preprint arXiv:2210.06155 (2022)
24. Powalski, R., Borchmann, Ł, Jurkiewicz, D., Dwojak, T., Pietruszka, M., Pałka, G.: Going full-tilt boogie on document understanding with text-image-layout transformer. In: Llados, J., Lopresti, D., Uchida, S. (eds.) ICDAR, vol. 12822, pp. 732–747. Springer, Heidelberg (2021). https://doi.org/10.1007/978-3-030-86331-9_47
25. Ren, S., He, K., Girshick, R., Sun, J.: Faster r-cnn: towards real-time object detection with region proposal networks. Adv. Neural Inf. Process. Syst. **28**, 1–9 (2015)
26. Su, J., Lu, Y., Pan, S., Murtadha, A., Wen, B., Liu, Y.: Roformer: enhanced transformer with rotary position embedding. arXiv preprint arXiv:2104.09864 (2021)
27. Vaswani, A., Shazeer, N., et al.: Attention is all you need. Adv. Neural Inf. Process. Syst. **30**, 1–11 (2017)

28. Wang, J., Jin, L., Ding, K.: Lilt: a simple yet effective language-independent layout transformer for structured document understanding. In: ACL, pp. 7747–7757 (2022)
29. Wu, C., Wu, F., Qi, T., Huang, Y.: Hi-transformer: hierarchical interactive transformer for efficient and effective long document modeling. In: ACL-IJCNLP, pp. 848–853 (2021)
30. Xu, Y., et al.: Layoutlmv2: multi-modal pre-training for visually-rich document understanding. In: ACL, pp. 2579–2591 (2021)
31. Xu, Y., Li, M., Cui, L., Huang, S., Wei, F., Zhou, M.: Layoutlm: pre-training of text and layout for document image understanding. In: KDD, pp. 1192–1200 (2020)
32. Yang, X., Yumer, E., Asente, P., Kraley, M., Kifer, D., Giles, C.L.: Learning to extract semantic structure from documents using multimodal fully convolutional neural network. IEEE Computer Society (2017)
33. Yang, Z., Yang, D., Dyer, C., He, X., Smola, A., Hovy, E.: Hierarchical attention networks for document classification. In: NAACL-HLT, pp. 1480–1489 (2016)
34. Zaheer, M., et al.: Big bird: transformers for longer sequences. Adv. Neural. Inf. Process. Syst. **33**, 17283–17297 (2020)
35. Zhang, X., Wei, F., Zhou, M.: Hibert: document level pre-training of hierarchical bidirectional transformers for document summarization. In: ACL, pp. 5059–5069 (2019)

NV-QALSH+: Locality-Sensitive Hashing Optimized for Non-volatile Memory

Zhili Yao, Yikai Huang, Zezhao Hu, and Jianlin Feng[✉]

School of Computer Science and Engineering, Sun Yat-sen University,
Guangzhou, China
fengjlin@mail.sysu.edu.cn

Abstract. Locality-Sensitive Hashing (LSH) is a well-known method to solve the Approximate Nearest Neighbor (ANN) search problem. Query-Aware LSH (QALSH), a state-of-the-art LSH method, is a disk-based algorithm and suffers from high latency of disk I/O, even though it exploits disk-friendly B+-Trees as index data structures. On the other hand, DRAM-based methods occupy large amounts of expensive DRAM space and have long index rebuilt time. To solve the hardware problems, a variant of QALSH called NV-QALSH was proposed to leverage non-volatile memory (NVM), which combines the advantageous features of DRAM and disks. In this paper, we first study the projection mechanism, the core of QALSH, and find the value-to-position property of the projection. According to that property, we propose Interpolation Search Array (ISA), a novel array-based data structure, which is more efficient than B+-Trees under NVM. Finally, we extend NV-QALSH to NV-QALSH+ by replacing B+-Trees with ISA. The experimental results show that NV-QALSH+ is 7–48× faster than the disk-based QALSH. Additionally, NV-QALSH+ occupies extremely low DRAM space and can recover instantly with a near-zero index rebuilt time. Furthermore, NV-QALSH+ outperforms NV-QALSH in query speed, DRAM occupancy, and index rebuilt time.

Keywords: Approximate Nearest Neighbor Search · Locality-Sensitive Hashing · Non-Volatile Memory

1 Introduction

Nearest neighbor (NN) search has wide applications in areas such as data mining, database and information retrieval [13]. But due to the curse of dimensionality [27], it is difficult to find the exact NN in high-dimensional space. Hence, people compromise to find an approximate nearest neighbor (ANN). Locality-Sensitive Hashing (LSH) and its variants [9,12,15,30] are the popular methods for solving ANN search problem in high-dimensional space.

Among varied LSH methods, QALSH [7,8] is a state-of-the-art method which has a high query performance, especially in high-dimensional space. In the tutorials of SIGKDD'19 [1], ICDE'21 [4] and VLDB'21 [3], QALSH is introduced as the representative of LSH methods. Recently, many variants of QALSH are proposed, including I-LSH [17], EI-LSH [18], R2LSH [20], PDA-LSH [28], VHP [21], NV-QALSH [32], etc., indicating that QALSH is still popular and widely studied. However, as a disk-based method, QALSH suffers from high latency of disk I/O, resulting in a limited query speed. Although DRAM-based LSH methods [2,14,22] place the whole index in DRAM for a fast query speed, they occupy large amounts of expensive DRAM space and have long index rebuilt time since data in DRAM is volatile and will be lost upon power failure.

Non-volatile memory (NVM) [10,11,29] can solve the above hardware problem, since it has both disk and DRAM features: 1) compared with DRAM, it has a lower price and a larger capacity, and most importantly a non-volatility feature; 2) compared with disk, it is byte-addressable and much faster.

QALSH projects all objects onto multiple random lines indexed by B+-Trees in the index phase. In the query phase, QALSH first locates the projected positions of the query object in all B+-Trees, and starts a range search for each B+-Tree to find the c-ANN. NV-QALSH [32] implements an NVM-based LSH which utilizes an NVM-optimized B+-Tree, the LB-tree [16], to combine QALSH and NVM. Note that B+-Trees organize data in I/O pages, i.e., the size of each node in a B+-Tree is fixed to the size of an I/O page, and the access granularity of a B+-Tree is a single node. Under NVM, however, data are accessed in byte rather than in disk page, indicating that B+-Trees may not be the best choice for QALSH under NVM. Hence, we propose NV-QALSH+, a better variant of QALSH optimized for NVM. We first study the projection mechanism of QALSH, and find an interesting property of value-to-position mapping: each projected random line can be easily divided into several approximate linear segments. Based on this finding, we propose a novel array-based data structure, named Interpolation Search Array (ISA). We find that, compared with B+-Trees, ISAs are more efficient for QALSH under NVM. Therefore, we use ISAs to replace LB-Trees in NV-QALSH, and get NV-QALSH+. Specifically, an ISA divides a random line into several linear segments by piece-wise linear approximation (PLA) [5,19,31] and applies interpolation search [24,25] to locate the projected position on each segment. Note that under DRAM or NVM, using B+-Trees to locate the projected position needs $O(\log n)$ memory accesses. However, interpolation search utilizes a more complex calculation to expect locating the projected position in $O(1)$ memory accesses. Intuitively, ISAs trade more CPU calculations for fewer memory accesses. Since the gap between the memory access speed of NVM and the calculation speed of CPU is very large, reducing NVM accesses will greatly improve the performance of NV-QALSH+.

Extensive experiments show that NV-QALSH+ is 7–48× faster than the disk-based QALSH. Since NVM is still 2–3× slower than DRAM, NV-QALSH+ has a slightly slower query speed compared with the state-of-the-art DRAM-based LSH methods. However, NV-QALSH+ occupies extremely low DRAM space and can recover instantly after the system reboots. More precisely, NV-QALSH+ saves up to 92%–99.8% DRAM space compared with the state-of-the-art DRAM-based LSH methods, and its index rebuilt time is near zero. Experimental results show that NV-QALSH+ outperforms NV-QALSH in query speed, DRAM occupancy, and index rebuilt time. The NVM characteristics and the experimental results indicate that NVM is a better hardware for LSH implementation.

The rest of this paper is organized as follows. Preliminaries are discussed in Sect. 2. Section 3 presents the detailed designs of NV-QALSH+. Experimental studies are shown in Sect. 4. Finally, we conclude our work in Sect. 5.

2   Preliminaries

2.1   c-ANN Search Problem

Given a dataset D containing n objects in a d-dimensional space R^d, let $dist(o, q)$ denote the distance between objects o and q, the c-ANN (c-approximate nearest neighbor) search problem is to find an object o such that $dist(o, q) \leq c \times dist(o^*, q)$, where q is any query object in R^d, c ($c > 1$) is the given approximation ratio, and o^* is the exact nearest neighbor of q. Similarly, the c-k-ANN search problem is to find k objects $o_i (1 \leq i \leq k)$ such that $dist(o_i, q) \leq c \times dist(o_i^*, q)$ for a given query q, where o_i^* is the exact i-th nearest neighbor.

2.2   QALSH

QALSH [7,8] is proposed to answer c-ANN queries for l_p distance with a rigorous theoretical guarantee and a favorable empirical performance. In this paper, we mainly focus on the l_2 (Euclidean) distance. QALSH adopts the query-aware LSH function family based on the p-stable distribution [2]: $H_a(o) = a \cdot o$, where o is the vector representation of an object, and a is a d-dimensional vector, in which each entry is drawn independently from the standard normal distribution $N(0, 1)$. Intuitively, this LSH function projects an object o onto a random line specified by a and the corresponding projected value $H_a(o)$ is regarded as the hash value for object o. An interesting property is that with a high probability, two close objects in the original space will remain close on the random line; and similarly, two far objects will stay far on the random line.

To find a c-ANN, QALSH projects the whole dataset onto m random lines indexed by m B+-Trees. When a query q arrives, QALSH projects q onto the m random lines using the same LSH functions. i.e., locating the position of projected value of q on each B+-Tree. Finally, QALSH performs collision counting [6] and virtual rehashing [6] for all the objects in the m B+-Trees, and an object with counting more than l ($l < m$) is chosen as a candidate. More precisely, QALSH performs a range search on each B+-Trees and counts the encountered objects. QALSH stops the range search when more than β (typically, β is 100 in [8]) candidates are chosen or one of the candidate is close enough to the query. QALSH returns the closest neighbor among candidates, and guarantees that with a constant probability, the returned object is a c-ANN. Similarly, for c-k-ANN search problem, QALSH terminates when more than $(\beta + k)$ candidates are chosen, or k of all candidates are found to be close enough to the query.

Besides QALSH, other disk-based LSH methods, such as C2LSH [6] and SRS [26], store the index structure on disks, and hence are sensitive to the high latency of the disk I/Os. DRAM-based LSH methods such as E2LSH [2], Multi-Probe LSH [22], LCCS-LSH [14], etc., enjoy a low query latency by placing the whole index in DRAM, resulting in substantial DRAM space overhead. When the whole index structure is larger than the memory capacity, we have to give up the DRAM-based methods. In addition, the whole index structure in DRAM will be lost upon power failure, and when the system reboots, it takes a lot of time to rebuild the index structure in DRAM. Generally, disk-based methods persist the LSH index structure on disks to minimize the usage of expensive DRAM space and benefit from a quick index rebuilt speed. DRAM-based methods place the LSH index structure in DRAM to achieve a low query latency. In this paper, we leverage NVM to combine the above advantages of disk-based methods and DRAM-based methods.

2.3 Non-volatile Memory and NV-QALSH

Non-Volatile Memory (NVM) [10,11] is a new storage technique. In this paper, we mainly target the Intel Optane DCPMM [10], a commercially available NVM solution, which has a higher capacity (128–512GB per DIMM) and a lower price than DRAM as well as the features of byte-addressability and data-durability. It has two configuration models, the Memory model and the App-Direct model. We leverage the App-Direct model for better performance and persistence. Precisely, when data are read from NVM to CPU caches in the App-Direct model, 256B data are first transferred from NVM storage to the XPBuffer in NVM, then 64B data among them are read to a CPU cache line [11,29].

In this paper, we use an NVM-aware file system to manage the Optane DCPMM, and memory-map the data into virtual addresses. In this manner, the application programs can manipulate the data in NVM by their virtual addresses like usual in-memory programming. As indicated in [29], Optane DCPMM is 2–3× slower than DRAM due to its longer media latency. Furthermore, Optane memory is more pattern-dependent than DRAM, and the gap between random and sequential reads is 20% for DRAM but 80% for Optane memory. Recently, many NVM-optimized B+-Trees [16,23] are proposed. Among them, LB-Tree [16] proposes some novel techniques such as entry moving and logless node split. NV-QALSH [32] utilizes LB-Trees to replace the disk-based B+-Tree in the original QALSH, and follows the selective persistence and 256B alignment optimizations, to improve the performance under NVM. Additionally, NV-QALSH optimizes the range search process according to the NVM characteristics.

2.4 Linear Approximation and Interpolation Search

Piece-wise Linear Approximation (PLA) [5,31] partitions a given curve into several linear segments, and generates a slope to indicate the trend for each segment.

For example, as illustrated in Fig. 1, given an array A consisting of all the projected values on a random line in ascending order, PLA divides the array into several linear segments, i.e., the sub-arrays $\{A_0, ..., A_k\}$, and generates a slope s_i for each sub-array $A_i (0 \leq i \leq k)$. Then, interpolation search [24,25] can be applied to predict the position in a sub-array A_i for the query's projected value v_q, i.e., the subscript of the array cell that stores v_q. More precisely, let v_0 denote the first value in the sub-array A_i. Interpolation search uses the following formula to calculate the predicted position $pos^*(v_q)$, s.t. $A_i[pos^*(v_q)] = v_q$.

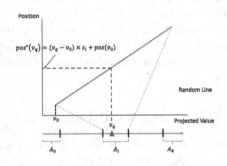

Fig. 1. An example of interpolation search on sub-array A_i.

$$pos^*(v_q) = (v_q - v_0) \times s_i + pos(v_0) \tag{1}$$

Shrinking-Cone segmentation [5], an efficient PLA algorithm, is proposed for sorted arrays. There are two main reasons why we use shrinking cone segmentation. First, it is a highly efficient one-pass algorithm with a time complexity of $O(n)$, i.e., it only needs to traverse the given array once to finish the segmentation. Second, it guarantees a maximal error in each segment. More specifically, given a value to be located, the error between the actual position of the given value and the predicted position calculated by Formula (1) is bounded by a parameter e. NV-QALSH+ leverages this theoretical guarantee to quickly locate the projected position of a query. Precisely, the projected position is the subscript of a sorted array composed of projected values.

The main idea behind Shrinking-Cone is that a new value can be added to a segment if and only if it does not violate the error constraint of any previous value in the segment. If a value added to the current segment violates the theoretical guarantee, Shrinking-Cone generates a new segment starting from this value. Please refer to [5] for more details about the Shrinking-Cone segmentation.

3 Design of NV-QALSH+

3.1 Overview of NV-QALSH+

NV-QALSH+ exploits a DRAM-NVM-disk hybrid storage architecture, which combines the advantages of DRAM, NVM, and disks. The projected arrays, the main part of the ISA, are placed in NVM for persistence since they contain all projected values of all random lines but only a portion of them are accessed during the query phase. As for the meta data of ISA, the space usage of them is small and they are frequently accessed during the range search in the query phase. Thus, they are stored in NVM for recovery, and after the system reboots, they are rebuilt in DRAM for fast access. More details about the ISA are presented in the Sect. 3.3. The m LSH functions are duplicated both in NVM for persistence and in DRAM for fast access the whole dataset, which is used to calculate distances in the original Euclidean space between the query object and the candidate objects, is placed in the cheaper disks. The space usage of them is enormous but only a tiny part of them are accessed for each query. With the DRAM-NVM-disk hybrid storage structure, NV-QALSH+ fully utilizes the lowest latency of DRAM, the persistence of NVM, and the economy of disks (Fig. 2).

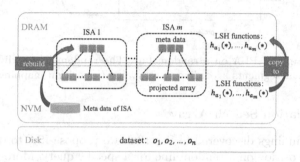

Fig. 2. Overview of NV-QALSH+.

3.2 Value-to-Position Mapping on Projected Arrays

As described in Sect. 3.1, the projected array is the main part of the ISA. We first sort all projected values and store them in the projected array. We conduct some experiments on the projected arrays for different datasets, and obtain some surprising findings. In brief, the projected array can be approximately divided into five linear segments by Piece-wise Linear Approximation (PLA). Specifically, we choose four real datasets. For each dataset, we project all the objects onto three random lines using LSH functions. As illustrated in Fig. 3, we plot the value-to-position mapping of the projected array for all datasets and all random lines. Intuitively, the shape of each mapping is like an "S". As illustrated in Fig. 4, such an "S"-shape curve can be easily divided into five approximate linear segments by PLA using the Shrinking-Cone segmentation introduced in Sect. 2.4. We also study the value-to-position mapping of the projected array in theory but gain little success. It is an open question that whether a random projection by query-aware LSH function based on p-stable distribution always results in an "S" mapping for any dataset. We will study this phenomenon thoroughly in the future work.

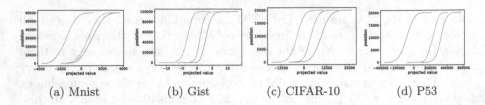

(a) Mnist   (b) Gist   (c) CIFAR-10   (d) P53

Fig. 3. Value-to-position mapping on projected arrays. For each of the four real datasets, we project all objects onto three random lines (the three curves in each sub-figure).

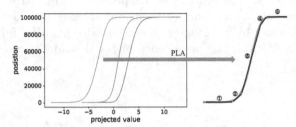

Fig. 4. An example of PLA on a random line. The random line is divided into five approximate linear segments according to the value-to-position mapping.

3.3 Interpolation Search Array

Based on the findings discovered in Sect. 3.2, we propose ISA to quickly locate the projected position on a random line for a specific query. More precisely, the original QALSH uses B+-Trees to locate the projected position with a time complexity of $O(\log n)$, and ISA aims to reduce this time complexity to approximate $O(1)$. Next, we first show how ISA is generated and how it works. Then, we explain why NV-QALSH+ with ISA can locate the projected position with a time complexity of approximate $O(1)$. As illustrated in Fig. 5, an ISA consists of two parts, a projected array and the corresponding meta data. In the index phase of QALSH, all objects in a dataset are projected onto multiple random lines. For each random line, we store the resulting projected values in ascending order in an array, forming a projected array. Then we apply Shrinking-Cone segmentation to segment each projected array logically. We do not divide the projected array into several sub-arrays physically, but keep them as a complete array for better sequential read performance. More specifically, for each resulting logical segment i, we record its slope s_i and the position pos_i of the first value of segment i. Precisely, pos_i is the subscript of the original projected array. Assuming the projected values in the array are $arr[0, ..., n-1]$ and the array is segmented to k parts logically, then the first projected value of segment i is $arr[pos_i]$. Note that s_i and pos_i constitute the meta data of a projected array.

The original disk-based QALSH builds a B+-Tree for each random line in the index phase. All the projected values are sorted first, and then are traversed once to bulk-load a B+-Tree. Note that Shrinking-Cone segmentation also traverses all

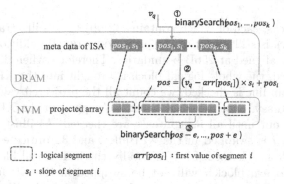

Fig. 5. Interpolation Search Array and an example of locating the position of a given value v_q.

the sorted projected values once to build an ISA. Therefore, the time complexity of building a B+-tree and that of building an ISA are both $O(n)$. In summary, NV-QALSH+ brings no extra overhead in the index phase.

In the query phase, ISA locates the actual position of a given query's projected value v_q in three steps as illustrated in Fig. 5 First, it performs a binary search to seek out which segment the given value belongs to. The time complexity of this procedure is $O(\log(k))$, where k is the number of the logical segments in the ISA generated by the Shrinking-Cone segmentation. Then, assuming the projected value v_q belongs to segment i, ISA applies interpolation search to calculate the predicted position of v_q in the original projected array. The calculation formula is similar to Formula 1: $pos^*(v_q) = (v_q - arr[pos_i]) \times s_i + pos_i$. The time complexity of this procedure is $O(1)$. Note that the theoretical guarantee of the Shrinking-Cone segmentation ensures that the error between the actual position and the predicted position is no more than e. Therefore, ISA finally performs a binary search between $[pos^*(v_q) - e, pos^*(v_q) + e]$ to find out the actual position of v_q in the original projected array. The time complexity of this procedure is $O(\log e)$. In summary, the time complexity of locating the actual position of a given value by ISA is $O(\log k + \log e)$.

To locate the projected position in a random line for a given query with a time complexity of approximate $O(1)$, the number of segments k in an ISA should be small, and the error e should be a constant. According to the discussion in Sect. 3.2, after applying Shrinking-Cone segmentation to the projected array of QALSH, each ISA is expected to contain approximate five segments regardless of the datasets or the LSH functions. When setting the error e to be a constant in Shrinking-Cone segmentation, ISA is able to locate the projected position of a given query with a time complexity of approximate $O(1)$. Actually, in experiments the bounded error e is set to be 128 for all datasets, and the resulting number of segments k is 13–27.

3.4 Range Search Optimization

We follow NV-QALSH to optimize the range search in NV-QALSH+ to improve the performance. NV-QALSH+ first divides the projected array in the ISA into

256B logical blocks, each of which contains 32 objects, as illustrated in Fig. 10. Each logical block is still a part of the projected array. In addition, each logical block is carefully aligned at 256B boundaries. Therefore, when data in a logical block are accessed, the whole logical block is brought into the XPBuffer. Then, When one random line is checked, we count all the projected values in a logical block. That is to say, once a logical block is swapped in the XPBuffer, it will not be swapped out until all its contents are checked. As illustrated in Fig. 6, the logical block i is swapped in the XPBuffer, and 32 projected values inside this logical block are checked. Since a totally checked logical block will not be checked again, logical block i will not be swapped in the XPBuffer again. In summary, to check all projected values in logical block i, the XPBuffer requires **at most one** data swapping in and out operation.

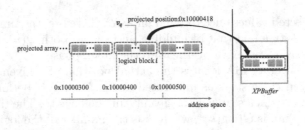

Fig. 6. Optimization of range search in NV-QALSH+.

4   Experiments

4.1   Experiment Setup

All methods are implemented in C++ and are compiled with gcc 8.3 using -O3 optimization. All experiments are conducted on a machine equipped with two Intel Xeon(R) Gold 5222 CPUs, a 128 GB DRAM, a 512 GB Optane DCPMM consisting of 4×128 GB Optane NVDIMM, and a 1TB NVMe SSD which is much faster than traditional disks. Thus, the disk-based QALSH achieves a better than usual performance. To remove the NUMA effect, we first place the NVM data on the node where CPU0 is located, and then bind CPU0 to the experimental program with *taskset* instruction.

Datasets and Queries. In the experiments, we use the following four real-world datasets to evaluate the performance of different LSH methods. For each dataset, we randomly choose 100 objects as queries.

- **Mnist**[1] has 60,000 objects. We follow [8] to only consider the top-50 dimensions with the largest variance.

---

[1] http://yann.lecun.com/exdb/mnist/.

- **Gist**[2] contains 100,000 960-dimensional feature vectors of Gist.
- **CIFAR**[3] contains 50,000 3,072-dimensional vectors of CIFAR-10.
- **P53**[4] contains 31,059 5408-dimensional vectors. We follow [8] to remove the incomplete objectsand normalize the coordinates to a range of [0,10000].

Evaluation Metrics. We use the following metrics to evaluate the performance of different methods when they achieve the same query accuracy.

- **Overall Ratio** is used to measure the query accuracy of different methods:

$$\frac{1}{k} \sum_{i=1}^{k} \frac{dist(o_i, q)}{dist(o_i^*, q)} \tag{2}$$

For a c-k-ANN query q, o_i is the searched i^{th} NN object and o_i^* is the exact i^{th} NN object. Intuitively, a smaller overall ratio means higher accuracy.
- **NVM to Disk Speedup Ratio** is the ratio of query speed of NVM-based LSH to that of disk-based ones. The larger Ratio means higher performance.
- **Query Time** is the time spent in the query phase for a given query.
- **DRAM Occupancy** is the size of DRAM space occupied by the LSH index.
- **Index Rebuilt Time** is the time cost for rebuilding the whole LSH index structure when the system reboots.

Benchmark Methods

- **LB-QALSH** combines QALSH and LB-Tree [16] without any optimization.
- **NV-QALSH** [32] optimizes the B+-Trees used in LB-QALSH and modifies the range search process to improve the query performance.
- **NV-QALSH+** is the proposed method further optimized for NVM, which replaces the B+-Trees with a novel array-based data structure, the ISA.
- **Disk-based QALSH** (simply **D-QALSH**)[5], the original version of QALSH [8], is a state-of-the-art disk-based LSH method.
- **DRAM-based QALSH** (simply **M-QALSH**)[6], the memory version of QALSH [8], is also a state-of-the-art DRAM-based method.
- **DRAM-based SRS** (simply **SRS**)[7], a state-of-the-art DRAM-based method.
- **LCCS-LSH**[8] LCCS-LSH [14] is a recently proposed LSH method, which is also a state-of-the-art DRAM-based method.

---

[2] http://corpus-texmex.irisa.fr/.
[3] http://www.cs.toronto.edu/~kriz/cifar.html.
[4] http://archive.ics.uci.edu/ml/datasets/p53+Mutants.
[5] https://github.com/HuangQiang/QALSH.
[6] https://github.com/HuangQiang/QALSH_Mem.
[7] https://github.com/DBWangGroupUNSW/SRS.
[8] https://github.com/1flei/lccs-lsh.

Parameters Setting. As for c-k-ANN search, k is set to 100 by default. We fine-tune all the LSH methods for the best query efficiency. For NV-QALSH+, the error e of Shrinking-Cone segmentation is set to 128 for all four datasets, and the resulting number of segments seg are 19, 27, 18, 13 for Mnist, Gist, CIFAR, and P53 respectively.

4.2 Results and Analysis

NVM to Disk Speedup Ratio. As shown in Fig. 7, NV-QALSH+ achieves a 7–48× speedup over D-QALSH, while NV-QALSH has a 5–40× speedup and LB-QALSH only achieves a 2–27× speedup. Generally, NV-QALSH+ has a 1.1–1.5× speedup over NV-QALSH because the ISA further accelerates the query phase in NV-QALSH+, indicating that the PLA and the interpolation search may be a better substitute for B+-Trees under NVM. Besides, NV-QALSH+ achieves a higher speedup ratio in high-dimensional space (CIFAR and P53 datasets) because the number of hash tables m increases as the dimension increases, and D-QALSH needs more disk I/Os to find the c-ANN. This also shows that NV-QALSH+ has more advantages in dealing with high-dimensional data.

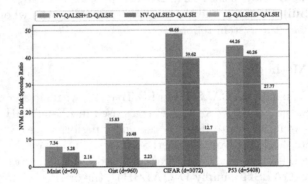

Fig. 7. NVM to Disk Speedup Ratio.

Query Time. As shown in Fig. 8, NV-QALSH+ always achieve a faster query speed than NV-QALSH and LB-QALSH which shows that NV-QALSH+ optimizes the query process of QALSH under NVM. As the query accuracy becomes higher, the query time gap between NV-QALSH+, NV-QALSH and LB-QALSH becomes larger, indicating that NV-QALSH+ is a better LSH method under NVM. Besides, as more random lines are used for query process, more data objects need to be checked to find the good enough candidates, which means a longer range search process in the query phase. A longer range search process causes a higher XPBuffer contention in LB-QALSH, resulting in a larger number of data swaps in XPBuffer and substantially degrades the NVM performance.

However, with the optimization of the range search modification, the XPBuffer contention in NV-QALSH+ still remains low, and the query time does not grow as fast as LB-QALSH. Note that NV-QALSH+ optimizes the process of locating the projected position of the query object in a single random line by ISA, which is faster than the B+-Trees in LB-QALSH and NV-QALSH. Therefore, the larger the number of random lines, the more obvious the optimization effect. In addition, NV-QALSH+ also optimizes the range search process, and the frequency of data swapping in and out of the XPBuffer still remains low for a high query accuracy. Since NVM is still 2–3× slower than DRAM, the query speed of NV-QALSH+ is slower than that of DRAM-based methods. But for high-dimensional datasets, i.e., for CIFAR and P53 datasets, NV-QALSH+ is faster than SRS or LCCS-LSH in some cases because it is based on QALSH, which is more beneficial in high-dimensional space than SRS and LCCS-LSH.

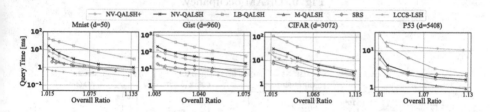

Fig. 8. Query time.

DRAM Occupancy. As illustrated in Fig. 9, although DRAM-based LSH methods have better query performance, they consume a lot of expensive DRAM space since they place the whole LSH index in DRAM. Among all the LSH methods, NV-QALSH+ has the lowest DRAM space occupancy, indicating that it is a practical LSH method. Specifically, the main DRAM-part data of M-QALSH are all random lines consisting of all the projected values, and hence occupy $O(mn)$ DRAM space, where m is the number of random lines (hash tables) and n is the size of dataset. For LCCS-LSH, the DRAM-part data are mainly composed of the Circular Shift Arrays (CSA), and each CSA contains $O(n)$ data. Since each hash table corresponds to a CSA, LCCS-LSH also occupies $O(mn)$ DRAM space. For LB-QALSH, the main DRAM-part data are the inner nodes of LB-Trees. There are totally m LB-Trees and the inner nodes of each LB-Tree occupy $O(n)$ DRAM space, resulting in $O(mn)$ DRAM space consumed by LB-QALSH. Besides, NV-QALSH also adopts the LB-Trees as the index structure, which occupies $O(mn)$ DRAM space. As for NV-QALSH+, it only stores the meta data of ISA in DRAM, and the time complexity of the meta data of each ISA is approximate to $O(1)$. Since there are m ISAs, NV-QALSH+ only occupies $O(m)$ DRAM space, implying that it greatly reduces the DRAM occupancy both in theory and in practice. As the query accuracy improves, the DRAM space required by all methods increases. However, NV-QALSH+ can save up to 90%-95% of DRAM space compared with DRAM-based methods. Note that the DRAM occupancy of NV-QALSH and LB-QALSH is also lower than that of DRAM-based methods, but is significantly higher than that of NV-QALSH+. Besides, for all datasets, NV-QALSH+ only occupies approximate 1 MB or less DRAM space, making it a cost-saving LSH method.

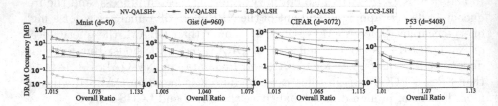

Fig. 9. DRAM occupancy.

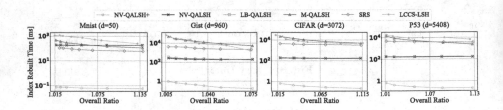

Fig. 10. Index rebuilt time.

Index Rebuilt Time. As shown in Fig. 10, the index rebuilt time of NV-QALSH+ is much shorter than that of other LSH methods. Note that the index rebuilt time is positively correlated to the DRAM occupancy since the process of rebuilding index is to restore the DRAM-part data. For DRAM-based methods, they need to rebuild the whole index data structure when the system reboots since they place the whole index in volatile DRAM. In general, they take long time to rebuild the LSH index. For LB-QALSH and NV-QALSH, the process of rebuilding index is mainly to recover the inner nodes of m LB-Trees. Specifically, all $O(n)$ leaf nodes are traversed to rebuild a single LB-Tree, and thus the index rebuilt time complexity is $O(mn)$. For NV-QALSH+, during the process of rebuilding the DRAM-part index, it only needs to read the meta data of m ISA from NVM and copies them into DRAM, resulting in a time complexity of $O(m)$. That is to say, the index rebuilt time of NV-QALSH+ is independent of the dataset size n, and is only positively correlated to the number of random lines m. Therefore, NV-QALSH+ has an extremely fast index rebuild speed. As the query accuracy is higher, the index rebuilt time becomes longer because the number of random lines becomes larger, and more data need to be restored. For a higher query accuracy, NV-QALSH+ is three to four orders of magnitudes faster than the DRAM-based LSH methods when rebuilding the LSH index structure. For all datasets, NV-QALSH+ can recover instantly within 1 millisecond, making it a more practical LSH method.

5 Conclusion

In this paper, we propose NV-QALSH+, which utilizes Interpolation Search Array, a novel array-based data structure, to accelerate the query performance under NVM. Experimental results show that NV-QALSH+ solves the problem of slow query speed suffered by disk-based QALSH, and meanwhile solves the problems of high DRAM occupancy and long index rebuilt time suffered by DRAM-based methods. The NVM characteristics and the experimental results indicate that NV-QALSH+ is a practical LSH method.

Acknowledgments. This work is partially supported by China NSFC under Grant No. 61772563.

References

1. Anastasiu, D.C., Rangwala, H., Tagarelli, A.: Are you my neighbor? bringing order to neighbor computing problems. In: KDD (2019)
2. Datar, M., Immorlica, N., Indyk, P., Mirrokni, V.S.: Locality-sensitive hashing scheme based on p-stable distributions. In: SCG, pp. 253–262 (2004)
3. Echihabi, K., Palpanas, T., Zoumpatianos, K.: New trends in high-d vector similarity search: Ai-driven, progressive, and distributed. VLDB **14**(12), 3198–3201 (2021)
4. Echihabi, K., Zoumpatianos, K., Palpanas, T.: High-dimensional similarity search for scalable data science. In: ICDE (2021)
5. Galakatos, A., Markovitch, M., Binnig, C., Fonseca, R., Kraska, T.: Fiting-tree: a data-aware index structure. In: COMAD, pp. 1189–1206 (2019)
6. Gan, J., Feng, J., Fang, Q., Ng, W.: Locality-sensitive hashing scheme based on dynamic collision counting. In: SIGMOD, pp. 541–552 (2012)
7. Huang, Q., Feng, J., Fang, Q., Ng, W., Wang, W.: Query-aware locality-sensitive hashing scheme for l_p norm. VLDB **26**(5), 683–708 (2017)
8. Huang, Q., Feng, J., Zhang, Y., Fang, Q., Ng, W.: Query-aware locality-sensitive hashing for approximate nearest neighbor search. VLDB **9**(1), 1–12 (2015)
9. Indyk, P., Motwani, R.: Approximate nearest neighbors: towards removing the curse of dimensionality. In: STOC, pp. 604–613 (1998)
10. Intel Optane Persistent Memory. https://www.intel.com/content/www/us/en/architecture-and-technology/optane-dc-persistent-memory.html
11. Izraelevitz, J., et al.: Basic performance measurements of the intel optane dc persistent memory module. arXiv:1903.05714 (2019)
12. Jafari, O., Maurya, P., Islam, K.M., Nagarkar, P.: Optimizing fair approximate nearest neighbor searches using threaded b+-trees. In: SISAP, pp. 133–147 (2021)

13. Jafari, O., Maurya, P., Nagarkar, P., Islam, K.M., Crushev, C.: A survey on locality sensitive hashing algorithms and their applications. arXiv:2102.08942 (2021)
14. Lei, Y., Huang, Q., Kankanhalli, M., Tung, A.K.: Locality-sensitive hashing scheme based on longest circular co-substring. In: SIGMOD, pp. 2589–2599 (2020)
15. Li, W., Zhang, Y., Sun, Y., Wang, W., Zhang, W., Lin, X.: Approximate nearest neighbor search on high dimensional data–experiments, analyses, and improvement. TKDE **32**, 1475–1488 (2016)
16. Liu, J., Chen, S., Wang, L.: LB+ trees: optimizing persistent index performance on 3dxpoint memory. Proc. VLDB Endow. **13**(7), 1078–1090 (2020)
17. Liu, W., Wang, H., Zhang, Y., Wang, W., Qin, L.: I-lsh: I/o efficient c-approximate nearest neighbor search in high-dimensional space. In: ICDE, pp. 1670–1673 (2019)
18. Liu, W., Wang, H., Zhang, Y., Wang, W., Qin, L., Lin, X.: EI-LSH: an early-termination driven i/o efficient incremental c-approximate nearest neighbor search. VLDB J. **30**(2), 215–235 (2021)
19. Liu, X., Lin, Z., Wang, H.: Novel online methods for time series segmentation. TKDE **20**(12), 1616–1626 (2008)
20. Lu, K., Kudo, M.: R2LSH: a nearest neighbor search scheme based on two-dimensional projected spaces. In: ICDE, pp. 1045–1056. IEEE (2020)
21. Lu, K., Wang, H., Wang, W., Kudo, M.: VHP: approximate nearest neighbor search via virtual hypersphere partitioning. Proc. VLDB Endow. **13**(9), 1443–1455 (2020)
22. Lv, Q., Josephson, W., Wang, Z., Charikar, M., Li, K.: Multi-probe LSH: efficient indexing for high-dimensional similarity search. In: VLDB, pp. 950–961 (2007)
23. Oukid, I., Lasperas, J., Nica, A., Willhalm, T., Lehner, W.: FPTree: a hybrid SCM-DRAM persistent and concurrent b-tree for storage class memory. In: COMAD, pp. 371–386 (2016)
24. Perl, Y., Itai, A., Avni, H.: Interpolation search-a log log n search. Commun. ACM **21**(7), 550–553 (1978)
25. Sandt, P.V., Chronis, Y., Patel, J.M.: Efficiently searching in-memory sorted arrays: revenge of the interpolation search? In: SIGMOD, pp. 36–53 (2019)
26. Sun, Y., Wang, W., Qin, J., Zhang, Y., Lin, X.: SRS: solving c-approximate nearest neighbor queries in high dimensional Euclidean space with a tiny index. Proc. VLDB Endow. (2014)
27. Weber, R., Schek, H.J., Blott, S.: A quantitative analysis and performance study for similarity-search methods in high-dimensional spaces. In: VLDB, vol. 98, pp. 194–205 (1998)
28. Yang, C., Deng, D., Shang, S., Shao, L.: Efficient locality-sensitive hashing over high-dimensional data streams. In: ICDE, pp. 1986–1989. IEEE (2020)
29. Yang, J., Kim, J., Hoseinzadeh, M., Izraelevitz, J., Swanson, S.: An empirical guide to the behavior and use of scalable persistent memory. In: FAST, pp. 169–182 (2020)
30. Zhang, S., Huang, J., Xiao, R., Du, X., Gong, P., Lin, X.: Toward more efficient locality-sensitive hashing via constructing novel hash function cluster. Concurr. Comput. Pract. Exp. **33**(20), e6235 (2021)
31. Zhao, H., Li, T., Chen, G., Dong, Z., Bo, M., Pang, C.: An online PLA algorithm with maximum error bound for generating optimal mixed-segments. JMLC **11**, 1483–1499 (2019)
32. Yao, Z., Zhang, J., Feng, J.: NV-QALSH: an nvm-optimized implementation of query-aware locality-sensitive hashing. In: DEXA, pp. 58–69 (2021)

A Novel Causal Discovery Model for Recommendation System

Guohao Sun, Huirong Hua, Jinhu Lu, and Xiu Fang[✉]

Donghua University, Shanghai, China
{ghsun,xiu.fang}@dhu.edu.cn, {2212651,1209126}@mail.dhu.edu.cn

Abstract. The recommendation system is now playing a more and more important role in our daily life. Recently, some scholars proposed that human behavior is governed by a complex web of causal models, and causal relationships are crucial in the recommendation process. Unfortunately, existing methods are limited in their ability to uncover hidden causal relationships because they may generate false causal relationships, which limits their recommendation performance. To address this issue, we propose a new recommendation model that leverages the causal relationship recommendation model and integrates a causal discovery module into the recommendation process. In this way, we can capture accurately the causal relationships underlying user behavior and generate more targeted recommendations. By fitting actual user behavior data, we can learn a cause-and-effect diagram that accurately reflects the real-world dynamics of the system. Extensive experiments conducted on two real-world datasets demonstrate that our method significantly outperforms the state-of-the-art approaches.

Keywords: Recommendation system · Causal discovery · Variational automatic encoder

1 Introduction

Recommendation systems play a crucial role in addressing the challenge of information overload and have garnered widespread attention. In recent years, numerous recommendation algorithms have been proposed, including collaborative filtering [18], content-based recommendation [15], and deep neural network-based recommendation [26]. Typically, these approaches recommend projects based on their correlation with other projects, aiming to identify those most likely to engage users. However, in real life, Human behavior is controlled by a series of potential causal models [10]. For example, in Fig. 1, users tend to buy Polaroid photo paper, battery, and album after purchasing Polaroid. This is because after purchasing Polaroid, users then have the idea of purchasing photo paper and other related items. Therefore, the Polaroid camera is the cause of purchasing other projects. From this example, we can see that causal relationship plays important role in the recommendation system. By understanding the causal relationship between different items, the recommendation system can make more precise and targeted suggestions, such that improving the overall effectiveness

© The Author(s), under exclusive license to Springer Nature Singapore Pte Ltd. 2024
X. Song et al. (Eds.): APWeb-WAIM 2023, LNCS 14334, pp. 261–276, 2024.
https://doi.org/10.1007/978-981-97-2421-5_18

of the recommendation system and enhancing the user experience. In addition, by applying causal relationships, people can easily understand how the system generates specific recommendations and can have a better understanding and trust in the recommendation system. Hence, learning about causality in the recommendation process is crucial.

Fig. 1. Potential causal effects in people's behavior

Currently, some scholars [7,23] have recognized the importance of causal relationships and proposed methods for learning and discovering hidden causal relationships. They use data analysis to obtain the probability of mutual occurrence between variables, and assume that variables with a higher probability of co-occurrence have a causal relationship. However, the causal relationships obtained by this method may be incorrect, which can even affect their recommendation performance. For example, in Fig. 2, when the weather gets hot, the sales of ice cream and the number of drowning people will increase at the same time. By the existing methods, some of them will erroneously conclude that there is a causal relationship between them.

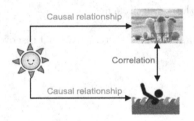

Fig. 2. Causal phenomena in nature

In order to solve this problem, we propose a new recommendation model called CDRS by applying causal relationships. In our new model, a causal structure learning method is proposed. We first postulate the causal diagram of the relevant items and filter out the irrelevant historical information. Then, we capture the causal relationship between user behaviors by fitting actual user behavior data and learning the causal diagram. With our new model, we can discover casual relationships more correctly and such that improve the recommendation performance.

The main contributions of this paper are summarized as follows:

- We propose a method to capture the causal relationship between user behavior to enhance interpretability and improve the performance of recommendations.

- We introduce a general recommendation framework called CDRS based on causal learning, which incorporates a causal discovery module into the recommendation model.
- We conduct experiments on two real-world datasets, demonstrating the efficacy of our framework in enhancing recommendation performance.

2   Related Work

Causal inference is one of the core issues in statistics and data science, which refers to the process of inferring causal relationships when a certain phenomenon has already occurred. Causal inference has extensive applications in biomedical research, economics, management, and social sciences. It enables researchers to reveal causal relationships between variables and uncover the underlying mechanisms behind observed phenomena. Currently, research on causal inference mainly involves two directions: causal discovery and causal effect estimation. Causal discovery aims to identify causal relationships among variables, while causal effect estimation aims to estimate the magnitude of causal effects. The recommendation systems that apply the causal inference are known as causal recommendations. In this paper, we aim to propose a new causal discovery method to improve recommendation performance. Therefore, in this section, we first introduce the work related to causal discovery and then the causal recommendation.

2.1   Causal Discovery

Causal discovery aims to uncover causal relationships among variables from complex data. It involves identifying a graphical network structure that accurately describes the causal relationships between variables. Typically, this graphical structure is a directed acyclic graph (DAG).

The problem of learning directed acyclic graphs(DAGs) from data has garnered significant interest in recent years. Various methods, such as differentiable DAG learning [3] and graph neural networks [16], have been proposed to address this problem. Initially, Constraint-based approaches [12] rely on a set of conditional independence tests to identify causal relationships between variables. However, these methods have limitations in distinguishing the structure of Markov equivalence classes. Then, Score-based methods [5,22] have been proposed, they use a score function to evaluate the fitting degree of different graphs to the data. These Score-based methods can only assess the impact of one variable on the dependent variable, and they cannot solve the real-world problem of multiple variables. To address these limitations, a hybrid method [14] was introduced which incorporates the likelihood framework into the causal function model to discover causal structures in multiple environments. All of the above-mentioned methods suffer the high complexity with the increase of searching of the graphs. To deal with the complexity problem, several methods have been proposed [4,29]. For example, NOTEARS [29] formulates the structure learning problem as a continuous optimization problem on a real matrix, which is easier to handle the models that contain continuous variables. DAG-GNN [27] extends the NOTEARS algorithm to support nonlinear relationships, but it may be limited in high-dimensional settings.DAG-GAN [6] proposes a method of generating antagonistic networks to simulate the causal generation mechanism, but it does not ensure non-circularity. CASTLE [13] learns the DAG structure of all input variables and uses the direct parent variable of the target variable as a predictor in regression or classification tasks.Overall, the existing methods still face challenges in accurately discovering the causal relationships among variables from complex data.

2.2 Causal Recommendation

Currently, there are three main types of work in the causal recommendation.

The first category aims to eliminate biases in recommendation systems, including popularity bias [1] and exposure bias [11] by using causal inference. The inverse propensity score (IPS) [19] method has been widely used and shown good performance, where it first estimates propensity scores based on some assumptions and then uses inverse propensity scores to reweight samples. CauseE [2] performs two rounds of MF on a large biased dataset and a small unbiased dataset, using L1 or L2 regularization to enforce the similarity between the two factorized embeddings. MCER [25] captures popularity bias as the direct causal effect of predicted scores and eliminates it by subtracting the direct popularity bias effect from the total causal effect.

The second category focuses on improving recommendation performance. For example, CARS [24] proposes a causal data augmentation framework to generate new data, address data sparsity issues, and improve the ability of sequential and top-N recommendations. CauseREC [28] models the distribution of counterfactual data to learn accurate and reliable user representations. These user representations are designed to be less sensitive to noisy behaviors and more trustworthy for fundamental behaviors.

The third category focuses on improving the interpretability of recommendation systems. For example, CountER [21] generates provider-side counterfactual explanations by finding a minimal set of user historical behaviors. [20] can specify the complexity and strength of the explanation and seek simple and effective explanations for model decisions.

3   Preliminary

In this section, we first show our target problem and then briefly introduce the related definitions.

3.1   Target Problem

Let $U = \{u_1, u_2, ..., u_n\}$ denote a set of users, $V = \{v_1, v_2, ..., v_m\}$ denote a set of items, $S = \left\{ \left(u_k, v_k^1, v_k^2, ..., v_k^{l_i} \right) \right\}_{k=1}^{N}$ denote the interaction set between users and items. The objective of our paper is to predict the next item $v_k^{l_i+1}$ to be selected. Here, $u_i \in R^{1 \times q^u}$ is a q^u-dimensional vector representing the user's latent characteristics, and $v_i \in R^{1 \times q^v}$ is a q^v-dimensional vector representing the item's latent characteristics. The prediction formula is given as follows.

$$p\left(v_k^{l_i+1} \mid \left(u_k, v_k^1, v_k^2, \ldots v_k^{l_i}\right)\right) \tag{1}$$

3.2 Relevant Definitions

Causal graphs, causal relationships, and causal inference are fundamental concepts in the field of causal reasoning. In addition, in our model, we employ a variational encoder for causal inference. Therefore, to better understand our paper, we will introduce the definitions of causal graphs, causal relationships, and variational encoder in this section.

Causal Graph. A causality diagram is a directed acyclic graph $G = (V, E)$, where V is a set of nodes and E is a set of edges, used to analyze and express the causal relationships between variables. The diagram captures the direction and strength of the causal relationships between variables. The cause-and-effect diagram consists of three basic configurations, as shown in Fig. 3. The first configuration is a chain, shown in Fig. 3(a). The two arrow lines in Fig. 3(a) between the three variables have the same direction and one variable acts as the "cause" of the other variable through the intermediary variable. The second configuration is a fork, shown in Fig. 3(b). The two arrow lines in Fig. 3(b) extend from the same variable as the source and point to the other two variables, which are the common causes of the outcome. The third configuration is the inverse fork, shown in Fig. 3(c). The arrow lines in Fig. 3(c) point to one variable and from another variable, and the variable is the common result of them.

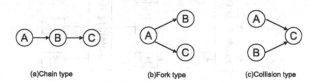

(a)Chain type   (b)Fork type   (c)Collision type

Fig. 3. Basic configuration of three components

Causal Relationship. Causality refers to the relationship between two variables (x and y) in statistical data, where the change of one variable (y) is caused by the change of the other variable (x). In this context, x is considered the cause and y the effect. The relationship between these two variables is known as a causal relationship. For instance, after buying a printer, it is likely that the consumer will also purchase printing paper and ink.

Causal Discovery. The objective of causal discovery is to identify and learn the underlying causal relationship structure among its internal variables from the observed dataset. Let $X = \{X_1, X_2, ..., X_m\}$ denote a set of random variable samples, and each sample is represented by a d-dimensional random vector, where $X \in R^{m \times d}$ and the causal relationships among them are represented by a causal graph G. $A \in R^{m \times m}$ is the weighted adjacency matrix of the DAG of m node. The definition of causal discovery is as follows: $A_{ij} = 1$ represents the edge from X_i to X_j and X_i is the reason for X_j. Conversely, $A_{ij} = 0$ represents no edge between the two variables. The purpose of causal discovery is to infer the structure A. In structure A, any two points can only have a unidirectional relationship, not a bidirectional one. That is, if m is the cause of n, it is impossible for n to be the cause of m.

Variational Automatic Encoder. As shown in Fig. 4, Variational Autoencoder (VAE) is a generative model that can learn the latent distribution of input data and generate new samples that are similar to it. Unlike traditional autoencoders, VAE not only learns the features of the data but also the underlying distribution, enabling it to generate new data samples. The core idea of VAE is to model the underlying distribution of input data by encoding and decoding the latent variables. In VAE, the encoder maps the input data $X = \{X_1, X_2, \cdots, X_n\}$, to the mean μ and variance δ^2 of the latent space and samples a latent vector $Z = \{Z_1, Z_2, \cdots, Z_n\}$ from this distribution. The decoder then maps the latent vector back to the input space to generate a new sample $\hat{X} = \left\{\hat{X}_1, \hat{X}_2, \cdots, \hat{X}_n\right\}$.

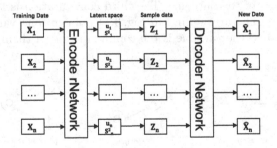

Fig. 4. Variational autoEncoder

4   Methodology

4.1   Overview of Model Structure

To recommend projects based on causal relationships, we propose a framework consisting of two modules: a causal discovery module and a causal recommendation module. The framework is shown in Fig. 5. Initially, user and project datasets are used as inputs for causal discovery analysis. Causal discovery is then performed using a variational encoder to generate reconstructed samples from the dataset. Finally, the reconstructed samples are used as new inputs, and a neural collaborative filtering network is applied for a recommendation, using reconstructed samples and negative samples to provide customized recommendations for potential causal relationships. The specific content is as follows.

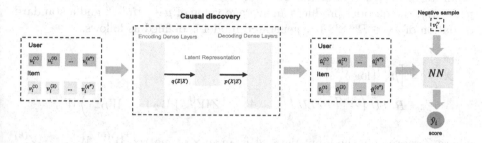

Fig. 5. The overall architecture of our method model

4.2 Causal Discovery

We propose two steps to achieve the goal of generating a new sample $\hat{X}$ that conforms to the same probability distribution as X. The first step is to extract data Z from the potential variable space that follows the same distribution as X. The second step is to generate a new sample $\hat{X}$ by sampling from Z. Next, we will introduce the detailed steps.

step1 (Coding Model): To ensure that the generated Z corresponds to the original X, we first establish a posterior distribution $p\left(Z|X^k\right)$ specific to X^k. We further assume it to be a normal distribution, i.e., $p\left(Z|X^k\right) = N(0,I)$. Because directly computing $p\left(Z|X^k\right) = N(0,I)$ is difficult, we use a variational posterior $q\left(Z|X^k\right)$ to approximate it, which is also a normal distribution. Finally, to derive the variational posterior distribution $q\left(Z|X^k\right)$, we establish a coding model for its implementation. The input data X^k is encoded by an encoder neural network into a latent variable Z with a density distribution of $q\left(Z|X^k\right)$. The variational posterior distribution $q\left(Z|X^k\right)$ is modeled as a Gaussian function with multiplicative coefficients. Specifically, the mean value of Z is multiplied by $\mu_Z \in R^{m \times d}$, and the standard deviation is expressed by multiplying $\delta_Z \in R^{m \times d}$. The module is shown below:

$$
\begin{aligned}
Z &:= \left[\mu_Z^{(k)} \mid \log \delta_Z^{(k)}\right] \\
&:= (I - A^T)\, ReLU \left(\cdots \left(ReLU \left(XW_1^{(k)}\right) W_2\right) \cdots W_{D-1}\right) W_D^{(k)}
\end{aligned}
\tag{2}
$$

where X represents input data, $W^{(k)}$ represents the weight of the neural network at level k, $ReLU$ represents the modified linear unit, D represents the depth of the neural network, and A^T represents the transposition of the matrix A.

step2 (Generation Model): Because $p\left(Z|X^k\right)$ belongs to X^k, we train a generation model $\hat{X}^k = g(Z)$, which restores X^k to $\hat{X}^k$ from Z^k sampled from distribution $p\left(Z|X^k\right)$. This model can accurately reconstruct each sample based on real user behavior data, encompassing the reconstruction of every element within X^k. By leveraging this approach, we can achieve a more precise and accurate reconstruction of the user behavior data. To calculate the likelihood $p\left(X^k|Z\right)$, the decoder produces an average value of $\mu_X \in R^{m \times d}$ and a standard deviation of $\delta_X \in R^{m \times d}$. The generated model is defined as follows.

$$
\begin{aligned}
\hat{X}^{(k)} &:= \left[\mu_X^{(k)} | \log \delta_X^{(k)} \right] \\
&:= ReLU\left(\cdots \left(ReLU\left((I - A^T)^{-1} Z W_1^{(k)} \right) W_2 \right) \cdots W_{D-1} \right) W_D^{(k)}
\end{aligned}
\tag{3}
$$

where Z represents the input data, A is an $m \times m$ matrix, $W_1^{(k)}, W_2, \ldots, W_D^{(k)}$ are parameters in the neural network that represent the weight matrix of each layer, D is the depth of the neural network, and $\hat{X}^{(k)}$ is the output result of the neural network.

After training the above model, we minimized the certain loss by the Evidence Lower Bound (ELBO) as the loss function. ELBO consists of two parts: reconstruction loss and Kullback-Leibler (KL) loss. We set the ratio of Reconstruction loss to KL loss to 1:1, which can not only ensure the quality of the generated samples, but also introduce some noise to make the generated samples have certain generalization ability. The loss function is defined as follows.

$$
L_{ELBO} = \frac{1}{n} \sum_{k=1}^{n} \left(L_{rec}^N + (-L_{KL}^N) \right)
\tag{4}
$$

where L_{rec}^N is the reconstruction loss and L_{KL}^N is the KL loss.

The reconstruction loss is a measure of the dissimilarity between the input and the output, which aims to minimize the reconstruction error. It can be expressed as follows: $L_{rec}^N = \frac{1}{n} \sum_{i=1}^{n} \left\| X - \hat{X} \right\|^2$, where X is the input data and $\hat{X}$ is the output from the model. On the other hand, the KL loss quantifies the divergence between the prior distribution and the approximate posterior distribution, which aims to regularize the latent space. The formula for kl loss is expressed as follows.

$$
L_{KL}^N = \frac{1}{2} \sum_{i=1}^{n} \sum_{j=1}^{d} (\delta_Z)_{ij}^2 + (\mu_Z)_{ij}^2 - 2\log (\delta_Z)_{ij} - 1
\tag{5}
$$

where n is the batch size, d is the dimension of the potential variable Z, and δ_Z and μ_Z represents the standard deviation and mean value of Z.

In addition, We need to ensure that the learned DAG is acyclic. Therefore, we add the following acyclic constraint:

$$tr\left[\left(I + \alpha\hat{A} \circ \hat{A}\right)^{n}\right] - n = 0 \tag{6}$$

where tr represents the trace of the matrix, $\hat{A}$ represents a normalized adjacency matrix, $\circ$ represents the Hadamard product (i.e., multiplication by element), n is the size of the adjacency matrix, α is a hyperparameter, and I is the identity matrix.

The interpretation of Eq. 6 can be understood in the following way: if the trace of the power of n of $(I + \alpha\hat{A} \circ \hat{A})$ equals n, then the DAG topology is legal and all nodes can be accessed. Otherwise, if the trace of the n power of $(I + \alpha\hat{A} \circ \hat{A})$ is less than n, it indicates that some nodes are unreachable or there are rings, which can lead to instability and overfitting of the model.

In summary, the overall objective of optimization is shown as follows.

$$L_{DAG} = -L_{ELBO} + \lambda \left(tr\left[\left(I + \alpha\hat{A} \circ \hat{A}\right)^{n}\right] - n\right) + \varsigma \left\|\left\{W_{d}^{(n)}\right\}_{d=1}^{D}\right\|_{2}^{2} \tag{7}$$

where ς is the regularization coefficient, and $W_{d}^{(n)}$ is the d-th weight matrix of the n-th layer. Overfitting is avoided by punishing large weight values.

4.3   Recommendation Based on Causality

In this section, we will present a method to enhance recommendation accuracy by utilizing causal preferences obtained through causal discovery. First, we conduct negative sampling that involves a characteristic of u_i for users, and compute the user's expected preference E. Next, we calculate the similarity between the user and other projects, and select the project v_j with the lowest similarity. Then, We replace the original user characteristics with data acquired through causal discovery as input. Finally, we input the negative samples along with the modified data into the model and train the neural network to obtain prediction results. To optimize the model, we use Bayesian personalized ranking loss to maximize user preference for positive samples over negative samples. The optimization objective can be formulated as follows.

$$L_1(O) = -\sum_{i=1}^{N} \delta\left(f(u, v) - f(u, v'')\right) \tag{8}$$

where u is a user, v is a positive sample, v'' is a negative sample and δ is the *sigmoid* activation function.

The formula after adding the causal discovery module can be converted into the following form:

$$L_2(O) = -\sum_{i=1}^{N} \delta\left(f(\hat{u}_i, \hat{v}_i) - f(\hat{u}_i, v_i'')\right) \tag{9}$$

where $\hat{u}_i$ and $\hat{v}_i$ are the reconstructed samples.

The loss function consists of two parts, one is the loss caused by causal discovery, and the other is the loss caused by prediction. The form is shown as follows.

$$L = L_{DAG} + L_2(O) \tag{10}$$

5   Experiments

In this section, we conduct experiments to show the effectiveness of the proposed framework. Our experiment aims to answer the following three essential questions:

RQ1. How does the performance of the CDRS compare to that of the baseline models?

RQ2. Does the application of the causal discovery module general framework to the recommendation system model have a certain effect on improvement?

RQ3. Can our methodology provide interpretability for the recommended results?

5.1   Experimental Settings

Datasets. We selected two real-world datasets, namely Cloud Theme Click and MovieLens10M, for our experiment. Cloud Theme Click is an e-commerce dataset collected from the Alibaba Taobao application, which captures click records from users and items in different purchase scenarios, such as 'what to bring for traveling' and 'how to dress for a party'. Cloud Theme Click includes user purchase history in the month preceding the promotion. MovieLens10M is a movie dataset containing ratings provided by users for each movie, as well as user characteristics such as gender, age, and occupation. Gender is represented as a binary feature, the occupation has 21 categories, and age is divided into 7 groups based on age range. Each movie is characterized by its ID, title, category (with 20 categories available), and year (which is divided into 18 categories) (Table 1).

Table 1. Statistics of datasets.

Dataset	Users	Items	Interaction
Cloud Theme Click	16452	9684	42356
MovieLens10M	7326	2443	418343

Baselines. To demonstrate the effectiveness of the causal discovery module in our model, we conducted experiments and compared it with six baseline models. BPR, GRU4Rec, and NeuMF models are commonly used standard models for recommendation systems, while IPS, DICE, and CauseE are standard models for causal reasoning in recommendation systems. The selection of these baselines allowed us to more accurately assess the additional value of our causal discovery module and to perform a more comprehensive comparison with state-of-the-art methods.

BPR [17] is a personalized sorting algorithm based on Bayesian posterior optimization.

NeuMF [8] is a popular deep learning-based recommendation algorithm. It combines traditional matrix decomposition with multi-layer perceptron and can simultaneously extract low-dimensional and high-dimensional features.

GRU4Rec [9] is a recommendation algorithm that uses session information and gated recurrent unit (GRU) to provide recommendations.

IPS [19] is a classical reverse probability weighting method for dealing with selection bias in observed data. The algorithm is based on the probability of sample selection by weighting the observed data so that the intervention effect can be estimated more accurately when dealing with causality.

CausE [2] is a machine learning-based causal inference method that uses probabilistic graphical models and machine learning methods to represent causal relationships as directed edges, learn the probability distribution of each node in the causal graph, and estimate the effect of the intervention using causal inference.

DICE [30] is a causal inference algorithm based on interpolation methods for inferring causal relationships from observed data. The algorithm uses the local structure of the causal graph to estimate the intervention effect by interpolation, and the global structure of the causal graph to control the accuracy of interpolation.

Evaluation Metrics. We have chosen two widely adopted evaluation metrics, namely Normalized Discounted Cumulative Gain(NDCG) and Recall, to assess the recommendation accuracy. Specifically, we computed the NDCG and Recall scores for the top 20 and top 25 recommended items, and the higher the score, the better the recommendation performance.

5.2 Overall Performance Comparison(RQ1)

Tables 2 and 3 present the overall performance of our proposed model and the six selected baselines. Table 2 displays the experimental results of the Cloud Theme Click dataset, while Table 3 presents the results of the MovieLens10M dataset. The best performance values are highlighted in bold in both tables.

The results demonstrate that our proposed method outperforms the other baselines in both datasets, exhibiting superior precision and NDCG scores, which validates the effectiveness of our framework.

Among the baselines that do not apply causal reasoning, i.e., BPR, NeuMF, and GRU4Rec, the BPR model performed the worst due to its shallow and simple structure; NeuMF model has improved performance compared to the BPR model because it incorporates a neural network in its model; and GRU4Rec model perform the best among them because the use of GRU. IPS, CausE, and DICE are all causal inference-based models for recommendation systems that address the bias inherent in traditional recommendation systems to achieve unbiased learning. Overall, these three methods achieve better performance compared with that of the methods that do not apply causal reasoning, demonstrating that incorporating causal inference into recommendation systems can improve the performance of the model. Our approach focuses on causal discovery, enabling the model to learn user causal preferences and ensuring that the training and test sets are independent and identically distributed, thus resulting in significant improvement in performance.

Table 2. Experimental Results on Cloud theme click Dataset.

Top-K	top20		top25	
	NDCG@$1e^{-2}$	Recall@$1e^{-2}$	NDCG@$1e^{-2}$	Recall@$1e^{-2}$
BPR	0.223	0.573	0.248	0.653
NeuMF	0.237	0.657	0.265	0.711
GRU4Rec	0.241	0.715	0.273	0.776
IPS	0.232	0.668	0.267	0.713
CausE	0.246	0.671	0.279	0.722
DICE	0.295	0.744	0.314	0.825
CDRS	**0.326**	**0.812**	**0.341**	**0.887**

Table 3. Experimental Results on MovieLens10M Dataset.

Top-K	top20		top25	
	NDCG@$1e^{-1}$	Recall@$1e^{-1}$	NDCG@$1e^{-1}$	Recall@$1e^{-1}$
BPR	0.238	0.341	0.359	0.404
NeuMF	0.254	0.372	0.374	0.439
GRU4Rec	0.276	0.416	0.387	0.470
IPS	0.313	0.413	0.404	0.462
CausE	0.273	0.385	0.374	0.433
DICE	0.368	0.463	0.452	0.508
CDRS	**0.398**	**0.491**	**0.474**	**0.534**

5.3   Effect of Causal Discovery Module(RQ2)

In order to study the effectiveness of our causal discovery module, we incorporated our proposed causal discovery module framework into two baseline methods, namely BPR and NeuMF, and conducted experiments on this basis, comparing the results with the baseline model. The dataset used is MovieLens10M.

NDCG and Precision indicators are used to evaluate the performance of recommended projects. The experimental results are shown in Fig. 6. The first figure shows the results of the NDCG and Recall of the BPR model and the BPR-CauD model configured with a causal discovery framework on top 20 and top 25, respectively. The second diagram is the result of NeuMF and NeuMF CauD model configured with a causal discovery framework. From the result, we can see that the performance of models configured with a causal framework has improved compared with that of the baseline model. The results show that the addition of the causal discovery module improves overall performance, verifying the effectiveness of our new module. Compared to traditional recommendation algorithms, causal discovery frameworks can better explain recommendation results and discover causal relationships between user behaviors.

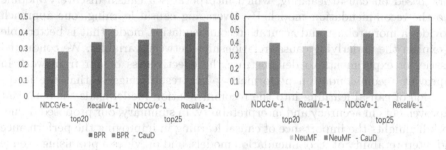

Fig. 6. Experimental results of two different models of BPR and NeuMF.

5.4   Interpretable Recommendation Structure(RQ3)

Our model provides reliable explanations because it considers the potential causal relationships that govern human behavior. Taking the movie dataset as an example, as shown in Fig. 7, when people choose movies, they often assume that the popularity of a movie influences their selection - the more popular the

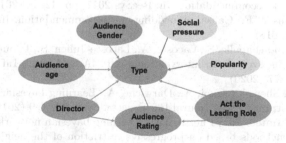

Fig. 7. Some causal structures exist in the movie dataset. The green part represents a false causal relationship, while the blue part represents an existing causal relationship. (Color figure online)

movie, the more likely it is to be chosen. However, this is a false causal relationship, as popular movies do not necessarily correspond to personal preferences. People's movie choices are determined by various factors, such as the director, the cast, and online ratings. Our model uses causal inference to discover and learn causal relationships, enabling it to provide reliable explanations and make accurate recommendations.

6 Conclusion

This paper proposes a novel approach to improve the performance and interpretability of recommendation models by capturing the causal relationships between user behavior. Specifically, we propose a general recommendation framework based on causal learning, which incorporates a causal discovery module into the recommendation model. By leveraging causal learning, our approach provides a more robust and accurate recommendation model that is better able to explain the underlying causal relationships between variables. We conducted a series of experiments to demonstrate the effectiveness of our framework in improving recommendation performance. The results suggest that the incorporation of causal learning into the recommendation model leads to significant improvements in accuracy and interpretability. In summary, our proposed framework highlights the importance of causal learning in improving the performance and interpretability of recommendation models, and provides a promising avenue for future research in this area.

Acknowledgements. This work was supported by Shanghai Science and Technology Commission (No. 22YF1401100), Fundamental Research Funds for the Central Universities (No. 22D111210, 23D111204), and National Science Fund for Young Scholars (No. 62202095).

References

1. Abdollahpouri, H., Burke, R., Mobasher, B.: Controlling popularity bias in learning-to-rank recommendation. In: RecSys 2017, pp. 42–46 (2017)
2. Bonner, S., Vasile, F.: Causal embeddings for recommendation. In: RecSys 2018, pp. 104–112 (2018)
3. Brouillard, P., Lachapelle, S., Lacoste, A., Lacoste-Julien, S., Drouin, A.: Differentiable causal discovery from interventional data. Adv. Neural. Inf. Process. Syst. **33**, 21865–21877 (2020)
4. Chen, E.Y.J., Shen, Y., Choi, A., Darwiche, A.: Learning bayesian networks with ancestral constraints. Adv. Neural Inf. Process. Syst. **29**, 1–9 (2016)
5. Gámez, J.A., Mateo, J.L., Puerta, J.M.: Learning bayesian networks by hill climbing: efficient methods based on progressive restriction of the neighborhood. Data Min. Knowl. Disc. **22**, 106–148 (2011)
6. Gao, Y., Shen, L., Xia, S.T.: DAG-GAN: causal structure learning with generative adversarial nets. In: ICASSP 2021, pp. 3320–3324. IEEE (2021)

7. Geng, C., Wu, H., Fang, H.: Causality and correlation graph modeling for effective and explainable session-based recommendation. arXiv preprint arXiv:2201.10782 (2022)
8. He, X., Liao, L., Zhang, H., Nie, L., Hu, X., Chua, T.S.: Neural collaborative filtering. In: WWW 2017, pp. 173–182 (2017)
9. Hidasi, B., Karatzoglou, A., Baltrunas, L., Tikk, D.: Session-based recommendations with recurrent neural networks. arXiv preprint arXiv:1511.06939 (2015)
10. Jenkins, J.M., Astington, J.W.: Theory of mind and social behavior: causal models tested in a longitudinal study. Merrill-Palmer Q. (1982-), 203–220 (2000)
11. Joachims, T., Swaminathan, A., Schnabel, T.: Unbiased learning-to-rank with biased feedback. In: WSDM 2017, pp. 781–789 (2017)
12. Koivisto, M., Sood, K.: Exact bayesian structure discovery in bayesian networks. J. Mach. Learn. Res. **5**, 549–573 (2004)
13. Kyono, T., Zhang, Y., van der Schaar, M.: Castle: regularization via auxiliary causal graph discovery. Adv. Neural. Inf. Process. Syst. **33**, 1501–1512 (2020)
14. Mooij, J.M., Magliacane, S., Claassen, T.: Joint causal inference from multiple contexts. J. Mach. Learn. Res. **21**(1), 3919–4026 (2020)
15. Musto, C.: Enhanced vector space models for content-based recommender systems. In: RecSys 2010, pp. 361–364 (2010)
16. Ng, I., Zhu, S., Chen, Z., Fang, Z.: A graph autoencoder approach to causal structure learning. arXiv preprint arXiv:1911.07420 (2019)
17. Rendle, S., Freudenthaler, C., Gantner, Z., Schmidt-Thieme, L.: BPR: bayesian personalized ranking from implicit feedback. arXiv preprint arXiv:1205.2618 (2012)
18. Schafer, J.B., Frankowski, D., Herlocker, J., Sen, S.: Collaborative filtering recommender systems. In: Brusilovsky, P., Kobsa, A., Nejdl, W. (eds.) The Adaptive Web: Methods and Strategies of Web Personalization, vol. 4321, pp. 291–324. Springer, Heidelberg (2007). https://doi.org/10.1007/978-3-540-72079-9_9
19. Schnabel, T., Swaminathan, A., Singh, A., Chandak, N., Joachims, T.: Recommendations as treatments: debiasing learning and evaluation. In: ICML, pp. 1670–1679. PMLR (2016)
20. Tan, J., et al.: Learning and evaluating graph neural network explanations based on counterfactual and factual reasoning. In: WWW 2022, pp. 1018–1027 (2022)
21. Tan, J., Xu, S., Ge, Y., Li, Y., Chen, X., Zhang, Y.: Counterfactual explainable recommendation. In: CIKM 2021, pp. 1784–1793 (2021)
22. Tsamardinos, I., Brown, L.E., Aliferis, C.F.: The max-min hill-climbing bayesian network structure learning algorithm. Mach. Learn. **65**, 31–78 (2006)
23. Wang, Z., Chen, X., Dong, Z., Dai, Q., Wen, J.R.: Sequential recommendation with causal behavior discovery. arXiv preprint arXiv:2204.00216 (2022)
24. Wang, Z., et al.: Counterfactual data-augmented sequential recommendation. In: SIGIR 2021, pp. 347–356 (2021)
25. Wei, T., Feng, F., Chen, J., Wu, Z., Yi, J., He, X.: Model-agnostic counterfactual reasoning for eliminating popularity bias in recommender system. In: KDD 2021, pp. 1791–1800 (2021)
26. Yu, F., Zhu, Y., Liu, Q., Wu, S., Wang, L., Tan, T.: TAGNN: target attentive graph neural networks for session-based recommendation. In: SIGIR 2020, pp. 1921–1924 (2020)
27. Yu, Y., Chen, J., Gao, T., Yu, M.: DAG-GNN: dag structure learning with graph neural networks. In: ICML, pp. 7154–7163. PMLR (2019)
28. Zhang, S., Yao, D., Zhao, Z., Chua, T., Wu, F.C.: Counterfactual user sequence synthesis for sequential recommendation. In: SIGIR 2021 (2021)

29. Zheng, X., Aragam, B., Ravikumar, P.K., Xing, E.P.: Dags with no tears: continuous optimization for structure learning. Adv. Neural Inf. Process. Syst. **31**, 1–12 (2018)

30. Zheng, Y., Gao, C., Li, X., He, X., Li, Y., Jin, D.: Disentangling user interest and conformity for recommendation with causal embedding. In: ACM Web Conference, pp. 2980–2991 (2021)

ECS-STPM: An Efficient Model for Tunnel Fire Anomaly Detection

Huansheng Song[1], Ya Wen[1], Xiangyu Song[2(✉)], ShiJie Sun[1], Taotao Cai[3], and Jianxin Li[4]

[1] Chang'an University, Xi'an, China
{hshsong,2021124096,shijieSun}@chd.edu.cn
[2] Swinburne University of Technology, Melbourne, Australia
x.song@deakin.edu.au
[3] Macquarie University, Sydney, Australia
taotao.cai@mq.edu.au
[4] Deakin University, Victoria, Australia
jianxin.li@deakin.edu.au

Abstract. The fire spreads rapidly in the tunnel due to the narrow space and high sealing, which makes rescue hard and threatens the citizen's lives. However, the lack of public fire datasets makes it challenging for networks to learn targeted representations of fire features, resulting in low detection accuracy. To tackle this problem, we construct a Tunnel Fire Anomaly Detection (TF-AD) dataset based on unsupervised training. This dataset contains 5200 high-resolution color images, including non-fire images for training and fire images with annotations for testing. Based on the TF-AD dataset, we propose an efficient tunnel fire anomaly detection model named ECS-STPM. ECS-STPM consists of a teacher and student network with identical EfficientNet-B1 structures. Additionally, considering the efficiency of adaptively assigning channel weights, we combine the convolutional kernel with channels to propose a novel attention mechanism, Efficient Kernel and Channel Attention (EKCA). EKCA replaces the Squeeze-and-Excitation (SE) networks in the MBConv module to prevent the loss of crucial information. Furthermore, we introduce the SPD-Conv module instead of the strided convolution layer to increase the detection accuracy in smaller fire areas. The experimental results on TF-AD dataset show that the pixel-level AUC-ROC and image-level AUC-ROC are up to 0.931 and 0.835, which verifies the effectiveness of our model.

Keywords: Tunnel Fire Anomaly Detection · TF-AD dataset · ECS-STPM · Unsupervised training · EKCA attention mechanism · SPD-Conv

1 Introduction

Highway traffic accidents caused by various reasons often lead to serious consequences, among which those caused by tunnel fires are the most hazardous. Tunnel fires are challenging to handle, as they can damage the monitoring, lighting,

© The Author(s), under exclusive license to Springer Nature Singapore Pte Ltd. 2024
X. Song et al. (Eds.): APWeb-WAIM 2023, LNCS 14334, pp. 277–293, 2024.
https://doi.org/10.1007/978-981-97-2421-5_19

and communication systems. Smoke can also reduce the visibility in the tunnel significantly, hindering fire fighting and personnel evacuation. Therefore, timely and effective fire detection in the tunnel is crucial.

The majority of current deep-learning approaches for fire detection require extensive training with large-scale annotated datasets of realistic fire scenarios to attain satisfactory detection outcomes. Bigger pre-training data improves performance logarithmically [19]. Limited tunnel fire data reduces model's detection and generalization skills. The network has low Precision and it confuses some flame-like objects such as headlights and tunnel lights with fire. Therefore, fire detection methods requiring many labeled fire data are unfit for tunnel fire anomaly detection.

Unsupervised learning-based anomaly detection networks improve greatly lately [17,30,32]. The tunnel fire is an example of an anomaly that deviates from the normal scenario of a tunnel. Wang *et al.* [23] proposes Student-Teacher Feature Pyramid Matching (STPM) for anomaly detection. The network achieves high accuracy by computing feature discrepancies between teacher and student network at different layers, and Fig. 1 shows tunnel fire results. As Fig. 1 illustrates, the feature maps of fire samples at various resolutions generated by STPM are contaminated with noise. Moreover, STPM confuses headlights with flames, this effect is more pronounced for low-resolution and small-scale fires.

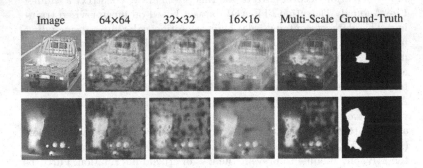

Fig. 1. STPM detection results at various resolutions

Anomaly detection networks can overcome the challenge of insufficient fire datasets for networks to learn fire features effectively. However, no public dataset for tunnel fire anomaly detection exists. We construct a Tunnel Fire Anomaly Detection (TF-AD) dataset, which consists of non-fire images for training and fire images for testing, based on realistic vehicle objects. Table 1 presents a comparison between the TF-AD dataset and the existing partial fire dataset [8].

As Table 1 shows, firstly they have low image resolution, high noise levels, and limited viewpoint diversity [29]. Secondly, the labeling is inconsistent, incomplete, and inaccurate. Lastly, they lack tunnel scenes. We introduce TF-AD, a novel dataset designed for unsupervised learning approaches that contains a large and diverse collection of images. The dataset includes fire and non-fire classes

Table 1. Existing fire datasets

Name	Year	Category	Scene	Label	Format	Drawback
BoWFire [18]	2014	Fire, non-fire	Incendiaries, non-fire scenarios	No	Images	Small number of pictures, low resolution images, no picture labeling
Fire-Dunnings-Dataset [22]	2018	Fire, non-fire	Complex scenarios	No	Images	Small sample size, low quality pictures
895 Fire Videos Data [5]	2020	Flame	Day, night	Yes	Videos	Unclear data labeling, data diversity
TF-AD (ours)	2023	Fire, non-fire	Traffic Scenes	Yes	Images	No smoke category

with pixel-wise labels, which can address the limitations of the above dataset in terms of traffic scenarios and data quality.

In this paper, we present ECS-STPM, an effective tunnel fire detection method based on the TF-AD dataset. ECS-STPM employs the unsupervised learning strategy for training. The backbone of the student-teacher network has been replaced with EfficientNet-B1 [21]. We propose Efficient Kernel and Channel Attention (EKCA) network as a replacement for the attention mechanism of the MBConv module. Furthermore, the SPD-Conv module [20] is adopted to enhance the accuracy of fire areas with small sizes and feature maps with low resolution.

We assess ECS-STPM on TF-AD and FDD dataset and contrast ECS-STPM with other anomaly detectors. The results demonstrate that ECS-STPM surpassed other networks in both image-level AUC-ROC and pixel-level AUC-ROC.

2   Related Work

2.1   Deep Learning Fire Detection

Many researchers propose various enhancements based on deep learning-based fire detection algorithms. Wang et al. [26] propose a convolutional neural network approach for fire detection by using data augmentation to train samples, which can avoid scattering due to insufficient datasets. Sharma et al. [16] develop a fire detection system by combining two pre-trained networks: VGG16 and Resnet50, but their dataset is unbalanced, containing more non-fire images than fire images. Wu et al. [28] create a forest fire benchmark and propose an improved tiny-Yolo. Zhang et al. [31] address the problem of data scarcity by synthesizing images for their dataset, but their network model had difficulty ensuring detection accuracy in real scenes. To conclude, existing deep learning-based fire detection methods face a trade-off between detection accuracy and speed. Moreover, they do not fully address the problem of tunnel fire dataset scarcity, which limits their applicability to tunnel scenarios.

2.2 Anomaly Detection

Anomaly detection can be categorized into two types: image-level and pixel-level. The former type involves classifying the whole image as normal or anomalous. Reiss *et al.* [14] introduce a representation learning approach for image anomaly detection that leverages neural networks to extract features from unlabeled data, but this approach suffers from challenging parameter tuning and high model complexity. Schlegl *et al.* [15] develop a Generative Adversarial Network (GAN)-based approach for anomaly detection, which detects anomalies by measuring reconstruction errors and feature mapping distances. This approach can handle high-dimensional, complex, and nonlinear data, but it demands a large amount of data and computational resources.

Pixel-level anomaly detection method can detect the exact locations of anomalies in a graph. Defard *et al.* [6] present an autoencoder (AE)-based approach for computing anomaly scores based on reconstruction errors. This approach is easy to implement but may fail to detect structural anomalies. Bergmann *et al.* [3] develop a student-teacher network-based approach that generates possible embeddings of normal data by training a teacher network and uses a student network to align the embedding space of the teacher network. It computes anomaly scores based on embedding distances and reconstruction errors.

STPM is an anomaly detection method that utilizes a student-teacher network, which can further transfer knowledge effectively from the pre-trained teacher network to the student network through the identical structured student-teacher network. It can detect anomalies of diverse sizes using the multi-scale feature matching strategy for acquiring different scale features in the feature pyramid. However, as mentioned in Sect. 1, STPM also has its disadvantages in the tunnel fire anomaly detection. We propose ECS-STPM to solve the problem of STPM for tunnel fire detection with better performance.

3 Methods

3.1 Tunnel Fire Anomaly Detection (TF-AD) Dataset

Statistical data indicate that self-ignition of the transported cargo accounts for 80% of the fire incidents on the highway, whereas traffic collisions cause the remaining 20% [2], which shows that vehicles are the main cause of highway fires. To solve the scarcity of public fire data in traffic scenarios, we build a Tunnel Fire Anomaly Detection (TF-AD) dataset with vehicle objects from real highway scenarios and public datasets. The dataset contains 5000 non-fire training images and 200 fire and non-fire images.

The training set is shown in (a) in Fig. 2. We assume that the tunnel fire video is unavailable and some interference factors in the tunnel can affect fire detection. The training set contains various vehicle from surveillance video under different conditions (daytime, nighttime, tunnels, and public), and vehicle in the public CoCo dataset [12], which can solve the poor accuracy caused by traffic scene changes. The training set can augment the suitability for tunnel traffic situations.

The test set contains both normal and fire samples, classified by scenario, as shown in (b) in Fig. 2. Figure 2 illustrates various tunnel fire scenarios with different types of vehicle objects in a), smaller fire samples and non-tunnel road fire scenarios in b), and vehicle object fire scenarios in the CoCo dataset, other object fire scenarios and non-fire scenarios in c). The test set are used to test the robustness of the model and to achieve distinction of tunnel fire samples.

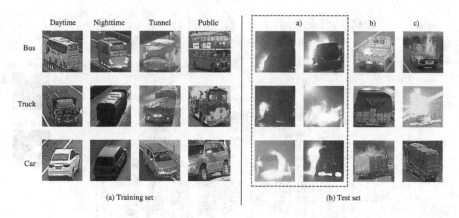

Fig. 2. Example TF-AD dataset diagram

The ground-truth is labeled with Labelme for the test set fire area. Since fire is a non-rigid object without a fixed shape, human impact exists when labeling, so the annotation rules are formulated: the salient features of fire are labeled to reduce the background information as much as possible. We supply pixel-level annotation for each fire region, and the dataset encompasses nearly 200 manually annotated regions.

3.2   ECS-STPM Network Structure

In this paper, we propose an efficient tunnel fire anomaly detection method called ECS-STPM. ECS-STPM can detect fire anomaly images with low resolution and small scale, and enhance the accuracy and speed of detection. The overall ECS-STPM network is shown in Fig. 3.

The ECS-STPM employs a student-teacher network architecture with identical structures. The teacher network is pre-trained on the image classification task (ImageNet on EfficientNet) and the student network is the untrained EfficientNet. In the training phase, features are extracted from a few successive layers of the teacher to guide the student network learning at the location of the knowledge distillation [24]. In the testing phase, the teacher network can effectively extract the fire features because of its high generalization ability. The student network is trained in non-fire samples and has poor representation of fire features. The two networks exhibit a significant discrepancy in feature extraction for the fire region across each layer. We take the fire feature maps obtained from three different feature extraction layers of the student-teacher network. These feature maps are bilinearly interpolated to conform to the dimensions of the input image, and subsequently multiplied to yield the ultimate fire feature maps.

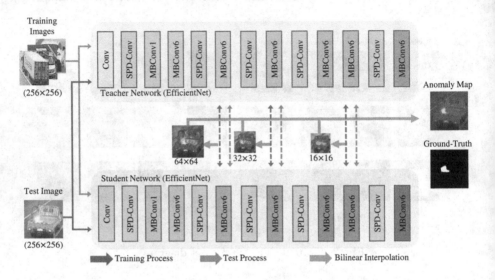

Fig. 3. ECS-STPM network structure

We adopt the same loss function as STPM. In the training phase, a set of non-fire images $\mathcal{D} = \{I_1, I_2, \ldots, I_n\}$ were input. $I_k \in R^{h \times w \times c}$ be the k-th input image, where h, w and c denote the height, width, and number of channels respectively. At l-th layer of the teacher and student networks, the feature vectors generated for image I_k are normalized at the feature map position (i, j) and denoted by $\hat{F}_t^l (I_k)_{ij}$ and $\hat{F}_s^l (I_k)_{ij}$ respectively. The loss at position (i, j) is given by the l_2-distance between the l_2-normalized feature vectors, as shown in Eq. 1. l_2-distance corresponds to the cosine similarity.

$$L^l (I_k)_{ij} = \frac{1}{2} \left\| \hat{F}_t^l (I_k)_{ij} - \hat{F}_s^l (I_k)_{ij} \right\|_{l_2}^2 . \tag{1}$$

In training, the student network learns the same feature outputs as the teacher network, so the loss function at l-th layer is defined as Eq. 2.

$$L^l (I_k) = \frac{1}{w_l h_l} \sum_{i=1}^{h_l} \sum_{j=1}^{w_l} L^l (I_k)_{ij} , \tag{2}$$

The feature map loss of different layers in the deep neural network is computed using Eq. 2, and the total loss is the weighted average of the losses of the different layers, as indicated by Eq. 3.

$$L\left(I_k\right) = \sum_{L=1}^{L} \alpha_l L^l\left(I_k\right), \quad \alpha_l \geq 0. \tag{3}$$

In testing, $J_k \in R^{h \times w \times c}$ represents the k-th input fire image, $\Omega^l(J)$ denotes the various fire feature map extracted in different feature layers of the student-teacher network, which follows Eq. 2. Finally, the feature maps are bilinearly interpolated to the same size as the input image, and the product of the fire feature maps yields the final fire feature anomaly map. An anomaly score for a test image J is computed based on the maximum value in the fire anomaly map, which indicates the presence of fire anomalous regions within the image.

Backbone Network. ECS-STPM network employs EfficientNet as the backbone networks for both the teacher and student networks. EfficientNet achieves model lightness and accuracy enhancement by balancing depth, width and resolution with a fixed scaling factor.

We select EfficientNet-B1 as the feature extraction network. EfficientNet-B1 is mainly composed of a stack of Mobile inverted Bottleneck Convolution (MBConv) structures. Figure 3 shows that each MBConv is followed by a number n, n represents the multiplicity factor. The network structure of MBConv is shown in Fig. 4.

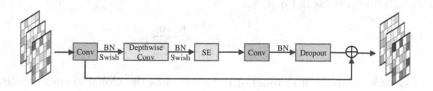

Fig. 4. MBConv module structure

When the input image passes through the MBConv structure, first it undergoes a 1×1 convolution (including BN and Swish) for dimensionality augmentation. Then it goes through a $k \times k$ Depthwise Convolution (containing BN and Swish) and applies the Squeeze-and-Excitation(SE) network [9] for weight update. Another 1×1 convolution is performed for dimensionality reduction, followed by a BN layer and Droupout. The final output feature map merges with the features of the input MBConv structure through shortcut connections to integrate the deeper feature information with the shallower image features.

Attention Mechanism. We introduce an Efficient Kernel and Channel Attention (EKCA) network, which replaces the SE attention mechanism in the MBConv structure. The EKCA network guides how to assign representations that focus on which convolution kernel to use through inter-channel dependencies of multi-scale extracted features. Figure 5 illustrates the structure of EKCA network, which consists of three parts: Split, Fuse, and Select.

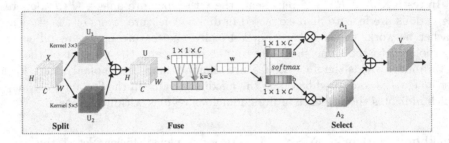

Fig. 5. EKCA network structure

Split: For the input feature $X \in R^{H \times W \times C}$, convolutional operations using two kernels with dimensions of 3×3 and 5×5 are performed, resulting in two distinct feature maps U_1 and U_2 that contain information with different scales.

Fuse: The global features U are obtained by summing the features U_1 and U_2, and then the global average information s is obtained by global average pooling. The calculation process is Eq. 4

$$s_c = F_{\text{gp}}(U_c) = \frac{1}{H \times W} \sum_{i=1}^{H} \sum_{j=1}^{W} U_c(i,j), \qquad (4)$$

where s_c denotes the c-th channel of s, F_{gp} denotes the global average pooling, and U_c represents the c-th channel of U. Cross-channel information exchange is achieved through Eq. 5 that assigns weights to each channel.

$$w = \sigma(W_k s), \qquad (5)$$

where σ is the Sigmoid activation function, W_k is the parameter matrix for computing channel attention, k denotes the coverage of local cross-channel interactions.

Select: After performing Fuse to obtain weight information at various spatial scales, the softmax function is applied to compute w and derive the weight a and b, the final characteristic V is calculated as in Eq. 6.

$$V = aU_1 + bU_2. \qquad (6)$$

SPD-Conv. ECS-STPM employs the SPD-Conv module to replace five strided convolution layer. The SPD-Conv module downsamples feature maps without losing learnable information, preserves more details, and improves the efficiency and accuracy of feature extraction. Figure 6 illustrates the SPD-Conv structure.

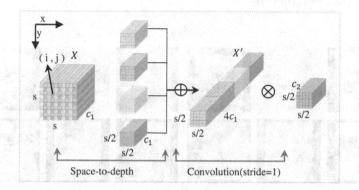

Fig. 6. SPD-Conv module structure

SPD-Conv is composed of a space-to-depth (SPD) layer and a non-strided convolution (Conv) layer. In the SPD, a feature map X of size $S \times S \times C_1$ can be scaled to form subgraphs $f_{x,y}$ consisting of $i + x$ and $j + y$, each of which is scaled to downsample X. In Fig. 6, the scale factor is 2, and each subgraph has the dimensions $\frac{S}{2} \times \frac{S}{2} \times C_1$. These four subgraphs are stitched in the channel dimension to obtain the feature map X', which has the dimensions $\frac{S}{2} \times \frac{S}{2} \times 2^2 C_1$. In the Conv, the SPD-Conv uses a convolutional layer with a step size of 1 to change the feature map size from X' to $\frac{S}{2} \times \frac{S}{2} \times C_2$, where $C_2 < 2^2 C_1$, which can retain as much critical information as possible.

4 Experiments

4.1 Experimental Details

In the experiments, ECS-STPM is tuned according to the loss function, and we choose the final weights when the loss converges. Three MBConv modules whose output feature map size is 64×64, 32×32 and 16×16 in EfficientNet-B1 are selected as knowledge distillation layers. The teacher network parameters are the pre-trained EfficientNet-B1 values on ImageNet, and the student network parameters are randomly initialized. The dataset is trained by ECS-STPM with 0.4 fixed learning rate, 0.9 momentum factor, 100 epoch, 32 batch size, and 256×256 input image resolution.

4.2 Dataset

We test ECS-STPM model on FDD dataset [7], which has fire and non-fire images from various scenes. We use non-fire images from six scenes as training set, fire images and non-fire images from the same scenes as test set. The dataset does not have tunnel scenes, but contains pedestrians and other objects. Figure 7 shows the dataset structure.

(a)   (b)

Fig. 7. Example FDD dataset diagram

In Fig. 7, from left to right (a) shows the non-fire images of six scenes in the training set, which are scenes 1~6. And (b) is the fire images of 6 scenes in the test set, where the non-fire images are not duplicated with the images in the training set.

4.3 Evaluation Metrics

This model is evaluated including frames per second (FPS), model size (S) and AUC-ROC (AR). The pixel-level AUC-ROC (AR_P) evaluates the fire region accuracy in the image, while the image-level AUC-ROC (AR_I) assesses the fire sample classification.

4.4 Experimental Results and Analysis

An Experimental Analysis of EfficientNet. We evaluate the backbone network against EfficientNet-B0 and B2, using three consecutive layers of Efficient-Net as feature extractors. The results are presented in Table 2.

As shown in Table 2, EfficientNet outperforms STPM in accuracy and parameters, indicating that the backbone network is highly robust. Among the three methods, Efficient-B1 achieves the best performance, with an improvement of 0.13 and 0.02 on AR_I and AR_P respectively, and a model size S of 63.4. Despite its small model size of 43.4, EfficientNet-B0 has relatively low expressive power. EfficientNet-B2 demonstrates comparable accuracy to our model, but less FPS

Table 2. Backbone network validation experiment

Model	AR_I	AR_P	FPS	S
STPM	0.65	0.83	100	104.8
EfficientNet-B0	0.74(+0.09)	0.84(+0.01)	**126.7(+26.7)**	**43.4(-61.4)**
EfficientNet-B2	0.779(+0.129)	0.846(+0.016)	108.6(+8.6)	74.0(−30.8)
EfficientNet-B1(ours)	**0.780(+0.13)**	**0.850(+0.02)**	119.3(+19.3)	63.4(−41.4)

increase and greater model size of 74. In conclusion, EfficientNet-B1 outperforms other networks and is more suitable as a backbone network. The model based on EfficientNet-B1 is named E-STPM.

An Experimental Analysis of EKCA Network. We replace SE network with EKCA, Efficient Channel Attention (ECA) [25], Convolutional Block Attention Module (CBAM) [27] and Selective Kernel (SK) network [11] in MBConv. Table 3 shows the results.

Table 3. Attention mechanism validation experiment

Model	AR_I	AR_P	S
E-STPM	0.78	0.85	63.4
E-STPM+ECA	0.817(+0.037)	0.872(+0.022)	**58.2(−5.2)**
E-STPM+CBAM	0.820(+0.04)	0.875(+0.025)	65.2(+1.8)
E-STPM+SK	0.826(+0. 046)	0.881(+0. 031)	61.4(−2.0)
E-STPM+EKCA(ours)	**0.830(+0.05)**	**0.894(+0.044)**	59.3(−4.1)

From Table 3, all attention mechanisms are superior to SE network. ECA improves AR_I by 0.037 respectively, and AR_P by 0.022 respectively. CBAM introduces additional parameters to the network in order to achieve an increase of 0.04 in AR_I and 0.025 in AR_P. SK network demonstrates an improvement of 0.046 in AR_I and 0.031 in AR_P. EKCA network achieves a significant increase of 0.05 in AR_I and 0.044 in AR_P, while simultaneously reducing the number of parameters and maintaining a high degree of accuracy. In conclusion, the EKCA network outperforms the student-teacher network, thereby leading to an improved model referred to as EC-STPM.

An Experimental Analysis of SPD-Conv Module. We present the experimental results of substituting five strided convolution layers with the SPD-Conv module in the EC-STPM model in Table 4.

From Table 4, the optimized model sacrifices the detection speed and model size, but the performance improvement in AR_I and AR_P are also very evident with 0.005 and 0.037. The refined model is referred to as ECS-STPM.

Table 4. SPD-Conv module verification experiment

Model	AR_I	AR_P	S	FPS
EC-STPM	0.830	0.894	**59.3**	**120.4**
ECS-SPTM (ours)	**0.835**	**0.931**	62.9	114.3

Validation Results of TF-AD Dataset. In this paper, we verify the model's accuracy at different resolutions of images, as shown in Table 5. The model results show that AR_I reaches 0.835 and AR_P reaches 0.931. It outperforms STPM at all resolutions and fire area detection.

Table 5. AR at different image resolutions

	Model	64×64	32×32	16×16	Multi-Scale
AR_I	STPM	0.529	0.597	0.638	0.65
	ECS-STPM(ours)	**0.757**	**0.799**	**0.815**	**0.835**
AR_P	STPM	0.756	0.791	0.824	0.83
	ECS-STPM(ours)	**0.872**	**0.883**	**0.923**	**0.931**

Figure 8 shows fire samples, STPM results, ECS-STPM results, and fire samples ground-truth, and illustrates different fire scenes: (a) a truck with a smaller fire, and (b) a tunnel with a larger fire.

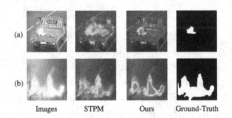

Fig. 8. STPM and ECS-STPM results on TF-AD dataset

The comparison results indicate that ECS-STPM have less noise and clearer fire edges than the STPM network's fire anomaly map under various fire sizes and illumination conditions. ECS-STPM enables more accurate fire area localization and enhanced fire detection performance on the image.

We test ECS-STPM on TF-AD test set and show results in Fig. 9. It contains fire image, fire anomaly maps at three resolutions and final fire area anomaly map with ground-truth.

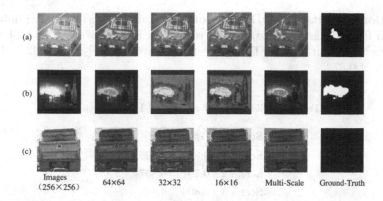

Fig. 9. ECS-STPM detection results at various resolution on TF-AD dataset

In Fig. 9(a) shows that ECS-STPM detects fire regions well in different resolutions and the final generated anomaly map approximate the ground-truth. Our model performs well in complex tunnel scenes with lighting and environment effects. (b) shows ECS-STPM detects fire regions in non-tunnel scenarios. (c) shows ECS-STPM distinguishes fire and non-fire areas well.

A Comparative Experiment of Existing Anomaly Detection Methods. We evaluate ECS-STPM against GANomaly [1], ITAE [10] and SPADE [4] on TF-AD dataset using image-level AR metrics. Table 6 presents the results. ECS-STPM surpasses others on all metrics, demonstrating superior accuracy, robustness and generalization for fire anomaly detection.

Table 6. Image level AR of different detection methods on TF-AD dataset.

Model	GANomaly	ITAE	SPADE	ECS-STPM (ours)
AR_I	0.672	0.747	0.751	**0.835**

We compare ECS-STPM with AnoGAN [15], CNN-Dict [13], and SPADE on pixel-level fire anomaly detection. Table 7 shows the results. ECS-STPM outperforms others and locates the fire region more accurately.

Table 7. Pixel level AR of different detection methods on TF-AD dataset.

Model	AnoGAN	CNN-Dict	SPADE	ECS-STPM (ours)
AR_P	0.769	0.804	0.923	**0.931**

Validation Results of FDD Dataset. We train ECS-STPM on training set based on FDD dataset. Table 8 shows results for six scenarios. ECS-STPM has high detection accuracy for FDN dataset, with average AR_I of 0.8092 and average AR_P of 0.8795.

Table 8. ECS-STPM results on FDD dataset

Scene	AR_I	AR_P
Scene 1	0.813	0.905
Scene 2	0.797	0.857
Scene 3	0.804	0.869
Scene 4	0.823	0.908
Scene 5	0.827	0.896
Scene 6	0.791	0.842
Average	0.8092	0.8795

Figure 10 illustrates visualization results of two scenes. Each row has input image, three fire anomaly maps at different resolutions, final anomaly map and ground-truth.

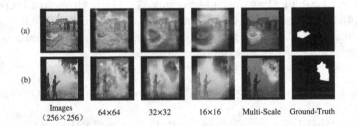

Fig. 10. ECS-STPM results on FDD dataset

Figure 10 shows ECS-STPM can locate fire regions with clear edges in fire anomaly maps from different feature extraction layers. ECS-STPM has high accuracy for image-level and pixel-level detection and can detect fire anomalies.

5   Conclusions

In this paper, we produce a novel dataset for tunnel fire anomaly detection called TF-AD, which contains abundant vehicle information. The dataset consists of non-fire images for training and fire images and non-fire images for testing. Utilizing this dataset, we present an efficient model for tunnel fire anomaly detection through unsupervised learning. This method only requires a large number of non-fire images for training and can effectively detect fire regions. Initially, we introduce a backbone network for the student-teacher network. Additionally, we propose a novel attention mechanism named EKCA. Moreover, we replace the strided convolution layer with the SPD-Conv module.

We propose an efficient model for detecting tunnel fire anomalies, named ECS-STPM, which is evaluated on the TF-AD dataset and compared to other existing networks. Our model achieves high performance ($AR_I = 0.835$, $AR_P = 0.931$), small model size (62.9 M), and fast speed (114.3 fps). Notably, ECS-STPM demonstrates excellent performance on the FDD dataset, thereby highlighting its ability to improve localization accuracy in complex environments.

Acknowledgement. This work was supported in part by the National Natural Fund Joint Fund Project of China under Grant U21B2041.

References

1. Akcay, S., Atapour-Abarghouei, A., Breckon, T.P.: GANomaly: semi-supervised anomaly detection via adversarial training. In: Jawahar, C.V., Li, H., Mori, G., Schindler, K. (eds.) ACCV 2018. LNCS, vol. 11363, pp. 622–637. Springer, Cham (2019). https://doi.org/10.1007/978-3-030-20893-6_39
2. Association NFP: Highway vehicle fires (2014-2016). NFPA Fire Analysis and Research Division (2018). https://www.nfpa.org/-/media/Files/News-and-Research/Fire-statistics-and-reports/Vehicle%20fires/osvehiclefires.pdf
3. Bergmann, P., Fauser, M., Sattlegger, D., Steger, C.: Uninformed students: Student-teacher anomaly detection with discriminative latent embeddings. In: Proceedings of the IEEE/CVF Conference on Computer Vision and Pattern Recognition, pp. 4183–4192 (2020)
4. Cohen, N., Hoshen, Y.: Sub-image anomaly detection with deep pyramid correspondences. arXiv preprint arXiv:2005.02357 (2020)
5. Company D: 895 fire videos data. https://www.datatang.ai/datasets/92?utm_source=PaperwithCode&utm_medium=dataset
6. Defard, T., Setkov, A., Loesch, A., Audigier, R.: PaDiM: a patch distribution modeling framework for anomaly detection and localization. In: Del Bimbo, A., et al. (eds.) ICPR 2021. LNCS, vol. 12664, pp. 475–489. Springer, Cham (2021). https://doi.org/10.1007/978-3-030-68799-1_35
7. Dunnings, A.J., Breckon, T.P.: Experimentally defined convolutional neural network architecture variants for non-temporal real-time fire detection. In: 2018 25th IEEE international conference on image processing (ICIP), pp. 1558–1562. IEEE (2018)
8. Fang, U., Li, M., Li, J., Gao, L., Jia, T., Zhang, Y.: A comprehensive survey on multi-view clustering. IEEE Trans. Knowl. Data Eng. (2023)
9. Hu, J., Shen, L., Sun, G.: Squeeze-and-excitation networks. In: Proceedings of the IEEE Conference on Computer Vision and Pattern Recognition, pp. 7132–7141 (2018)
10. Huang, C., Ye, F., Cao, J., Li, M., Zhang, Y., Lu, C.: Attribute restoration framework for anomaly detection. arXiv preprint arXiv:1911.10676 (2019)
11. Li, X., Wang, W., Hu, X., Yang, J.: Selective kernel networks. In: Proceedings of the IEEE/CVF Conference on Computer Vision and Pattern Recognition, pp. 510–519 (2019)

12. Lin, T.-Y., et al.: Microsoft COCO: common objects in context. In: Fleet, D., Pajdla, T., Schiele, B., Tuytelaars, T. (eds.) ECCV 2014. LNCS, vol. 8693, pp. 740–755. Springer, Cham (2014). https://doi.org/10.1007/978-3-319-10602-1_48

13. Napoletano, P., Piccoli, F., Schettini, R.: Anomaly detection in nanofibrous materials by cnn-based self-similarity. Sensors 18(1), 209 (2018)

14. Reiss, T., Cohen, N., Horwitz, E., Abutbul, R., Hoshen, Y.: Anomaly detection requires better representations. In: Computer Vision–ECCV 2022 Workshops: Proceedings, Part IV, pp. 56–68. Springer (2023). https://doi.org/10.1007/978-3-031-25069-9_4

15. Schlegl, T., Seeböck, P., Waldstein, S.M., Schmidt-Erfurth, U., Langs, G.: Unsupervised anomaly detection with generative adversarial networks to guide marker discovery. In: Niethammer, M., et al. (eds.) IPMI 2017. LNCS, vol. 10265, pp. 146–157. Springer, Cham (2017). https://doi.org/10.1007/978-3-319-59050-9_12

16. Sharma, J., Granmo, O.C., Goodwin, M., Fidje, J.T.: Deep convolutional neural networks for fire detection in images. In: Engineering Applications of Neural Networks: 18th International Conference, EANN 2017, pp. 183–193. Springer (2017). https://doi.org/10.1007/978-3-319-65172-9_16

17. Song, X., Li, J., Cai, T., Yang, S., Yang, T., Liu, C.: A survey on deep learning based knowledge tracing. Knowl.-Based Syst. 258, 110036 (2022)

18. Gustavo Botelho de Sousa, R.d.S.T.: Bowfire. https://bitbucket.org/gbdi/bowfire-dataset/downloads/

19. Sun, C., Shrivastava, A., Singh, S., Gupta, A.: Revisiting unreasonable effectiveness of data in deep learning era. In: Proceedings of the IEEE International Conference on Computer Vision, pp. 843–852 (2017)

20. Sunkara, R., Luo, T.: No more strided convolutions or pooling: A new cnn building block for low-resolution images and small objects. arXiv preprint arXiv:2208.03641 (2022)

21. Tan, M., Le, Q.: Efficientnet: rethinking model scaling for convolutional neural networks. In: International Conference on Machine Learning, pp. 6105–6114. PMLR (2019)

22. University, D.: Fire-dunnings-dataset. https://collections.durham.ac.uk/files/r2d217qp536#.YG_uqsgzqkL

23. Wang, G., Han, S., Ding, E., Huang, D.: Student-teacher feature pyramid matching for unsupervised anomaly detection. arXiv preprint arXiv:2103.04257 (2021)

24. Wang, L., Yoon, K.J.: Knowledge distillation and student-teacher learning for visual intelligence: a review and new outlooks. IEEE Trans. Pattern Analy. Mach. Intell. (2021)

25. Wang, Q., Wu, B., Zhu, P., Li, P., Zuo, W., Hu, Q.: Eca-net: efficient channel attention for deep convolutional neural networks. In: Proceedings of the IEEE/CVF Conference on Computer Vision and Pattern Recognition, pp. 11534–11542 (2020)

26. Wang, Y., Dang, L., Ren, J.: Forest fire image recognition based on convolutional neural network. J. Algorithms Comput. Technol. 13, 1748302619887689 (2019)

27. Woo, S., Park, J., Lee, J.-Y., Kweon, I.S.: CBAM: convolutional block attention module. In: Ferrari, V., Hebert, M., Sminchisescu, C., Weiss, Y. (eds.) ECCV 2018. LNCS, vol. 11211, pp. 3–19. Springer, Cham (2018). https://doi.org/10.1007/978-3-030-01234-2_1

28. Wu, S., Zhang, L.: Using popular object detection methods for real time forest fire detection. In: 2018 11th International Symposium on Computational Intelligence and Design (ISCID), vol. 1, pp. 280–284. IEEE (2018)

29. Xu, C., Zhao, W., Zhao, J., Guan, Z., Song, X., Li, J.: Uncertainty-aware multiview deep learning for internet of things applications. IEEE Trans. Industr. Inf. **19**(2), 1456–1466 (2022)
30. Zhang, P.F., Luo, Y., Huang, Z., Xu, X.S., Song, J.: High-order nonlocal hashing for unsupervised cross-modal retrieval. World Wide Web **24**, 563–583 (2021)
31. Zhang, Q.x., Lin, G.h., Zhang, Y.m., Xu, G., Wang, J.j.: Wildland forest fire smoke detection based on faster r-cnn using synthetic smoke images. Proc. Eng. **211**, 441–446 (2018)
32. Zhou, X., Delicato, F.C., Wang, K.I.K., Huang, R.: Smart computing and cyber technology for cyberization. World Wide Web **23**, 1089–1100 (2020)

ACE-BERT: Adversarial Cross-Modal Enhanced BERT for E-Commerce Retrieval

Boxuan Zhang[1], Chao Wei[1], Yan Jin[1], Cai Xu[2(✉)], Weiru Zhang[1], Haihong Tang[1], and Ziyu Guan[2]

[1] Alibaba Group, Hangzhou, China
{boxuan.zbx,weichao.wc,yan.jinyan,weiru.zwr,piaoxue}@alibaba-inc.com
[2] Xidian University, Xi'an, China
{cxu,zyguan}@xidian.edu.cn

Abstract. Nowadays on E-commerce platforms, products usually contain multi-modal descriptions. To search related products from user-generated texture queries, most previous works learn the multi-modal retrieval models by the historical query-product interactions. However, the tail products with fewer interactions tend to be neglected in the learning process. Recently, product pre-training methods have been proposed to solve this problem. Motivated by these works, we aim to solve two challenges in this area: (1) Existing works share the same network for the user-generated texture query and the product title, which ignores the semantic gap between them; (2) The irrelevant backgrounds in products images would disturb the cross-modal alignment. In this work, we propose the Adversarial Cross-modal Enhanced BERT (ACE-BERT) for E-commerce multi-modal retrieval. In the pre-training stage, ACE-BERT learns multi-modal Transformers by product title, image and additional Hot Query. Specifically, ACE-BERT constructs Hot Query for each product to learn the correlations of products and queries in the pre-training stage. In addition, ACE-BERT detects the significant objects and removes irrelevant backgrounds of the product image, then takes them as image representation. In the fine-tuning stage, ACE-BERT performs semantic matching and adversarial learning tasks to better align the representations of queries and products. Experimental results demonstrate that ACE-BERT outperforms the state-of-the-art approaches on both a public dataset and a real-world application. It is remarkable that ACE-BERT has already been deployed in the search engine, leading to a 1.46% increase in revenue.

Keywords: E-Commerce · Cross Modal Retrieval · Adversarial Learning · BERT

1 Introduction

With the rapid development of the Internet, lots of people are used to purchasing on global E-commerce platforms, such as Amazon, Aliexpress and eBay. In the E-

X. Song et al. (Eds.): APWeb-WAIM 2023, LNCS 14334, pp. 294–311, 2024.
https://doi.org/10.1007/978-981-97-2421-5_20

commerce retrieval systems, a customer usually first submits a textural query to the search engines, then the search engines would retrieve the candidate products from the product database. Most retrieval systems are still based on textual matching [2], which calculates the similarities of query and product titles by their matching scores. Recently, many deep learning methods are applied to retrieval systems to mine the high-level semantic information of the queries and product titles. For example, Embedding-Based Retrieval (EBR) [16] first learns high-level embeddings of query and product title by the deep neural network, then converts the retrieval problem into a nearest neighbor search [20] problem. In essence, these deep learning-based retrieval systems still rely on the matching of queries and product titles.

Fig. 1. The retrieval result of the textual matching-based retrieval system: the products are returned to a user looking for a "red dress". The red bounding boxes show the title and image of a mismatched product. (Color figure online)

In E-commerce platforms, the products usually consist of multiple modalities, e.g. textual title and image. The textual matching-based retrieval systems emphasize the matching in textual modal, leading to the *cross-modal mismatch* problem. An example is shown in Fig. 1. When a user is looking for a "red dress", a product (in the bounding boxes) with a title containing "red dress" is presented to the user. This is an apparent mismatch between query and product image, which would seriously impact the user's experience.

The multi-modal retrieval methods are proposed to solve the cross-modal mismatch problem [3,5,30]. These methods usually first fuse the multi-modal contents of the product to learn comprehensive product embedding, then match the embedding with query embedding. Most multi-modal retrieval methods use historical query-product interactions for model training. However, with millions of products in the E-commerce platforms, users tend to provide feedback for a very small set of them, causing a power-law distribution in E-commerce. The *tail products* with fewer interactions tend to be neglected in the learning process.

In order to fully consider the tail products, product pre-training methods [11,42] have been proposed recently. These methods pre-train the multi-modal Transformers [26,27] via large-scale multi-modal product data. By learning the multi-modal matching in the pre-training stage, these methods achieve state-of-the-art performance on downstream retrieval tasks. However, there still exist two challenges: (1) Existing works share the same network for the user-generated texture query and the product title, which ignores the semantic gap between them; (2) The irrelevant backgrounds in product images would disturb the cross-modal alignment. We notice that product images are different from the images in the general domain. They usually contain significant objects and obvious background noises. As shown in Fig. 2, it is easy to locate the significant Regions of Interests (RoIs) of product images, while hard for the images in the general domain[1]. Therefore, we dedicate to precisely extracting the significant objects in product images and facilitating the cross-modal alignment.

In this paper, we propose a new product pre-training based multi-modal retrieval method, named Adversarial Cross-modal Enhanced BERT (ACE-BERT). In the pre-training stage, ACE-BERT learns multi-modal Transformers by product title, image and additional Hot Query. The Hot Query is constructed for each product and would help to learn the correlation between product and query in the pre-training stage. In addition, ACE-BERT detects the "object-level" RoI (such as the right part of Fig. 2(a)) of the product image and takes it as pixel feature. ACE-BERT constructs complementary high-level patch features and low-level pixel features as image representation. In the fine-tuning stage, we perform the semantic matching task to achieve cross-modal retrieval. We also employ the adversarial learning strategy to better align the representation of the query and product.

To summarize, the contributions are as follows: (1) we point out that the crucial problem in E-commerce pre-training, i.e., the user-generated texture query should be considered in the pre-training since it has the semantic gap with the product title. We also construct the hot query for product query pre-training to solve this problem; (2) We find the product images usually contain significant objects and obvious background noises. Accordingly, We propose to extract the significant "object-level" RoI in product images to facilitate the cross-modal alignment; (3) Experiments on both a public dataset and the real-world E-commerce plateform (offline and online) verify ACE-BERT achieves the state-of-the-art performance.

---

[1] http://cocodataset.org.

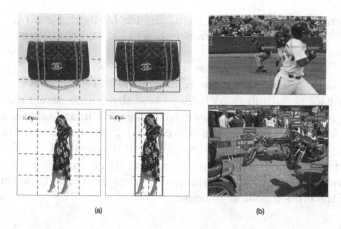

(a)   (b)

Fig. 2. (a) The RoIs of product images from the E-commerce domain. The left part cuts images into patches with the dotted bounding boxes, the right part detects the Region of Interest (RoIs) with red bounding boxes before cutting into patches. (b) The RoIs of MSCOCO images from the general domain. (Color figure online)

2   Related Work

Pre-training aims to learn models by leveraging data from other related tasks. It has been widely adopted in Computer Vision (CV) at first. With the success of AlexNet [25] in ImageNet classification [8], researchers subsequently find that the Convolutional Neural Network (CNN) [35] pre-trained on the large-scale image corpus show superior performance on a variety of down-stream tasks [10]. The knowledge embedded in original data [4,12,13,17,18] is captured by pre-training. Then, a Transformer-based pre-training model, BERT [9] is widely used in Natural Language Processing (NLP). It learns a universal language encoder by pre-training with large-scale unlabeled data. In recent years, there are research works [33,41] on pre-training the generic representation for NLP tasks. Inspired by these studies, many multi-modal pre-training models for vision-language have been proposed. Different from those multi-modal pre-training models in general domain, we need to consider the user-generated texture query in the product pre-training. Therefore, we construct the hot query to solve this problem.

Multi-modal Retrieval searches single/multi-modal contents according to single/multi-modal queries [21,24]. It utilizes the consistent and complementary information of multi-modal data [22,23,31,32,34,36,37] to promote retrieval. Existing methods aim to maximize the similarity between the samples from different modalities in a common subspace [18]. Earlier works concentrated on correlation [15] maximize the canonical correlations of the representations of different modals. Another line is based on generative adversarial networks (GANs) [14]. These methods [7,39,40] introduce the discriminator to distinguish samples from different modalities. The above methods usually adopt historical query-product interactions to train the models. However, with millions of products on E-commerce platforms, users tend to provide feedback for a very small set of them. The tail products with fewer interactions tend to be neglected in the training process.

Recently, researchers extend the multi-modal pre-training for cross-modal task and achieve great success. The representative methods contain ViLBERT [28], VisualBERT [27], Unicoder-VL [26]. These algorithms extract multiple RoIs or split image into multiple patches from images as image tokens. In fashion domain, FashionBERT [11] and Kaleido-BERT [42] use patch based methods to obtain image tokens. In this paper, we follow this research line. We point out that product images in E-commerce context usually contain only one significant object. Therefore, we extract the significant "object-level" RoI in product images to facilitate the cross-modal alignment.

3   Adversarial Cross-Modal Enhanced BERT

In this section, we first define the multi-modal retrieval task, then present the pre-training and fine-tuning stages of ACE-BERT in detail.

3.1   Notations and Task Definition

In the multi-modal retrieval task, suppose we are given a dataset with a set of texture queries $\mathcal{Q} = \{q_1, q_2, ..., q_S\}$ and products $\mathcal{A} = \{a_1, a_2, ..., a_K\}$. Each product a_k contains texture title a_k^T and image a_k^I. Given a set of training tuples $\{q_i, a_i\}_{i=1}^{W}$ (historical query-product interactions), the goal is to develop a model that can accurately predict the clicked products for a given query.

Despite we can collect millions of users' click-through pairs ($\{q_i, a_i\}$) from search logs for training our matching model, users tend to provide feedback for a very small set of products. The tail products with fewer interactions tend to be neglected in the learning process. In addition, different from the long product title, the semantic modeling of short queries is challenging. Therefore, we propose to learn the cross-modal relations of product title, product image and query in the pre-training task. For the fine-tuning task, we aim to the solve retrieval task.

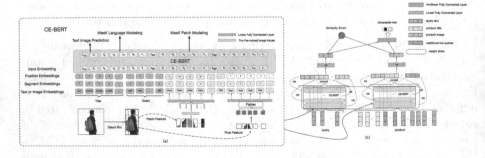

Fig. 3. The architecture of ACE-BERT. The left part is the pre-training model named CE-BERT. The right part is the fine-tuning model for E-commerce retrieval task.

3.2 Pre-training

The overall architecture of ACE-BERT is shown in Fig. 3. It consists of pre-training model named CE-BERT and fine-tuning model for E-commerce retrieval tasks. In the pre-training stage, we train the multi-modal Transformers, CE-BERT, by product titles $\{a_k^T\}$, images $\{a_k^I\}$ and additional hot queries. The Hot Query is constructed for each product and would help to learn the correlation of product and query in the pre-training stage. We train CE-BERT by all products. For clarify, we omit subscript k in this subsection. In addition, we detect the "object-level" RoI of the product image and takes it as pixel feature. We construct high-level patch features and low-level pixel features as image representation. Finally, we pre-train CE-BERT with three unsupervised tasks, Masked Language Modeling (MLM), Masked Patch Modeling (MPM) and Text and Image Prediction (TIP).

Pre-training Model Pipeline

In this subsection, we elaborate the pipeline of CE-BERT.

Input Product Title Representation: We conduct the standard pre-processing of BERT [9] on product title a^T. The input token sequence of product title, $T_a = \{t_a\}$, is the sum of initial word embeddings, segment embeddings and position embeddings, where t_a denotes the embedding of a specific token (word).

Input Product Image Representation: Different from the images in general domain, the product images usually contain significant objects and little background noises. It is easy to locate the significant RoIs of product images. Therefore, we resize the image $\{a^I\}$ to a fixed-size resolution and then detect the object-level RoI from images by Edge Detection Method. Next, we cut the object-level RoI into multiple patches. Then we send the patch representations into ResNet to obtain the high-level semantic embeddings. We add these embedding and corresponding segment embeddings, as well as position embeddings to obtain the input high-level patch sequence of product image, $\{i_a^O\}$.

Most of the current multi-modal pre-training methods only use the high-level patchs as the input. This ignores that different modals may have different levels of correlation. Therefore, we regard the raw image feature as another input of CE-BERT, which can be regarded as low-level feature. In detail, we split the detected RoI into a sequence of pieces with the same pixel number to obtain the fixed-length representation. For each piece, we reshape it into a sequence of flattened 1D vector. Specifically, the pixel features provide a priori knowledge without any inductive bias and are beneficial to cross-modal learning. Finally, the position embedding and the segment embedding are added to the pixel embedding to obtain the input low-level patch sequence of product image, $\{i_a^R\}$. i_a^R contains the low-level information of a specific part of the detected RoI. We assigned a special [SEP] token separate them $\{i_a^R\}$ and $\{i_a^O\}$. The final input patch sequence of product image is $I_a = \{\cdots, i_a^O, \cdots, [SEP], \cdots, i_a^R, \cdots\}$. We further verify the effectiveness of the two kinds patch embeddings of product image in Experiments.

Input Hot Query Representation: Existing pre-training methods [11,42] use the product titles and images to learn the model. For the retrieval task, these methods take the same sub-model to handle product titles and texture queries. This ignores the semantic gap between them. The queries are generated by users and even have various descriptions for a same target. In addition, the user-generated queries are usually short, which makes semantic modeling even more challenging. It is difficult to directly learn query embeddings by Transformer structure. Therefore, we propose the Hot Query to help semantic modeling and let the query-product pair interact with each other during training.

For each product a, we collect its interacted queries from the historical query-product interactions. Then we count the number of clicks for each inter-acted query. We select the (at most) top F frequent queries as the Hot Query for product a. These frequent queries are concatenated to construct the initial token embeddings of query. Similar to Text Representation mentioned above, the sum of initial token embeddings of query, the segment embeddings and position embeddings are regard as input token sequence of query, $H_a = \{h_a\}$.

CE-BERT: As shown in Fig. 3, the input token sequence of product title T_a, input patch sequence of product image I_a and input token sequence of query, H_a are concatenated as the input of CE-BERT. Similar to BERT, the special token [CLS] and separate token [SEP] are added in the first position and between these token sequences, respectively.

We adopt pre-trained standard BERT as the backbone network of CE-BERT. The information of product tokens and query tokens thus interact freely in mul-tiple self-attention layers. CE-BERT outputs the final representations of each token or patch. Here we denote v_{T_a}, v_{I_a} and v_{H_a} as the output token sequence of product title, output patch sequence of product image and output token sequence of query, respectively.

Pre-training Loss Function

We pre-train ACE-BERT with three tasks, Masked Language Modeling (MLM), Masked Patch Modeling (MPM) and Text and Image Prediction (TIP). The MLM and MPM tasks are similar to the mask task in BERT pre-training. We randomly mask the text tokens and image patches, which stimulates CE-BERT to learn the multi-modal correlations. Specifically, for the MLM task, we mask some tokens of T_a or H_a. Next we obtain the output tokens of this masked tokens (corresponding tokens in v_{T_a} or v_{H_a}) by CE-BERT. Finally we use the output token to predict the word in this token. Suppose $\hat{y}^{MLM}/y^{MLM}$ denote the predicted/true label of this word, the loss of MLM task is:

$$\mathcal{L}_{MLM} = -\sum_i y_i^{MLM} \log(\hat{y}_i^{MLM}). \tag{1}$$

Similar to MLM, we mask out certain patch (denoted as i_a) in input patch sequence of product image I_a in the MPM task. It would be set to zero. The corresponding output patch is v_{i_a}. When i_a is masked-out, the distributions of v_{i_a} is $q(v_{i_a})$, otherwise is $p(v_{i_a})$. The target of the MPM task is to minimize the KL-divergence over the two distributions:

$$\mathcal{L}_{MPM} = \sum p(v_{i_a}) log \frac{p(v_{i_a})}{q(v_{i_a})}. \tag{2}$$

The TIP task is to predict whether the image and text are from one same product. We concatenate the outputs of the special token [CLS] of all layers, and fed it into a binary classifier to predict whether the multi-modal inputs are matched. The predicted label is denoted as $\hat{y}^{TIP}$. For a positive example in the train dataset, the text (product title and hot query) and product image are extracted from the same product. Therefore the label $y^{TIP} = 1$. Otherwise for one negative sample, the text and image are randomly selected from different products ($y^{TIP} = 0$). The objective of the TIP task is:

$$\mathcal{L}_{TIP} = -\sum [y^{TIP} \log \hat{y}^{TIP} + (1 - y^{TIP}) log(1 - \hat{y}^{TIP})]. \tag{3}$$

In the pre-training stage, we jointly optimize the three tasks to learn ACE-BERT. We elaborate the implementation details in the Appendix.

3.3 Fine-Tuning

In this section, we will detail how to fine-tune the ACE-BERT for our multi-modal retrieval task. Figure 3(b) illustrates the overall architecture of ACE-BERT. It consists of two pre-trained CE-BERTs, which are desinged for query and product respectively. We encode query q_s and product a_k by the pre-trained CE-BERTs to obtain the corresponding query and product embeddings v_{q_s}, v_{a_k}. Next, we compute the conditional probability by the similarity of query and product pair:

$$P(v_{a_s}|v_{q_s}) = \frac{exp(\gamma cos(v_{q_s}, v_{a_s}))}{\sum_{k=1}^{K} exp(\gamma cos(v_{q_s}, v_{a_k}))}, \tag{4}$$

where γ denotes a smoothing parameter.

The main loss for the retrieval task is the the Semantic Matching Loss, which minimizes the following bidirectional log loss:

$$\mathcal{L}_{main} = -\frac{1}{W}(\sum_{s=1}^{W} log P(v_{a_s}|v_{q_s}) + \sum_{s=1}^{W} log P(v_{q_s}|v_{a_s})), \tag{5}$$

where W is the number of training tuples.

To further minimize the divergence of paired query and product embeddings, we design adversarial learning in fine-tuning. Specifically, we add a discriminator D to classify the domain sources (query or product). The objective of adversarial loss is:

$$\mathcal{L}_{adv} = -\frac{1}{W}\sum_{i=s}^{W}[log D(v_{q_i}; \theta_D) + log(1 - D(v_{a_s}; \theta_D))], \tag{6}$$

where D is fully connected neural network and θ_D denotes the parameters of D. Similar to Generative Adversarial Network [14], we conduct a minimax two-player game is conducted between the discriminator and encoder. We incorporate $\mathcal{L}_{main}$ and $\mathcal{L}_{adv}$ to jointly train the model:

$$\min_{\theta_E} \max_{\theta_D} (\mathcal{L}_{main} - \mathcal{L}_{adv}), \tag{7}$$

where θ_E denotes the encoder including CE-BERT and subsequent neural network. We elaborate the network architecture details in the Appendix. In the fine-tuning stage, we iteratively update the θ_E and θ_D with the other is fixed until convergence.

4 Experiments

In this section, we conduct experiments on both a public dataset and the real-world E-Commerce application (offline and online) to verify the effectiveness of ACE-BERT.

4.1 Datasets

We conduct offline experiments on two datasets in this paper. The first one is the public fashion product dataset, Fashion-Gen[2]. The second one is collected from E-Commerce platform. This dataset will be released soon on Tianchi Datasets[3].

- The Fashion-Gen dataset contains 67,666 fashion products and 293,008 images. We collect 53,826 products which consist of text descriptions and images for training and 6,706 products for testing.
- For pre-training on the E-Commerce platform dataset, we collect about 4.7 million high-quality products with titles and main images as positives (called PT dataset). The ratio of positive products (users' clicked products) to negative products is set as 1:3, which is determined by cross-validation in preliminary offline experiments. For fine-tuning, we collect over 25 million users' click-through information from search logs (called FT dataset). In particular, we sample the users' clicked products as positives. Negative products are sampled online in each mini-batch updated during the model training. For the offline evaluation on the downstream retrieval task, we sample the test set from the users' click-through records in the next day.

4.2 Baselines

The baselines in this experiment are as follows:

---

[2] https://fashion-gen.com.
[3] https://tianchi.aliyun.com.

- **TwinBERT** [29] has twin-structured BERT-like encoders to represent query text and product title respectively. It has been demonstrated as the basic baseline of the BERT-based model in retrieval.
- **FashionBERT-Finetune** [11] only adopts patches as image features and achieves state-of-the-art performance in fashion domain. We fine-tune it for the retrieval task.
- **CE-BERT-Finetune** considers the user-generated texture query and constructs high-level patch features and low-level pixel features as image representation. We also fine-tune it for the retrieval task.
- **ACE-BERT** is different from CE-BERT-Finetune. It is equipped with the adversarial learning for better cross-modal matching.

4.3   Evaluation Metrics

For offline experiments, we adopt the AUC and Recall@K (k=10, 50, 100) as evaluation metric. In detail, we randomly sample a query set. For each query, we construct a target set $T = \{t_1, t_2, ..., t_M\}$. And an evaluation set $E = \{e1, e2, ..., e_K\}$ is returned by the model. The Recall@K is defined as:

$$Recall@K = \frac{\sum_{i=1}^{K} f(e_i)}{M}, \qquad f(e_i) = \begin{cases} 1 & e_i \in T \\ 0 & else, \end{cases}$$

where M is the size of T. We evaluate on two datasets in offline experiments. For our platform's dataset, the query set and target set have 100,000 randomly sampled queries and 3.4 million clicked products, respectively. The target set is constructed by considering user click log. In particular for calculating the Recall@K (k=10, 50, 100), the embeddings of queries and products are generated by models. And then for each query, the ANN search is used to find the top-k products.

Similar to above, for Fashion-Gen dataset, we collect 6,706 product of test set. And for each product, we use the product image which is photographed from 1 to 6 different angles as target set. We adopt ANN search to find the top-k (k=10, 50, 100) image among 293,008 image for each product text description.

For online experiments, we use click-through rate (CTR) and Revenue which is defined as the money received from normal business operations (e.g. pay-per-click in this paper) as the metrics. Furthermore, we also calculate the p-value according to Fisher's exact test, in order to assess the significance of improvements in online experiments.

4.4   Performance Comparison

Evaluation on the Fashion-Gen Dataset. We compare the two pre-trained models with the offline experiment on the Fashion-Gen dataset. As demonstrated in Table 1, CE-BERT performs better than FashionBERT, which indicates the CE-BERT is effective in extracting the information from the image. The additional pixel patches can help improve the model performance. We further verify this in Sect. 4.6.

4.5 Implementation Details

Pre-training. In pre-training, we reuse the pre-trained parameters from the 4-layer BERT-mini. The setting of hyper-parameters are as follows: (1) The maximum text sequence length is set to 512 and the maximum image sequence length is set to 16 and 49 for patch features and pixel features, respectively. (2) For patch features, ResNeXt101 is adopted to extract a 2048d embedding for each patch. For pixel features, the image is resize to 224×224 and each patch has 32×32 input size. (3) Adam is used as the optimizer with the learning rate set to $2e-5$. (4) The training will stop when the max number of epochs is reached either early stop to avoid overfitting. Other hyper-parameters are adopted from BERT pre-training.

Fine-Tuning. In fine-tuning, we initialize the parameters of FashionBERT and CE-BERT with the parameters of the pre-trained 4-layer BERT-mini and train them on the PT dataset with the settings of hyper-parameters mentioned above. And then, we fine-tune above models with the following hyper-parameters: (1) the mini-batch size is set to 384 and Adam is used as the optimizer, with learning rate set to $5e-5$ (2) The embedding size of query and product are set to 128. (3) the training will stop when the max number of epochs is reached either early stop to avoid overfitting. Above training is done on five P100 GPUs.

In the online setting, the large-scale online A/B tests are conducted to compare those methods. Totally over 100 million search impressions are collected in our A/B tests and search traffic is equally split into the control and treatment groups.

Table 1. Offline comparison results on the Fashion-Gen dataset.

ID	Description	AUC	Recall@10	Recall@50	Recall@100
1	FashionBERT	0.9383	1.58%	4.49%	6.70%
2	CE-BERT	0.9470	2.12%	5.88%	8.52%

Evaluation on the E-Commerce Platform Dataset. In this section, we compare the four models mentioned in Sect. 4.2 on the downstream E-commerce retrieval task. The results are summarized in Tables 2 and 3. We can obtain that: (1) Multimedia data helps to improve E-commerce retrieval performance by comparing FashionBERT-Finetune to TwinBERT; (2) CE-BERT-Finetune benefits more from patch feature and pixel feature than from only patch feature compared with FashionBERT-Finetune; (3) ACE-BERT achieves the best performance on both AUC and Recall@K metric. The reason should be that, although the multi-modal features help to understand the product in depth, the downstream multi-modal retrieval task will leave a representation gap between query and product. We apply the adversarial learning on ACE-BERT and thus improve its performance.

Table 2. Offline comparison results on the E-Commerce dataset.

ID	Description	AUC	Recall@10	Recall@50	Recall@100
1	TwinBERT	0.7695	3.62%	10.16%	14.92%
2	FashionBERT-Finetune	0.7768	3.70%	10.33%	15.16%
3	CE-BERT-Finetune	0.7917	3.90%	10.89%	15.90%
4	ACE-BERT	0.7966	5.21%	13.61%	19.29%

Table 3 shows the online comparison results. We can see both CE-BERT-Finetune and ACE-BERT outperform the FashionBERT. Compared with Twin-BERT, ACE-BERT boots the CTR of products and the revenue of the platform by 0.78% and 1.46%, respectively. The p-value of 0.00001 (significant when significance level=0.05) also indicates this is a very significant improvement. While the improvement of FashionBERT-Finetune over TwinBERT are 0.07% and 0.08%, its p-value shows no significance.

Table 3. Offline comparison results on the E-Commerce platform.

Control	Treatment	CTR	Revenue	P-Value	Total Impressions
TwinBERT	FashionBERT-Finetune	+0.07%	+0.08%	0.617	247,706,406
TwinBERT	CE-BERT-Finetune	+0.32%	+0.49%	0.00001	174,239,400
TwinBERT	ACE-BERT	+0.78%	+1.46%	0.00001	68,409,368

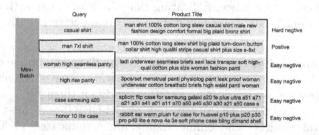

Fig. 4. The mixed negative sampling in mini-batch. The products' titles in black bounding boxes are training instances for "main 7xl shirt".

4.6   Analysis

Negative Sampling. In this section we will introduce the mixed negative sampling in mini-batch. A sample is shown in Fig. 4. The mini-batch consists of positive query and product pairs. For the query of each positive pair in a mini-batch, we use the other product samples as the negatives of the query. This provides a large amount of negatives whose size is coupled with the mini-batch size.

Image Representation. As mentioned above, CE-BERT-Finetune outperforms FashionBERT-Finetune on both AUC and Recall@K metrics. Here we analyze detailed results to find out whether the high-level patch features and the low-level pixel features are truly work.

We compare the pre-trained multi-modal BERTs with different image representations. For the sake of fairness, we pre-train a series of multi-modal BERTs by using different image representations and fine-tune them on the same FT dataset with the same hyper-parameters. Finally, we evaluate the effectiveness of these models on AUC and Recall@100 metrics.

Table 4. Results of different image representations.

Image Representation	AUC	Recall@100
Original Patch	0.7768	15.16%
RoI Patch	0.7816	15.30%
RoI Patch and Pixel	0.7917	15.90%

The evaluation results are shown in Table 4. Compared with the patch feature used in FashionBERT, the RoI-based patch feature achieves the better performance. In addition, the RoI-based patch and pixel feature used in CE-BERT achieves the best performance, indicating that the way of extracting the image feature used in our model is meaningful for E-commerce retrieval.

Hot Query and Adversarial Learning. Both offline and online experiments confirm that ACE-BERT helps to improve the performance of the retrieval task. Thus in this section, we conduct an ablation study on them.

In particular, we reuse the parameters of CE-BERT and fine-tune a series of the retrieval models. The offline results are summarized in Table 5. The model with the additional hot query in Setting 2 achieves significant Recall@100 improvements compared to Setting 1, which proves its superiority in the retrieval task. Furthermore, the Setting 3 with adversarial learning achieves significant gains over Setting 2, which indicates that the fine-tuning step can bridge the representation gap across domains by introducing adversarial learning.

Table 5. Ablation study results.

ID	Description	AUC	Recall@100
1	No Other Optimization	0.7917	15.90%
2	Add Hot Query	0.7934	18.85%
3	Add Discriminator	0.7966	19.29%

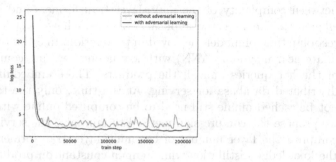

Fig. 5. The development of the Semantic Matching Loss (i.e. Eq. 5) during the two training procedures (i.e. with adversarial learning *vs.* without adversarial learning).

Figure 5 shows the real value of the semantic matching loss (i.e., Eq. 5) during the training procedure. We can find that the values of the loss with the adversarial learning converge smoothly and has a lower loss. To further explore the effectiveness of the adversarial learning, we visualize the distribution of query and product embeddings returned by our trained model through the t-SNE [38] tool. In detail, we sample 8000 points for query and product embeddings, respectively. These points belong to 8 specific categories. Compared with the points in Fig. 6(b), the points in Fig. 6(a) with the same color (i.e. the same category) easily appear in the same cluster. This indicates that the adversarial learning

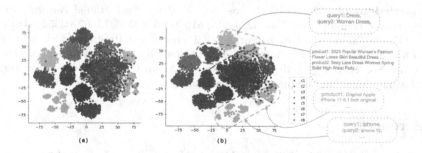

Fig. 6. The t-SNE visualization: (a) without adversarial learning, (b) with adversarial learning and the points in the dotted bounding boxes represent they are close each other. The color denotes the different categories (e.g. {c1, c2,...} denotes {Dresses, Outdoor Fun&Sports, ...}) in E-commerce.

helps to keep the embeddings from different domains (i.e. query and product) well aligned.

4.7 Online System Deployment

Serving over 100 million cross-border active users is not a trivial task. In deployment, we encounter some challenges. One of the main concern is to balance between complexity of the multi-layer transformers and online computational efficiency. The industry applications always have a strict constraints on the response time of models deployed in production. In order to perform the approximate near neighbor (ANN) with low latency, we pre-compute the embeddings of the hot queries and all the products. These embeddings are indexed in a distributed database for serving. At run-time, only the tail queries which can not be cached offline still need to be computed online with high latency. Thus, we propose the compression method to minimize the serving cost. In detail, we compress the layer number of the multi-layer transformers on query side with the knowledge distillation while remain constant on product side.

5   Conclusion

In this paper, we focus on the query text and multi-modal product matching in the E-commerce retrieval. There are long-term benefits by introducing the multiple modalities into the retrieval system to boost its performance. We propose ACE-BERT, a novel adversarial cross-modal enhanced BERT for the E-commerce retrieval. In particular, we make use of the patch-level and pixel-level image features for efficient image extracting. We also incorporate the Hot Query into the pre-training for better modeling semantics of the short query. Furthermore, we adopt the adversarial learning to bridge the representation gap across domains. We perform offline experiments and online A/B tests to verify the state-of-the-art performance of ACE-BERT.

Acknowledgment. This research was supported by the National Natural Science Foundation of China (Grant Nos. 62133012, 61936006, 62103314, 62203354, 61876144, 61876145, 62073255), the Key Research and Development Program of Shaanxi (Program No. 2020ZDLGY04-07), and Innovation Capability Support Program of Shaanxi (Program No. 2021TD-05), the Open Project of Anhui Provincial Key Laboratory of Multimodal Cognitive Computation, Anhui University, No. MMC202105".

References

1. Andrew, G., Arora, R., Bilmes, J., Livescu, K.: Deep canonical correlation analysis. In: International Conference on Machine Learning, pp. 1247–1255. PMLR (2013)
2. Baeza-Yates, R., Ribeiro-Neto, B., et al.: Modern information retrieval, vol. 463, ACM press (1999)

3. Bogolin, S., Croitoru, I., Jin, H., Liu, Y., Albanie, S.: KCross modal retrieval with querybank normalisation. In: Proceedings of the IEEE/CVF Conference on Computer Vision and Pattern Recognition, pp. 5194–5205 (2022)
4. Wang, X., et al.: Efficient subgraph matching on large rdf graphs using mapreduce. Data Sci. Eng. **4**, 24–43 (2019)
5. Cheng, M., et al.: ViSTA: vision and scene text aggregation for cross-modal retrieval. In: Proceedings of the IEEE/CVF Conference on Computer Vision and Pattern Recognition, pp. 5184–5193 (2022)
6. Bradley, A.P.: The use of the area under the roc curve in the evaluation of machine learning algorithms. Pattern Recogn. **30**, 1145–1159 (1997)
7. Chi, J., Peng, Y.: Dual adversarial networks for zero-shot cross-media retrieval. In: IJCAI, pp. 663–669 (2018)
8. Deng, J., Dong, W., Socher, R., Li, L.-J., Li, K., Fei-Fei, L.: Imagenet: a large-scale hierarchical image database. In: 2009 IEEE Conference on Computer Vision and Pattern Recognition, pp. 248–255. IEEE (2009)
9. Devlin, J., Chang, M.-W., Lee, K., Toutanova, K.: Bert: Pre-training of deep bidirectional transformers for language understanding, arXiv preprint arXiv:1810.04805 (2018)
10. Donahue, J., et al.: Decaf: a deep convolutional activation feature for generic visual recognition. In: International Conference on Machine Learning, pp. 647–655. PMLR (2014)
11. D. Gao, D., et al.: Fashionbert: text and image matching with adaptive loss for cross-modal retrieval. In: Proceedings of the 43rd International ACM SIGIR Conference on Research and Development in Information Retrieval, pp. 2251–2260 (2020)
12. Yin, H., Song, X., Yang, S., Li, J.: Sentiment analysis and topic modeling for covid-19 vaccine discussions. World Wide Web **25**(3), 1067–1083 (2022)
13. Wang, L., Ren, J., Xu, B., Li, J., Luo, W., Xia, F.: Model: Motif-based deep feature learning for link prediction. IEEE Trans. Comput. Soc. Syst. **7**(2), 503–516 (2020)
14. Goodfellow, I.J.: Generative adversarial networks, arXiv preprint arXiv:1406.2661 (2014)
15. Zhao, W., et al.: Telecomnet: tag-based weakly-supervised modally cooperative hashing network for image retrieval. IEEE Trans. Pattern Analy. Mach. Intell., 7940–7954 (2022)
16. Huang, J.-T., et al.: Embedding-based retrieval in facebook search. In: Proceedings of the 26th ACM SIGKDD International Conference on Knowledge Discovery & Data Mining, pp. 2553–2561 (2020)
17. Song, X., Li, J., Lei, Q., Zhao, W., Chen, Y., Mian, A.: Bi-CLKT: Bi-graph contrastive learning based knowledge tracing. Knowl.-Based Syst., 108274 (2022)
18. Yang, S., et al.: Robust cross-network node classification via constrained graph mutual information. Knowl.-Based Syst., 109852 (2022)
19. Huang, P.-S., He, X., Gao, J., Deng, L., Acero, A., Heck, L.: Learning deep structured semantic models for web search using clickthrough data. In: Proceedings of the 22nd ACM International Conference on Information & Knowledge Management, pp. 2333–2338 (2013)
20. Johnson, J., Douze, M., Jégou, H.: Billion-scale similarity search with gpus. IEEE Trans. Big Data (2019)

21. Ma, H.: Ei-clip: entity-aware interventional contrastive learning for e-commerce cross-modal retrieval. In: Proceedings of the IEEE/CVF Conference on Computer Vision and Pattern Recognition, pp. 18051–18061 (2022)
22. Xu, C., Liu, H., Guan, Z., Wu, X., Tan, J., Ling, B.: Adversarial incomplete multiview subspace clustering networks. IEEE Trans. Cybernet. 52(10), 10490–1503 (2021)
23. Yin, H., Yang, S., Song, X., Liu, W., Li, J.: Deep fusion of multimodal features for social media retweet time prediction. World Wide Web 24, 1027–1044 (2021)
24. Lu, H., Fei, N., Huo, Y., Gao, Y., Lu, Z., Wen, J.: COTS: Collaborative two-stream vision-language pre-training model for cross-modal retrieval. In: Proceedings of the IEEE/CVF Conference on Computer Vision and Pattern Recognition, pp. 15692–15701 (2022)
25. Krizhevsky, A., Sutskever, I., Hinton, G.E.: Imagenet classification with deep convolutional neural networks. Adv. Neural. Inf. Process. Syst. 25, 1097–1105 (2012)
26. Li, G., Duan, N., Fang, Y., Gong, M., Jiang, D.: Unicoder-vl: a universal encoder for vision and language by cross-modal pre-training. In: Proceedings of the AAAI Conference on Artificial Intelligence, vol. 34, pp. 11336–11344 (2020)
27. Li, L.H., Yatskar, M., Yin, D., Hsieh, C.-J., Chang, K.-W.: Visualbert: a simple and performant baseline for vision and language, arXiv preprint arXiv:1908.03557 (2019)
28. Lu, J., Batra, D., Parikh, D., Lee, S.: Vilbert: pretraining task-agnostic visiolinguistic representations for vision-and-language tasks, arXiv preprint arXiv:1908.02265 (2019)
29. Lu, W., Jiao, J., Zhang, R.: Twinbert: distilling knowledge to twin-structured bert models for efficient retrieval, arXiv preprint arXiv:2002.06275 (2020)
30. Lu, X., Zhu, L., Liu, L., Nie, L., Zhang, H.: Graph convolutional multi-modal hashing for flexible multimedia retrieval. In: Proceedings of the 29th ACM International Conference on Multimedia, pp. 1414–1422 (2021)
31. Zhao, W., Xu, C., Guan, Z., Liu, Y.: Multiview concept learning via deep matrix factorization. IEEE Trans. Neural Netw. Learn. Syst. 32(2), 814–825 (2020)
32. Xu, C., Guan, Z., Zhao, W., Wu, Q., Yan, M., Chen, L., Miao, Q.: Recommendation by users' multimodal preferences for smart city applications. IEEE Trans. Industr. Inf. 17(6), 4197–4205 (2020)
33. Radford, A., Narasimhan, K., Salimans, T., Sutskever, I.: Improving language understanding by generative pre-training (2018)
34. Xu, C., Zhao, W., Zhao, J., Guan, Z., Song, X., Li, J.: Uncertainty-aware multiview deep learning for internet of things applications. IEEE Trans. Industr. Inf. 19(2), 1456–1466 (2022)
35. Simonyan, K., Zisserman, A.: Very deep convolutional networks for large-scale image recognition, arXiv preprint arXiv:1409.1556 (2014)
36. Xu, C., Guan, Z., Zhao, W., Wu, H., Niu, Y., Ling, B.: Adversarial incomplete multi-view clustering. IJCAI 7, 3933–3939 (2019)
37. Xu, C., Guan, Z., Zhao, W., Niu, Y., Wang, Q., Wang, Z.: Deep multi-view concept learning. IJCAI, 2898–2904 (2018)
38. Van der Maaten, L., Hinton, G.: Visualizing data using t-sne. J. Mach. Learn. Res. 9 (2008)
39. Wang, B., Yang, Y., Xu, X., Hanjalic, A., Shen, H.T.: Adversarial cross-modal retrieval. In: Proceedings of the 25th ACM International Conference on Multimedia, pp. 154–162 (2017)
40. Xu, X., He, L., Lu, H., Gao, L., Ji, Y.: Deep adversarial metric learning for cross-modal retrieval. World Wide Web 22, 657–672 (2019)

41. Yang, Z., Dai, Z., Yang, Y., Carbonell, J., Salakhutdinov, R., Le, Q.V.: Xlnet: generalized autoregressive pretraining for language understanding, arXiv preprint arXiv:1906.08237 (2019)
42. Zhuge, M.: Kaleido-bert: vision-language pre-training on fashion domain. In: Proceedings of the IEEE/CVF Conference on Computer Vision and Pattern Recognition, pp. 12647–12657 (2021)

Continual Few-Shot Relation Extraction with Prompt-Based Contrastive Learning

Fei Wu[1], Chong Zhang[1], Zhen Tan[2(✉)], Hao Xu[1], and Bin Ge[1]

[1] Laboratory for Big Data and Decision, National University of Defense Technology, Changsha, China
{wufei21,chongzhang,xuhao,gebin}@nudt.edu.cn
[2] Science and Technology of Information Systems Engineering Laboratory, National University of Defense Technology, Changsha, China
tanzhen08a@nudt.edu.cn

Abstract. Continual relation extraction (CRE) aims to continually learn new relations while maintaining knowledge of previous relations in the data streams. Recently, continual few-shot relation extraction (CFRE) is introduced, in which only the first step of training has enough data, while the later steps have only a handful of samples, since emerging relations are often difficult to obtain enough training data quickly in realistic scenarios. Some previous work has proved that storing samples from previous tasks can alleviate the problem of catastrophic forgetting in continual learning. However, such approaches rely heavily on memory size and fail to fully exploit the knowledge associated with memorized samples in the pre-trained language model (PLM). To solve these problems, we propose a novel Prompt-based Contrastive Learning method, namely **PCL** for CFRE with two optimizations. Firstly, to make better use of the knowledge in the limited number of training samples, we propose to obtain better representations of the relations by a prompt-based encoder with the assistance of a well-designed template. A mix-up data augmentation strategy is harnessed to increase the robustness of the model and prevent few-shot tasks from over-fitting. Secondly, to alleviate forgetting previous knowledge, in addition to replaying samples in the memory, our method maintains the stability of instance embeddings and retains the knowledge of the learned relations by an instance-level distillation loss. With extensive experiments on two benchmarks, we demonstrate that our model significantly outperforms state-of-the-art baselines and increases the generalization and robustness of the CFRE model on the imbalanced dataset.

Keywords: Continual Learning · Prompt-based Encoder · Contrastive Learning

1 Introduction

Relation extraction(RE) plays a vital role in knowledge extraction and management, which serves for a series of tasks in natural language processing, such

X. Song et al. (Eds.): APWeb-WAIM 2023, LNCS 14334, pp. 312–327, 2024.
https://doi.org/10.1007/978-981-97-2421-5_21

as question and answer, text understanding, etc. Given a sentence x with the annotated entities pairs e_1 and e_2, the task of RE aims at identifying the relation between the entity pair by classifying it into a pre-defined collection of relations R_k. Traditionally, RE models [1,2] require a pre-defined fixed set of relations, which can only classify the relation between the pair of entities in a sentence into one of the categories in the fixed relation set. However, since the real world is open [3], the emergence of new data gives rise to a continually increasing number of relations. In response to this scenario, the paradigm of continual relation extraction (CRE) [4,5] is proposed.

In existing research, most CRE models assume that sufficient labeled data is available. However, in the real world, most of relations have insufficient labeled data, and this problem becomes more noteworthy for newly emerged relations. As a result, considering it is expensive to obtain enough labeled data, Qin et al. [6] proposed Continual Few-shot Relation Extraction (CFRE), which requires the model to learn new relations from very few training examples without forgetting the old ones. Current CFRE models all face two challenges: catastrophic forgetting of previous knowledge [7] and insufficient training samples of emergent relations [8].

To address the catastrophic forgetting problem, current methods fall into three broad categories: regularization-based methods [9], architecture-based methods [10] and memory-based methods [4,5]. Recent research [11] has shown that memory-based methods are most effective in natural language processing applications, where typical samples from previous tasks are stored in memory and replayed together with the current tasks. Successful memory-based methods include EMAR [4], CML [5], RP-CRE [12], CRL [13] and so on. As for the second challenge, Qin et al. [6] enrich the training dataset with additional relevant data via a self-supervised manner. Chen et al. [14] employ the nearest neighbor scheme to improve representation learning for few-shot novel classes.

To meet the challenges, we propose a CRE model based on prompt-based contrastive learning, namely PCL with several optimization. Facing the dilemma of catastrophic forgetting, we adopt the memory-based method to avoid forgetting previous relational knowledge. At the same time, we use contrastive learning to learn more discriminative representations for instances of diverse relations and harness knowledge distillation to keep the stability of these representations along the task sequence. To handle the problem of few-shot training samples, different from Qin et al. [6] which augments the training samples with external data, we employ a prompt-based encoder to learn relational representations for instances. In CRE and CFRE, all of these models proposed to use a language model and fine-tuning it with continuous data flow. However, due to the gap between the pre-trained task and the fine-tuning task, the knowledge of the pre-trained model is not fully exploited. Based on the observation that prompt learning has achieved great success in low-resource scenarios, we propose a prompt-based contrastive learning method to learn robust representations for few-shot samples. Besides, we propose a data augmentation approach by mixuping features of the masked position in the prompt. By this way, we can expand the training set by constructing soft samples.

The main contributions of our work can be summarized as follows:

- We propose a Prompt-based Contrastive Learning and Data Augmentation method for CFRE, which uses prompt-based contrastive learning and knowledge distillation to alleviate catastrophic forgetting for continual learning and few training samples for the few-shot scenario.
- Prompt-based Contrastive Learning can make better use of PLM, samples in the memory and knowledge of external prompt templates, which is effective to obtain better relation representation and make sure the stability of embedding space disturbed by new tasks. Besides, prompt-based method is more suitable for few-shot setting and mixing features increases the generalization and robustness ability of the model.
- Extensive experiments results on FewRel and TACRED datasets show remarkable superiority of our model compared to existing ones and alleviate catastrophic forgetting effectively.

2   Related Work

2.1   Continual Learning

Continual learning is a learning paradigm that requires a model to continually learn new tasks on new dataset without forgetting how to perform well in the previously learned tasks. The main challenge of continual learning is to solve the problem of catastrophic forgetting [7]. There are three methods currently used to address the problem of catastrophic forgetting: (1) Regularization-based methods [9] impose constraints on the weights that are important for learned tasks during network parameter updating; (2) Architecture-based methods [10] dynamically adjust the neural resources to change the architecture of the model and learn new knowledge without forgetting what has already been learned; (3) Memory-based methods [4,5] keep a representative portion of data in the previously learned task in memory and train them with the dataset of new tasks to review the old knowledge that has been learned.

For CRE, memory-based methods have been proven the most promising, the key to which is to obtain a consistent representation through replay samples in memory. Inspired by the success of memory-based methods in NLP, we use a memory replay framework to review old knowledge while learning continually emerging relations, which helps a lot to solve the problem of catastrophic forgetting.

2.2   Few-Shot Learning

Few-shot learning [15–17] is dedicated to enabling a model to learn a new task quickly with only a small number of samples, and is of great practical importance, especially in special areas with low-resource data. There are three main approaches to solve this problem: (1) Data augmentation methods [18] use prior knowledge to expand or synthesize more data to avoid model overfitting; (2)

Model fine-tuning methods [19] pre-train the model on large-scale data, and fine-tune the parameters of the fully connected layers or the top layers of the neural network model on the target small sample dataset to obtain the fine-tuned model; (3) Transfer-based learning methods [20] use old knowledge to learn new knowledge and the main goal is to quickly transfer the learned knowledge to a new domain, such as metric learning, meta-learning, etc.

In this paper, we use the sentence mix-up method to construct soft samples to increase the generalization and robustness ability of the model.

2.3   Contrastive Learning

The goal of contrastive learning is to learn an encoder that enhances the representation of data by reducing the distance between samples of the same label while expanding the distance between samples of different labels. Liang et al. [21] introduce a memory contrastive learning algorithm that can help the feature extractor to optimize the feature representation capability. Gao et al. [22] propose an active and contrastive Learning framework that can discover the category model without pre-defining semantic categories.

Considering the excellent performance of contrast learning in few-shot learning, we employ contrast learning to obtain a better representation for each relation, which is effective for relation classification tasks.

2.4   Prompt Learning

Fueled by the birth of GPT-3, prompt-based learning has been widely used in NLP downstream tasks [23,24]. Different from the conventional fine-tuning method needing extra task-specific heads for downstream tasks and having a different objective from pre-training, prompt-based learning can better exploit knowledge in PLMs by reformulating the objective of downstream tasks as cloze tasks, whose task objective is the same as pre-training. Besides, recent research [25] has demonstrated its effectiveness in the low-data regime.

In this work, we use a prompt-based encoder to learn relational representations for instances, which enables the model to obtain more semantic knowledge and perform well in the few-shot tasks.

3   Methodology

3.1   Problem Definition

In contrast to traditional relation extraction tasks, continual relation extraction does not have a pre-defined relation set, and the number of relations increases as tasks progress. Formally, given a stream of tasks $\mathcal{T} = \{T_1, T_2, ..., T_n\}$, the k-th task T_k has its own training set D_{train}^k, test set D_{test}^k, and relation set R_k. Each dataset D has some samples $\{(x_i, y_i)\}_{i=1}^N$, where x_i is the input sentence and a pair of entities in the sentence and $y_i \in R_k$ is the relation between the entity

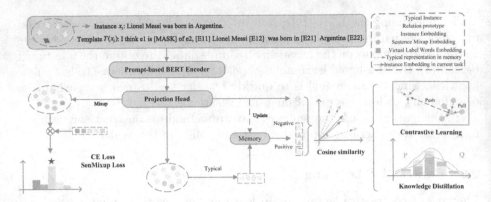

Fig. 1. Framwork of Prompt-based Contrastive Learning

pair. The goal of CFRE is to learn the relations R_k in T_k without forgetting the relations $\hat{R}_{k-1} = \cup_{i=1}^{k-1} R_i$, which are already learned in the historical tasks. Each task in $\mathcal{T}$ is a multi-relation classification task. Unlike CRE, CFRE assumes that only the first task has enough samples and the subsequent tasks have few samples, e.g., each relation in the first task has 100 samples, but the subsequent tasks have only 5 samples. The CFRE setting is in line with the realistic scenario, because, for the already existing relations, there are generally enough labeled data, while for the newly emerged relations, there are few labeled data.

The CFRE model Θ is continually trained on a series of datasets $\{D_{\text{train}}^i\}_{i=1}^k$ corresponding to the stream of tasks $\{T_i\}_{i=1}^k$, where D_{train}^1 is the large-scale training dataset and $D_{\text{train}}^k (k > 1)$ is the few-shot training datasets of newly emerged relations, noting that only D_{train}^k is available at the k-th training task. After training on D_{train}^i, Θ is tested to recognize all encountered classes in $\{T_i\}_{i=1}^k$ on the $\hat{D}_{test}^k = \cup_{i=1}^k D_{test}^i$.

The biggest challenge for CFRE is to make the model learn new relations while maintaining the classification ability of learned relations. In this regard, we use the memory model $\mathcal{M} = \{M_1, M_2, ..., M_n\}$ to store samples that appeared in the previous tasks. When the model learns the k-th task, the samples stored in $\{M_1, ..., M_{k-1}\}$ are merged into D_{train}^k for training together to solve the catastrophic forgetting problem. Since there is no fixed number of tasks under continual learning, there is a limit to the number of memory samples in general. Due to the few-shot setting of new tasks and few training samples, we adopt the previous setting of storing only one sample per class of relations in memory as Qin et al. [6] do.

3.2 Framework

The learning of PCL in a new task is shown in Fig. 1, which contains three major steps:

Initial Training: The parameters of the pre-trained language model BERT are fine-tuned with the training samples in dataset D_{train}^k.

Sample Selection: For each relation $r_i \in R_k$, we use the k-means algorithm to cluster samples labeled with label r_i and select the closest sample to the centroid and store it in the memory.

Better Representation Learning: In order to retain the knowledge of learned relations and the knowledge between relations in the previous tasks, we use contrastive replay and knowledge distillation to obtain better representation of relations.

3.3 Represent Sentence with the Prompt

The key of PCL is to help PLM adapt to the domain-specific answer distribution and obtain better representation of relations by designing sophisticated templates, which provide external knowledge and help to exploit desired knowledge in the PLM better.

Given the excellent performance in extracting contextual representation of text, we use BERT [26] as a pre-trained language model. Recent study [27] has shown that entity information can help encoders understand relation information and obtain better sentence representations in relational extraction tasks. Therefore, we use some special markers $[E_{11}]$, $[E_{12}]$, $[E_{21}]$ and $[E_{22}]$ at the beginning and end of the entity to augment sentence information for a better representation, where e_1 and e_2 are the head entity and tail entity in this sentence.

$$x_{aug} = \{w_1, ..., [E_{11}], e_1, [E_{12}], ..., [E_{21}], e_2, [E_{22}], ..., w_x\}. \tag{1}$$

At the same time, we construct a template mapping function $T(\cdot)$ to obtain the prompt input, which can spire the PLM to generate better representations for instances. We obtain x_{prompt} by taking x_{aug} as an input to $T(\cdot)$.

$$x_{prompt} = T(x_{aug}) = \text{I think } e_1 \text{ is [MASK] of } e_2 \text{ [SEP] } x_{aug}. \tag{2}$$

Then, feed the new token sequence x_{prompt} to BERT to generate the presentation of [MASK] position and the probability distribution over the label set $\mathcal{Y}$ as:

$$\begin{aligned} p(y \mid x_{prompt}) &= p\left([\text{MASK}] = \mathcal{M}(y) \mid x_{prompt}\right) \\ &= \frac{\exp\left(\mathbf{v} \cdot \mathbf{h}_{[\text{MASK}]}\right)}{\sum_{\mathbf{v}_i \in \mathcal{V}} \exp\left(\mathbf{v}_i \cdot \mathbf{h}_{[\text{MASK}]}\right)} \end{aligned} \tag{3}$$

where $y \in \mathcal{Y}$ is the target label word, $\mathbf{v}$ is the embedding of the token $v = \mathcal{M}(y)$ in the BERT.

Finally, to obtain denser representations, we use a projection head $Proj$ to further extract the feature and map it to a low-dimensional embedding space.

$$\tilde{\mathbf{z}} = Proj(\mathbf{h}_{[\text{mask}]}) \tag{4}$$

In this model, we use the hidden representation for relation classification and normalize the hidden representation to obtain $\mathbf{z} = \tilde{\mathbf{z}}/\|\tilde{\mathbf{z}}\|$ for comparison learning.

3.4   Initial Training for New Task

According to the general assumption of CFRE, all relations that have been represented in historical tasks are not available in the new task T_k. Therefore, at time step k, we first initialize memory bank M_{all} with the normalized embedding z in T_k.

$$M_{all} \leftarrow M_k = \cup_{r \in R_k} M_r \tag{5}$$

where M_{all} is the memory bank of all tasks, R_k is the relation set of T_k, and $M_r = \{(x_i, y_i)\}_{i=1}^{O}$ is the memory set of relation r, O is the number of samples stored in the memory bank for each relation. In this paper, we set $O = 1$ as Qin et al. [6] do.

We fine-tune the model Θ on the training dataset D_{train}^k by minimizing loss $\mathcal{L}_{all}$ that consists of a cross entropy loss, a sentence mixup loss and a contrastive loss to optimize the parameters in the model.

The **cross entropy loss** portrays the similarity between the actual output probability and the desired output probability, which is commonly used in relation classification tasks. The formula of the cross entropy loss function is shown as follows:

$$\mathcal{L}_{CE}(\Theta) = \sum_{i=1}^{|D_{train}^k|} -\log P\left(y = y_i \mid \mathbf{z}_i, y_i\right) \tag{6}$$

where y_i is the true label of sample i in D_{train}^k.

Mix-up is a data augmentation method that generates more training samples by using linear interpolation, which has demonstrated its excellent performance in few-shot classification tasks and the ability of increasing the generalization and robustness for few-shot tasks. In this model, we use the sentence mixup method.

In **senMix-up loss**, a pair of hidden representation $\tilde{z}$ and their corresponding class labels y are interpolated linearly as follows:

$$\tilde{\mathbf{z}}_{ij} = \lambda \tilde{\mathbf{z}}_i + (1 - \lambda)\tilde{\mathbf{z}}_j \tag{7}$$

$$\tilde{y}_{ij} = \lambda y_i + (1 - \lambda)y_j \tag{8}$$

The hidden representation vector $\tilde{\mathbf{z}}_{ij}$ is used to obtain the distribution over the possible target labels through the softmax layer. Then, we use multi-class cross entropy loss for training:

$$\mathcal{L}_{SM} = -\lambda \log P(y = y_i | \tilde{\mathbf{z}}_i, y_i) - (1 - \lambda) \log P(y = y_j | \tilde{\mathbf{z}}_j, y_j) \tag{9}$$

The **contrastive loss** attempts to reduce the distance between samples with the same relation label while expanding the distance between samples with different relation labels. The contrastive loss function is shown below.

$$\mathcal{L}_{CL} = \sum_{i=1}^{n} \frac{-1}{|P(i)|} \sum_{p \in P(i)} \log \frac{\exp\left(\mathbf{z}_i \cdot \mathbf{z}_p / \tau\right)}{\sum_{j \in S_I} \exp\left(\mathbf{z}_i \cdot \mathbf{z}_j / \tau\right)} \tag{10}$$

where n is the number of samples in a batch of D_{train}^k, $P(i)$ is the set with the same relation to the training sample x_i, $|P(i)|$ is its cardinality, z_i is the representation of the training sample x_i, and S_i is the set of some samples in M_{all}.

The overall training loss can be written as:

$$\mathcal{L}_{All} = \mu_1 \cdot \mathcal{L}_{CE} + \mu_2 \cdot \mathcal{L}_{SM} + \mu_3 \cdot \mathcal{L}_{CL} \tag{11}$$

where $\mu_1 = 0.1$, $\mu_2 = 0.1$, $\mu_3 = 1.0$.

We minimize the overall training loss $\mathcal{L}_{All}$ for fast adaption on new tasks.

3.5   Selecting Typical Samples for Memory

After training the model with Eq. (11), in each new task, to enable the model memorize the knowledge of relations in previous tasks, we save a typical sample for each relation in memory. For each relation r_i, we use k-means to cluster embedding of its training samples in D_{train}^k and save the closest one in the memory M_{all}.

3.6   Better Representation Learning

This module is used to retain more knowledge of learned relations when replaying samples in memory. In continual learning, the parameters of the model will change when learning new relations in a new task, which means, the embedding space of the learned relations in the historical tasks will be disturbed by the learning of new relations to some extent.

Unlike just replaying samples in memory for recovery, we use two replay strategies to make the encoder not forget the knowledge of learned relations while learning new relations, which are contrastive replay and knowledge distillation respectively.

When finishing training a new task, we use Eq. (10) to replay samples in memory to review learned knowledge. Note that S_I here is not a part of the samples in the memory, but all the samples in the memory. By replaying all the samples in memory, our model regains a stable understanding of learned relations as well as consolidates the relations learned in the current task.

However, since storing only one memory sample per relation can easily cause overfitting and the distribution of relations will be disturbed in a new task, we use knowledge distillation to retain the knowledge between relations and samples in the previous task. The model hopes to keep the similarity between samples and its corresponding relation prototypes and the dissimilarity between samples and other relation prototypes as much as possible while maintaining the stability of the embedding space. We obtain the prototype of each relation by averaging the samples per relation saved in the memory.

$$\mathbf{P}_c = \sum_{i=1}^{O} \mathbf{z}_i^c \tag{12}$$

The similarity between sample $\mathbf{z}_i$ and relation prototype $\mathbf{p}_j$ is calculated using cosine similarity:

$$\alpha_{ij} = \frac{\mathbf{z}_i^T \mathbf{p}_j}{\|\mathbf{z}_i\| \, \|\mathbf{p}_j\|} \tag{13}$$

where α_{ij} is the cosine similarity between sample $\mathbf{z}_i$ and relation prototype $\mathbf{p}_j$.

After training new tasks, the parameters of the model will change, including the encoder. Therefore the embedding of samples saved in memory is dynamic. Whenever a new task is learned, we use KL divergence to keep the similarity and dissimilarity between each sample and all relation prototypes in the last task as consistent as possible, with what it can preserve the knowledge between this step and last step for those learned relations.

$$\mathcal{L}_{KL} = \sum_i KL\left(P_i \| Q_i\right) \tag{14}$$

where $P_i = \{p_{ij}\}_{j=1}^{|\hat{R}_k|}$ is the metric distribution of similarity between samples and others relation prototypes before training a new task, and $p_{ij} = \frac{\exp(\alpha_{ij}/\tau)}{\sum_j \exp(\alpha_{ij}/\tau)}$. Similarly, $Q_i = \{q_{ij}\}_{j=1}^{|\hat{R}_k|}$ is the metric distribution of similarity similarity between samples and others relation prototypes after training a new task, and $q_{ij} = \frac{\exp(\tilde{\alpha}_{ij}/\tau)}{\sum_j \exp(\tilde{\alpha}_{ij}/\tau)}$. By the way, $\tilde{\alpha}_{ij}$ is the cosine similarity between sample $\mathbf{z}_i$ and prototype $\mathbf{p}_j$ calculated by their dynamic embedding.

3.7   Inference

During inference, giving an input sentence $x_i \in \hat{D}_{test}^k$, the probability of labeling the relation with class $y \in \mathcal{Y}$ is computed by Eq. (3). The relation with the max probability are selected as the predicted relation.

4   Experiment

4.1   Datasets

We use two datasets which are widely used in the field of relation extraction under continual learning as our benchmark datasets.

FewRel [28] is constructed using Wikipedia as a corpus and a knowledge graph, and is currently the largest finely labeled dataset in the field of few-shot relation extraction. 80 classes of relations are available in FewRel, each with more than hundreds of instances. Following the experimental settings of CFRE by Qin et al. [6], 80 classes of relations are divided into 8 tasks, each with 10 classes of relations (10 way in a task). According to the settings of CFRE, the first task has enough data and the subsequent tasks have a small amount of data. In the experiment, each relation in the first task has 100 samples, while for the remaining subsequent tasks, we chose 5 and 10 samples per relation for the 5-shot and 10-shot experiments, respectively.

TACRED [29] is a large-scale relation extraction dataset with 106,264 examples built over newswire and web text. It has 42 relations including no_relation. Following the experimental settings of CFRE by Qin et al. [6], only 41 relations outside of no_relation are used in the experiments, and they are divided into 8 tasks, of which the first task has 6 relations and each task of the subsequent tasks have 5 relations. Similar to the FewRel dataset mentioned above, the first task has 100 samples per relation, while the subsequent tasks have 5 and 10 samples per relation for the 5-shot and 10-shot experiments.

4.2 Evaluation Metrics

With the gradual progress of tasks, more and more relations will appear. Some models use an unstrict evaluation strategy, which means, for relations that have appeared, the authors only compare them with the corresponding negative labels in current task for judging whether the pred label is correct or not. Unlike the evaluation criteria adopted by ERDA [6], we use the accuracy of relation classification in all learned relations which had appeared in previous tasks as the evaluation criterion for our proposed model, i.e., the accuracy of classifying all the relations $\hat{R}_k = \cup_{i=1}^k R_i$ that have emerged after learning T_k on the test dataset $\hat{D}_{test}^k = \cup_{i=1}^k D_{test}^i$. This evaluation metric can be a good response to how the model performs on previous tasks after learning a new task and can respond to whether the model is effective in mitigating the catastrophic forgetting problem associated with continual learning.

4.3 Baseline

We evaluate PCL and several baselines on benchmarks for comparison:

1. EMAR [4] proposed episodic memory activation and reconsolidation to alleviate catastrophic forgetting.
2. CML [5] introduce a curriculum-meta learning method to tackle catastrophic forgetting as well as order-sensitivity problems.
3. RP-CRE [12] selects and memorizes typical samples under each relation to generate a prototype representation of the relation and uses it for refining sample embeddings.
4. CRL [13] proposed a consistent representation learning method to maintain the stability of the relation embedding.
5. ERDA [6] proposed embedding space regularization and data augmentation to alleviate catastrophic forgetting in CFRE.

4.4 Experimental Settings

For a fair comparison, only one sample per relation is allowed to store in the memory for all baselines and our model. We conduct experiments with the baselines on the datasets used by ERDA [6]. Due to the order-sensitivity problem, we use 6 different random seeds to obtain 6 different orders of relations occurrence as Qin et al. [6] do. We average the results of these 6 rounds as final results. Table 1 is the results of 6 rounds on Fewrel 5-shot dataset.

Table 1. Accuracy (%) of six different runs with different task order on Fewrel benchmark for 10-way 5-shot setting.

Run index	Task index							
	T1	T2	T3	T4	T5	T6	T7	T8
1	95.40	84.75	73.43	73.23	73.10	70.10	68.86	66.90
2	94.90	82.00	81.17	78.60	74.22	71.05	68.03	67.11
3	94.60	86.65	79.93	80.23	75.70	70.22	69.17	67.34
4	95.30	86.90	78.80	73.05	70.06	71.05	68.17	66.75
5	95.00	85.20	79.33	74.10	74.08	71.60	69.54	66.75
6	95.10	87.55	78.50	72.65	71.10	68.47	66.66	66.68
Average	**95.05**	**85.51**	**78.53**	**75.31**	**73.04**	**70.42**	**68.41**	**66.92**

4.5 Main Results

We compare the performance of PCL and baselines with the same setting as ERDA. By the way, our rigorous evaluation method increases the difficulty of CFRE, which brings more difficulties to baselines and our model. Table 2 shows the results of the proposed methods on two datasets, where the reported scores are the average of 6 rounds. From Table 2, we can observe that:

1. Our proposed model PCL is significantly better than baselines in CFRL setting. Although EMR, CML, and ERDA adopt unstrict evaluation criteria, our model receives the highest scores with strict evaluation criteria, which demonstrates that PCL is better at retaining learned knowledge when learning new tasks and is more effective to mitigate catastrophic forgetting problems.
2. PCL outperforms baselines by making better use of memory, PLM and relation knowledge, including knowledge between relations and knowledge of external prompts, to better maintain the stability of the embedding space.
3. All the models on the TACRED dataset do not score as well as the FewRel dataset, and the main reason for this is that TACRED is an unbalanced dataset. In particular, in the TACRED 10-shot experiment, the accuracy of our last task is 19.09% higher than the best score in the baselines while only 5.75% higher in the Fewrel 10-shot experiment, which shows that our model is more robust to scenarios with class-imbalanced. We attribute this to our better utilization of PLM information compared to baselines by proposing a prompt-based contrastive learning method, as well as data augmentation of training samples using feature mixup, which help to learn better representations for few-shot sample and improve the generalization and robustness of our model.
4. Due to the few-shot setting in CFRE, only one sample per relation are allowed to store in the memory, which leads to a significant decrease in the results of baselines. It can be seen that the number of samples allowed to store in the memory and the availability of sufficient training data have a significant

Table 2. Accuracy (%) on all appeared relations (including relations in the previous tasks) at the stage of T_k. The method marked by † represents the results generated from unstrict evaluation criteria.

FewRel								
Model	T1	T2	T3	T4	T5	T6	T7	T8
Few-shot Setting: 5-shot								
EMAR†	71.35	56.4	47.74	41.09	38.48	33.69	30.03	28.41
CML†	85.40	68.45	58.10	46.40	41.40	37.92	35.97	35.64
RP-CRE	94.01	71.51	69.64	68.34	66.36	62.38	60.06	58.27
CRL	**95.58**	79.90	70.51	66.79	65.09	61.79	59.20	57.73
ERDA†	91.60	79.40	70.60	64.78	61.82	57.92	55.40	53.32
PCL	95.05	**85.51**	**78.53**	**75.31**	**73.04**	**70.41**	**68.40**	**66.92**
Few-shot Setting: 10-shot								
EMAR†	69.50	55.91	47.48	42.47	37.55	33.79	34.41	29.73
CML†	87.70	74.00	66.57	59.85	52.16	48.73	44.33	37.75
RP-CRE	94.23	77.31	75.42	73.91	67.92	64.39	63.70	62.35
CRL	**95.75**	79.51	73.52	68.82	65.94	64.16	62.09	60.05
ERDA†	92.55	81.11	71.02	64.84	61.17	58.70	54.66	53.43
PCL	95.55	**84.73**	**83.11**	**79.48**	**73.96**	**70.44**	**69.54**	**68.10**
TARCED								
Model	T1	T2	T3	T4	T5	T6	T7	T8
Few-shot Setting: 5-shot								
EMAR†	49.88	40.27	31.25	27.43	22.79	19.83	18.27	16.75
CML†	68.92	43.62	35.11	27.63	22.23	24.43	21.51	15.84
RP-CRE	88.03	71.27	64.51	55.27	49.65	48.12	45.34	43.01
CRL	81.43	60.12	52.48	46.73	52.60	51.66	46.63	47.04
ERDA†	68.85	46.31	37.16	34.36	30.47	27.59	23.26	22.23
PCL	**88.77**	**79.13**	**76.80**	**73.03**	**65.59**	**61.86**	**59.08**	**57.59**
Few-shot Setting: 10-shot								
EMAR†	43.42	42.79	35.84	30.22	24.99	18.56	18.82	17.01
CML†	71.43	53.49	41.89	35.67	28.47	25.93	16.57	16.43
RP-CRE	88.03	71.27	64.51	55.27	49.65	48.12	45.34	43.01
CRL	87.03	71.21	64.28	52.39	52.98	47.49	44.47	41.66
ERDA†	75.90	54.29	41.35	35.81	31.04	28.43	22.97	21.94
PCL	**88.45**	**84.87**	**78.68**	**73.93**	**69.10**	**67.27**	**65.22**	**62.10**

impact on the results. We overcome these conditions and the proposed model is more suitable for realistic emergent relations with insufficient labeled data and limited storage space, which is of great practical application and facilitates rapid deployment of projects.

4.6   Ablation Study

We conduct ablations studies to investigate the effect of different components in our model. The experiments are performed on the TARCED(10-shot),which are shown in Table 3. From it, we can observe that all components are important to our model, especially the prompt-based encoder. Comparing PCL with PCL(w.o. prompt), not using the prompt-based encoder to learn relational representations for instances leads to a significant decrease in scores from the second task, the score for the second task is down 12.96% from the first task, which demonstrates the importance of the prompt-based encoder in mitigating knowledge forgetting.

Table 3. Ablations on TARCED benchmark (10-shot).

Model	T1	T2	T3	T4	T5	T6	T7	T8
TARCED 10-shot								
PCL	**88.45**	**84.87**	**78.68**	**73.93**	**69.10**	**67.27**	**65.22**	**62.10**
w.o.L_{CE}	88.42	83.41	75.09	69.02	64.51	60.17	58.60	56.72
w.o.L_{SM}	88.51	71.27	69.33	69.85	62.31	61.07	60.11	58.01
w.o.prompt	87.36	74.40	68.85	61.30	56.76	51.15	51.81	49.85

4.7   Effect of Prompt-Based Contrastive Learning

We use t-SNE, a commonly used dimension reduction algorithm, to represent the dimension reduction relation and show the embedding space of ERDA, RP-CRE and PCL. We randomly choose 6 relations to visualize samples in their corresponding test dataset, four of which (Id 26, 19, 29 and 32) are selected in the first task and two of which (Id 20 and 14) in the emerging tasks. We can see from Fig. 2 that in the TARCED 10-shot dataset, compared with other models, our model has larger interclass distance and better intra-class compactness, which indicate the effectiveness of our model for representing relations after learning new tasks and retaining learned knowledge.

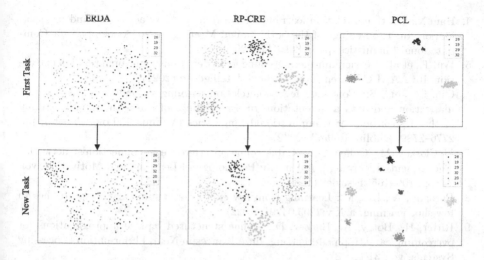

Fig. 2. t-SNE visualization of ERDA, RP-CRE and PCL at two stages in TARCED 10-shot

5   Conclusion

In this paper, we propose a novel prompt-based contrastive learning and data augment method for CFRE, which can better exploit the knowledge of PLM, samples saved in the memory, and sophisticated templates. Through comparative learning and knowledge distillation, we retain the knowledge that has been learned and make better use of the knowledge in the memory sample when learning new tasks. Specifically, the prompt-based encoder helps to obtain a better representation of relation, which is effective for few-shot samples. Besides, constructing soft samples increase the generalization and robustness of our model. Extensive experimental results and analysis on two benchmark datasets show that PCL has great advantages in maintaining learned knowledge while keep learning new knowledge and performs well in alleviating catastrophic forgetting problem. In the future, we will continue to study the application of the prompt-based contrastive approach under continual learning in NLP.

References

1. Pang, N., Tan, Z., Zhao, X., Zeng, W., Xiao, W.: Domain relation extraction from noisy Chinese texts. Neurocomputing **418**, 21–35 (2020)
2. Huang, P., Fang, Y., Zhu, H., Xiao, W.: End-to-end knowledge triplet extraction combined with adversarial training. J. Comput. Res. Develop. **56**, 2536–2548 (2019)
3. Sennrich, R., Haddow, B., Birch, A.: Neural machine translation of rare words with subword units. In: Proceedings of the 54th Annual Meeting of the Association for Computational Linguistics (Volume 1: Long Papers), pp. 1715–1725. Berlin, Germany (2016)

4. Han, X.: Continual relation learning via episodic memory activation and reconsolidation. In: Proceedings of the 58th Annual Meeting of the Association for Computational Linguistics, pp. 6429–6440 (2020)

5. Wu, T., et al.: Curriculum-meta learning for order-robust continual relation extraction. In: AAAI Conference on Artificial Intelligence (2021)

6. Qin, C., Joty, S.: Continual few-shot relation learning via embedding space regularization and data augmentation. In: Proceedings of the 60th Annual Meeting of the Association for Computational Linguistics (Volume 1: Long Papers), pp. 2776–2789. Dublin, Ireland (2022)

7. McCloskey, M., Cohen, N.J.: Catastrophic interference in connectionist networks: The sequential learning problem. In: Psychology of Learning and Motivation, vol. 24, pp. 109–165. Elsevier (1989)

8. Parnami, A., Lee, M.: Learning from few examples: a summary of approaches to few-shot learning. arXiv:2203.04291 (2022)

9. Ritter, H., Botev, A., Barber, D.: Online structured laplace approximations for overcoming catastrophic forgetting. In: Advances in Neural Information Processing Systems, vol. 31 (2018)

10. Mallya, A., Davis, D., Lazebnik, S.: Piggyback: adapting a single network to multiple tasks by learning to mask weights. In: Proceedings of the European Conference on Computer Vision (ECCV), pp. 67–82 (2018)

11. Sun, F., Ho, C., Lee, H.: LAMOL: language modeling for lifelong language learning. In: 8th International Conference on Learning Representations, ICLR 2020. Addis Ababa, Ethiopia (2020)

12. Cui, L., et al.: Refining sample embeddings with relation prototypes to enhance continual relation extraction. In: Proceedings of the 59th Annual Meeting of the Association for Computational Linguistics and the 11th International Joint Conference on Natural Language Processing (Volume 1: Long Papers), pp. 232–243 (2021)

13. Zhao, K., Xu, H., Yang, J., et al.: Consistent representation learning for continual relation extraction. In: Findings of the Association for Computational Linguistics: ACL 2022, pp. 3402–3411. Dublin, Ireland (2022)

14. Chen, K., Lee, C.G.: Incremental few-shot learning via vector quantization in deep embedded space. In: International Conference on Learning Representations (2021)

15. Pang, N., Tan, Z., Hao, X., Xiao, W.: Boosting knowledge base automatically via few-shot relation classification. Front. Neurorobot. 14, 584192 (2020)

16. Pang, N., Zhao, X., Wang, W., Xiao, W., Guo, D.: Few-shot text classification by leveraging bi-directional attention and cross-class knowledge. Sci. China Inf. Sci. 64, 1–13 (2021)

17. Fei, J., Zeng, W., Zhao, X., Li, X., Xiao, W.: Few-shot relational triple extraction with perspective transfer network. In: Proceedings of the 31st ACM International Conference on Information & Knowledge Management, pp. 488–498 (2022)

18. Zhang, L., Deng, Z., Kawaguchi, K., Ghorbani, A., Zou, J.: How does mixup help with robustness and generalization? arXiv:2010.04819 (2020)

19. He, K., Girshick, R., Dollár, P.: Rethinking imagenet pre-training. In: Proceedings of the IEEE/CVF International Conference on Computer Vision, pp. 4918–4927 (2019)

20. Jingyao, W., Zhao, Z., Sun, C., Yan, R., Chen, X.: Few-shot transfer learning for intelligent fault diagnosis of machine. Measurement 166, 108202 (2020)

21. Tian, R., Shi, H.: Momentum memory contrastive learning for transfer-based few-shot classification. In: Applied Intelligence, pp. 1–15 (2022)

22. Gao, B., Zhao, X., Zhao, H.: An active and contrastive learning framework for fine-grained off-road semantic segmentation. In: IEEE Transactions on Intelligent Transportation Systems (2022)
23. Chen, X., et al.: Knowprompt: Knowledge-aware prompt-tuning with synergistic optimization for relation extraction. In: Proceedings of the ACM Web Conference 2022, pp. 2778–2788 (2022)
24. Liu, P., Yuan, W., Jinlan, F., Jiang, Z., Hayashi, H., Neubig, G.: Pre-train, prompt, and predict: a systematic survey of prompting methods in natural language processing. ACM Comput. Surv. 55(9), 1–35 (2023)
25. Scao, T.L., Rush, A.M.: How many data points is a prompt worth? arXiv:2103.08493 (2021)
26. Devlin, J., Chang, M.W., Lee, K., Toutanova, K.: Bert: pre-training of deep bidirectional transformers for language understanding. arXiv:1810.04805 (2018)
27. Soares, L.B., FitzGerald, N., Ling, J., Kwiatkowski, T.: Matching the blanks: Distributional similarity for relation learning. arXiv:1906.03158 (2019)
28. Han, X., et al.: Fewrel: a large-scale supervised few-shot relation classification dataset with state-of-the-art evaluation. arXiv:1810.10147 (2018)
29. Zhang, Y., Zhong, V., Chen, D., Angeli, G., Manning, C.D.: Position-aware attention and supervised data improve slot filling. In: Conference on Empirical Methods in Natural Language Processing (2017)

An Improved Method of Side Channel Leak Assessment for Cryptographic Algorithm

Fuxiang Lu[1], Weijian Li[2(✉)], Zanyu Huang[2], Chuanlu Chen[1], and Peng Chen[1]

[1] China Quality Certification Center South China Laboratory, Guangzhou, China
[2] Guangdong Polytechnic Normal University, Guangzhou, China
weijianlee@126.com

Abstract. As a standard method of side channel leak assessment, TVLA is a popular research direction of side channel attack. TVLA mainly conducts leak assessment based on Welch's t-test or paired t-test, but in some evaluation scenarios, different leak assessment results may appear in the two hypothesis tests. Firstly, the modified relationship between the two kinds of t-tests is studied in theory, and then it is pointed out that the paired t-test is better when there is a positive correlation between the two groups of sampled populations to be evaluated, while the correcting two-sample t-test is needed when there is an independent or negative correlation. Secondly, it is verified by experiments that in common experimental scenarios, such as noise of common measuring equipment, temperature fluctuation and the choice of cross-input plaintext sequence will lead to positive correlation between samples, especially in the first two scenarios, the paired t-test can play a fast, efficient and robust role. Theoretical analysis and experimental results show that the method of mutual correction by two hypothesis tests ensures the accuracy and efficiency of side channel leakage assessment.

Keywords: TVLA · paired t-test · correcting two-sample t-tests · assessment of side channel · side channel attack

1 Introduction

Since Paul Kocher first proposed Timing Attack in 1996, cryptographers have conducted extensive research on side-channel attack techniques [1]. The gradual development of side-channel attacks poses a serious threat to the secure implementation of cryptographic algorithms, so the ability to resist side-channel attacks must be considered in the design process of cryptographic algorithm implementation. In order to ensure the secure application of cryptographic chips, side-channel leakage evaluation has become an essential part of cryptographic chip security evaluation and certification.

At present, ISO/IEC 15408 (Common Criteria, CC), which adopts an "attack-oriented" evaluation method [2], is mainly used internationally for information security general evaluation criteria. Since the implementation of CC requires mastering the implementation details of cryptographic algorithms and coping with various side-channel attack methods, it leads to high cost, long time consumption, and can only evaluate existing side-channel attack methods. In 2013, CRI company (Cryptography Research, Inc) took TVLA (Test Vector Leakage Assessment) as a standard method for side-channel leakage evaluation [3], which is suitable for black box model evaluation and industrial application with low operation requirements, easy implementation and high efficiency. In 2015, Tobias Schneider *et al.* [4] proposed a side-channel leakage evaluation based on non-specific t-test and high-order test as well as efficient implementation methods, laying a theoretical foundation for side-channel leakage evaluation. In 2016, Adam Ding *et al.* [5] proposed that the side-channel leakage evaluation method based on paired t-test has advantages such as fastness, efficiency and robustness. In 2017, S. Guilley *et al.* [6] found that selecting good experimental equipment to collect power consumption and reasonable statistical tools can improve the evaluation efficiency. In 2019, O. Bronchain *et al.* [7] pointed out the main factors affecting the results of side-channel leakage evaluation and proposed a side-channel leakage evaluation method based on D-test. In 2021, Yang and Jia et al. [22] proposed a black-box leakage detection method based on analysis of variance (ANOVA), which extended the binary evaluation of TVLA to multi-class evaluation, and pointed out that this method can use less power to detect more leakage. WANG et al. [23] proposed a side-channel leakage evaluation method based on Bartlett and multi-class F-test (Bartlett-F test) to solve the problem of false negative in evaluation. In 2022, Wang et al. [24]introduced a metric observed power to evaluate the effect of T-test and proposed the Levene-test based-leakage assessment (LTBLA) to solve the unreliable result in T-test.

The choice of statistical tools has a crucial impact on TVLA. In addition to the main methods of Welch's t-test [3,4,8–11] and paired t-test [5,9,12], other literature attempts to use tests such as ρ-test [13], χ^2-test [14], and Hotelling's T^2-test [7]. The use of t-tests is the most widespread and researched. The evaluation schemes based on other hypothesis tests are less studied and not yet mature. The use of each hypothesis testing tool requires strict conditions to be met. If the conditions are not met, a corrected t-test can be used. The optimized scheme designed in this paper is based on the modified two-sample t-test proposed by Zimmerman D W [15]. Under the optimized scheme proposed in this paper, side-channel leakage evaluation based on paired t-tests is more efficient, faster and robust than traditional Welch's t-tests. This paper has conducted relevant research on this topic.

1. The effects of measurement noise, temperature fluctuations and cross-order selection of power consumption acquisition devices on the side-channel leakage measurements were analyzed and verified from the measurements. It is found that the side-channel leakage measurements based on paired t-test and Welch's t-test in some scenarios show different results for the same sampling point.

2. Theoretically analyze the similarities and differences between Welch's tt-test and paired t-test and the correction relationship between them, i.e., the sampling points that do not satisfy the paired t-test combined with the correlation coefficient will be corrected to a two-sample t-test. The preferred method of the two under different conditions is proposed and applied to side-channel leakage measurement, which can improve the efficiency and reliability of side-channel leakage measurement.

2   Related Background

The power analysis attack mainly exploits the correlation leakage between the power consumption generated by the encryption device during encryption and the encryption operation of the device [16], so the leakage of the encryption operation of the device can be measured by two sets of power consumption traces. TVLA determines whether the two sets of power consumption traces come from the same overall by collecting two sets of power consumption generated by the device under test using the same key for two different sets of plaintext encryption [17]. If not, the device has side-channel leakage. The two sets of power consumption traces represent a sampling of the two overall classes, and the two sets of power consumption traces are denoted by L_F and L_R, representing the power consumption traces generated by the encryption process of the fixed plaintext group and the random plaintext group, respectively. The power trajectory model is as follows [5]:

$$L_F = V(k, x_F) + r_F + r_E,$$
$$L_R = V(k, x_R) + r_R + r_E \tag{1}$$

where $V(k, x_F)$ and $V(k, x_R)$ denote the power consumption data caused by intermediate values corresponding to fixed plaintext x_F and random plaintext x_R, respectively, which are Gaussian distributed and independent of each other. r_F and r_R denote the measurement noise generated by the measurements, which are mutually independent Gaussian noises with mean 0 variances σ_F^2 and σ_R^2. r_E is Gaussian noise with mean 0 variances σ_F^2 caused by a common environment. The sample size is n_F and n_R and the mean and variance are $\bar{L}_F$, $\bar{L}_R$ and s_F^2, s_R^2, respectively, and the degrees of freedom are denoted by ν. The Welch's t-test statistic is expressed as [4].

$$t_u = \frac{\bar{L}_F - \bar{L}_R}{\sqrt{\frac{s_F^2}{n_F} - \frac{s_R^2}{n_R}}}, \nu = \frac{\left(\frac{s_F^2}{n_F} + \frac{s_R^2}{n_R}\right)^2}{\frac{\left(\frac{s_F^2}{n_F}\right)^2}{n_F - 1} - \frac{\left(\frac{s_R^2}{n_R}\right)^2}{n_R - 1}} \tag{2}$$

where μ_F and μ_R denote the expected values of the overall F and R. The original and alternative hypotheses of Welch's t-test are:

$$H_0 : \mu_F = \mu_R, H_1 : \mu_F \neq \mu_R.$$

When paired t-test is used as the evaluation tool, the difference between L_F and L_R can eliminate part of the influence of environmental noise and improve the efficiency of evaluation. When $n_F = n_R = N$, $\nu = 2N - 2$, the design of paired samples $D = L_F - L_R$ two adjacent encryptions correspond to L_F, L_R. And finally, n pairs of D are obtained as $D_1, D_2, \cdots, D_N$, $\bar{D}$ and s_D^2 denote the mean and variance of D, respectively, and the test statistic of paired t-test is expressed as [5].

$$t_p = \frac{\bar{D}}{\sqrt{\frac{s_D^2}{n}}}, \nu = n - 1 \tag{3}$$

Here, μ_D represents the expected value of population D. The null and alternative hypotheses of paired t-test are as follows:

$$H_0 : \mu_D = 0, H_1 : \mu_D \neq 0.$$

Through t distribution cumulative function, t value and degree of freedom n the side-channel leakage evaluation calculates the probability of this situation when the hypothesis accepts H_0 and determines whether it is less than the significance level α. Generally, the value of α is 0.00001, and a corresponding fixed threshold C is set [4]. In the actual process of side-channel leakage evaluation, it is found that paired t-test has more advantages than Welch's t-test in most cases. First, the paired t-test is simpler to calculate, and the sparse matrix may appear in the construction of paired samples D to make the calculation of t_p more efficient. Secondly, the implementation of TVLA requires a large number of power traces, and the collection of power consumption is difficult to complete in a short period of time. Keeping environmental temperature and other factors unchanged for a long time requires a high cost. Statistical methods can eliminate the influence caused by changes in environmental factors, and the use of TVLA based on paired t-test reflects this advantage [5]. Finally, in the process of side-channel leakage evaluation, some objective factors cause t_p to change more rapidly and the evaluation efficiency is higher than t_u.

3   Theoretical Analysis and Evaluation Design

3.1   T-Test for Related Samples

In order to improve the efficiency of side-channel leakage evaluation and enhance the robustness, Adam Ding et $al.$ [5] introduced paired t-test. Let $\tilde{\sigma}_F^2 = \sigma_F^2 + var\,[V\,(k, x_F)]$ and $\tilde{\sigma}_R^2 = \sigma_R^2 + var\,[V\,(k, x_R)]$ denote that the variance of L_F and L_R consists of two parts: the variance of the power consumption caused by the intermediate value and the variance of the measurement noise. Since the implementation of the encryption algorithm is very fast, two adjacent encryptions are considered to be done in the same experimental environment, and the difference between L_F and L_R can eliminate r_E. The test statistics of Welch's t-test and paired t-test can be expressed as follows, respectively.

$$t_p = \frac{\overline{L_F} - \overline{L_R}}{\sqrt{\frac{\tilde{\sigma}_F^2 + \tilde{\sigma}_R^2 + 2\sigma_E^2}{n}}}, t_u \approx \frac{\overline{L_F} - \overline{L_R}}{\sqrt{\frac{\tilde{\sigma}_F^2 + \tilde{\sigma}_R^2}{n}}} \tag{4}$$

In theory, $|t_p|$ is larger than $|t_u|$ if there is common noise. When evaluating, $|t_p|$ changes more rapidly as the power trace increases, enabling faster detection of leaks. However, the sample correlation will affect the results of the t-test. This paper tries to give a more reasonable theoretical analysis of this problem. When the correlation coefficient of two sets of samples X_1 and X_2 is ρ, the standard deviation of the difference between their means is $\sigma_{\overline{X}_1 - \overline{X}_2} = \sqrt{\sigma_{\overline{X}_1}^2 + \sigma_{\overline{X}_2}^2 - 2\rho\sigma_{\overline{X}_1}\sigma_{\overline{X}_2}}$. Let $\widetilde{\sigma_F^2} + \sigma_F^2 = \sigma_F'^2$ and $\widetilde{\sigma_R^2} + \sigma_R^2 = \sigma_R'^2$, so $\sigma_D^2 = \sigma_F'^2 + \sigma_R'^2 - 2\rho\sigma_F'^2\sigma_R'^2$ and $|t_p|$ is written as follows.

$$t_p \approx \frac{\overline{L_F} - \overline{L_R}}{\sqrt{\frac{\tilde{\sigma}_F^2 + \tilde{\sigma}_R^2 + 2\sigma_E^2 - 2\rho\sigma_F'\sigma_R'}{n}}} \tag{5}$$

We have $t_p = t_u/\sqrt{1-\rho}$ if the variances of the two samples are unknown but equal. In the actual construction of paired samples, the variances may not be equal, and the corresponding equality relationship is often obtained by analogy in many literatures [15,18].

$$\frac{t_p}{t_u} \approx \frac{1}{\sqrt{1-\rho}} \tag{6}$$

According to Formula (6), there is an equality relationship of $t_p = t_u/\sqrt{1-\rho}$ between paired t-test and Welch's t-test. When there is correlation between samples, t_u should be corrected to t_u', that is, $t_u' = t_u/\sqrt{1-\rho}$, which is called modified two-sample t-test [15,19], so $t_p = t_u'$. By analyzing Formula (6), the following conclusions can be obtained:

1. When the correlation coefficient between groups of sampling points $\rho > 0$, the paired t-test can ensure that the significance level corresponding to the fixed threshold C is closer to α, so that the evaluation results are more accurate (when the sample size is greater than 100). Otherwise, Welch's t-test with fixed threshold will produce incorrect results due to significance level distortion. The common experimental equipment noise and temperature fluctuation in the evaluation cause positive correlation samples. In order to prove the advantages of paired t-test under this condition, comparative experiments are carried out in Sect. 3 evaluation scenarios 1) and 2).

2. When the correlation coefficient between groups of sampling points $\rho < 0$, the advantages of paired t-test cannot be played. In addition, paired t-test is not appropriate because it reduces the test power and increases the type II error rate because it reduces the number of degrees of freedom by half. Different crossover methods affect the correlation coefficient of samples, and the correlation of some sampling points will affect the results of hypothesis testing. In this case, the test statistics of the two are similar but the different

degrees of freedom may lead to different hypothesis test results. Therefore, in the evaluation scenario 3) in Sect. 4, a comparative experiment under this condition is designed.

3. In order to achieve accurate and efficient side-channel leakage evaluation based on TVLA, the following optimizations can be made in the process of t-test calculation and selection:

   (a) Firstly, formula (3) is used to quickly and efficiently calculate the value of paired t-test t_p ;

   (b) Secondly, it was divided into two cases according to the size of ρ. If $\rho > 0$, the paired t-test was performed; otherwise, the modified two-sample t-test was performed according to Formula (6) and t'_u.

   (c) Finally, the overall results of side-channel leakage evaluation are obtained.

3.2 Design of Evaluation Scenarios

The optimization scheme as shown in Fig. 1 in this paper makes the TVLA based on t-test more efficient and accurate, and has stronger robustness to adapt to different evaluation scenarios. In this paper, a variety of common test scenarios are carefully designed for verification. Firstly, the test scenario 1) common measurement equipment noise and the test scenario 2) fluctuating ambient temperature will cause positive correlation of samples under such conditions. At this time, the paired t-test can be used to complete the test efficiently and robustly. Secondly, in practical applications, the positive and negative correlations of samples may not be completely consistent, so using paired t-test or Welch's t-test alone will lead to wrong evaluation conclusions. In this scheme, t_p is calculated by paired t-test, and then the positive or negative correlation is judged. If the correlation is positive, the decision of paired t-test is continued; if the correlation is negative or not, t_p is corrected and then the decision of modified two-sample t-test is performed. Finally, the overall evaluation results are obtained, which can reflect the advantages of accurate, efficient and simple calculation of the evaluation results.

3.3 Choosing the Crossover Order

TVLA is to obtain two groups of samples by sampling, and use the information provided by the samples to infer the characteristics and relationships of the two populations. In side-channel leakage testing and evaluation, if the power consumption of the device under test is collected in a fixed crossover order, the fixed crossover order will produce sampling deviation because the internal state of the device under test will affect the power consumption trajectory of the next encryption [4]. When the deviation reaches a certain level, it will cause the hypothesis test of small probability event occurrence to lose statistical significance, and if the sampling deviation is not controlled, it may reduce the credibility of the hypothesis test. Reducing the sampling bias makes the sample more representative of the population characteristics, and a more suitable crossover order should be selected. The order of the crossed input plaintext refers

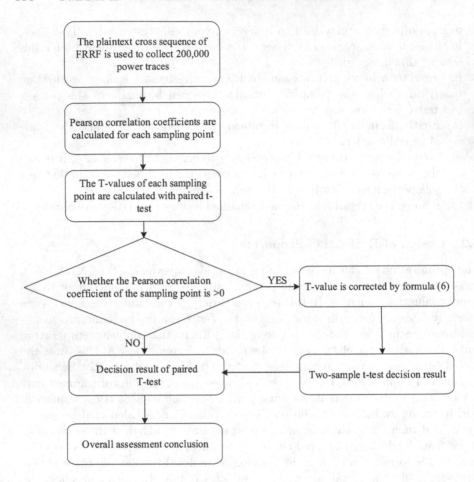

Fig. 1. Flowchart of optimized evaluation scheme

to the alternating order of the plaintext types (such as fixed class and random class) of the input in each encryption, which is simply referred to as "crossed order" in this paper.

In the process of TVLA usage and optimization, it is found that the order of cross-input plaintext has an impact on hypothesis testing results and efficiency [5,8,20]. Cross order $FRFR$ is pointed out that the cross order may lead to invalid intervals in the test results, such as the test results of the first round of AES [4,5]. In the crossover order proposed in literature [8], the sequence in which F and R are encrypted twice in each group appears randomly with equal probability. In reference [5], $FRRF$ crossover order was proposed. Theoretically, this method is more suitable for constructing paired t-test and can avoid invalid interval problem. In order to verify the influence of the crossover order on the t-test results and the equality relationship of Formula (6), the influence of the three crossover orders on the side-channel leakage evaluation was compared and analyzed [21].

4 Experimental Results and Analysis

SAKURA-G side-channel analysis platform and Agilent DSO 3034T oscilloscope are the main experimental equipment in this paper Fig. 2. In order to compare the influence of different experimental devices on the experimental results, two different measurement devices, called device 1 and Device 2, are designed. For device 1, the oscilloscope channel 2 is connected by SMA to BNC wire, and the Mini-Circuits BLP-50+ low-pass filter is connected in series at the BNC end to collect the power consumption generated by encryption. For device 2, the connection mode of channel 2 in device 1 is changed to passive probe N2943A and the low-pass filter is removed, which has greater device noise, especially high-frequency noise [6,7]. In this paper, we evaluate the side-channel leakage of the unprotected AES encryption implementation on SAKURA-G in different scenarios. In scenario 1) and scenario 2), it is the positive correlation samples caused by the main environmental factors in the evaluation, and in scenario 3) it is the impact of three different crossover orders on the evaluation.

Fig. 2. Experimental platform

4.1 Positively Correlated Samples Suitable for Paired T-Test

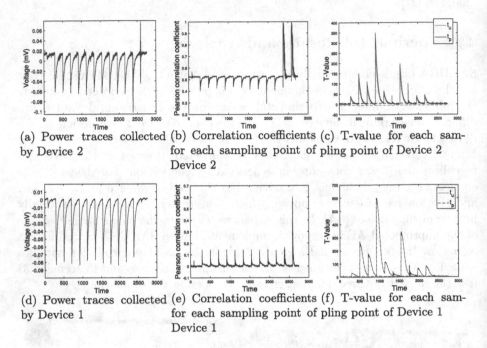

(a) Power traces collected by Device 2

(b) Correlation coefficients for each sampling point of Device 2

(c) T-value for each sampling point of Device 2

(d) Power traces collected by Device 1

(e) Correlation coefficients for each sampling point of Device 1

(f) T-value for each sampling point of Device 1

Fig. 3. Impact of equipment 1 and equipment 2 on the evaluation results

Evaluation Scenario 1: Common Noise Caused by Experimental Equipment

In this paper, device 1 and device 2 are used for comparative experiments. Device 2 is used to encrypt and collect 200,000 power traces in the cross-order of FRRF, and the side-channel leakage test results are shown in Fig. 3.

Figure 3(b) and (e) comparison found that device 2 caused each sampling point to be a positive correlation sample due to the greater common noise generated by the device. The positive correlation leads to a larger $|t_p|$ than $|t_u|$ in Fig. 3 (c), while Fig. 3(f) makes $|t_p|$ and $|t_u|$ basically equal due to the small number of relations at most sampling points. The experimental results show that the common noise caused by the experimental equipment causes each sample point to be a positive correlation sample, which can detect the leakage faster and reflect the advantages of paired t-test.

Evaluation Scenario 2: Fluctuating Ambient Temperature

The change of experimental environment temperature has a great impact on the side-channel leakage evaluation. Device 1 was used in an experimental environment with fluctuating temperature to encrypt and collect 200 000 power consumption traces in FRRF cross order. Two kinds of t-test are used to evaluate

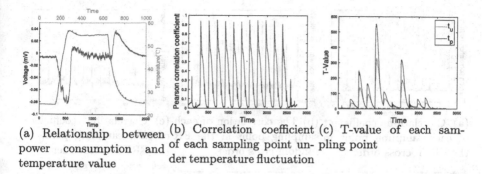

(a) Relationship between power consumption and temperature value

(b) Correlation coefficient of each sampling point under temperature fluctuation

(c) T-value of each sampling point

Fig. 4. Impact of fluctuating temperature environments on the evaluation results

the side-channel leakage respectively, and the experimental results are shown in Fig. 4.

Figure 4(a) Diagram of power consumption (taking any sampling point) and temperature fluctuation. By comparing Fig. 4(b) and (c) with Fig. 3(e) and (f), it is found that since the power consumption generated by two adjacent encryption is affected by the temperature change trend at the same time, each sampling point in Fig. 4(b) is a positive correlation sample, which is more suitable for using paired t-test. As can be seen in Fig. 4(c), because the positive correlation coefficient makes $|t_p|$ larger than $|t_u|$, the paired t-test can play its advantages.

4.2 Effect of Crossover Order on T-Test

Evaluation Scenario 3: Selection of Crossover Order
In this paper, device 1 is used to collect 200,000 power traces each under three crossover orders, and the correlation coefficients, $|t_p|$ and $|t_u|$ under the three crossover orders are calculated respectively, and the results are shown in Fig. 5.

A sampling point in the first round of AES whose correlation coefficient varies greatly due to the crossover order is used for analysis. Among the three cross orders, FRFR sequence has the best independence among the several test scenarios to cross the input plaintext to obtain the correlation between samples. However, all the test results of the first round of AES encryption process are wrong, as shown in Fig. 5(f).

The second crossover order induces the largest correlation coefficient and has the greatest impact on hypothesis testing. By observing the correlation coefficient, t value change and size of the sampling points caused by the three crossover orders, it is found that $FRRF$ is the most appropriate crossover order for side-channel leakage evaluation.

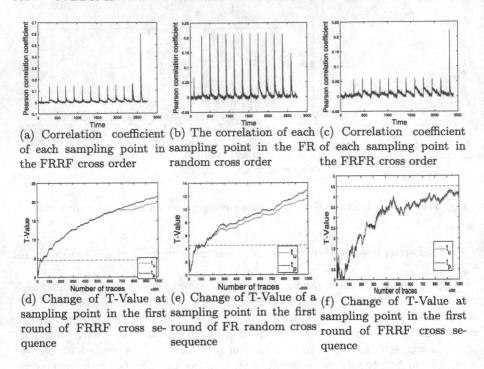

(a) Correlation coefficient of each sampling point in the FRRF cross order

(b) The correlation of each sampling point in the FR random cross order

(c) Correlation coefficient of each sampling point in the FRFR cross order

(d) Change of T-Value at sampling point in the first round of FRRF cross sequence

(e) Change of T-Value of a sampling point in the first round of FR random cross sequence

(f) Change of T-Value at sampling point in the first round of FRRF cross sequence

Fig. 5. Evaluation results of three cross-sequence input plaintexts

4.3   Errors and Corrections in the Evaluation

In order to verify that the misuse of the two t-tests will lead to wrong evaluation results, this paper selected some obvious sampling points for analysis. By increasing the number of power traces, the changes of correlation coefficient, $|t_p|$ and $|t_u|$ of sampling points were observed. Finally, the accuracy of the two t-tests in different cases and the correction relationship between them were verified, and the results were shown in Fig. 6 and Fig. 7

Figure 6 shows the case of error II in Welch's t-test caused by the presence of some sampling points for $\rho > 0$ in the evaluation scenarios 1) and 2). The correlation coefficients corresponding to the three sampling points in Fig. 6(d) are all positive, and the correlation coefficients are gradually stable with the increase of the power consumption trace. The $|t_p|$ and $|t_u|$ corresponding to these sampling points can be clearly distinguished by the set threshold C, and the corresponding hypothesis testing decisions are different.

Figure 7 shows that A sample point with $\rho \leq 0$ in the evaluation scenario 3) after accurate calculation, and then comparing with α, it is found that the paired t-test shows the II error. Figure 7(a) shows the random occurrence of the sequence of two encrypted plaintext inputs of F and R in the evaluation scenario 3) with equal probability, which leads to type II error in the paired t-test due to $\rho \leq 0$. The correlation coefficient of the sampling point in Fig. 7(b) is negative

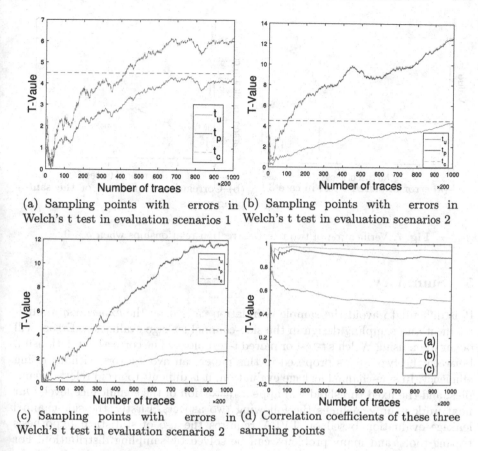

(a) Sampling points with errors in Welch's t test in evaluation scenarios 1

(b) Sampling points with errors in Welch's t test in evaluation scenarios 2

(c) Sampling points with errors in Welch's t test in evaluation scenarios 2

(d) Correlation coefficients of these three sampling points

Fig. 6. Verification of two t-test correction relationships for positively correlated samples

but small after stabilization, so there is no obvious difference between $|t_p|$ and $|t_u|$ in Fig. 7(a). With the increase of the power trace, the t value is basically stable near the threshold C, but the two t-test pairs appear different decisions after calculation. The significance level $alpha = 0.00001$ is set and $C \approx 4.4173$ is calculated by combining the sample size of the experiment in this paper. Because the degrees of freedom are reduced by half, the probability values of the two t-tests calculated when the value of t is near the value of C are 1.0004621e-5 and 9.999383e-6 are located on both sides of α, and the two t-tests will make different decisions at this time. Because the paired t-test has fewer degrees of freedom, it reduces the power of the test so using the paired t-test for $\rho \leq 0$ can lead to error II.

340   F. Lu et al.

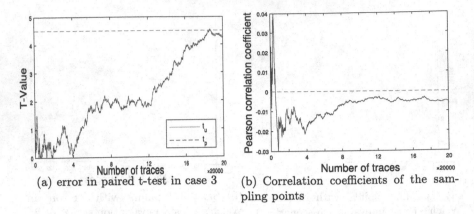

(a) error in paired t-test in case 3   (b) Correlation coefficients of the sampling points

Fig. 7. Verification of two t-test correction relationships when $\rho \leq 0$

5   Summary

It is difficult to avoid the sample correlation caused by the experimental environment and sampling design in the side-channel leakage evaluation. Errors will occur when using Welch's t-test or paired t-test alone. The correction relationship between the two t-tests proposed in this paper can avoid errors without losing efficiency. In side-channel leakage evaluation, it is difficult to avoid relevant samples, and the fixed threshold will appear distortion of the significance level. Our next work mainly includes the following two aspects. Firstly, the side-channel leakage evaluation based on TVLA requires the use of statistical hypothesis testing tools, and many problems can be solved by sampling distribution. For instance, accuracy can be solved by the selection of significance level and the influencing factors of sampling process, while efficiency can be achieved by the complexity of sampling samples. Secondly, the evaluation of the countermeasure implementations of cryptographic algorithm needs to be improved according to the characteristics of implementations.

Acknowledgement. This work was supported by China Quality Certification Center science and technology project under Grant No. 2022CQC18-GZ, Guangdong Provincial Department of Education Key Field Project for Ordinary Universities under Grant No. 2023ZDZX1008 and No. 2020ZDZX3059, Guangzhou key field research and development plan project under Grant No. 202206070003, General Project of the National Natural Science Foundation of China under Grant No. 62072123.

References

1. Kocher, P.C.: Timing attacks on implementations of diffie-hellman, RSA, DSS, and other systems. In: Koblitz, N. (ed.) CRYPTO 1996. LNCS, vol. 1109, pp. 104–113. Springer, Heidelberg (1996). https://doi.org/10.1007/3-540-68697-5_9

2. Hongsong, S.H.I., Jinping, G.A.O., Wei, J.I.A., et al.: Analyse of the security architecture and policy model in the common criteria. J. Tsinghua Univ. (Sci. Technol.) 56(5), 493–498 (2016)

3. Becker G, Cooper J, Demulder E, et al. Test vector leakage assessment (TVLA) methodology in practice. In: International Cryptographic Module Conference, pp. 1001-1013. Springer, Cham (2013)

4. Schneider, T., Moradi, A.: Leakage assessment methodology. In: International Workshop on Cryptographic Hardware and Embedded Systems, pp. 495–513. Springer, Cham (2015)

5. Ding, A.A., Chen, C., Eisenbarth, T.: Simpler, faster, and more robust t-test based leakage detection. In: Standaert, F.-X., Oswald, E. (eds.) COSADE 2016. LNCS, vol. 9689, pp. 163–183. Springer, Cham (2016). https://doi.org/10.1007/978-3-319-43283-0_10

6. Merino del Pozo, S., Standaert, F.-X.: Getting the most out of leakage detection. In: Guilley, S. (ed.) COSADE 2017. LNCS, vol. 10348, pp. 264–281. Springer, Cham (2017). https://doi.org/10.1007/978-3-319-64647-3_16

7. Bronchain, O., Schneider, T., Standaert, F.-X.: Multi-tuple leakage detection and the dependent signal issue. IACR Trans. Cryptographic Hardware Embed. Syst. 2019, 318–345 (2019)

8. Bilgin, B., Gierlichs, B., Nikova, S., Nikov, V., Rijmen, V.: Higher-order threshold implementations. In: Sarkar, P., Iwata, T. (eds.) ASIACRYPT 2014. LNCS, vol. 8874, pp. 326–343. Springer, Heidelberg (2014). https://doi.org/10.1007/978-3-662-45608-8_18

9. Masoum, M.: A highly efficient and secure hardware implementation of the advanced encryption standard. J. Inf. Secur. Appl. 48, 102371–102384 (2019)

10. Bache, F., Plump, C., Wloka, J., et al.: Evaluation of (power) side-channels in cryptographic implementations. It-Inf. Technol. 61(1), 15–28 (2019)

11. Jiazhe, C., Hexin, L., Yanan, W., et al.: Evaluating side-channel information leakage in 3DES using the t-test. J. Tsinghua Univ. (Sci. Technol.) 65(5), 499–503 (2016)

12. Shahverdi, A., Taha, M., Eisenbarth, T.: Lightweight side channel resistance: threshold implementations of Simon. IEEE Trans. Comput. 66(4), 661–671 (2017)

13. Durvaux, F., Standaert, F.-X.: From improved leakage detection to the detection of points of interests in leakage traces. In: Fischlin, M., Coron, J.-S. (eds.) EURO-CRYPT 2016. LNCS, vol. 9665, pp. 240–262. Springer, Heidelberg (2016). https://doi.org/10.1007/978-3-662-49890-3_10

14. Moradi, A., Richter, B., Schneider, T., et al.: Leakage detection with the x2-test. IACR Trans. Cryptographic Hardware Embed. Syst. 2018, 209–237 (2018)

15. Zimmerman, D.W.: Correcting Two-Sample "z" and "t" tests for correlation: an alternative to one-sample tests on difference scores. Psicologica: Int. J. Methodol. Exp. Psychol. 33(2), 391–418 (2012)

16. Bronchain, O., Hendrickx, J.M., Massart, C., Olshevsky, A., Standaert, F.-X.: Leakage certification revisited: bounding model errors in side-channel security evaluations. In: Boldyreva, A., Micciancio, D. (eds.) CRYPTO 2019. LNCS, vol. 11692, pp. 713–737. Springer, Cham (2019). https://doi.org/10.1007/978-3-030-26948-7_25

17. Standaert, F.-X.: How (Not) to use welch's t-test in side-channel security evaluations. In: Bilgin, B., Fischer, J.-B. (eds.) CARDIS 2018. LNCS, vol. 11389, pp. 65–79. Springer, Cham (2019). https://doi.org/10.1007/978-3-030-15462-2_5

18. Hsu, H., Lachenbruch, P.A.: Paired t test. Wiley encyclopedia of clinical trials, pp. 1–3 (2007)

19. Zimmerman, D.W.: Teacher's corner: a note on interpretation of the paired-samples t test. J. Educ. Behav. Stat. **22**(3), 349–360 (1997)

20. Mather, L., Oswald, E., Bandenburg, J., Wójcik, M.: Does my device leak information? An *a priori* statistical power analysis of leakage detection tests. In: Sako, K., Sarkar, P. (eds.) ASIACRYPT 2013. LNCS, vol. 8269, pp. 486–505. Springer, Heidelberg (2013). https://doi.org/10.1007/978-3-642-42033-7_25

21. David, H.A., Gunnink, J.L.: The paired t test under artificial pairing. Am. Stat. **51**(1), 9–12 (1997)

22. Yang, W., Jia, A.: Side-channel leakage detection with one-way analysis of variance. Secur. Commun. Netw. **2021**, 1–13 (2021)

23. Yaru, W.A.N.G., Ming, T.A.N.G.: Side channel leakage assessment with the Bartlett and multi-classes F-test. J. Commun. **42**(12), 35–43 (2021)

24. Wang, Y., Tang, M., Wang, P., et al.: The Levene test based-leakage assessment. Integration **87**, 182–193 (2022)

Fusing Global and Local Interests with Contrastive Learning in Session-Based Recommendation

Su Zhang[1], Ye Tao[2], Ying Li[3(✉)], and Zhonghai Wu[3]

[1] Center for Data Science, Peking University, Beijing, China
zachzsu@pku.edu.cn
[2] School of Software and Microelectronics, Peking University, Beijing, China
tao.ye@pku.edu.cn
[3] National Engineering Center of Software Engineering and Key Lab of High Confidence Software Technologies (MOE), Peking University, Beijing, China
{li.ying,wuzh}@pku.edu.cn

Abstract. Session-based Recommendation aims at predicting the next item based on anonymous users' behavior sequences. During a session, both the user's global interest and local interest influence the decisions. However, due to their discrepancy, most existing works fail to fully leverage this comprehensive information. In this paper, we propose a contrastive learning based interest fusion model, which provides a more reasonable method to fuse global and local interests in session-based recommendation. Specifically, we design two session-level feature encoders to extract them respectively. Instead of fusing interests at the representation level, we take the prediction results from different encoders as supervision signals to enhance their ability in extracting global and local interests. Experiments on real-world datasets show that the proposed method achieves state-of-the-art performance.

Keywords: Session-based recommendation · Contrastive learning · Global and local interests

1 Introduction

Recommender systems have been playing crucial roles in alleviating information overload and providing personalized advice. Unlike traditional recommenders built with users' profiles and historical behaviors, session-based recommendation (SBR) is adopted in services where only actions during an ongoing session can be exploited. For a user's behavior decision, both global long-term interest and local short-term interest are crucial and mutually complementary. It's difficult to predict a user's current purpose solely from one side, *e.g.*, in e-commerce applications, a user usually starts a session with a specific shopping intention (global interest) but may also show interest in other captivating items within a short timeframe (local interest). As a result, both global long-term interests and local short-term interests should be taken into account for recommendation.

However, the existing methods lack in-depth thinking on utilizing users' global and local interests. Many works capture only one of them and can't achieve

X. Song et al. (Eds.): APWeb-WAIM 2023, LNCS 14334, pp. 343–358, 2024.
https://doi.org/10.1007/978-981-97-2421-5_23

satisfying results for lack of users' comprehensive interest information [4,16]. Meanwhile, works on fusing information from the two sides basically obtain a unified representation [8,11,18], *e.g.*, using the linear combination result of global and local interest representations, or learning a session representation without explicitly distinguishing them. However, the unified representation neglects the discrepancy between global and local interests and can't get rid of the mutual interference between them [25]. Thus these methods fail to make rational use of the comprehensive information from different interests.

To solve these problems, we propose to apply contrastive learning (CL) to fuse the information from global and local interests. We build auxiliary tasks by means of mining supervision information from the model itself. Due to the discrepancy between global and local interests, we design two session-level feature encoders to extract them separately. Instead of fusing interests at the representation level, we obtain the prediction results with the two encoders separately and use them as supervision signals to enhance the model's ability to extract global and local interests.

Specifically, for the global interest encoder, the positive and negative instances with high confidence are sampled according to the results of local interest encoder. They are used as pseudo-labels to optimize the representation from global interest encoder, and vice-versa. Considering the discrepancy between global and local interests, CL can provide explicit supervision signals to guide the encoder to adaptively introduce information from the other side instead of combining them directly. Note that the importance of global and local interests varies with scenarios, *e.g.*, for music recommendation, users may prefer to choose songs with the same genre as the latest liked song, while in e-commerce apps, users' behaviors in a session often revolve around the primary static intention. So we choose either global or local interest encoder as the main encoder for prediction according to specific application scenarios.

This paper analyses the discrepancy between global and local interests based on our experiments, *i.e.*, they focus on different aspects of information and can lead to mutual interference when simply combined together. So it's necessary to make rational use of the discrepancy and avoid the mutual interference which counteracts the effectiveness of the comprehensive information. Thus we proposed the contrastive learning based method to fuse global and local interests. Extensive experiments prove that our method can utilize the two types of interests efficiently, achieving outstanding performance. To summarize, our contributions come from these following aspects:

- We analyse the discrepancy between the information from global long-term interests and local short-term interests, and show the significance of reasonably extracting and exploiting them in session-based recommendation.
- We propose a novel interest fusion model for session-based recommendation and design a contrastive learning method between different encoder views which can extract and fuse the global and local interests efficiently.
- We conduct extensive experiments on real-world datasets and prove our superiority to the existing methods. The code of our model is released here.

2   Related Work

2.1   Session-Based Recommendation

The common idea for SBR is to utilize item transitions and temporal information from sessions. Early methods are mainly based on Markov Chain, *e.g.*, FPMC [14] combines it with matrix factorization to make predictions. With the rise of deep learning, RNNs are applied to capture sequential information and achieve great success [4,8,11], along with the attention mechanism introduced to learn session representations. Recently, graph neural networks have shown their capability in modeling graph data and are introduced to session-based recommendation, exhibiting great performance [1,16,18,19]. These works build global graphs or session-specific graphs to capture information from item transitions. SR-GNN [18] and TAGNN [22] apply GGNN [10] to learn item embeddings and use attention mechanism to generate session representations. Later, more graph models are introduced as the backbone to this task, including WGAT by FGNN [13] and SGAT by LESSR [1]. More recently, some works focus on digging information from the sessions by constructing complicated graph structures, like hypergraphs which captures high-order correlations [19] and item graphs at different levels [16,20], and become new sota methods.

On the task of session-based recommendation, early works utilize either one of global and local interests to make predictions, which omits some part of information [4,16]. Recently more works give attention to both of them. STAMP [11] uses the Hadamard product of their representation as the session embedding. Other works [8,18] combine them using linear transition. However, the unified representation can be less expressive due to the interference caused by the discrepancy between global and local interests.

2.2   Contrastive Learning in Recommender Systems

Recently, contrastive learning [5] (CL), as a novel learning paradigm to alleviate data sparsity and enhance the ability in feature extraction, is introduced to recommender systems [24]. Contrastive Learning mines supervision signals from data or the model itself and can build auxiliary tasks to enhance the models. In recommendation, works utilizing CL are usually based on different views generated using data augmentation techniques.

For non-sequential recommendation tasks, SGL [17] utilizes edge and node dropout as the method of producing new views and MHCN [23] generates positive data by means of subgraph sampling. The negative pairs are usually sampled from different items or gained by adding feature noises, *e.g.*, feature shuffling. As for the sequential scenarios, CL4SRec [21], S^3-Rec [26] and H^2SeqRec [9] uses item masking, reordering or cropping methods to generate new sequences. Besides, S^2-DHCN [19] and COTREC [20] build item-level and session-level graphs as different views to conduct contrastive learning. These are the prior works to adopt CL auxiliary tasks in session-based recommendation, achieving outstanding performance.

3   Analysis: Global and Local Interests in Session-Based Recommendation

3.1   Preliminaries

Problem Formulation. Let $V = \{v_1, v_2, \ldots, v_m\}$ denote the set of the unique items involved in all session data, where m is the number of items. Each anonymous session is represented by a sequence of clicked items as $s = [v_{s,1}, v_{s,2}, \ldots, v_{s,t}]$ which are ordered by timestamps. And we denote the set of all sessions as S. Formally, session-based recommendation aims at predicting the next possible click (*i.e.*, $v_{s,t+1}$) for a given session s. The output of the model should be the predicted probabilities among the set of all items V, *i.e.*, $\hat{y} = \{\hat{y}_1, \hat{y}_2, \ldots, \hat{y}_m\}$.

Typical Procedure of Session-Based Recommender. For a typical session-based recommender, we extract item-level features and then learn session representations with a session-level encoder. Finally the scores are computed with the representations of sessions and items. In this procedure, the session-level encoder is used to capture users' preferences and reflect their different interests.

Later in this section, we first introduce the two types of session representation encoders designed in our work, *i.e.*, global interest encoder and local interest encoder. After that, we present the difference between information extracted by the two encoders and give further analysis on global and local interests.

3.2   Global and Local Interest Encoder

Given the item features, session encoders learn a representation for each session based on items in it. Formally, given a session $s = [v_{s,1}, v_{s,2}, \ldots, v_{s,t}]$ and embeddings of corresponding items $[h_{s,1}, h_{s,2}, \ldots, h_{s,t}]$, the interest encoder outputs the session representation θ_s. We design the two-phase session interest encoder consisting of two modules: Session Aggregation and Gated Combination.

Session Aggregation. Here the item embeddings in a session are aggregated into one representation and we use different aggregation methods for global and local interest encoders.

- **Mean Aggregation for global interest encoder.** The global static interest reflects the initial motivation or basic characteristics of a user. It focuses on users' global intention, having weak correlation with the item sequence. Thus it's often expressed in the form of weighted sum. Here we simply adopt a mean pooling layer:

$$h_s^{mean} = \mathrm{MEAN}([h_{s,1}, h_{s,2}, \ldots, h_{s,t}])$$

- **GRU Aggregation for local interest encoder**. The local interest reveals users' preference in a short period. It changes over time and pays more attention to sessions' sequential patterns. So we use the RNN model GRU to capture it, taking the mean pooling result as the initial hidden state:

$$h_s^{gru} = \text{GRU}([h_{s,1}, h_{s,2}, \ldots, h_{s,t}], h_s^{mean})$$

Gated Combination (GC). After the aggregation, we introduce the information from the last item in a session to the result. A gated mechanism is designed to control its weight. The final result can be formulated as:

$$\alpha = \text{MLP}(h_s^*, h_{s,t}) = \sigma(W_{GC}[h_s^*; h_{s,t}] + b_{GC})$$
$$\theta_s = \alpha \circ h_s^* + (1 - \alpha) \circ h_{s,t} \circ h_s^*$$

where σ is the sigmoid activation function and W_{GC}, b_{GC} are the parameters of MLP. h_s^* represents the aggregation result which is either h_s^{mean} or h_s^{gru}.

Finally, based on different session aggregation methods, we get the global interest encoder (**global-IE**) and local interest encoder (**local-IE**). We denote their outputs as θ_s^{global} and θ_s^{local} respectively. In the sections that follow, we also use global-IE and local-IE as a symbol of the model using them respectively.

3.3   Analysis on Global and Local Interest Based Predictions

In order to reveal the influence of global and local interests on recommendation separately, we complete the overall procedure for session-based recommendation using different session interest encoders and compare their results.

Overall Difference of Global and Local Interests. Figure 1a shows the ranks of true labels for each test session on Tmall dataset given by the two models. The x- and y-coordinates are the ranks given by the global and local encoder respectively, *e.g.*, a point (2, 10) means the true label of the session is ranked No.2 by global-IE while No.10 by local-IE. As is seen, quite a considerable number of points deviate from the line $y = x$, which means the two encoders do make different predictions to a large extent. To be specific, Fig. 1b presents the difference of ranks for each true label given by global-IE and local-IE. And 33% of true next items are given a higher probability rank by the global encoder while 29% are predicted better by the local one.

The Preference for Global or Local Interests Has Relation to the Length of Sessions. Though the preference varies in different situations and is hard to predict, we can still find some relation between the performance of two types of interests and the length of sessions. We classified the sessions in Tmall dataset according to their lengths and analyzed the average rank of the ground-truth labels. The relative difference of the ranks given by global-IE and local-IE is shown in Fig. 1c. According to the results, the importance of global

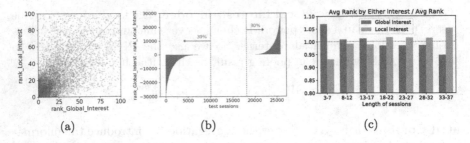

Fig. 1. Discrepancy between results given by either interest on Tmall dataset. (a) Rank comparison between global and local interests. (b) Rank difference between global and local interests. (c) Average rank by either interest on sessions of different lengths (shown after divided by the average rank of two interests).

interests, on the whole, is increasing as the length of sessions increases. When sessions become longer, local interests are likely to deviate from users' global preference and global interests can play a bigger role in users' decisions. And in cases where sessions are quite short, the global representation can be incapable of steadily and clearly capturing users' global preference, while the local interests can achieve better performance by reason of revealing users' current preference.

These results can further explain the difference between global and local interests in sessions of different lengths, reflecting that the two types of information focus on different user interests. Therefore, they are both important and indispensable for recommendation. However, due to the discrepancy, the global and local interests may interfere with each other when combining them or directly integrating their results. So we utilize their results as pseudo-labels for mutual supervision and provide guidance for the learning of session representations. In this way the representations from two encoders can contain each other's information while preserving their own semantic information.

4 Methodology

In this section, the proposed **C**ontrastive **L**earning based **I**nterest **F**usion model (**CLIF**) for session-based recommendation is described in detail. We use graph neural network (GNN) as the item-level feature encoder and the procedure of the primary SBR task is divided into four phases: (1) Graph construction. (2) Item Representation Learning. (3) Session Interest Encoding. (4) Prediction. In addition, an auxiliary contrastive supervision task is employed to enhance the primary task. The overview of the procedure can be seen in Fig. 2.

4.1 Graph Construction

In order to utilize GNN as the feature encoder, the first step is to construct a meaningful graph using all sessions. We treat each item $v_i \in V$ as a node in

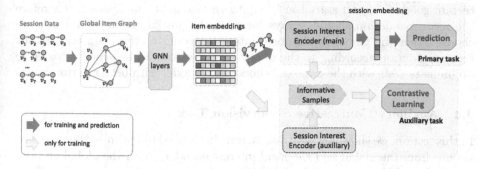

Fig. 2. The Overall Procedure of CLIF

the graph. And for a session $s = [v_{s,1}, v_{s,2}, \ldots, v_{s,t}] \in S$, each contiguous pair of clicked items $(v_{s,j}, v_{s,j+1})$ forms a directed edge in the graph. In this way, we build a global directed graph and the weight for each edge (v_i, v_j) counts how many times the adjacent clicks $[v_i, v_j]$ appear in all sessions. Specially, we set the weight of all self-loop edges to 1 because we observe many repeated clicks which can be disadvantageous to the item representation learning.

4.2   Item Representation Learning

Inspired by the idea [3] that feature transformation and nonlinear activation modules can be useless and even harmful in recommendation tasks, we adopt GCN [7] without the these modules on the homogeneous directed graph. Given the incoming adjacent matrix $A \in \mathbb{R}^{m \times m}$ and the item embeddings from the l-th layer $H^{(l)} \in \mathbb{R}^{m \times d}$ (d is the dimension of the item representations), the information propagation in one graph convolution layer can be formalized as:

$$H^{(l+1)} = D_{(in)}^{-\frac{1}{2}} A D_{(out)}^{-\frac{1}{2}} H^{(l)} \tag{1}$$

where $D_{(in)}$ and $D_{(out)}$ are the in-degree and out-degree matrix for the graph respectively. That is, $D_{(in)} = \sum_i A_{ij}$ and $D_{(out)} = \sum_j A_{ij}$. After L layers of forward propagation, we average the item embeddings obtained from each layer to be the final item embeddings $H = \frac{1}{L+1} \sum_{i=0}^{L} H^{(i)}$.

4.3   Session Interest Encoder

We design two distinct encoders to capture global and local interests separately, which are described in detail in Sect. 3.2. It's worth noting that their importance for recommendation varies with scenarios. In some services user's preference changes over time while in other services user behaviors often revolve around the primary static intention. For example, for music or video recommendation, users are likely to choose items with the same type or genre as the latest enjoyed one. And when it comes to e-commerce, users usually have demand for some certain goods first and start the shopping service as a consequence. Therefore the recommendation should always focus on this initial global interest. So we choose either the global or the local interest encoder as the main encoder for recommendation according to the application scenario. To be specific, we make preliminary tests with the data and choose the encoder with better performance.

4.4 Auxillary Contrastive Supervision Task

In this section, we show how contrastive learning is used to optimize the representations from the global and the local interest encoder. With the global encoder global-IE and the local encoder local-IE, we first mine supervision signals from their results and then compute the contrastive learning loss.

Mining Supervision Signals. After obtaining the global interest representation θ_s^{global} and the local interest representation θ_s^{local}, we use them to predict the probabilities among the whole item set to be the next item:

$$\hat{y}_{i,s}^* = \text{Softmax}(score(h_i, \theta_s^*)), \quad score(h_i, \theta_s^*) = \frac{h_i^\top \theta_s^*}{\|h_i\|\|\theta_s^*\|} \quad (2)$$

where $*$ means either *global* or *local* and $\hat{y}_{i,s}$ denotes the probability of item i to be recommended. The scores are computed in the form of cosine similarity.

We select items with the top-K highest probabilities as positive samples. And as for negative samples, we randomly select K items from those that are ranked top-η but after $4K$. These hard difficult samples can provide more meaningful information than those sampled on the whole item set. The sets of positive samples and negative samples are denoted as C_+^* and C_-^* respectively.

Contrastive Learning. The auxiliary supervision task is built with the pseudo-labels and refines the model through a contrastive learning objective. For global-IE, we hope it to learn dynamic information from local-IE, so its scores on C_+^{local} should be far higher than scores on C_-^{local}, and vice versa. For each specific session s, we follow InfoNCE [12] to design our optimization objective:

$$\mathcal{L}_{CL} = -\log \frac{\sum_{i \in C_+^{local}} \psi\left(h_i, \theta_s^{global}\right)}{\sum_{i \in C_+^{local}} \psi\left(h_i, \theta_s^{global}\right) + \sum_{i \in C_-^{local}} \psi\left(h_i, \theta_s^{global}\right)}$$
$$-\log \frac{\sum_{i \in C_+^{global}} \psi\left(h_i, \theta_s^{local}\right)}{\sum_{i \in C_+^{global}} \psi\left(h_i, \theta_s^{local}\right) + \sum_{i \in C_-^{global}} \psi\left(h_i, \theta_s^{local}\right)} \quad (3)$$

where $\psi(h_i, \theta_s^*) = \exp(score(h_i, \theta_s^*)/\tau)$ and τ is the temperature to control the impact of hard negative samples [15].

In this way, both encoders can get guided with the positive and negative samples from the other view which can reflect different interests of users. Thus, the two views can exchange information and the representations can be refined. Compared with the combined representation for the two types of interests, our method can truly introduce global or local interests into the other representation while keeping the original semantic information from being interfered.

4.5 Recommendation and Model Optimization

Given a session s and an item v_j, we obtain the score using the cosine similarity between h_j and the session representation θ_s^* from the main encoder. Therefore, during training as well as inference, we conduct L_2-normalization on the representations:

$$\tilde{h}_j = \frac{h_j}{\|h_j\|} \qquad \tilde{\theta}_s^* = \frac{\theta_s^*}{\|\theta_s^*\|} \tag{4}$$

Then the cosine similarity score can be obtained from their inner product and a softmax function is applied to compute the probabilities for all items:

$$\hat{y}_j = \frac{\exp(\beta \tilde{h}_j^\top \tilde{\theta}_s^*)}{\sum_{i=1}^m \exp(\beta \tilde{h}_i^\top \tilde{\theta}_s^*)} \tag{5}$$

As shown in [2], the cosine similarities are restricted to $[-1, 1]$ and the softmax loss can get saturated at high values for the training set. So we use a scaling factor β to amplify the discrimination and achieve better convergence. Then the learning objective of the primary SBR task is formulated as a cross entropy loss:

$$\mathcal{L}_{SBR} = -\sum_{i=1}^m y_i \log(\hat{y}_i) \tag{6}$$

where y denotes the one-hot vector of the ground truth labels. Finally, along with the auxiliary contrastive learning task, we formulate the total loss as(λ is the hyperparameter controlling the magnitude of the auxiliary task):

$$\mathcal{L} = \mathcal{L}_{SBR} + \lambda \mathcal{L}_{CL} \tag{7}$$

5 Experiments

5.1 Experimental Setup

Datasets. We conduct experiments on three real-world datasets, *i.e.*, *Tmall*[1], *Diginetica*[2] and *RetailRocket*[3]. Tmall dataset, released in IJCAI-15 competition, contains anonymous users' shopping logs on Tmall app. Diginetica comes from CIKM Cup 2016, and only its transactional data is used. RetailRocket is an e-commerce dataset and contains anonymous users' browsing activities within six months. Following [8,18], we filter out sessions of length 1 and items appearing less than 5 times for each dataset. Sessions during the last week are divided as test data and previous sessions compose the training set. Besides, a sequence splitting method is used to augment the data. Specifically, we obtain a series of sequences and labels $([v_{s,1}], v_{s,2}), ([v_{s,1}, v_{s,2}], v_{s,3}), \ldots, ([v_{s,1}, v_{s,2}, \ldots, v_{s,t-1}], v_{s,t})$ from the session $s = [v_{s,1}, v_{s,2}, \ldots, v_{s,t}]$. The statistics of the datasets are listed in Table 1.

---

[1] https://tianchi.aliyun.com/dataset/dataDetail?dataId=42.
[2] https://competitions.codalab.org/competitions/11161.
[3] https://www.kaggle.com/retailrocket/ecommerce-dataset.

Table 1. Statistics of the datasets.

Dataset	Tmall	Diginetica	RetailRocket
# training sessions	351,268	719,470	433,643
# testing sessions	25,898	60,858	15,132
# items	40,728	43,097	36,968
average length	6.69	5.12	5.43

Table 2. Paramter settings for datasets.

Parameters	Tmall	Diginetica	RetailRocket
number of GNN layers (L)	1	2	1
number of samples (K)	2	2	2
negative sampling range (η)	50	50	50
main encoder	global-IE	local-IE	local-IE
weight for CL task (λ)	0.25	0.05	0.02

Evaluation Metrics. We adopt two widely used metrics, *i.e.*, Hit Rate (HR@k) and Mean Reciprocal Rank (MRR@k), for top-k recommendation where $k = 20$.

Baselines. To evaluate the recommendation performance, We compare our method with the following representative baselines:

- **FPMC** [14] is a classic hybrid model combing matrix factorization and Markov chain for next-basket recommendation. We ignore the user latent representations in session-based recommendation.
- **GRU4REC** [4] is an RNN based method that uses GRU to model session sequences.
- **NARM** [8] introduces attention mechanism to model the session representation on the basis of a RNN-based item-level feature encoder.
- **STAMP** [11] employs attention mechanism to capture long-term interest and regard the last item as short-term interest. They are combined into the final representation using Hadamard product.
- **SR-GNN** [18] adopts gated graph neural network to obtain item embeddings and employs soft-attention mechanism to generate session embeddings.
- **S^2-DHCN** [19] constructs two hypergraphs to learn inter- and intra- session information respectively and use CL to enhance the primary model.
- **GCE-GNN** [16] constructs two types of graphs to learn item representation from local and global contexts and uses both of them to make predictions.
- **COTREC** [20] constructs an item-level hypergraph and a session-level global garph as two views and build CL tasks to enhance the primary model on the item view.

Table 3. Performance of all comparison methods on three datasets. The best and the second best results of each column are boldfaced and underlined respectively.

Models	Tmall		Diginetica		RetailRocket	
	HR@20	MRR@20	HR@20	MRR@20	HR@20	MRR@20
FPMC	16.06	7.32	26.53	6.95	32.37	13.82
GRU4REC	10.93	5.89	29.45	8.33	44.01	23.67
NARM	23.30	10.70	49.70	16.17	50.22	24.59
STAMP	26.47	13.36	45.64	14.32	50.96	25.17
SR-GNN	27.52	13.72	50.73	17.59	50.32	26.57
S^2-DHCN	31.42	15.05	53.18	18.44	53.66	27.30
GCE-GNN	33.42	15.42	_54.22_	**19.04**	_56.11_	_29.14_
COTREC	_36.35_	_18.04_	_54.18_	**19.07**	_56.17_	**29.97**
CLIF	**40.41**	**19.61**	**54.91**	_18.95_	**58.01**	_29.11_

Parameter Settings. Following previous works, we set the batch size for mini-batch to 100, the embedding size to 100, and the L_2 regularization factor to 10^{-5}. We use Adam [6] as the optimizer with an initial learning rate of 0.001, which is scheduled to decay every 3 epochs with the rate of 0.1. The temperature τ of CL task and the scaling factor β are fixed to 0.2 and 12 for all datasets. Other parameters are shown in Table 2.

5.2   Performance Comparison

The performance of different methods on three datasets are reported in Table 3. From the results we can draw the following conclusions:

Compared with traditional methods, deep learning based methods generally achieve better performance. Despite performing worse on Tmall, GRU4REC still outperforms FPMC on other datasets, demonstrating the capability of RNN in modeling sequences. In addition, this result also implies that local interests are less referential for the scenario of Tmall, which agrees to our observation.

GNN based models outperform sequential models in general, demonstrating that graph structure is more suitable for modeling session data. Among them GCE-GNN achieves better results than SR-GNN, which indicates the importance of capturing information from different levels (intra- and inter- session).

Finally, we give an analysis on methods involving the fusion of different information. STAMP and SR-GNN take users' both global and local interests into account, but their results are not satisfying for lack of reasonable approaches to fusing information. CL based methods (S^2-DHCN, COTREC) extract information from different levels and build auxiliary tasks accordingly, but the self-supervision signals can be less informative and need the process of self-discrimination to avoid mode collapse. Overall, the proposed CLIF method can make rational use of the comprehensive information from global and local interests, leading to consistent better performance.

Table 4. Comparisons of different fusion methods

Methods	Tmall		Diginetica		RetailRocket	
	HR@20	MRR@20	HR@20	MRR@20	HR@20	MRR@20
global-IE	39.19	19.22	53.09	18.46	55.45	28.19
local-IE	36.62	17.82	54.56	18.67	57.72	28.71
global-IE-CLIF	**40.41$^\ominus$**	**19.61$^\ominus$**	52.08	18.14	53.99	28.16
local-IE-CLIF	37.47	18.34	**54.91$^\ominus$**	**18.95$^\ominus$**	**58.01$^\ominus$**	**29.11$^\ominus$**
AVE	39.23	19.28	54.46	18.73	57.70	28.78
LIN	39.26	19.13	54.46	18.65	57.64	28.83
AVE-CLIF	40.16	19.42	54.58	18.88	57.75	28.95
LIN-CLIF	40.09	19.34	54.62	18.82	57.73	29.05

5.3 Analysis on CL Based Interest Fusion Method

Comparisons of Different Fusion Methods. To reveal the improvements brought by the designed contrastive learning method, we compare the performance of using the original encoders and encoders with different fusion methods. The results are reported in Table 4. **Global-IE** and **local-IE** mean the primary task using global or local interest encoder respectively. The suffix **-CLIF** denotes the one using CLIF framework and the symbol $\ominus$ means it's used as the main encoder. Besides, we provide the results of two other combination methods: (1) Averaging (**AVE**). (2) Linear transformation (**LIN**). To verify the effectiveness of the proposed contrastive supervision module, we conduct the same contrastive learning procedure on encoders in these methods and also use the suffix **-CLIF** to note the results. From the results we have the following observations:

CLIF brings significant improvements to the main encoders consistently. On Tmall, global-IE achieves particularly great progress as the main encoder. Besides, though the auxiliary encoders are trained without ground-truth labels, they still maintain quite good performance with the supervision from the main encoder.

Directly combining global and local interests shows little improvements. Meanwhile, it's observed that applying our contrastive supervision module can improve the performance of these fusion methods. This demonstrates that our CL module can help bridge the gap between global and local interests, which lessens the mutual interference when they are combined together.

On all datasets, the proposed CLIF method performs best compared to other interest fusion methods, even if they are enhanced by the CL module. This implies that the crude combination leads to mutual entanglement between global and local interests. In contrast, simply using their results as supervision signals is a more rational way to utilize the discrepancy and obtain comprehensive information, ensuring our method to have the consistent best performance.

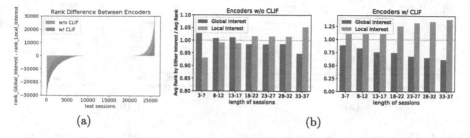

Fig. 3. Rank difference of true labels without and with CLIF on Tmall dataset. (a) Rank difference between global and local interests. (b) Average rank by either interest without (left) and with (right) CLIF on sessions of different length.

Information Discrepancy With & Without CLIF. In this part we try to find out whether and how the main encoder learns information from the auxiliary one. We measure the discrepancy of the results between encoders without and with CL training by the difference of true labels' rank. The visualization result is shown in Fig. 3a. To reveal the specific improvements brought by CLIF, we analyse the performance on sessions of different lengths and report the results in Fig. 3b.

As can be seen in Fig. 3a, cases where the main encoder(*i.e.* global-IE on Tmall) gives poorer ranks than the auxiliary encoder decrease a lot. Considering that the auxiliary encoder in CLIF still performs well, we conclude that the main encoder can truly obtain interest information from the other encoder, which is the basic reason for the consistent better performance.

Specifically in Fig. 3b, we take a look into the performance on sessions of various lengths and check how the method helps to improve the main encoder. Compared with the performance without CLIF, we can find that the main encoder (global-IE) keeps the tendency of having higher ranks when sessions become longer. Moreover, its performance on shorter sessions exceeds those of local-IE. This result indicates that global-IE can get improved by the information of local interests and meanwhile the original information of global interests is well preserved, showing the effectiveness of the proposed CLIF method.

5.4 Analysis on Hyper-parameters for CL

In this section, we focus on the details of the sampling process. We conduct extensive experiments on Tmall and Diginetica datasets and check the impacts of sampling number K and the range of negative sampling η.

Impact of the Sampling Number. To investigate the impact of the sampling number K, we range it in $\{1, 2, 3, 5, 10\}$ and report the results in Fig. 4a. Basically we can notice that the tendencies of the two metrics are different. When the sampling number increases, hit rate increases first and then decreases while MRR keeps dropping. The possible reason could be that fewer positive

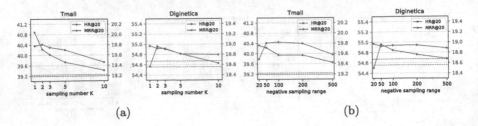

Fig. 4. The impact of (a) sampling number K and (b) negative sampling range η. Performance of the main encoders without CLIF are shown in dashed line.

samples can focus on more targeted information and promote the rank of the next-click item in some cases, leading to higher MRR. And more samples can cover more comprehensive information and thus improve the overall hit rate. So it might be proper to choose less sampling numbers when rankings of top few items are more important. When the sampling number further increases, the samples can be less positive and informative, leading to worse performance.

Impact of the Negative Sampling Range. To study the effect of the negative sampling range η, we conduct experiments with a set of representative η values $\{20, 50, 100, 200, 500\}$. The results are shown in Fig. 4b. On the whole, the tendency when η increases is similar with the case of the sampling number, but the metrics show more stability on both datasets. When the sampling range is quite small, the negative samples might actually be the true positive items, leading to lower hit rate. But meanwhile the true negative samples can be more difficult ones and provide more targeted information, which can help promote the rank of the next-click item. When the range becomes much wider, the negative samples are likely to be easier ones which provide less effective information, resulting in the dropping of metrics. Nevertheless, the performance declines fairly slow, due to the high quality information provided by positive samples.

6   Conclusion

In this paper, we illustrate the significance of making rational use of users' both global long-term interests and local short-term interests in SBR and propose a novel interest fusion framework based on contrastive learning. We conduct experiments to show the discrepancy between users' global and local interests, which can lead to mutual interference when they are combined together. Thus we propose to capture them separately and use their results as supervision signals to utilize information from the discrepancy. Extensive experiments demonstrate the effectiveness of our framework and show its superiority over other methods.

References

1. Chen, T., et al.: Handling information loss of graph neural networks for session-based recommendation. In: Proceedings of the 26th ACM SIGKDD International Conference on Knowledge Discovery & Data Mining, pp. 1172–1180 (2020)
2. Gupta, P., et al.: Niser: Normalized item and session representations to handle popularity bias. arXiv preprint arXiv:1909.04276 (2019)
3. He, X., Deng, K., Wang, X., et al.: Lightgcn: simplifying and powering graph convolution network for recommendation. In: SIGIR, pp. 639–648 (2020)
4. Hidasi, B., et al.: Session-based recommendations with recurrent neural networks. arXiv preprint arXiv:1511.06939 (2015)
5. Hjelm, R.D., et al.: Learning deep representations by mutual information estimation and maximization. In: 7th International Conference on Learning Representations, ICLR 2019 (2019)
6. Kingma, D.P., Ba, J.: Adam: a method for stochastic optimization. arXiv preprint arXiv:1412.6980 (2014)
7. Kipf, T.N., Welling, M.: Semi-supervised classification with graph convolutional networks. arXiv preprint arXiv:1609.02907 (2016)
8. Li, J., Ren, P., Chen, Z., Ren, Z., Lian, T., Ma, J.: Neural attentive session-based recommendation. In: CIKM, pp. 1419–1428 (2017)
9. Li, Y., et al.: Hyperbolic hypergraphs for sequential recommendation. In: CIKM, pp. 988–997 (2021)
10. Li, Y., et al.: Gated graph sequence neural networks. In: 4th International Conference on Learning Representations, ICLR 2016 (2016)
11. Liu, Q., et al.: Stamp: short-term attention/memory priority model for session-based recommendation. In: Proceedings of the 24th ACM SIGKDD International Conference on Knowledge Discovery & Data Mining, pp. 1831–1839 (2018)
12. Van den Oord, A., Li, Y., Vinyals, O.: Representation learning with contrastive predictive coding. arXiv e-prints pp. arXiv–1807 (2018)
13. Qiu, R., Li, J., Huang, Z., Yin, H.: Rethinking the item order in session-based recommendation with graph neural networks. In: CIKM, pp. 579–588 (2019)
14. Rendle, S., Freudenthaler, C., Schmidt-Thieme, L.: Factorizing personalized Markov chains for next-basket recommendation. In: Proceedings of the 19th International Conference on World Wide Web, pp. 811–820 (2010)
15. Wang, F., Liu, H.: Understanding the behaviour of contrastive loss. In: CVPR, pp. 2495–2504 (2021)
16. Wang, Z., et al.: Global context enhanced graph neural networks for session-based recommendation. In: SIGIR, pp. 169–178 (2020)
17. Wu, J., Wang, X., Feng, F., et al.: Self-supervised graph learning for recommendation. In: SIGIR, pp. 726–735 (2021)
18. Wu, S., Tang, Y., Zhu, Y., Wang, L., Xie, X., Tan, T.: Session-based recommendation with graph neural networks. In: AAAI, vol. 33, pp. 346–353 (2019)
19. Xia, X., Yin, H., Yu, J., et al.: Self-supervised hypergraph convolutional networks for session-based recommendation. In: AAAI, vol. 35, pp. 4503–4511 (2021)
20. Xia, X., Yin, H., Yu, J., Shao, Y., Cui, L.: Self-supervised graph co-training for session-based recommendation. In: CIKM, pp. 2180–2190 (2021)
21. Xie, X., et al.: Contrastive learning for sequential recommendation. In: ICDE, pp. 1259–1273. IEEE (2022)
22. Yu, F., Zhu, Y., Liu, Q., Wu, S., et al.: TAGNN: target attentive graph neural networks for session-based recommendation. In: SIGIR, pp. 1921–1924 (2020)

23. Yu, J., Yin, H., Li, J., Wang, Q., Hung, N.Q.V., Zhang, X.: Self-supervised multi-channel hypergraph convolutional network for social recommendation. In: Proceedings of the Web Conference 2021, pp. 413–424 (2021)
24. Yu, J., Yin, H., Xia, X., Chen, T., Li, J., Huang, Z.: Self-supervised learning for recommender systems: a survey. arXiv preprint arXiv:2203.15876 (2022)
25. Zheng, Y., et al.: Disentangling long and short-term interests for recommendation. In: Proceedings of the ACM Web Conference 2022, pp. 2256–2267 (2022)
26. Zhou, K., et al.: S3-rec: self-supervised learning for sequential recommendation with mutual information maximization. In: CIKM, pp. 1893–1902 (2020)

Exploring the Effectiveness of Student Behavior in Prerequisite Relation Discovery for Concepts

Jifan Yu[1], Hanming Li[1], Gan Luo[2], Yankai Lin[3], Peng Li[1], Jianjun Xu[4], Lei Hou[1(✉)], and Bin Xu[1]

[1] Tsinghua University, Haidian District, Beijing, China
{yujf21,lhm22}@mails.tsinghua.edu.cn,
{houlei,xubin}@tsinghua.edu.cn
[2] Meituan Inc., Beijing, China
[3] Renmin University of China, Beijing, China
[4] Beijing Caizhi Technology Co., Ltd., Beijing, China
xjj@czkj1010.com

Abstract. What knowledge should a student grasp before beginning a new MOOC course? To answer this question, it is essential to automatically discover prerequisite relations among course concepts. Although researchers have devoted intensive efforts to detecting such relations by analyzing various types of information, there are few explorations of utilizing student behaviors in this task. In this paper, we investigate the effectiveness of student behaviors in prerequisite relation discovery. Specifically, we first construct a novel dataset to support the study, and then formally define four typical student behavior patterns based on analysis of the dataset. Moreover, We explore how this behavior information can be utilized to enhance existing methods, including feature-based and graph-based, via extensive experiments. Experimental results demonstrate that proper modeling of student behaviors can significantly improve the performance of the methods on this task. We hope our study could call for more attention and efforts to explore student behavior for prerequisite relation discovery.

Keywords: Prerequisite relation discovery · Student behavior · MOOCs

1 Introduction

Since the first edition of Robert Gagne's *Principles of Instructional Design* [6] came out, many efforts from pedagogy have suggested that students should grasp prerequisite knowledge before moving forward to learn subsequent knowledge. Such prerequisite relations are described as the dependence among knowledge concepts and are crucial for students to learn, organize, and apply knowledge [19].

J. Yu and H. Li—Equal contribution.

Figure 1 shows an example of the prerequisite relations in Massive Open Online Courses (MOOCs). For a student who wants to learn the concept "Convolutional Neural Network" (CS224:video18), it is beneficial to previously grasp the prerequisite concepts ("Gradient Descent" and then "Back Propagation Algorithm") for better understanding.

Prerequisite relations play an essential role in intelligent educational applications such as curriculum planning [1], reading list generation [7], etc. Figure 1 illustrates an example where explicit prerequisite relations among concepts (represented in red) enable the recommendation of a coherent and reasonable learning sequence to the student (represented in green or blue).

However, as the quantity of educational resources grows rapidly in the era of MOOCs, it is expensive and ineffective to obtain fine-grained prerequisite relations by expert annotations as before [4]. Therefore, automatically discovering prerequisite relations becomes a rising topic in recent years.

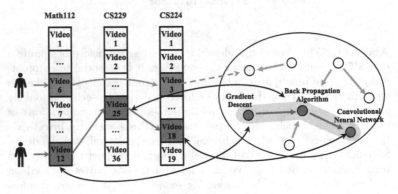

Fig. 1. An application example of prerequisite relations. A student who wants to study "Convolutional Neural Network" can be suggested to follow the prerequisite chain (green path) to learn "Gradient Descent" and "Back Propagation Algorithm" first. (Color figure online)

This task is previously defined as **Prerequisite Relation Discovery of Concepts in MOOCs**. Earlier works on this topic include extracting such relations from the content of MOOC videos [15,17] and the preset orders of MOOC resources [16,21]. Recently, graph-based approaches are introduced into the task, either relying solely on graph [9] or taking advantage of both content- and graph-based methods [27]. However, despite these attempts on this topic, the current methods are still inadequate for direct application in practical scenarios, primarily due to the following challenges

First, unlike the factual relations of general entities (e.g., *Bill Gates* has the relation **founder** with *Microsoft Inc.*), prerequisite relations are more cognitive than factual, which makes it rarely mentioned explicitly in texts and challenging to be captured from MOOC corpus. Second, existing MOOC resources considered prerequisite clues are noisy, e.g., the order of MOOC videos. As a video usually teaches several concepts, it is common that some of these concepts are not prerequisites to the ones in later videos. Therefore, it is crucial to explore more resources to help the discovery of prerequisite relations.

Inspired by the idea from educational psychology that students' learning behaviors are highly related to the cognitive structure of knowledge [3], we investigate leveraging the *video watching behaviors of students* in the task of discovering prerequisite concepts in MOOCs. For supporting the investigation, we collect student behavior data from real MOOC courses and organize expert annotations to construct a novel dataset of prerequisite concept relations. With careful analysis of the data, we then summarize four typical behavior patterns.

Furthermore, to explore the use of behavior patterns to enhance existing feature- and graph-based methods, we design features for each of these patterns, the effectiveness of which is examined through experiments. Results show that incorporating student behaviors effectively enhances existing representative models. We also provide several empirical suggestions for research on related topics.

Our contributions include:

- An investigation on how to leverage student behaviors to extract prerequisite relations of concepts;
- Validation of the effect that existing methods can be enhanced significantly combined with behaviors information;
- A novel benchmark dataset using real courses from MOOC websites[1].

2   Related Work

Our work mainly follows the efforts in discovering *prerequisite relations* among course concepts, aiming to detect the dependence of concepts from MOOC resources. The task of identifying prerequisite relations originates from educational data mining, which could help in automatic curriculum planning [19] and other educational applications [20]. In the area of education, early works discover general prerequisite structures from students' test performance [8,23,25], and these early efforts have mainly focused on discovering the dependence among courses or knowledge units. [24] and [14] further propose to learn more fine-grained prerequisite relations, i.e., the prerequisite relations among concepts. In recent years, detecting prerequisite concepts from courses (especially online courses) has become a rising research topic. Researchers explore various kinds of methods from matrix optimization [16], feature engineering [17], neural networks [21] to the recent trend of graph-based models [9,13,27] to consider the static information of MOOCs (course/video) as indispensable clues for discovering such relations.

There are also some attempts to extract prerequisite relations from other resources, e.g., paper citation networks [7] and textbooks' unit sequences and titles [11]. Recently, the user clickstream of Wikipedia pages [22] is also proven to

---

[1] https://github.com/HwHunter/Prerequisite-Relation-Discovery-in-MOOCs.

indicate concept dependence. This inspires us to improve prerequisite prediction by considering the user behaviors in MOOCs, which contain more behavioral details and are more relevant to the cognitive learning process.

3   Problem Formulation

In this section, we give fundamental definitions and present the problem formulation for discovering prerequisite relations among course concepts in MOOCs.

A MOOC corpus is composed of courses from MOOCs, denoted as $\mathcal{M} = \{\mathcal{C}_i\}_{i=1}^{|\mathcal{M}|}$, where $\mathcal{C}_i$ indicates the i-th course. Each course includes a sequence of videos, i.e., $\mathcal{C}_i = [v_{ij}]_{j=1}^{|\mathcal{C}_i|}$, where v_{ij} refers to a video with its subtitles from the course. And the **Student Behavior** that we use in this paper is the Video Watching Behaviors $\mathcal{S} = \{(u, v, t)\}$, where each behavior records student $u \in U$ started to watch the video v at time t, and U is the set of all students.

Course dependence is defined as a prerequisite relation between courses [16], denoted as $\mathcal{D} = \{(\mathcal{C}_i, \mathcal{C}_j) \,|\, \mathcal{C}_i, \mathcal{C}_j \in \mathcal{M}\}$, which indicates that course $\mathcal{C}_i$ is a prerequisite course of $\mathcal{C}_j$. This information is often provided by the teachers when setting up new courses.

Course Concepts refer to the subjects taught in a course (e.g., "LSTM" is a concept of the Deep Learning course). We respectively denote the concepts of a certain video, a course, and the whole MOOC corpus as $\mathcal{K}^v$, $\mathcal{K}^c$ and $\mathcal{K}$. The video concepts $\mathcal{K}_{ij}^v = \left\{c_1, ..., c_{|\mathcal{K}_{ij}^v|}\right\}$ is the concepts taught in course video v_{ij}. As a course consists of several videos, the course concept $\mathcal{K}_i^c = \mathcal{K}_{i1}^v \cup ... \cup \mathcal{K}_{i|\mathcal{C}_i|}^v$. And all the concepts of the MOOC corpus is $\mathcal{K} = \mathcal{K}_1^c \cup ... \cup \mathcal{K}_{|\mathcal{M}|}^c$.

Discovering prerequisite relation of course concepts in MOOCs is formulated as follows: Given the MOOC corpus $\mathcal{M}$, course dependence $\mathcal{D}$, student behaviors $\mathcal{S}$, and the corresponding course concepts $\mathcal{K}$, the goal is to learn a function $\mathcal{L}: \mathcal{K}^2 \rightarrow \{0, 1\}$ that maps a pair of concepts (c_a, c_b), where $c_a, c_b \in \mathcal{K}$, to a binary class indicating whether c_a is a prerequisite of c_b.

4   Data Collection and Annotation

Although there are a few datasets for mining prerequisite relations from online courses [12,17], the information on student behaviors is rarely studied due to the lack of accessible data. To support our investigation, we collect data on courses, videos, and student behaviors from XuetangX[2], a large MOOC website, and organize multi-level annotations to construct a novel prerequisite concept dataset in MOOCs.

Due to the sparsity of prerequisite relationships, it is challenging to ensure data quality while keeping low annotation cost [16]. Hence, we consider two issues in data construction: (1) the connectivity of course concepts, and (2) the effectiveness of annotations. In particular, with the protection of user privacy in mind, we collect and annotate the student behavior data in three stages after constructing the annotation team.

---

[2] https://www.xuetangx.com/.

Annotation Team: The group of annotators consists of 1 experienced CS professor, 2 CS Ph.D.s, and 8 Students who enrolled in these courses. We also employ the teachers of our selected courses as consultants to deal with disagreements in the annotation process.

Table 1. Statistics of our dataset. As course concepts and students may overlap in different courses, their total number is not a simple numerical addition.

		CS	PL	AI	ALL
#Course		4	4	4	12
#Video		312	222	233	767
#Concept		369	227	377	700
#Pair	+pos	672	673	267	1,612
	−neg	1,258	539	218	2,015
#Student		12,094	12,014	3,541	17,587
#Behavior		430,769	337,953	39,136	807,858
Kappa		0.765	0.737	0.769	0.754

Stage 1: MOOC Information Collection. We select 12 sample courses in three domains to collect information on MOOCs, including "Basic Knowledge of Computer Science" (**CS**), "Programming Languages" (**PL**), and "Artificial Intelligence" (**AI**). These courses are selected because their concepts are highly relevant. Then we collect course and student data in three steps: (1) downloading all course materials which include the video orders and subtitles; (2) obtaining the video watching logs of students who participated in these courses during 2017–2019 as user behavior data source, which could help us to infer a student's learning frequency, coverage, watching duration, and other information of a particular video; (3) annotating the dependence of courses. The annotation requires each course pair to be labeled by two students. If there exist conflicts, two Ph.D. students will further label the result. And the final results are confirmed by the professor and corresponding teacher.

Stage 2: Data Processing. Regarding all the subtitles of selected courses as the MOOC corpus, we employ a concept extraction method [18] that is widely used in MOOC-related tasks to obtain concept candidates. For each candidate, two annotators label it as "not course concept" or "course concept", and the disagreements are confirmed by the teacher. Each labeled concept's Wikipedia abstract is also dumped as side information for the reproduction of baseline methods.

Considering the protection of user privacy, we strictly abide by the agreement between the platform and the users, remove sensitive personal information, and anonymize the users into UserIDs. Especially, to alleviate the noises introduced by users' random operation, we filter out the student behavior data by following steps: (i) Remove students who have less than 2 elective courses and watch less than 10 videos. (ii) Delete behavior records that the student watched less than 30% of the video.

Stage 3: Prerequisite Relation Annotation. For all the labeled course concepts, we manually annotate the prerequisite relations among them. A critical challenge in the annotation is the giant quantity and sparsity. If the concept number is n, the candidate pair number is $n(n-1)/2$, which necessitates extensive human labeling efforts. Therefore, we present a two-step strategy to reduce the workload:

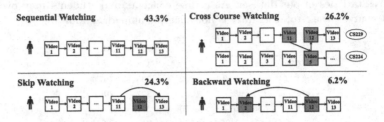

Fig. 2. Four typical patterns when a student watches course videos. The figure shows the proportion of video pairs that match each pattern in all behaviors.

• Step 1: The teacher of the corresponding course leads the annotators to cluster the concepts into several groups, which may maintain possible prerequisite relations. After this step, we get 28 clusters of 700 course concepts, where the largest contains 210 concepts and the smallest contains 14 concepts. We organize the following annotation within these concept clusters.

• Step 2: We generate the candidate concept pairs within the clusters, and sample a small scale of them as golden standard (300). Then we employ them to train existing baselines (i.e., MOOC-RF [18], GlobalF [15], PREREQ [21] and CPR-Recover) as candidate filters. To ensure the Recall, we only filter out the pair if none of the classifiers predict it to be a prerequisite, and preserve the remaining pairs in the annotation.

In the annotation process, two annotators from corresponding domains are asked to label whether concept A is a prerequisite of concept B, i.e. the annotator needs to answer the question of "whether A helps understand B". A pair is labeled as positive only when the two annotators agree. The statistics of our dataset are shown in Table 1, where *#Course, #Video, #Concept, #Pair, #Student* are the number of corresponding items and *#Behavior* is the number of video watching records. The *Kappa* statistics of the inter-annotator agreement is 0.754, showing the reliability of the annotation results[3].

5   Student Behavior Patterns and Prerequisite Features

To make full use of the student behavior information, a closer view of the characteristic of data is necessary. In this section, four typical patterns of student

---

[3] From our experience, such annotation is not easy to determine a perfect standard. E.g. "Stack" and "Queue": Grasping one of them indeed help the understanding of the other, but someone may think these two concepts are not prerequisite. We finally control the Kappa over 0.7 with the help of corresponding teachers, which indicates a good quality of the final presented dataset.

behavior during watching videos in MOOCs are summarized based on our observation and analysis of the dataset. Furthermore, we present several features to model student behaviors, which will be validated with thorough experiments in the following section. The design of our approach is based on the following cognitive learning hypothesis:

Hypothesis 1. *Students tend to follow the prerequisite cognitive structure to learn new knowledge.*

The hypothesis was proposed in educational psychology [3], and was widely applied in prerequisite-driven instructional design [19,20]. We extend this hypothesis by analyzing the clues of the prerequisite concepts implied in student learning orders. Surprisingly, although MOOCs preset the video order, our observation of student behavior data indicates that students often learn MOOC videos in their orders. As shown in Fig. 2, we summarize four typical behavior patterns, and the out-of-pre-order learning behaviors are even more than half of the total (56.72%). To leverage student behaviors in this task, we analyze the causes of these patterns and build several features to model prerequisite relations from student behaviors.

To model the video watch behavior, we first construct a sequence S_u for each user u from student behavior record S, where the video watch behaviors are sorted in time order. Comparing the preset video order of each C_i with S_u, we summarize four typical patterns of students' video-watching behaviors: *Sequential Watching, Cross Course Watching, Skip Watching* and *Backward Watching* as shown in Fig. 2. Before introducing the modeling details, we will begin by providing definitions for the behavior patterns as follows.

Definition 1 (Behavior Pattern). *A behavior pattern P is formed by one or more video pairs. A video pair (v_i, v_j) belongs to a pattern P when it matches the conditions of the pattern.*

As the student behavior patterns are at *video* level, we infer the prerequisite features of a concept pair $c_a \in K_i^v$ and $c_b \in K_j^v$ by considering videos as bags of course concepts, where K_i^v, K_j^v correspond to the concepts taught in v_i, v_j, and a concept may be taught in more than one videos.

Over 72.5% of the students' behavior records contain all the four typical patterns in Fig. 2, indicating that they are not accidental. Therefore, we attempt to speculate the causes of these four patterns from the cognitive perspective, and build prerequisite features f^P to model them.

Sequential Watching Pattern. Sequential watching indicates that a student watches videos in the course's preset video order, which indicates that the concepts taught in these videos are in accordance with the prerequisite cognitive structure. To leverage this pattern, we assign prerequisite feature f_1^P for the concepts c_a and c_b as:

$$f_1^P(c_a, c_b) = \sum_{u \in U} \sum_{v_i, v_j \in S_u} \alpha^{|j-i|} \cdot \frac{\mathsf{Seq}(u, v_i, v_j)}{\max(|K_i^v|, |K_j^v|)}, \tag{1}$$

where function $\mathtt{Seq}(u, v_i, v_j) = 1$ holds when 1) v_i, v_j are the i-th, j-th videos of a student's watching record $\mathcal{S}_u$ and are in the same course, $j > i$; 2) $c_a \in \mathcal{K}_i^v$ and $c_b \in \mathcal{K}_j^v$ (Otherwise $\mathtt{Seq}(u, v_i, v_j) = 0$).

Considering there are multiple concepts taught in each video, we employ $\max(|\mathcal{K}_i^v|, |\mathcal{K}_j^v|)$ to normalize the feature of a certain concept pair. Furthermore, since the distance between watching videos corresponds to their relatedness, we employ an attenuation coefficient of $\alpha \in (0, 1)$ to capture distant dependence from long sequences in this pattern.

Cross Course Watching Pattern. Besides watching in one course, there is a phenomenon that some students choose to watch videos in other courses before continuing on the present study. The main reason is that the knowledge provided by other courses' videos is helpful to study this course. Hence, cross course watching behavior could reflect the dependence between concepts from different courses. The prerequisite feature f_2^P for c_a and c_b is calculated as:

$$f_2^P(c_a, c_b) = \sum_{u \in U} \sum_{v_i, v_j \in \mathcal{S}_u} \alpha^{|j-i|} \cdot \frac{\mathtt{Crs}(u, v_i, v_j)}{\max(|\mathcal{K}_i^v|, |\mathcal{K}_j^v|)}, \tag{2}$$

where function $\mathtt{Crs}(u, v_i, v_j) = 1$ holds when 1) v_i, v_j are the i-th, j-th videos of a student's watching record $\mathcal{S}_u$ and are in the different course; 2) $c_a \in \mathcal{K}_i^v$ and $c_b \in \mathcal{K}_j^v$ (Otherwise $\mathtt{Crs}(u, v_i, v_j) = 0$).

Skipping Watching Pattern. An abnormal student behavior is skipping some videos when learning a course, which drops a hint that the "skipped videos" are not so necessary for the latter videos' comprehension. Given a student behavior sequence $\mathcal{S}_u$ and course video orders $\mathcal{C} = [v_1..v_i...]$, we can detect the skipped video pairs and assign a negative f_3^P for the concept pair c_a and c_b as:

$$f_3^P(c_a, c_b) = -\sum_{u \in U} \sum_{v_i, v_j \in \mathcal{S}_u} \alpha^{|j-i|} \cdot \frac{\mathtt{Skp}(u, v_i, v_j)}{\max(|\mathcal{K}_i^v|, |\mathcal{K}_j^v|)}, \tag{3}$$

where function $\mathtt{Skp}(u, v_i, v_j) = 1$ holds when 1) v_i, v_j are the i-th, j-th videos of a same course, and $i < j$; 2) v_j is watched by user u but v_i is not watched; 3) $c_a \in \mathcal{K}_i^v$ and $c_b \in \mathcal{K}_j^v$ (Otherwise $\mathtt{Skp}(u, v_i, v_j) = 0$).

Backward Watching Pattern. This pattern means a student goes back to a video that he/she watched before. A possible explanation is he/she jumps back to a video to re-learn prerequisite knowledge of the current video. Based on this assumption, we adjust the equation for the feature f_4^P between c_a and c_b.

$$f_4^P(c_a, c_b) = \sum_{u \in U} \sum_{v_i, v_j \in \mathcal{S}_u} \alpha^{|j-i|} \cdot \frac{\mathtt{Bck}(u, v_i, v_j)}{\max(|\mathcal{K}_i^v|, |\mathcal{K}_j^v|)}, \tag{4}$$

where function $\mathtt{Bck}(u, v_i, v_j) = 1$ holds when 1) v_i, v_j are the i-th, j-th videos of a student behavior record $\mathcal{S}_u$, and $i < j$; 2) v_i is watched again after v_j; 3) $c_a \in \mathcal{K}_i^v$ and $c_b \in \mathcal{K}_j^v$ (Otherwise $\mathtt{Bck}(u, v_i, v_j) = 0$).

6 Explore Graph-Based Modeling of Student Behavior

Building *concept graphs* is a common idea in concept mining tasks, including concept extraction and expansion [18,26]. Since the prerequisite relations among concepts are transitive, i.e., if $a{\to}b, b{\to}c$ then $a{\to}c$, previous works also often employ a directed graph to describe the dependence on a set of concepts [5,7]. Moreover, in the past few years, works like [13,27] tend to treat prerequisite relations as edges in a graph. Thus the task is formalized as link prediction and can be tackled with graph-based methods. All this progress inspires us to leverage student behaviors better by building a *concept graph*, defined as:

Definition 2 (Concept Graph). *A concept graph* $\mathcal{G} = (\mathcal{K}, E)$ *is a weighted directed graph, whose nodes are course concepts* $\mathcal{K}$ *and each edge* $e = (c_a \to c_b) \in E$ *is associated with a weight* w_e.

Regarding the prerequisite relation learning as a link prediction problem in a graph, we can leverage the student behavior better by utilizing Graph Convolutional Networks (GCNs) [10] to model information propagation of the concepts. Meanwhile, as several types of MOOC information have been applied to detect prerequisite concepts in previous research, including course dependence [16,21], video order [17], we also design similar concept graphs for these resources. By comparing the model performance of different graphs, we can explore the role of student information more fairly (excluding the factors of the graphical modeling).

In the following part of this section, we introduce the construction of concept graphs and how to conduct prerequisite relation learning on them. We also present experimental results to give insights into the effectiveness of the student behavior in this task.

6.1 Concept Graph Construction

As shown in Fig. 3, we design a concept graph $\mathcal{G}^s$ based on student behaviors as while as $\mathcal{G}^c$ based on course dependence graph and $\mathcal{G}^v$ based on video order graph. As the nodes of these concept graphs are the same course concept $\mathcal{K}$, the only difference is the setting of their edges.

Our graph construction stage's main idea is to assign edge weight for each concept pair in these graphs. After calculating all edges' weights in a graph, we only preserve the edges with positive weights, for they are helpful for relation reasoning.

Concept Graph Based on Student Behavior. To build a concept graph from student behaviors, a straightforward idea is to model the prerequisite clues by combining the extracted features in Sect. 5. Hence, we assign the weight w_e^s for the edge $e = (c_a \to c_b)$ in this graph $\mathcal{G}^s$ as:

$$w_e^s = \sum_{i=1}^{4} f_i^P(c_a, c_b) \times \frac{\log(|U|)}{|U|}, \tag{5}$$

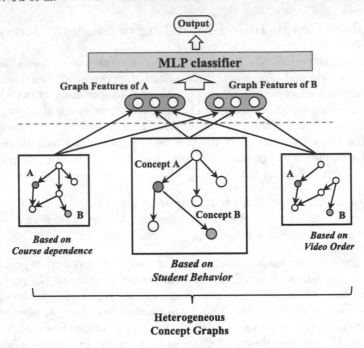

Fig. 3. The framework of our graph-based model. The nodes of each graph are course concepts. Features of concepts are the concatenation of the corresponding node embeddings learned by GCN.

where $f_i^{\mathcal{P}}(i = 1, 2, 3, 4)$ denotes the features of the concept c_a and c_b from the four behavior patterns. $log(|U|)/|U|$ is used to normalize the weight to combine with the other two user-independent graphs.

Except for student behaviors, we also build concept graphs for existing static MOOC prerequisite clues through similar methods, including the dependence among courses and the preset order of videos. By modeling this information, we can more fairly compare the contribution of these clues in graphs and explore whether they can be integrated to further enhance the model.

Concept Graph Based on Course Dependence. Course dependence is widely used in prerequisite learning. When a course is certain to be a prerequisite course of another one, there must be dependent relations between some of its concepts. So we build a concept graph $\mathcal{G}^c$ based on course dependency to exploit this information. Suppose c_a and c_b are respectively concepts of course $\mathcal{C}_i$ and $\mathcal{C}_j$, for an edge $e = (c_a \rightarrow c_b)$ of this concept graph, we can calculate its weight w_e^c as:

$$w_e^c = \sum_{\mathcal{C}_i, \mathcal{C}_j \in \mathcal{M}} \frac{\text{CD}(\mathcal{C}_i, \mathcal{C}_j)}{\max(|\mathcal{K}_i^c|, |\mathcal{K}_j^c|)}, \tag{6}$$

where function $\text{CD}(\mathcal{C}_i, \mathcal{C}_j) = 1$ only when pair $(\mathcal{C}_i, \mathcal{C}_j)$ is in course dependence set $\mathcal{D}$ (otherwise $\text{CD}(\mathcal{C}_i, \mathcal{C}_j) = 0$). We also use $\max(|\mathcal{K}_i^c|, |\mathcal{K}_j^c|)$ to normalize such information to concept-level.

Concept Graph Based on Video Order. Video order indicates the dependence between videos. In general, the previous videos in a course are helpful for the latter ones [21] and such dependence is stronger when two videos are closer. Based on this assumption, when calculating the weight for the concept graph $\mathcal{G}^v$ based on video order, we also apply the attenuation coefficient α to obtain edge weight w_e^v for the edge e between concept c_a and c_b:

$$w_e^v = \sum_{u \in U} \sum_{v_i, v_j \in \mathcal{S}_u} \alpha^{|j-i|} \cdot \frac{\mathrm{VO}(u, v_i, v_j)}{\max(|\mathcal{K}_i^v|, |\mathcal{K}_j^v|)}, \tag{7}$$

where function $\mathrm{VO}(v_i, v_j) = 1$ only when 1) v_i, v_j are the i-th, j-th videos of a same course; 2) $c_a \in \mathcal{K}_i^v$ and $c_b \in \mathcal{K}_j^v$ (otherwise $\mathrm{VO}(v_i, v_j) = 0$).

6.2   Prerequisite Relation Learning

After building concept graphs $\mathcal{G}^c$, $\mathcal{G}^v$, and $\mathcal{G}^s$, we utilize GCNs to reason prerequisite relations in these graphs. We first train the embedding representation of all concept nodes via GCNs. And then we feed the graph embeddings of the concept pair (c_a, c_b) into an MLP classifier to predict whether concept c_a is a prerequisite concept of c_b.

In particular, we initialize the adjacency matrix A of the graph and the feature matrix X of the concept nodes for each graph. The adjacency matrix A, with a size of $|\mathcal{K}|^2$, can be derived from edge weights, e.g., for the adjacency matrix A^s of the student behavior graph $\mathcal{G}^s$, we have $A_{ij}^s = w_e^s$, where w_e^s is the weight of edge $e = (c_i \rightarrow c_j)$. And the $|\mathcal{K}| \times d$ sized feature matrix X of the concept nodes in all graphs is initialized by a pre-trained d-dimension language model, i.e., X_i is the word embedding of the text concept c_i.

The training of GCNs on our directed concept graphs follows the propagation rule shown below, which is an adapted version for directed graphs:

$$Z = \hat{D}^{-1} \hat{A} X \Theta, \tag{8}$$

where Θ is a matrix of filter parameters, Z is the convolved signal matrix, $Z_i = h_i$ is the graph embedding of concept c_i, $\hat{D}_{ii} = \sum_j \hat{A}_{ij}$ and the Laplacian is $\hat{A} = I_N + A$. The other settings are the same with [10].

After the graph-training stage, we input the graph embeddings h_a, h_b of a concept pair (c_a, c_b) into a two-layer MLP followed with a sigmoid function to do classification:

$$\Pr(\mathcal{L}(c_a, c_b) = 1) = \sigma(\max(0, (h_a \oplus h_b) \mathbf{W_1}) \mathbf{W_2}), \tag{9}$$

where $\Pr$ is the probability, $\sigma(\cdot)$ is the sigmoid function, $\mathbf{W_1} \in \mathcal{R}^{2d \times d}$ and $\mathbf{W_2} \in \mathcal{R}^{d \times 1}$ are trainable matrices, and $\oplus$ denotes vector concatenation.

6.3   Experiment

To have a clear assessment of the effectiveness of the graph-based modeling of student behavior, we conduct experiments on the newly presented dataset in

Table 2. Overall performance. P, R and $F1$ represent *precision, recall,* and *F1 score* respectively, and Δ represents the improvement of F1 score after adding student behavior features. $^{+sf}$: enhanced with student behavior features. For graph-based models, $^{+cv}$: $\mathcal{G}^c$ and $\mathcal{G}^v$ are used; $^{+s}$: only $\mathcal{G}^s$ is used; $^{+cvs}$: all $\mathcal{G}^c$, $\mathcal{G}^v$ and $\mathcal{G}^s$ are used.

	P	R	$F1$	Δ
MOOC-RF	0.749	0.584	0.656	–
MOOC-RF^{+sf}	0.755	0.639	0.691	**+3.5**
GlobalF	0.679	0.631	0.650	–
GlobalF^{+sf}	0.710	0.657	0.680	**+3.0**
PREREQ	0.468	0.792	0.567	–
PREREQ^{+sf}	0.511	0.712	0.595	**+2.8**
Seq2Seq	0.706	0.743	0.723	–
Seq2Seq^{+sf}	0.707	0.736	0.720	–0.3
GCN^{+cv}	0.789	0.792	0.790	–
GCN^{+s}	0.762	0.784	0.772	–
GCN^{+cvs}	**0.792**	**0.814**	**0.802**	**+1.2**

Sect. 4. Besides, supplemental experiments are carried out on other typical methods with different features and structures to further verify the general accordance of student behaviors and prerequisite structure. The models we select for the supplemental experiments are described as follows:

- **MOOC-RF**: A widely-used method [17], which extracts seven features from the video and subtitle corpus of MOOCs. We reproduce this method and select Random Forest as the classifier to match its claimed best performance.
- **GlobalF**: This method [15] extracts the graph-based features and text-based features for a certain concept pair. The graph-based features are based on Wikipedia Anchor Links, and the text features are based on the description of concepts. We reproduce this method as a typical baseline.
- **PREREQ**: This method [21] utilizes course dependence and video orders to find prerequisite relations through a siamese network.
- **Seq2Seq**: Several efforts have been put forth to utilize neural models to extract prerequisite relations from text, and we reproduce the sequence-to-sequence model in [2] to encode the concepts' texts as prerequisite features.

For enhancing the above models, we concatenate the student behavior features $f_k^{\mathcal{P}}$ ($k = 1, 2, 3, 4$) with original features and then utilize the same classifiers in the respective papers to obtain experimental results. During all the experiments, we apply 10-fold cross-validation and balance the training set by oversampling the positive instances. Table 2 presents a summary of the comparative outcomes obtained from different methods. We analyze the performance in the following aspects: (1) **Student behaviors are effective in prerequisite relation discovery.** MOOC-RF, GlobalF, and PREREQ gain significant improvement after adding the extracted student behavior features. It preliminarily proves that the student behaviors imply clues of prerequisite concepts and are useful to prerequisite relation discovery. (2) **The effectiveness of Student Behavior in Graph Modeling.** GCN^{+s} performs better than the Seq2Seq model and has a competitive performance among all baselines. Further, GCN^{+cvs} performs better than GCN^{+cv}, which indicates that except for the improvement of graph-based modeling, student behavior is still beneficial in advanced attempts of prerequisite relation discovery.

6.4 Analysis of Graph Modeling

We also analyze different components of the graph-based modeling, including:

Necessity of Edge Weights. We set all the edge weights to 1 to convert three concept graphs into unweighted ones and present the corresponding results in Table 3. The performance of all three GCN-based models declines severely, especially the most competitive GCN^{+cvs}, indicating the necessity of edge weights.

Table 3. Absolute performance drops without the edge weights of the concept graphs.

	P	R	F_1
GCN^{+cv}	−8.5	−11.1	−9.9
GCN^{+s}	−7.7	−11.8	−9.8
GCN^{+cvs}	**−10.5**	**−14.5**	**−12.6**

Analysis of Different Behavior Patterns. We further investigate the impact of four behavior patterns by only using some patterns when building graph $\mathcal{G}^s$, the performance changes are shown in Table 4, which provides some insights for understanding the student behaviors: (1) Sequential watching covers high-quality prerequisite concept pairs, resulting in a significant improvement of the precision (P). However, such a pattern is not so effective for discovering prerequisite relations that do not match with the preset order of courses, resulting in a relatively small improvement of recall (R); (2) The other three patterns, which do not follow the preset order of courses, improve recall significantly, indicating that they are effective for discovering prerequisite relations those not covered by sequential watching; (3) Therefore, the four behavior patterns are complementary, and all of them help to discover prerequisite concepts.

Table 4. Performance improvement of GCN^{+cvs} compared with GCN^{+cv} when only using one student behavior pattern while building the concept graph $\mathcal{G}^s$. Seq: sequential watching, Crs: cross course watching, Bck: backward watching, and Skp: skip watching.

Pattern	P	R	F_1
Seq.	**+0.6**	+0.5	+0.5
Crs.	+0.3	+2.1	+1.2
Skp.	+0.1	+2.3	+1.1
Bck.	+0.1	**+2.6**	**+1.3**

7   Conclusion and Future Work

In this work, we perform an investigation on employing the students' video-watching behaviors in prerequisite relation discovery of concepts in MOOCs. To support the study, we collect student behaviors and conduct data annotations to build a novel dataset. After analyzing the typical patterns, we design features for each of them and experimentally verify the student behaviors' effectiveness in enhancing existing models, both feature-based and graph-based.

Based on our investigation, we present several promising future directions, including 1) A more detailed analysis of the relationship between user behavior and prerequisite concepts, e.g., dividing the typical patterns into a finer-grained level for analysis. 2) Prerequisite-driven intelligent applications on real MOOCs. And it is also beneficial to develop interactive applications to collect more kinds of user behavior for prerequisite relation discovery.

Acknowledgement. This work is supported by the National Key R&D Program of China (2020AAA0105203). It also got partial support from the National Natural Science Foundation of China (No. 62277033), National Engineering Laboratory for Cyberlearning and Intelligent Technology, and Beijing Key Lab of Networked Multimedia.

References

1. Agrawal, R., Golshan, B., Papalexakis, E.: Datadriven synthesis of study plans. Data Insights Laboratories (2015)
2. Alzetta, C., et al.: Prerequisite or Not Prerequisite? That's the Problem! An NLP-Based Approach for Concept Prerequisites Learning (2019)
3. Ausubel: Educational Psychology: A Cognitive View (1968)
4. Bergan, J.R., Jeska, P.: An examination of prerequisite relations, positive transfer among learning tasks, and variations in instruction for a seriation hierarchy. Contemp. Educ. Psychol. **5**, 203–215 (1980)
5. Brunskill, E.: Estimating Prerequisite Structure from Noisy Data (2011)
6. Gagne, R.M., Briggs, L.J.: Principles of Instructional Design. Holt, Rinehart & Winston, Austin (1974)

7. Gordon, J., Zhu, L., Galstyan, A., Natarajan, P., Burns, G.: Modeling concept dependencies in a scientific corpus. In: Proceedings of the 54th Annual Meeting of the Association for Computational Linguistics, vol. 1: Long Papers, pp. 866–875 (2016)
8. Huang, X., Yang, K., Lawrence, V.B.: An efficient data mining approach to concept map generation for adaptive learning. In: Industrial Conference on Data Mining (2015)
9. Jia, C., Shen, Y., Tang, Y., Sun, L., Lu, W.: Heterogeneous graph neural networks for concept prerequisite relation learning in educational data. In: Proceedings of the 2021 Conference of the North American Chapter of the Association for Computational Linguistics: Human Language Technologies, pp. 2036–2047 (2021)
10. Kipf, T.N., Welling, M.: Semi-supervised classification with graph convolutional networks. In: International Conference on Learning Representations (ICLR) (2017)
11. Labutov, I., Huang, Y., Brusilovsky, P., He, D.: Semi-supervised techniques for mining learning outcomes and prerequisites. In: SIGKDD. ACM (2017)
12. Li, I., Fabbri, A.R., Tung, R.R., Radev, D.R.: What should i learn first: introducing lecturebank for NLP education and prerequisite chain learning. In: Proceedings of the AAAI Conference on Artificial Intelligence, vol. 33, pp. 6674–6681 (2019)
13. Li, I., Fabbri, A.R., Hingmire, S., Radev, D.: R-VGAE: relational-variational graph autoencoder for unsupervised prerequisite chain learning. In: Proceedings of the 28th International Conference on Computational Linguistics, pp. 1147–1157 (2020)
14. Liang, C., Wu, Z., Huang, W., Giles, C.L.: Measuring prerequisite relations among concepts. In: Proceedings of the 2015 Conference on Empirical Methods in Natural Language Processing, pp. 1668–1674 (2015)
15. Liang, C., Ye, J., Wang, S., Pursel, B., Giles, C.L.: Investigating active learning for concept prerequisite learning. In: Thirty-Second AAAI Conference on Artificial Intelligence (2018)
16. Liang, C., Ye, J., Wu, Z., Pursel, B., Giles, C.L.: Recovering concept prerequisite relations from university course dependencies. In: Thirty-First AAAI Conference on Artificial Intelligence (2017)
17. Pan, L., Li, C., Li, J., Tang, J.: Prerequisite relation learning for concepts in MOOCs. In: Proceedings of the 55th Annual Meeting of the Association for Computational Linguistics, vol. 1: Long Papers, pp. 1447–1456 (2017)
18. Pan, L., Wang, X., Li, C., Li, J., Tang, J.: Course concept extraction in MOOCs via embedding-based graph propagation. In: Proceedings of the Eighth International Joint Conference on Natural Language Processing, vol. 1: Long Papers, pp. 875–884 (2017)
19. Parkay, F.W., Hass, G.: Curriculum Planning: A Contemporary Approach. Allyn & Bacon Incorporated, Boston (1999)
20. Romero, C., Ventura, S.: Educational data mining: a survey from 1995 to 2005. Expert Syst. Appl. **33**, 135–146 (2007)
21. Roy, S., Madhyastha, M., Lawrence, S., Rajan, V.: Inferring concept prerequisite relations from online educational resources. In: Proceedings of the AAAI Conference on Artificial Intelligence, vol. 33, pp. 9589–9594 (2019)
22. Sayyadiharikandeh, M., Gordon, J., Ambite, J.L., Lerman, K.: Finding prerequisite relations using the wikipedia clickstream. In: Companion Proceedings of the 2019 World Wide Web Conference, pp. 1240–1247 (2019)
23. Scheines, R., Silver, E., Goldin, I.M.: Discovering prerequisite relationships among knowledge components. In: EDM, pp. 355–356 (2014)

24. Talukdar, P., Cohen, W.: Crowdsourced comprehension: predicting prerequisite structure in wikipedia. In: Proceedings of the Seventh Workshop on Building Educational Applications Using NLP, pp. 307–315 (2012)
25. Vuong, A., Nixon, T., Towle, B.: A method for finding prerequisites within a curriculum. In: EDM (2011)
26. Yu, J., et al.: Course concept expansion in MOOCs with external knowledge and interactive game. In: Proceedings of the 57th Annual Meeting of the Association for Computational Linguistics, pp. 4292–4302 (2019)
27. Zhu, Y., Zamani, H.: Predicting prerequisite relations for unseen concepts. In: Proceedings of the 2022 Conference on Empirical Methods in Natural Language Processing, pp. 8542–8548 (2022)

Wasserstein Adversarial Variational Autoencoder for Sequential Recommendation

Wenbiao Liu, Xianjin Rong, Yingli Zhong, and Jinghua Zhu[✉]

School of Computer Science and Technology, Heilongjiang University, Harbin 150000, China
2211938@s.hlju.edu.cn, zhujinghua@hlju.edu.cn

Abstract. Variational autoencoders (VAEs) have shown unique advantages as a generative model for sequence recommendation. The core of VAEs is the reconstruction of error targets through similarity metrics to provide a supervised signal for training. However, VAE reconstruction tends to generate non-realistic outputs, which severely affects the accuracy of sequential recommendation. To solve the above problem, in this paper, we propose a new framework called Wasserstein Adversarial Variational Autoencoder (WAVAE) for Sequential Recommendation. In WAVAE, the VAE first combines with the Generative Adversarial Network (GAN) network to differentiate the true and false samples by introducing an adversarially trained discriminative network, and then the learning feature representation in the discriminative network is used as the basis for the VAE reconstruction target so that the VAE tends to generate true samples. We further used Wasserstein loss to optimise the training process, with the aim of avoiding the gradient disappearance problem that occurs when the above-mentioned adversarial network is trained on discrete data, ensuring that the VAE can obtain accurate reconstruction targets through adversarial learning. In addition, we concatenate the original samples with their labels as input to control the generated content and thus control the generated sample tendency. Finally, we conduct experiments on several real datasets to evaluate the model, and the experiment results show that our model outperforms the state-of-the-art baselines significantly.

Keywords: Sequential Recommendation · Variational Autoencoder · Adversarial Learning · Wasserstein Loss

1 Introduction

Sequential Recommender Systems (SRS) [18] dominate a variety of online services. Unlike traditional Collaborative Filtering (CF), SRS considers sequential dependencies while treating user-item interactions as dynamic sequences to capture users' current and recent preferences for more accurate recommendations. In order to make the best use of sequence information to improve SRS, a number of methods have been proposed to capture sequence patterns. Factorizing

X. Song et al. (Eds.): APWeb-WAIM 2023, LNCS 14334, pp. 375–389, 2024.
https://doi.org/10.1007/978-981-97-2421-5_25

Personalised Markov Chain (FPMC) is one of the most classical algorithms for the sequential recommendation, which combines the user's interest and sequence information to predict the next item, but FPMC only captures short-term sequence information and ignores long-term information. GRU4Rec[+] [6] can solve this problem by using a Gated Recurrent to learn the correlation between items in a sequence. In this way, GRU4Rec[+] can effectively capture both long and short-term user behaviour, further improving the performance of the recommendation system. Some other models use Convolutional Neural Networks (CNNs) for sequential recommendations. The classic one is Caser [15] takes the items in each window of a sequence and inputs them into a convolutional layer to obtain an embedding, which is then used to predict the classification of the next item or the next few items. A more direct advantage of CNN over RNN is that it can obtain overall information. SASRec [8] and Bert4Rec [14] alleviate the RNN long-range dependency problem by using the attention mechanism when extracting sequence information from the global, thus obtaining better prediction results.

Recently, Generative Adversarial Networks (GAN) and Variational Autoencoders (VAE) have been very successful as powerful generative models in Computer Vision (CV) and Natural Language Processing (NLP), and they also have achieved great progress in the field of recommender systems. The VAE-based recursive version of the Sequential Variational Autoencoder (SVAE) [11] has received a lot of attention in recent years for its remarkable performance in SRS, and SVAE has powerful sequence reconstruction capabilities with good prediction results. ACVAE [20] introduces adversarial learning into the Adversarial Variational Bayes (AVB) [10] framework to obtain high-quality latent variables through adversarial training, addressing the limitations of the posterior distribution. However, the two state-of-the-art SRS-based VAE models mentioned above share a common problem. The main objective of the VAE generation model is to optimise the KL divergence between the true joint distribution and the generative joint distribution. In traditional VAE, assume the posterior, prior and generative distributions to be independent Gaussian distributions, so the KL divergence between the two joint distributions cannot theoretically be zero, and letting distributions approximate each other will only yield a rough and average result which will lead to the reconstruction of VAE tending to generate unrealistic outputs, resulting in lower quality of the generated samples.

To deal with the above problems, we propose the Wasserstein Adversarial Variational Autoencoder (WAVAE) in this paper. The model solves the above problem by jointly training a VAE and a GAN, where the VAE consists of an encoder and a decoder. The decoder is very much like the generator in GAN. The discriminator in GAN can discriminate the authenticity of the samples, thus helping VAE to generate more realistic samples. Because the data are discrete in recommender systems, adversarial networks are prone to gradient disappearance during training. Wasserstein, with its superior smoothing properties compared to KL divergence [16] and JS divergence, can effectively solve the problem of gradient disappearance in adversarial networks and can guarantee adversarial training, so we use Wasserstein loss [1] to train the adversarial networks. Furthermore, to control the generated content, invoking the idea of Conditional Variational Autoencoder (CVAE) [13], we concatenate the original samples with the corresponding labels to control the generated content. We can generate the desired data by specifying its label when generating the data. Finally, we conducted experiments on three publicly available real datasets to evaluate the performance of our model and to experimentally validate the importance of each component. Our main contributions are listed as follows:

- We propose to combine VAE and GAN for sequential recommendation to improve the quality of generated samples, reduce reconstruction loss and improve recommendation accuracy through adversarial training.
- To avoid the disappearance of gradients in adversarial networks when training discrete data, we propose to use Wasserstein loss training to assist the adversarial network, ensuring that the generator and discriminator are in equilibrium.
- We concatenate the original samples with labels to control the generated content and make the generated samples more realistic.
- The experimental results show that our model outperforms the state-of-the-art baseline on several real datasets. Our extensive ablation study of WAVAE confirms its success stems from all its components.

2   Related Work

2.1   Sequential Recommendation

Sequential recommendation refers to using various modelling methods to mine sequential patterns in user interactions in chronological order and use them to support the recommendation of one or more items at the next moment. RNNs are a common approach to sequence modelling and are widely used for various sequence modelling. GRU4Rec+ [6] introduces RNNs to session-based recommender systems, where interactions within a session are used as sequence history for sequence modelling. Caser [15] points out that RNN-based models can only model point-level sequential patterns, not union-level patterns, while CNN can solve this problem well. The above approach only models sequence interactions but does not store them. When the sequence is long, some past interactions may be forgotten, and RUM [3] introduces a user memory module to store information about sequence interactions. In addition, Transformer has significantly improved NLP tasks, and large pre-trained models such as Bert have been proposed based on Transformer. Bert4Rec [14] introduces such structural ideas to recommender systems. In conclusion, with the successful development of deep learning, especially recurrent neural networks, sequential recommender systems have been developed rapidly.

2.2   VAE for Recommendation

The VAE [5] model is a generative model that uses neural network training to obtain, also known as inferential and generative networks that generate data not contained in the input data. VAE has many variants, such as β-VAE [7], that enhance the ability of the VAE model to represent untangling. InfoVAE [22] can effectively improve the quality of variable follow-up without resorting to the flexibility of decoding distributions. Multi-VAE [9] applies the VAE successfully to the CF problem. WVAE [4] uses the Wasserstein distance to measure the similarity of the distribution of potential attributes and shows its superiority over the evidence of lower bound of KL divergence (ELBO) under mild conditions. SVAE [12] processes the input sequence of items by GRU and finally outputs the candidate probability distributions of the items. All of the above models have made some improvements based on VAE, but none consider the effect of reconstruction error on model performance.

2.3   GAN for Recommendation

The GAN is an important generative model in deep learning, where two networks (generator and discriminator) are trained at the same time and compete in a minimax algorithm. IRGAN [17] brings the GAN, a big hit in CV, to information retrieval for the first time. WGAN [1] introduces the Wasserstein distance to solve the problem of gradient disappearance. CFGAN [2] solves the problems of traditional IRGAN through a series of methods and also optimises collaborative filtering. VAEGAN [21] is a new CF framework base on adversarial VAE. Although many recommendation models are based on GAN networks, most are applied in CF, and further research is needed in the sequential recommendation.

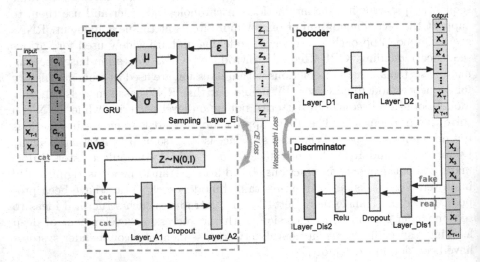

Fig. 1. WAVAE model diagram. It consists of four parts: Encoder, Decoder, Adversarial Variational Bayes network, and Auxiliary Discriminative.

3   Method

In this section, we first introduce the main formulation of the classical VAE and then present our approach, the WAVAE. Figure 1 shows the structure of WAVAE. The WAVAE consists of four main parts: Encoder, Decoder, Adversarial Variational Bayes network, and Auxiliary Discriminative.

3.1   Problem Formulation

$U = \{u_1, u_2, u_3 \cdots\cdots u_n\}$ denotes the set of all users, $V = \{v_1, v_2, v_3 \cdots v_m\}$ denotes the set of all items, where n denotes the number of users and m denotes the number of items. $X_u = \{x_{u,1}, x_{u,2}, x_{u,3} \cdots\cdots x_{u,T_u}\}$ denotes the sequence of u user interactions, $x_{u,i}$ denotes the i-th item that u-user interacts with, T_U denotes the length of the sequence of u user interactions, and the task of the sequence recommendation model is to predict the next item x_u of interest to the user based on the historical interaction information $x_{u,T_{u+1}}$.

3.2   Sequential Variational AutoEncoders

In VAE, the focus is on latent encoding variational inference, and thus VAE provides a suitable framework for latent variable learning and effective Bayes inference. To perform a particular inference, the joint distribution $P_\theta(x, z)$ between the input x and the latent variable z needs to be found. $P_\theta(x, z)$ is the distribution of the input data and its attributes. $P_\theta(x)$ can be calculated from the edge distribution. According to the Bayes theorem, we can obtain:

$$P_\theta(x) = \int P_\theta(z \mid x)P(x)dz. \tag{1}$$

However, the posterior distribution $p_\theta(z \mid x)$ is often difficult to compute. To make $p_\theta(z \mid x)$ easy to handle, VAE introduces a variational inference model (encoder). The goal of VAE is to find an estimable distribution that approximates $p_\theta(z \mid x)$, that an estimate of the conditional distribution of the potential encoding z given an input x. So introducing $q_\phi(z \mid x)$, $q_\phi(z \mid x)$ provides a good estimate of $q_\theta(z \mid x)$. It is both parameterisable and easy to handle. We can obtain:

$$\max_{\theta, \phi} \mathbb{E}_{x \sim p_D(x)}[\mathbb{E}_{z \sim q_\phi(z|x)}[\log p_\theta(x \mid z)] - \mathrm{KL}\left(q_\phi(z \mid x)\|p(z)\right)], \tag{2}$$

where $p_D(x)$ is the data distribution and θ is the generative model parameter, ϕ is the inference model parameter, which is often referred to as evidence lower bound (ELBO).

SVAE is the recursive version of VAE that can be used to model time-aware user preferences, with latent variables capable of expressing temporal dynamics, thus causal and dependency relationships between users' historical preferences. We can model temporal dependencies by conditionally relating each event to previous events. In sequential recommendation tasks, it is important to model temporal dependencies between items. Time dependence can be modelled using conditional probabilities. For a given time step t, the sequential model predicts the $(t+1)$-th term based on the items numbered from 1 to t. We can obtain the SVAE [11]:

$$z_{u(t)} \sim \mathcal{N}(0, I_K), \quad \pi(z_{u(t)}) \sim \exp\{f_\theta(z_{u(t)})\}$$
$$x_{u(t)} \sim \text{Multi}\,(1, \pi(z_{u(t)})), \tag{3}$$

where N denotes the Gaussian distribution, $f_\theta()$ is the neural network, and θ is the parameter.

3.3 Adversarial Variational Bayes

AVB is a technique for training VAE with an arbitrary inference model. We revisit the maximum-likelihood problem by building an auxiliary discriminative model. The inference model and the discriminant model are like two competitors. So AVB essentially combines VAEs and GANs (As shown in the lower left part of Fig. 1). The core equation for the recursive version of VAE is as follows:

$$\max_{\theta,\phi} \mathbb{E}_{x_{u,t} \sim p_D(x_{u,t})} [\mathbb{E}_{z_{u,t} \sim q_\phi(z_{u,t}|x_{u,[1:t]})} [\log p_\theta(x_{u,t+1} \mid z_{u,t})]$$
$$- \text{KL}(q_\phi(z_{u,t} \mid x_{u,[1:t]}) \| p(z_{u,t}))]. \tag{4}$$

The inference model $q_\phi(z_{u,t} \mid x_{u,[1:t]})$ in the above equation is very flexible with respect to the user interaction information x and very restrictive with respect to the latent variable z, a factor that also limits the effectiveness of the generated model. In the AVB, we can avoid using the inference model $q_\phi(z_{u,t} \mid x_{u,[1:t]})$, which is the expression shown, and we train the black box model with the adversarial form to obtain a more flexible expressive power. For KL divergence, the approximate posterior distribution is difficult to handle. However, the KL divergence can be reformulated as:

$$\text{KL}(q_\phi(z_{u,t} \mid x_{u,[1:t]}) \| p(z_{u,t})) = \log q_\phi(z_{u,t} \mid x_{u,[1:t]}) - \log p(z_{u,t}). \tag{5}$$

To compute the Eq. 5, a discriminant network $T^*(x_u, z_u)^{(t)}$ is introduced, which takes the entire sequence x_u and the latent variable z_u and outputs a sequence of values T_u with length [20]. The purpose of the discriminant model is to distinguish whether (x_u, z_u) comes from $p_D(x_u)p(z_u)$ or from our current inference model $p_D(x_u)q_\phi(z_u \mid x_u)$. $p_D(x_u)p(z_u)$ is the true value and $p_D(x_u)q_\phi(z_u \mid x_u)$ obtained from the model. The $p(z)$ true exists in the distribution behind the data, which we cannot obtain, and we try to approximate $p(z_u)$ by $q_\phi(z_u \mid x_u)$. The discriminant network $T^*(x_u, z_u)^{(t)}$ objectives are as follows:

$$\max_{\Psi} \sum_{t=1}^{T_u} [\mathbb{E}_{x_u \sim P_D(x_u)} \mathbb{E}_{z_u \sim Q_\phi(z_u|x_u)} \log(\sigma(T_\Psi(x_u, z_u)^{(t)}))$$
$$+ \mathbb{E}_{x_u \sim P_D(x_u)} \mathbb{E}_{z_u \sim P(z_u)} \log(1 - \sigma(T_\Psi(x_u, z_u)^{(t)}))], \tag{6}$$

where σ is the sigmoid and Ψ is the network parameter. To make the backpropagation work smoothly, we use the reparameterization technique [20]. Specifically, we define the function $z_\phi(x, \epsilon) = f(f(x, \epsilon_1), \epsilon_2))$ where f denotes a non-linear function and ϵ_1, ϵ_2 is sampled from the Gaussian distribution. Thus the Eq. (4) can be expressed as:

$$\max_{\theta, \phi} \sum_{t=1}^{T_u} \mathbb{E}_{x_{u,t} \sim p_\mathcal{D}(x_{u,t})} \mathbb{E}_{\varepsilon \sim \mathcal{N}(0,I)} [\log p_\theta \left(x_{u,t+1} \mid z_{u,t} \right) \qquad (7)$$
$$- T_\Psi^* \left(x_u, z_\phi(x_u, \epsilon) \right)^{(t)}].$$

3.4   Auxiliary Discriminative Network

We combine VAE and GAN and introduce an auxiliary discriminative network D for adversarial training to reduce reconstruction losses in the user-item interaction sequence. A nice feature of the GAN is that its discriminator network implicitly learns sample similarity measures to distinguish them from fake samples. We, therefore, propose to exploit this observation to translate the properties of the samples learned by the discriminator into the more abstract reconstruction error of the VAE. This combines the advantages of GAN as a high-quality generative model with the benefits of VAE as a method of encoding data into a latent space. We view VAE as the generator in a GAN network for generating the most similar and plausible samples to the real data and D as the discriminator that attempts to distinguish the real samples from the generated samples as much as possible. The generator and discriminator are in a state of mutual play to generate samples closer to the real data.

Specifically, we input x_u as true samples and x'_u from the decoder as false samples into the discriminator and use the gradient from D to guide the VAE to reduce the reconstruction loss between true and false samples to obtain a more accurate reconstruction result (As shown in the lower right part of Fig. 1). The click probability of interactive items is predicted to be 1, and the click probability of non-interactive items is set to 0. In a recommendation system, we should pay attention to the recommendation of non-interactive items. For VAE to have a certain prediction probability for non-interactive items, we set the reconstruction prediction probability of non-interactive items to 0 before discriminator D for true-false samples input. By setting this, the discriminator D can distinguish true-false samples according to the interactive items, and VAE does not get the gradient of non-interactive items from D, ensuring that VAE has a certain prediction probability for non-interactive items. The auxiliary discriminant network D objective function is as:

$$\max_{\xi} \sum_{t=1}^{T_u} [\mathbb{E}_{x_u \sim P_\mathcal{D}(x_u)} \mathbb{E}_{z_u \sim Q_\phi(z_u|x_u)} \log(\sigma(D_\xi \left(x_u \mid z_u \right)^{(t)}))$$
$$+ \mathbb{E}_{x'_u \sim P_\theta(x_u|z_u)} \mathbb{E}_{z_u \sim Q_\phi(z_u|x_u)} \log(1 - \sigma(D_\xi \left(x'_u \mid z_u \right)^{(t)}))], \qquad (8)$$

where ξ is a discriminative network parameter, which can be obtained by training. The objective function of the generator G is expressed as:

$$\min_{\theta,\xi} \mathbb{E}_{\boldsymbol{x'}_u \sim P_\theta(\boldsymbol{x}_u|z_u)} \mathbb{E}_{z_u \sim Q_\phi(z_u|\boldsymbol{x}_u)} \log(1 - \sigma(D_\xi\left(\boldsymbol{x'}_u \mid z_u\right)^{(t)}))]. \tag{9}$$

3.5 Wasserstein Loss

We can equivalently transform the loss of the generator to minimize the JS divergence before the true distribution and the generated distribution. When the discriminator reaches optimality, the JS divergence between the true and generative distributions is minimized [1]. However, suppose there is no or negligible overlap between the two distributions. In that case, this will make the JS divergence a constant, which means that the gradient of the generator approximates zero, causing the gradient to vanish. The dataset of users and items in the recommendation task is discrete, the generation of discrete data needs to be sampled according to probability and the sampling process is not derivable, which can lead to the discriminator not being able to Backpropagation (bp) the gradient to the generator. So the model in this paper suffers from the same problem of gradient disappearance. To solve this problem, we introduce the Wasserstein distance, defined as follows:

$$W\left(P_r, P_g\right) = \inf_{\gamma \sim \Pi(P_r, P_g)} \mathbb{E}_{(\boldsymbol{x},\boldsymbol{y})\sim\gamma}[\|\boldsymbol{x} - \boldsymbol{y}\|], \tag{10}$$

where $\gamma \sim \Pi\left(P_r, P_g\right)$ denotes the set of joint distributions of the combination of the true distribution P_r and the generated distribution P_g, for each possible joint distribution γ, the true sample $\boldsymbol{x}$ and the generated sample $\boldsymbol{y}$ can be sampled, and the distance between the pair is calculated. The expected value $\mathbb{E}_{(\boldsymbol{x},\boldsymbol{y})\sim\gamma}[\|\boldsymbol{x} - \boldsymbol{y}\|]$ of the sample pair under the joint distribution γ is then calculated, and a lower bound is taken on the expected value, which is the Wasserstein distance. Wasserstein distance ensures that adversarial training of discrete data can continue. So Eq. (8) can be expressed as:

$$\max_{\xi} \sum_{t=1}^{T_u} [\mathbb{E}_{\boldsymbol{x}_u \sim P_D(\boldsymbol{x}_u)} \mathbb{E}_{z_u \sim Q_\phi(z_u|\boldsymbol{x}_u)} D_\xi\left(\boldsymbol{x}_u \mid z_u\right)^{(t)}$$
$$+ \mathbb{E}_{\boldsymbol{x'}_u \sim P_\theta(\boldsymbol{x}_u|z_u)} \mathbb{E}_{z_u \sim Q_\phi(z_u|\boldsymbol{x}_u)}(1 - D_\xi\left(\boldsymbol{x'}_u \mid z_u\right)^{(t)})]. \tag{11}$$

3.6 Conditional Data

One of the problems with generating data with VAE is that we do not have any control over the type of data generated, which means that items may be generated that do not belong in the item list. This results in the generated sample being too far removed from the true sample, thus reducing the accuracy of the prediction. To generate items that match the candidate items, then a conditional input needs to be added to the VAE. We were inspired by CVAE [13] to concatenate the generated samples with their labels, making the samples generated by the variational autoencoder controllable and thus improving the recommendation accuracy.

Specifically, the original data x_u is processed by $one-hot$ encoding to obtain the label c_u [19]. Then the original data and the label are concatenated to obtain the input data, fed into the encoder. Thus the Eq. (8) can be expressed as:

$$\max_{\theta,\phi} \sum_{t=1}^{T_u} \mathbb{E}_{x_{u,t}\sim p_{\mathcal{D}}(x_{u,t})}\mathbb{E}_{\varepsilon\sim\mathcal{N}(0,I)}[\log p_\theta\,(x_{u,t+1}\mid z_{u,t})$$
$$-T_\Psi^*\,(\{x_u,c\},z_\phi(\{x_u,c\}\epsilon))^{(t)}], \qquad (12)$$

where $\{\ \}$ denotes the concatenate. It should be noted that c_u is not involved in the loss calculation.

3.7 Objective Functions

Overall, the ultimate objectives of WAVAE are as follows:

$$J^G : \max_{\theta,\phi,\xi} \sum_{t=1}^{T_u} \mathbb{E}_{x_{u,t}\sim p_{\mathcal{D}}(x_{u,t})}\mathbb{E}_{\varepsilon\sim\mathcal{N}(0,I)}$$
$$[(\log P_\theta\,(x_{u,t+1}\mid z_{u,t}) - \alpha\cdot T_\phi^*\,(\{x_u,c\}z_\phi(\{x_u,c\},\varepsilon)))]$$
$$+\,\mathbb{E}_{x'_u\sim P_{\mathcal{D}}(x_u)}\mathbb{E}_{z_u\sim Q_\xi(z_u\mid x_u)}\beta\cdot(D\,(x'\mid z_u)),$$

$$J^T : \max_{\Psi} \sum_{t=1}^{T_u}[\mathbb{E}_{x_u,\sim P_{\mathcal{D}}(x_u)}\mathbb{E}_{z_u\sim Q_\phi(z_u\mid x_u)}\log(\sigma(T_\Psi\,(\{x_u,c\},z_u)^{(t)}))$$
$$+\,\mathbb{E}_{\{x_uc\}\sim P_{\mathcal{D}}(x_u)}\mathbb{E}_{z_u\sim P(z_u)}\log(1-\sigma(T_\Psi\,(x_u,z_u)^{(t)}))],$$

$$J^D : \max_{\xi} \sum_{t=1}^{T_u}[\mathbb{E}_{x_u\sim P_{\mathcal{D}}(x_u)}\mathbb{E}_{z_u\sim Q_\phi(z_u\mid x_u)}D_\xi\,(x_u\mid z_u)^{(t)}$$
$$+\,\mathbb{E}_{x'_u\sim P_\theta(x_u\mid z_u)}\mathbb{E}_{z_u\sim Q_\phi(z_u\mid x_u)}(1-D_\xi\,(x'_u\mid z_u)^{(t)})], \qquad (13)$$

where J^G denotes the objective function of the encoder and decoder, J^T denotes the objective function of the Adversarial Variational Bayes, and J^D denotes the auxiliary discriminative network. α and β are hyper-parameters that control the weights of the Adversarial Variational Bayes and auxiliary discriminative network, respectively. We experimentally investigate these two parameters in Sect. 4.6.

4 Experiments

4.1 Datasets

To demonstrate the validity of WAVAE, we conduct experiments on three real datasets of different sizes.

- **MovieLens Latest (ML-latest)**: This contains information on 1,682 movies rated by 943 users, including the latest movie ratings and detailed times-tamps.
- **MovieLens 1m (ML-1m)**: This contains approximately 1 million ratings for 3,900 movies from 6,040 unique users, a common dataset for sequential recommendation systems.
- **MovieLens 10m (ML-10m)**: This is the larger version of ML-1m and con-tains 10 million ratings from 7,200 users for 10,000 movies.

For each user, we partition the interaction sequence into 8:2 for training and testing. As forecasts, we use the three most widely used top-K valuation metrics, NDCG, Recall and MRR.

4.2 Baselines

We compare our approach with nine representative SR models:

- **POP**: POP ranks users based on the number of interactions and recommends the most popular interactions.
- **FPMC**: FPMC combines the user's interests and sequence information to make a prediction about the next item.
- **Caser** [15]: Caser uses CNNs to extract information about short-term sequences.
- **GRU4Rec+** [6]: GRU4Rec+ application of RNN (GRU) to session recom-mendation, the core idea is that the user's behaviour of clicking on a series of items in a session is seen as a sequence.
- **BERT4Rec** [14]: BERT4Rec uses deep bi-directional self-attention to model sequences of user behaviour.
- **CFGAN** [2]: Introducing collaborative filtering to adversarial networks for recommender systems.
- **Mult-VAE** [9]: VAE for collaborative filtering, work on learning user repre-sentations in recommender systems using VAE.
- **SVAE** [12]: Sequence Variable Autoencoder use GRU and VAE to generate the target sequence.
- **ACVAE** [20]: ACVAE introduces adversarial training for sequence genera-tion with an Adversarial Variational Bayes (AVB) framework.

4.3 Implementation Details

We implement WAVAE using PyTorch and determine the structure and param-eters based on its test metrics. Specifically, we apply dropout at the input layer with a probability of 0.5 to prevent overfitting. We optimise the VAE using Adam and the AVB and auxiliary discriminant network using SGD. If it is less than a fixed length, add 0 to the end of the sequence. We train the model on ML-latest for 350 cycles, on ML-1m for 250 cycles and ML-10m for 150 cycles. Other key hyper-parameters are discussed further below. The source code of WAVAE is available on GitHub[1].

---

[1] https://github.com/WAVAE/WAVAE-PyTorch.

4.4 Performance Comparison

We compare WAVAE with baselines, and the experimental results for each baseline are shown in Table 1, where WAVAE outperforms the other methods in all evaluation metrics.

Table 1. Comparison of WAVAE with other baselines on top-k, $k \in \{5,10,20\}$. Results are in bold for the highest scores and underlined for the highest scores other than the highest (in percentages).

Dataset	Metric	POP	FPMC	Caser	GRU4Rec$^+$	CFGAN	Mult-VAE	BERT4REC	SVAE	ACVAE	WAVAE
ML-latest	N@5	5.69	6.83	4.91	7.12	6.51	6.34	5.64	6.52	_8.11_	**8.86**
	N@10	5.51	7.73	6.04	7.58	7.00	6.66	5.75	7.43	_9.21_	10.19
	N@20	6.23	9.13	7.74	8.61	8.11	80.5	6.63	8.93	_10.72_	11.48
	R@5	2.83	4.43	2.71	3,94	3.01	2.94	3.83	4.36	_5.21_	**5.56**
	R@10	4.51	7.73	5.82	6.54	6.03	6.05	5.93	7.53	_9.13_	9.33
	R@20	7.03	12.06	10.71	10.54	9.84	11.16	8.73	12.03	_14.52_	15.22
	M@5	11.43	13.21	9.75	13.63	12.31	12.85	11.25	12.35	_15.53_	16.48
	M@10	12.31	15.12	11.61	15.13	13.93	14.93	12.22	13.94	_17.01_	18.14
	M@20	13.00	16.15	12.93	16.01	14.89	15.43	12.94	14.81	_17.86_	18.76
ML-1m	N@5	7.77	13.33	13.89	12.52	7.31	6.02	10.41	15.29	_18.75_	19.53
	N@10	7.62	14.13	14.62	12.71	7.93	6.93	11.03	15.73	_19.19_	19.82
	N@20	8.71	15.82	16.42	14.13	8.43	8.55	12.78	17.68	_21.22_	21.76
	R@5	2.43	6.51	6.66	6.45	3.32	2.91	5.38	7.76	_9.48_	9.83
	R@10	4.31	11.33	11.44	10.45	6.23	6.31	9.23	12.93	_15.81_	16.21
	R@20	8.81	18.03	18.63	16.53	11.13	11.43	15.41	20.74	_24.33_	24.75
	M@5	14.12	24.46	24.81	22.81	14.02	11.63	19.58	26.74	_31.99_	33.43
	M@10	15.55	26.71	27.13	24.73	16.04	13.91	21.83	28.93	_33.91_	35.36
	M@20	16.81	27.73	28.13	25.69	17.25	15.23	23.01	29.89	_34.83_	36.28
ML-10m	N@5	5.61	9.03	6.54	12.23	6.91	8.23	12.93	17.33	_17.73_	18.11
	N@10	6.03	10.11	7.24	13.28	7.43	9.41	14.41	18.39	_18.96_	19.23
	N@20	6.92	12.31	8.83	15.27	8.43	11.75	16.94	20.79	_21.53_	21.83
	R@5	3.21	5.81	3.74	8.03	3.91	4.89	8.44	10.93	_11.22_	11.64
	R@10	5.53	10.09	6.72	13.06	6.78	9.21	14.36	17.22	_18.03_	18.76
	R@20	8.88	16.93	11.82	19.99	10.76	16.63	22.84	26.01	_27.31_	27.89
	M@5	10.21	16.31	12.51	21.43	12.81	15.23	22.31	29.53	_30.21_	31.13
	M@10	11.43	18.31	14.26	23.21	14.31	17.35	24.46	31.61	_32.15_	32.85
	M@20	12.23	19.51	15.38	24.16	15.34	18.61	25.54	32.51	_33.07_	33.73

Firstly, there is no doubt that almost all models outperforms the traditional models POP and FPMC. This is because deep learning-based models are more capable of learning and are better suited to recommender systems. FPMC can also be found to exceed POP, as POP is not a sequential recommendation method. This proves that sequential behavioural information is more helpful in predicting user interests.

Secondly, GRU4Rec$^+$ and Caser are both more classical deep learning-based methods. Gru4Rec$^+$ can effectively capture both long and short-term information about users. Caser is a CNN-based model with the advantage of obtaining global information through experimental results. We found through experiments that WAVAE outperforms both GRU4Rec$^+$ and Caser. Compared with BERT4Rec, WAVAE achieve better results despite considering only one-way information and not utilizing global attention mechanisms. All of which demonstrates the strong predictive power of the generative model. Furthermore, CFGAN uses the GAN network structure for collaborative filtering, which cannot capture the dynamic changes in user interest, so the experimental results are inferior to our model.

Finally, our model also achieves significant improvements compared to the VAE-based approach. In the dataset ML-latest, we improve (mean value) 38% NDCG, 25% Recall, and 30% MRR relative to SVAE. SVAE introduces GRU to generate target sequences but does not consider the effect of high-quality latent variables on the model. Both our model and ACVAE experiments outperform SVAE, which could demonstrate the effect of high-quality latent variables on the model. However, SVAE and ACVAE do not consider the effect of reconstruction error on model performance. Our model improves recommendation performance by addressing the VAE model's reconstruction tendency to produce non-real data. We outperform the strongest baseline ACVAE in all metrics across all datasets (up to 11% improvement and down to 3% improvement). This demonstrates that accurate reconstruction propensity is the key to good results in sequence-oriented generative models.

4.5 Ablation Study

To verify the impact of the individual modules on the overall performance, an ablation study was carried out on WAVAE. We chose NDCG@10 as the evaluation metric.

Effect of Conditional Data. Figure 2 shows the effect of original samples and label concatenate on the overall model. The results for the model without concatenating labels were worse than for WAVAE in all metrics. This demonstrates that concatenating labels can control generative convergence.

Effect of Auxiliary Discriminative Network. We remove the discriminant network without affecting the training of the VAE model. We can see from Fig. 2 that the model performs better with adding the auxiliary discriminant network D. The discriminant network D uses a black-box neural network to accurately measure the distance between the input data distribution and the reconstructed data. Through adversarial training, the two distributions get closer and closer, letting VAE converge to generate real data.

Effect of AVB. We remove the AVB part without changing the rest of the structure. From Fig. 2, WAVAE obtains better results than the model without AVB. This is because AVB uses a flexible black-box inference model and adversarial training to enhance the expressiveness of the model, resulting in a method closer to the true posterior distribution.

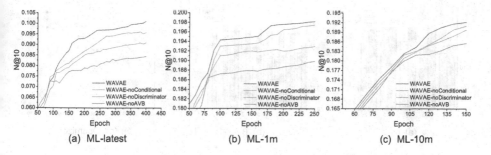

Fig. 2. Ablation study on ML-latest, ML-1m and ML-10m.

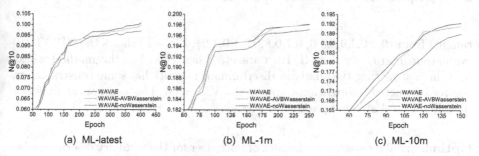

Fig. 3. Ablation study (effect of Wasserstein loss) on ML-latest, ML-1m and ML-10m.

Effect of Wasserstein Loss. From Fig. 3, we can see that adding Wasserstein loss on the auxiliary adversarial network works better, which indicates that in our model, introducing Wasserstein loss prevents not only the gradient of the adversarial network from vanishing but also ensures the diversity of the generated samples. In addition, the experiments show that there is no significant effect of using Wasserstein in AVB, because there is a fusion of real data between the true and false samples input to the AVB discriminator, so the distribution between the true and false samples will have overlapping parts, so there is no gradient disappearance problem.

4.6   Parameter Effects

In this section, we focus on the hyperparameters and optimisers in the approach of this paper.

Hyper-parameters α. To further investigate the effect of AVB on the overall model, we tested the hyper-parameters α ranging from $\{0.0,0.05,0.1,0.15,0.2\}$, Fig. 4 (a) shows the NDCG@10 evaluation metrics in the ML-1m dataset. Our optimal model effect is achieved when $\alpha = 0.05$, it is shown that adversarial training can constrain the expression of latent variables and thus prevent overfitting. The effect worsens when α more than 0.05 or smaller than 0.05.

Hyper-parameters β. We also do a corresponding study on the weight coefficients of the auxiliary discriminant network, and we test the hyper-parameters β

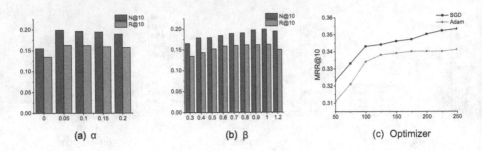

Fig. 4. (a) and (b) show the performance of different settings of hyper-parameters α and hyper-parameters β. (c) shows the performance of different optimizers.

ranging from {0.3,0.4,0.5,0.6, 0.7,0.8,0.9,1.0,1.2}, Fig. 4(b) shows the NDCG@10 evaluation metrics on the ML-1m dataset. When $\beta = 1.0$, the method works best, indicating that the auxiliary discriminant network has some constraints on the reconstructed information to make the VAE true. This component may not be fully utilised at other parameter settings.

Optimizer. We research the impact of optimizer for the auxiliary discriminative network and the effect of SGD and Adam optimizers on the model separately. The experimental results Fig. 4(c) show that using a momentum-based algorithm such as Adam is unsuitable for discriminative networks with unstable loss gradients. At the same time, SGD is suitable for the case of unstable gradients.

5   Conclusions

In this paper, we propose a novel sequential recommendation model WAVAE. The inference model is based on the reconstruction loss of GAN to provide training signals for VAE, which solves the characteristics that the reconstruction of VAE models for sequential recommendation tends to be non-realistic. To prevent the gradient of the adversarial network from vanishing, we introduce the Wasserstein loss. In addition, we splice the samples with labels to control the generated content. Experiments show that WAVAE achieves a significant improvement compared with state-of-the-art models.

Acknowledgements. This work was supported by the Natural Science Foundation of Heilongjiang Province of China, LH2022F045.

References

1. Adler, J., Lunz, S.: Banach wasserstein gan. Adv. Neural Inf. Process. Syst. (2018)
2. Chae, D.K., Kang, J.S., Kim, S.W.: CFGAN: a generic collaborative filtering framework based on generative adversarial networks. In: CIKM, pp. 137–146 (2018)

3. Chen, X., et al.: Sequential recommendation with user memory networks. In: WSDM, pp. 108–116 (2018)

4. Chen, Z., Liu, P.: Towards better data augmentation using wasserstein distance in variational auto-encoder. arXiv preprint arXiv:2109.14795 (2021)

5. Ezukwoke, K., Hoayek, A., Batton-Hubert, M., Boucher, X.: GCVAE: generalized-controllable variational autoencoder (2022)

6. Hidasi, B., Karatzoglou, A., Baltrunas, L., Tikk, D.: Session-based recommendations with recurrent neural networks. arXiv preprint arXiv:1511.06939 (2015)

7. Higgins, I., et al.: Beta-vae: learning basic visual concepts with a constrained variational framework (2016)

8. Kang, W.C., McAuley, J.: Self-attentive sequential recommendation. In: ICDM, pp. 197–206. IEEE (2018)

9. Liang, D., Krishnan, R.G., Hoffman, M.D., Jebara, T.: Variational autoencoders for collaborative filtering. In: Proceedings of the 2018 World Wide Web Conference, pp. 689–698 (2018)

10. Mescheder, L., Nowozin, S., Geiger, A.: Adversarial variational bayes: unifying variational autoencoders and generative adversarial networks. In: ICML, pp. 2391–2400. PMLR (2017)

11. Miladinović, D., Shridhar, K., Jain, K., Paulus, M.B., Buhmann, J.M., Allen, C.: Learning to drop out: an adversarial approach to training sequence VAEs (2022)

12. Sachdeva, N., Manco, G., Ritacco, E., Pudi, V.: Sequential variational autoencoders for collaborative filtering. In: WSDM, pp. 600–608 (2019)

13. Sohn, K., Lee, H.: Learning structured output representation using deep conditional generative models. Adv. Neural Inf. Process. Syst. (2015)

14. Sun, F., et al.: Bert4rec: sequential recommendation with bidirectional encoder representations from transformer. In: CIKM, pp. 1441–1450 (2019)

15. Tang, J., Wang, K.: Personalized top-n sequential recommendation via convolutional sequence embedding. In: WSDM, pp. 565–573 (2018)

16. Van Den Oord, A., Kalchbrenner, N.: Pixel recurrent neural networks. In: International Conference on Machine Learning, pp. 1747–1756. PMLR (2016)

17. Wang, J., et al.: A minimax game for unifying generative and discriminative information retrieval models. In: SIGIR Proceedings (2018)

18. Wang, S., Hu, L., Wang, Y., Cao, L., Sheng, Q.Z., Orgun, M.A.: Sequential Recommender Systems - Challenges, Progress and Prospects, pp. 6332–6338 (2019)

19. Wang, X., et al.: Normative modeling via conditional variational autoencoder and adversarial learning to identify brain dysfunction in Alzheimer's disease (2022)

20. Xie, Z., Liu, C., Zhang, Y., Lu, H., Wang, D., Ding, Y.: Adversarial and contrastive variational autoencoder for sequential recommendation. In: Proceedings of the Web Conference 2021, pp. 449–459 (2021)

21. Yu, X., Zhang, X., Cao, Y., Xia, M.: Vaegan: a collaborative filtering framework based on adversarial variational autoencoders. In: Twenty-Eighth International Joint Conference on Artificial Intelligence IJCAI-19 (2019)

22. Zhao, S., Song, J., Ermon, S.: Infovae: Balancing learning and inference in variational autoencoders, pp. 5885–5892 (2019)

Fine-Grained Category Generation
for Sets of Entities

Yexing Du[1], Jifan Yu[2], Jing Wan[1(✉)], Jianjun Xu[3], and Lei Hou[2]

[1] Beijing University of Chemical Technology, Beijing 100029, China
wanj@mail.buct.edu.cn
[2] Tsinghua University, Beijing 100084, China
[3] Beijing Caizhi Technology Co., Ltd., Beijing 100081, China

Abstract. Category systems play an essential role in knowledge bases by groupings of semantically related entities. Category generation task aims to produce category suggestions which can help knowledge editors to expand a category system. Most past research has focused on solving coarse-grained problems, not fine-grained scenarios. In this paper, we propose a two-stage framework to generate fine-grained categories for sets of entities. In the category generation stage, we extract conceptual texts from the context of entities and then employ the Seq2Seq model to generate candidate categories. In the category selection stage, we cluster the entities and design discrete patterns using entity names for prompt ranking, which are further ensembled to preserve the final categories. We construct a new fine-grained category generation dataset based on Wikipedia. Experimental results demonstrate the effectiveness of the framework over the state-of-the-art abstractive summarization methods.

Keywords: Category generation · Fine-grained · Prompt

1 Introduction

Category systems play a crucial role in building knowledge bases, and the constant emergence of new entities leads to the need for constructing new categories. In contrast to coarse-grained categories, We focus on fine-grained categories, which have longer and more complex names (such as *Minesweepers of the United Kingdom* vs. *Minesweeper*). These categories can reflect human classifications and entity relations, and have irreplaceable value for various tasks, including knowledge acquisition [19] and entity set expansion [34]. However, manual updating of a category system by knowledge editors can be costly. Therefore, there is a clear need for automatic generation of category suggestions.

The fine-grained category generation task aims to produce category suggestions for a set of entities along with their surrounding context. Currently, a large portion of relevant research is based on fine-grained entity typing [5,7,21], which assumes that suitable categories already exist. However, this assumption does not always hold true in practical scenarios. The problem of creating new categories for existing category systems has not received sufficient attention.

© The Author(s), under exclusive license to Springer Nature Singapore Pte Ltd. 2024
X. Song et al. (Eds.): APWeb-WAIM 2023, LNCS 14334, pp. 390–405, 2024.
https://doi.org/10.1007/978-981-97-2421-5_26

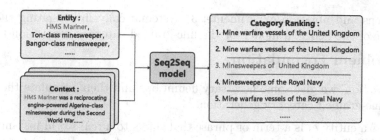

Fig. 1. Given a set of entities along with context (surrounding text), we aim to generate category suggestions. The blue labels indicate the categories of entities in Wikipedia. (Color figure online)

Inspired by recent Seq2Seq models [13,23,32], the task of generating new categories can be viewed as an abstract summarization task [33]. Figure 1 shows an example of fine-grained category generation. However, this process suffers from three challenges: (1) **Noise** - Category creation should not be solely based on individual entities. Directly feeding raw data into the model will introduce much noisy information about individual entities. (2) **Drift** - Entities and context are mixed input into the model, which can lead to ineffective utilization of entity features. As a result, the category suggestions may deviate from the entities. (3) **Repetition** - The model relies on a search algorithm to rank the candidate. Sampling strategies like beam search [24] can result in repetition, where the same category is generated multiple times.

To address the above problem, we design a two-stage category generation framework that includes three new modules:

1) **Conceptual Text Mining:** This module is aimed at reducing noise by extracting common features, making it easier for the model to capture the essential parts of the input. After extracting the conceptual text, it is fed into the Seq2Seq model to generate candidate categories.

2) **Prompt Ranking:** To address the drift problem caused by entity mixing, we design a prompt ranking module to associate entities with categories. Inspired by prompt paradigm, we construct patterns and use prompt [4,15] to rank the categories by perplexity [10]. This helps to ensure that the category suggestions are more closely related to the entities.

3) **Ensemble Ranking:** This module alleviates repetition problems by integrating multiple models' parsing abilities. Ensemble ranking [27] combines the results of candidate ranking and prompt ranking, employing a dynamic rule design that preserves diversity while alleviating repetition.

For evaluation, we construct a new dataset from Wikipedia, which consists of 23.8M entities and context, 1.1M categories, and 5.3K fine-grained categories. We compare it with five abstractive summarization methods. Additionally, we conduct ablation experiments to evaluate the contribution of three modules.

Contributions. Our contributions include: 1) proposing a two-stage framework that uses a multi-model ensemble for fine-grained category generation; 2) constructing a new dataset based on Wikipedia and conducting five baseline evaluations and ablation experiments; 3) systematically investigating the performance of the Seq2Seq model for the fine-grained category generation task.

2   Problem Formulation

In this section, we give some necessary definitions and then formulate the problem of fine-grained category generation.

Entity. An entity e_i is a term or phrase that refers to a real-world instance, e.g. entity *HMS Peterhead (J59)* is a minesweeper built for the Royal Navy during the Second World War. A set of entities of the same category constitutes an entity set, denoted as $E = \{e_i\}_{i=1}^{n}$.

Context. The context a_i refers to a piece of text that contains the entity e_i. A set of context text constitutes a text set, denoted as $\mathcal{A} = \{a_i\}_{i=1}^{n}$.

Category. A category c_i is the set of entities with particular shared characteristics. Fine-grained Categories usually have complex names compared to coarse-grained categories, which contain more features, such as people, items, attributes, and semantic relations.

Fine-Grained Category Generation. Given a set of Entities E and its contexts $\mathcal{A}$, this task is to generate a ranked list of fine-grained categories suggestions $\mathcal{C}^e$.

3   Method

To generate fine-grained categories, we need to address two crucial problems: 1) How to effectively generate enormous fine-grained candidate categories (for

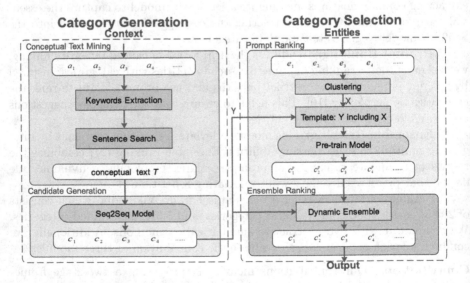

Fig. 2. Our framework for fine-grained category generation which adds three modules: (1) **Conceptual Text Mining**; (2) **Prompt Ranking**; (3) **Ensemble Ranking**. Y is the candidate category, and X is the entity after clustering.

recall)? 2) How to choose high-quality categories with language model probing (for precision)? As shown in Fig. 2, we introduce our two-stage framework:

Category Generation: To generate a large number of candidate categories, we first filter data by extracting conceptual texts among context $\mathcal{A}$ and then employ the Seq2Seq language model to generate a candidate list $\mathcal{C}$.

Category Selection: To select high-quality categories from the candidate list. We design discrete templates to get prompt ranking $\mathcal{C}^p$. These generated results are further ensemble ranked to preserve final results $\mathcal{C}^e$.

3.1   Category Generation

The category generation stage can be considered as a type of text summarization process. Recent studies have shown that proper extraction of input texts can greatly enhance the final generation results [17]. Therefore, we use a pre-processing technique called conceptual text mining to filter out texts. Then, the conceptual text is inputted into a Seq2Seq model to generate the category.

Conceptual Text Mining. This pre-processing module aims to reduce noise in the contextual text. The given context set $\mathcal{A}$ may contain descriptive text that pertains only to specific entities, which can introduce noise when generating categories. Meanwhile, the large amount of texts may exceed the input capacity of existing language models. To address these issues, we propose a method for filtering out low-value text using unsupervised metrics.

We follow the idea of **Information Density** [1], an evaluation metric sourced from pedagogy, to preserve the text $\mathcal{T}$ with valuable information. Specifically, we count all the keywords in the context set $\mathcal{A}$ and filtered out the sentences that only contain high-frequency keywords, which can be formally presented in the following metric:

$$\mathcal{T} = \left(\sum s_{i,j} \mid k_l \in s_{i,j}, f_{k_l,\mathcal{A}} > \Theta \right), \tag{1}$$

where k_l is the keyword of $\mathcal{A}$, $s_{i,j}$ denotes the j-th sentence in the i-th context. $k_l \in s_{i,j}$ means that k_l is in the sentence $s_{i,j}$. Θ is a is a predefined threshold for keyword occurrence frequency[1] for filtering, and $f_{k_l,\mathcal{A}}$ indicates the frequency of keywords k_l in $\mathcal{A}$. In particular, we conduct this conceptual text mining stage with two steps:

1) Keyword extraction: Given a set of context text $\mathcal{A}$, we employ a keyword extraction algorithm, such as TF-IDF [25], to extract the keywords of each context. Then we count the number of occurrences for each keyword and filter out the ones that appear less frequently.

2) Sentence search: We search for sentences in a given context text that contain specific keywords. We merge the sentences into a text $\mathcal{T}$, which is called as conceptual text.

---

[1] In the experiment, the best result is achieved on 0.15.

Candidate Generation. In the task of summarization, the dataset typically consists of (document, summary) pairs, while for category generation training, we use a dataset where each example (x, y) is a (conceptual text $\mathcal{T}$, category) pair. During the Seq2Seq fine-tuning process, the model is trained using the standard cross-entropy loss [30]:

$$L_{Data} = -\sum_{t=1}^{T} \log p(y_{t+1} \mid y_{1:t}, x), \tag{2}$$

where T is the target sequence length and p is the model's predicted probability for the correct word.

For the generation of candidate categories, we employ beam search as the search algorithm, which is used to find the most probable textual sequence in the set of candidate sequences. In the algorithm, each candidate sequence has a corresponding probability value, referred to as the "confidence" of the sequence. Specifically, given an input sequence x and a target sequence y, the sequence confidence s is defined as:

$$s = logP(y|x) = \sum_{n=1}^{T} logp(y^{<t>}|x, y^{<1>}, ..., y^{<t-1>}), \tag{3}$$

where T is the target sequence length and $y^{<t>}$ is the candidate category's t-th word. In this stage, we record the confidence s of all candidate categories $\mathcal{C} = \{c_i\}_{i=1}^{n}$ for use in the final ensemble ranking.

3.2   Category Selection

We consider the category selection stage as a reranking process and aim to improve precision. To achieve this, we propose a lightweight approach that leverages the parsing capabilities of multiple models. First, we create a template to rerank category suggestions based on perplexity using another pre-trained model. Next, we use ensemble ranking to combine candidate and prompt ranking results through a special rule design.

Prompt Ranking. This module is designed to address the drift problem where the entity name included in the context is not effectively utilized. We propose a process that involves constructing discrete patterns and computing perplexity for ranking. This helps to ensure that the suggested categories are more closely related to the entities.

- **Entity Clustering:** Entity clustering [18] is a key step in our work. The purpose of clustering is to select the most appropriate entity X.
  In particular, we use K-means [8] algorithm to select the X from entity set E. We randomly initialize K cluster centers, and select the entity closest to the center with the maximum cluster distance as the representative entity X.

- **Template:** The key for the prompt paradium is to construct a template that is adapted to the downstream task. Based on the work of [9], we construct five hand-built Hearst templates (as illustrated in Table 1):
  Where X is an entity e_i, and Y is a category c_i in C. We manually build the template with the same X and different Y. We create templates to enhance the connection between entity and category.

Table 1. Hearst templates.

input(X)	input(Y)	Template
Entity	Category	*class Y : X*
		X a special case of Y
		X is a Y that
		Y including X
		X and some other Y

- **Perplexity (ppl):** Perplexity is one of the most common metrics for evaluating language models. The large pre-trained model can determine the degree of sentence fluency based on the value of ppl. If we have a tokenized sequence $\mathcal{X} = \{x_i\}_{i=1}^{n}$ then the perplexity of $\mathcal{X}$ is:

$$PPL(\mathcal{X}) = exp\left\{ -\frac{1}{t} \sum_{i}^{t} \log p_\theta(x_i \mid x_{<i}) \right\}, \tag{4}$$

where $logp_\theta(x_i|x_{<i})$ is the log-likelihood of the i-th token conditioned on the preceding tokens $x_{<i}$.

We feed the template into a pre-trained model and calculate the perplexity of category Y, which is then used to re-rank the candidate categories and obtain the prompt ranking, denoted as $C_p = \{c_i^p\}_{i=1}^{n}$. The smaller the perplexity of the template, the better the result in the prompt ranking.

Ensemble Ranking. This module is designed to alleviate the problem of repetition. In last stage, we already obtain the candidate generation and prompt ranking results. We aim to combine them to create the ensemble ranking $C_e = \{c_i^e\}_{i=1}^{n}$, where c_i^e represents a category.

We set a threshold λ^2 by comparing beam search confidence s to dynamically adjust the contribution of the two ranking groups in the ensemble ranking C^e. If the beam search score s of a category suggestion c_j in the candidate generation results is greater than the threshold, it is directly included in the ensemble ranking. Otherwise, we take the corresponding category c_i^p in prompt ranking to obtain the final category C_e in the ensemble ranking:

---

[2] We determined a threshold of 0.85 which provided the best result.

$$\mathcal{C}^e = \sum_{i=1}^{n}(\lceil(s-\lambda)\rfloor c_i + \lceil(\lambda-s)\rfloor c_i^p), \tag{5}$$

where $(s-\lambda)$ varies in the range of $(-1,1)$, $\lceil\rfloor$ means ROUNDUP function (Fig. 3).

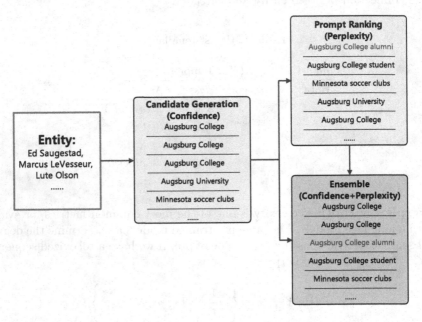

Fig. 3. Ensemble ranking. Ed Saugestad is the name of an individual. The top 2 results of the ensemble ranking come from confidence, while the rest come from perplexity. The blue labels indicate the categories of entities in Wikipedia. (Color figure online)

4   Experimental Evaluation

4.1   Experiment Settings

Datasets. Since there is no publicly available dataset for fine-grained category generation, we collect data from Wikipedia[3] and construct a new dataset called Wiki-category. Wikipedia is a multilingual online encyclopedia and the largest reference work in history. Our dataset is constructed through a three-stage process.

In the first stage, we gather information from Wikipedia on entities, context texts, categories, and their relationships. We normalize this information into three sets: 1) the entity and its category; 2) the entity and its corresponding context; 3) the category and its parent categories. In the second stage, we use

---

[3] https://www.wikipedia.org/.

the category relationships to construct a hierarchical tree of categories and select specific layers as fine-grained categories. Finally, we filter out all subcategories under the fine-grained categories, retaining only entities.

We collect 23.8M entities and context, and 1.1M categories from Wikipedia and construct a dataset of 5.3K fine-grained categories. The average length of a category is 28 characters, and on average, each category contains 21 entities. In our task, we use 5.3K fine-grained categories as the dataset. The dataset is divided into three parts: training, validation, and testing sets, with a split ratio of 80/10/10, respectively.

Baselines. We compare our model with five typical methods:

- **Pointer-generator** [26] proposes an architecture that augments the Seq2Seq attention model. It can copy words from the source text via pointing and uses coverage to discourage repetition.
- **NATS** [28] is an abstractive text summarization approach. It uses the pointer-generator network as the base model and equips it with the coverage mechanism and unknown word replacement.
- **Bart** [13] uses a standard Seq2Seq translation architecture with a bidirectional encoder (like BERT [11]) and a left-to-right decoder (like GPT [22]). BART is particularly effective when fine-tuned for text generation.
- **T5** [23] is an encoder-decoder model in which each task is converted to text-to-text format, using the same model,function and training process for all NLP tasks.
- **Pegasus** [32] is a large Transformer-based encoder-decoder model in which important sentences are removed/masked from an input document and are generated from the remaining sentences, similar to an extractive summary.

Basic Settings. The experiments are conducted on a Linux server equipped with an AMD EPYC 7402 CPU, 256 GB RAM, and an NVIDIA GeForce RTX 3090. The proposed models are implemented using PyTorch 1.10.1 in Python 3.7. We fine-tune the model with five typical methods. The pointer-generator method is based on the configuration of the paper[4]. For NATS, we use their publicly released toolkit[5]. For Seq2Seq models(T5, PEGASUS, Bart), we choose the tokenizer and model from huggingface[6].

Evaluation Metric. We use the **rouge** and **text similarity** as our evaluation metric, which is the metric in text generation for evaluating ranked lists.

1) Rouge [14] is a set of metrics for evaluating automated abstracts and machine translations. This metric primarily relies on the concept of recall to

---

[4] https://github.com/abisee/pointer-generator.
[5] https://github.com/tshi04/NATS.
[6] https://huggingface.co/.

evaluate the degree of similarity between the generated summary and the reference summary. We report the scores for ROUGE-1, ROUGE-2 and ROUGE-L. We select the result with the highest rouge score among the top 5 results.

2) Text similarity is used to compare sequence differences. This metric primarily relies on the concept of the longest common subsequence(LCS)[7].

$$ratio = 2 * M/T \qquad (6)$$

M = matches, T = total number of elements in both sequences. To reduce noise, we compare the text are only retained English characters which are converted to lowercase.

4.2   Overall Evaluation

As shown in Table 2, we conduct a comprehensive evaluation of the performance of commonly used Seq2Seq models on fine-grained category generation tasks. To ensure a fair comparison, all models are trained and tested using conceptual text T. Our method is also optimized and improved based on the Bart-large model, resulting in state-of-the-art performance. Moreover, we notice several interesting observations in our experiments.

For Different Evaluation Metrics. we find that the Rouge evaluation method could not effectively distinguish between similar fine-grained categories, which prompted us to design a text similarity evaluation method. Furthermore, we aim to generate categories that match the titles of corresponding Wikipedia pages, and as such, the text similarity evaluation was designed to better assess our models' performance in this regard. *E.g.*, T5-large performs well in rouge evaluation but only achieves mediocre results in text similarity evaluation.

For Pre-train Models. We compare the performance of pre-trained models with Seq2Seq methods such as PG and NATS, which performed poorly. Our hypothesis is that these methods rely on pointwise copying of words from context. However, conceptual texts T do not contain explicit category information. Therefore, the generated categories are likely derived from model pre-training.

For Different Baselines. We compare our methods with five fine-turned models and a prompt-based zero-shot model and observe Bart achieves the best performance among the baseline models. Our method is based on Bart and has the same $Prec@1$ and $Dist@1$, but our approach significantly improves the performance of $Prec@5$. Furthermore, we observe that larger-scale models with the same structure tend to produce better results.

---

[7] We calculate text similarity by using SequenceMatcher.ratio(). $Prec@1$ and $Prec@5$ indicate whether the top1 and top5 results include a ratio value of 1. $Dist@1$ and $Dist@5$ indicate the largest ratio among the top1 and top5 results.

Table 2. Rouge and text similarity evaluation of different methods. GPT-2‡ indicates the zero-shot result prompt. Results in bold are optimal, and the results underlined are suboptimal.

Model	Rouge			Text Similarity			
	1	2	L	Prec@1	Prec@5	Dist@1	Dist@5
PG	15.92	5.72	15.83	0.28	0.56	15.37	38.69
NATS	12.16	4.48	11.80	0.28	0.56	29.01	32.86
T5-small	27.48	13.86	27.31	2.78	6.39	35.95	48.58
T5-base	30.25	17.80	30.05	5.56	8.91	40.72	51.13
T5-large	36.41	21.87	36.06	5.56	10.83	41.46	55.76
pegasus-xsum	37.42	23.16	37.13	6.94	13.06	42.49	56.27
pegasus-large	38.74	25.18	38.51	8.33	14.72	44.29	_56.89_
Bart-base	35.52	20.85	35.36	6.11	11.11	41.09	54.27
Bart-large	_39.83_	_26.91_	_39.73_	**11.11**	16.39	**46.17**	56.29
GPT-2‡	39.02	26.33	38.93	6.67	_17.22_	43.90	56.35
Ours	**41.07**	**28.62**	**40.89**	**11.11**	**20.28**	**46.17**	**57.20**

4.3 Result Analysis

Ablation Study. We conduct ablation experiments to verify the efficacy of conceptual text mining, prompt ranking, and ensemble ranking. As can be seen from Table 3, the use of conceptual text mining successfully extracts relevant information from context, effectively reducing noise.

Regarding the Prompt and ensemble ranking methods, They doesn't alter the top-1 result, hence yielding the same values for $Prec1$ and $Dist1$. For prompt ranking, the rouge evaluation remained basically the same, while text-similarity improved, demonstrating that entity name information can promote category ranking. For ensemble ranking, both rouge and text-similarity are improved. Prompt integrates the relationships between entities and categories, while ensemble ranking combines the parsing capabilities of multiple models.

Table 3. Ablation study.

Model	Rouge			Text Similarity			
	1	2	L	Prec@1	Prec@5	Dist@1	Dist@5
w/o mining	−8.29	−5.44	−8.27	−5.44	−10.50	−5.60	−8.24
w/o prompt	0.07	0.02	0.16	/	−0.83	/	−0.06
w/o ensemble	−1.31	−1.42	−1.31	/	−3.06	/	−0.85

Implementation Observation. We input the dataset to generate the top10 results and record the beam search scores of categories. We comprehensively consider the performance of *Prec@1* and *Prec@5*. In Fig. 4, there are a series of hyperparameter settings:

1) For the maximum constraint of the generation length, we set the value to 16 or more; 2) No-repeat size is the similarity penalty set. It refers to the size of the n-gram that is considered when generating new text to ensure that no repeated sequence of words is produced. We set the value to 3; 3) For the ensemble ranking scale, we compare the effects of fixed and dynamic scales. The scale of the model and prompt is [4:1 to 1:4]. The *Prec@5* is highest when using the dynamic scale setting.

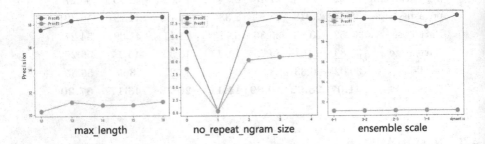

Fig. 4. Hyperparameter settings.

Case Study. This section shows two real cases and compares our approach with Bart for case studies. It can be seen in Table 4.

1) Case 1 demonstrates the effectiveness of the prompt method in reducing semantic drift using entity information. *E.g.*, we input the names of three persons, but Bart preferentially generates categories related to an organization, prompt can select a category related with people based on the given entity information. This approach helps generate categories that are more relevant and aligned with the input entity information.

2) Case 2 illustrates how ensemble ranking can alleviate the problem of repetition in specific situations. Specifically, we observe that the Seq2Seq model tends to generate similar results due to the constraints of the search algorithm. Furthermore, when using the Bart model to generate categories, we find that the term "backup" was frequently included in the generated results. However, by combining the two sets of ranked results using ensemble ranking, we can increase the diversity of the generated categories and reduce the problem of repetition.

Table 4. Case studies between Bart and Ours. The ground truth category is marked in blue, and the entity is marked in red.

Case 1	Augsburg College alumni: [Ed Saugestad, Marcus LeVesseur, Lute Olson]
Bart	Augsburg College, Augsburg College, Augsburg College, Augsburg University, Minnesota soccer clubs
Ours	Augsburg College, Augsburg College, Augsburg College alumni, Augsburg Student, Minnesota soccer clubs
Case 2	Disk computer storage: [Parallel ATA, Disk read-and-write head, Seagate ST1, M.2, Hard disk drive, Serial ATA, RAID]
Bart	Computer backup, Computer backup, Disk computer backup, Computer backu, Personal computer backup
Ours	Computer backup, Computer backup, Disk computer backup, Computer backu, Disk computer storage

5   Related Work

Abstract Text Summarization. Our work is more closely related to the abstract text summarization task persisting salient information of source articles. There are two broad approaches to summarization: extraction and abstraction. Extractive summarization extracts key sentences and keywords from source documents. Abstract summarization allows the generation of new words and phrases to compose summaries.

Point-generator network [26] can copy words from source text via pointing and use coverage to discourage repetition. NATS [28] supplies a toolkit of RNN-based Seq2seq models. BART [13] is an encoder-decoder model and has shown superior performance in abstract summarization. T5 [23] is a unified framework that converts all text-based language problems into a text-to-text format and adapts for text summarization. Pegasus [32] is a pre-trained model tailored for abstractive text summarization. However, these works do not work well in fine-grained category generation tasks where severe noise and drift problems.

Fine-Grained Entity Typing. For category generation, a large amount of previous work has focused on assigning entities to categories, usually called fine-grained entity typing, which aims to classify marked entities from input sequences into specific types in a pre-defined label set. Ultra-fine types [5] introduce a new entity typing task to predict a set of free-form phrases that describe the entity's role in a given sentence. [20] utilizes the type description available from Wikipedia to build a distributed semantic representation of the types. [35] requires no annotated data and can flexibly identify newly defined types. [6] investigates the application of prompt-learning on fine-grained entity typing in fully supervised, few-shot, and zero-shot scenarios. This task generates pre-defined categories and can not expand new categories.

Prompt-Based Learning. Our work also benefits from prompt-learning, which is based on language models that model the probability of text directly. In this paradigm, a model with a fixed architecture is pre-trained as a language model (LM), predicting the probability of observed textual data. GPT-2 [22] proposes to use unsupervised pre-trained models for supervised tasks. The seminal work that stimulates the development of prompt-learning is the birth of GPT-3 [3], which uses hand-crafted prompts for tuning and achieves very impressive performance on various tasks. Pattern design can be divided into discrete [29,31] and continuous [12]. Inspired by Hearst patterns [9], we construct different templates and show stable performance. There are related works [2,16] about prompt, which are very similar to the standard pre-training and fine-tuning paradigm, but adding prompt can provide additional guidance at the beginning of model training. In our work, we mainly use prompt-learning to model the probability of templates and rerank candidate categories.

6   Conclusion and Future Work

The category systems present in large-scale knowledge bases are unique and valuable resources that can be leveraged to accomplish a wide range of information access tasks. However, these systems are currently created and maintained manually by editors. To address this issue, this paper proposes an approach that utilizes the Seq2Seq model to generate categories based on a set of entities and their context as input. However, this task faces several challenges, including noise, drift, and repetition.

To overcome these challenges, we propose a novel approach that utilizes a two-stage category generation framework. We first utilize the Seq2Seq model, which produces a large number of fine-grained categories. Next, we apply language model probing to select high-quality categories. To reduce the noise in the original text, we use the conceptual text mining approach in the first stage to extract common text from context. The second stage includes two modules, prompt ranking and ensemble ranking. Prompt ranking is based on a manually constructed template, which includes the names of categories and entities. We utilize perplexity to re-rank categories, which helps to improve the connection between entities and categories and reduce the drift phenomenon. Ensemble ranking is designed to alleviate repetition issues. In this module, we use dynamic rule algorithms combined with candidate categories and prompt ranking results to improve the diversity of generated categories.

To evaluate the effectiveness of our proposed framework, we develop a dataset dedicated to fine-grained category generation based on Wikipedia. We systematically investigate the performance of the Seq2Seq model for the fine-grained category generation task. Our experimental results, based on five baseline evaluations, demonstrate the efficacy of our approach, while the ablation study validates the effectiveness of the three modules. Moving forward, we plan to explore the construction of continuous patterns for category generation and apply fine-grained categories to downstream tasks.

Acknowledgments. This work is supported by National Key R&D Program of China (2020AAA0105203).

References

1. Aiken, E.G., Thomas, G.S., Shennum, W.A.: Memory for a lecture: effects of notes, lecture rate, and informational density. J. Educ. Psychol. **67**(3), 439 (1975)
2. Ben-David, E., Oved, N., Reichart, R.: Pada: a prompt-based autoregressive approach for adaptation to unseen domains. arXiv preprint arXiv:2102.12206 (2021)
3. Brown, T., et al.: Language models are few-shot learners. Adv. Neural. Inf. Process. Syst. **33**, 1877–1901 (2020)
4. Carlsson, F., Öhman, J., Liu, F., Verlinden, S., Nivre, J., Sahlgren, M.: Fine-grained controllable text generation using non-residual prompting. In: Proceedings of the 60th Annual Meeting of the Association for Computational Linguistics (Volume 1: Long Papers), pp. 6837–6857 (2022)
5. Choi, E., Levy, O., Choi, Y., Zettlemoyer, L.: Ultra-fine entity typing. In: Proceedings of the 56th Annual Meeting of the Association for Computational Linguistics (Volume 1: Long Papers), pp. 87–96 (2018)
6. Ding, N., et al.: Prompt-learning for fine-grained entity typing. arXiv preprint arXiv:2108.10604 (2021)
7. Han, X., et al.: Cross-lingual contrastive learning for fine-grained entity typing for low-resource languages. In: Proceedings of the 60th Annual Meeting of the Association for Computational Linguistics (Volume 1: Long Papers), pp. 2241–2250 (2022)
8. Hartigan, J.A., Wong, M.A.: Algorithm as 136: a k-means clustering algorithm. J. Roy. Stat. Soc. Ser. C (Appl. Stati.) **28**(1), 100–108 (1979)
9. Hearst, M.A.: Automatic acquisition of hyponyms from large text corpora. In: COLING 1992 Volume 2: The 14th International Conference on Computational Linguistics (1992)
10. Jelinek, F., Mercer, R.L., Bahl, L.R., Baker, J.K.: Perplexity-a measure of the difficulty of speech recognition tasks. J. Acoust. Soc. Am. **62**(S1), S63–S63 (1977)
11. Kenton, J.D.M.W.C., Toutanova, L.K.: Bert: pre-training of deep bidirectional transformers for language understanding. In: Proceedings of NAACL-HLT, pp. 4171–4186 (2019)
12. Lester, B., Al-Rfou, R., Constant, N.: The power of scale for parameter-efficient prompt tuning. In: Proceedings of the 2021 Conference on Empirical Methods in Natural Language Processing, pp. 3045–3059 (2021)
13. Lewis, M., et al.: Bart: denoising sequence-to-sequence pre-training for natural language generation, translation, and comprehension. In: Proceedings of the 58th Annual Meeting of the Association for Computational Linguistics, pp. 7871–7880 (2020)
14. Lin, C.Y.: Rouge: a package for automatic evaluation of summaries. In: Text Summarization Branches Out, pp. 74–81 (2004)
15. Liu, P., Yuan, W., Fu, J., Jiang, Z., Hayashi, H., Neubig, G.: Pre-train, prompt, and predict: a systematic survey of prompting methods in natural language processing. ACM Comput. Surv. **55**(9), 1–35 (2023)
16. Liu, X., et al.: GPT understands, too. arXiv preprint arXiv:2103.10385 (2021)

17. Liu, Y., Shen, S., Lapata, M.: Noisy self-knowledge distillation for text summarization. In: Proceedings of the 2021 Conference of the North American Chapter of the Association for Computational Linguistics: Human Language Technologies, pp. 692–703 (2021)
18. Min, E., Guo, X., Liu, Q., Zhang, G., Cui, J., Long, J.: A survey of clustering with deep learning: from the perspective of network architecture. IEEE Access **6**, 39501–39514 (2018)
19. Mota, T., Sridharan, M.: Commonsense reasoning and knowledge acquisition to guide deep learning on robots. In: Robotics: Science and Systems (2019)
20. Obeidat, R., Fern, X., Shahbazi, H., Tadepalli, P.: Description-based zero-shot fine-grained entity typing. In: Proceedings of the 2019 Conference of the North American Chapter of the Association for Computational Linguistics: Human Language Technologies, Volume 1 (Long and Short Papers), pp. 807–814 (2019)
21. Pang, K., Zhang, H., Zhou, J., Wang, T.: Divide and denoise: learning from noisy labels in fine-grained entity typing with cluster-wise loss correction. In: Proceedings of the 60th Annual Meeting of the Association for Computational Linguistics (Volume 1: Long Papers), pp. 1997–2006 (2022)
22. Radford, A., Wu, J., Child, R., Luan, D., Amodei, D., Sutskever, I., et al.: Language models are unsupervised multitask learners. OpenAI Blog **1**(8), 9 (2019)
23. Raffel, C., et al.: Exploring the limits of transfer learning with a unified text-to-text transformer. J. Mach. Learn. Res. **21**(140), 1–67 (2020)
24. Reddy, D.R., et al.: Speech understanding systems: a summary of results of the five-year research effort. Department of Computer Science. Camegie-Mell University, Pittsburgh, PA, vol. 17, p. 138 (1977)
25. Salton, G., Buckley, C.: Term-weighting approaches in automatic text retrieval. Inf. Process. Manag. **24**(5), 513–523 (1988)
26. See, A., Liu, P.J., Manning, C.D.: Get to the point: summarization with pointer-generator networks. In: Proceedings of the 55th Annual Meeting of the Association for Computational Linguistics (Volume 1: Long Papers), pp. 1073–1083 (2017)
27. Shen, J., Wu, Z., Lei, D., Shang, J., Ren, X., Han, J.: Setexpan: corpus-based set expansion via context feature selection and rank ensemble. In: Ceci, M., Hollmén, J., Todorovski, L., Vens, C., Dzeroski, S. (eds.) Joint European Conference on Machine Learning and Knowledge Discovery in Databases, pp. 288–304. Springer, Cham (2017). https://doi.org/10.1007/978-3-319-71249-9_18
28. Shi, T., Keneshloo, Y., Ramakrishnan, N., Reddy, C.K.: Neural abstractive text summarization with sequence-to-sequence models. ACM Trans. Data Sci. **2**(1), 1–37 (2021)
29. Shin, T., Razeghi, Y., Logan IV, R.L., Wallace, E., Singh, S.: Autoprompt: eliciting knowledge from language models with automatically generated prompts. In: Proceedings of the 2020 Conference on Empirical Methods in Natural Language Processing (EMNLP), pp. 4222–4235 (2020)
30. Shleifer, S., Rush, A.M.: Pre-trained summarization distillation. arXiv preprint arXiv:2010.13002 (2020)
31. Wallace, E., Feng, S., Kandpal, N., Gardner, M., Singh, S.: Universal adversarial triggers for attacking and analyzing NLP. In: Proceedings of the 2019 Conference on Empirical Methods in Natural Language Processing and the 9th International Joint Conference on Natural Language Processing (EMNLP-IJCNLP), pp. 2153–2162 (2019)
32. Zhang, J., Zhao, Y., Saleh, M., Liu, P.: Pegasus: pre-training with extracted gap-sentences for abstractive summarization. In: International Conference on Machine Learning, pp. 11328–11339. PMLR (2020)

33. Zhang, S., Balog, K., Callan, J.: Generating categories for sets of entities. In: Proceedings of the 29th ACM International Conference on Information & Knowledge Management, pp. 1833–1842 (2020)
34. Zhang, Y., Shen, J., Shang, J., Han, J.: Empower entity set expansion via language model probing. In: Proceedings of the 58th Annual Meeting of the Association for Computational Linguistics, pp. 8151–8160 (2020)
35. Zhou, B., Khashabi, D., Tsai, C.T., Roth, D.: Zero-shot open entity typing as type-compatible grounding. In: Proceedings of the 2018 Conference on Empirical Methods in Natural Language Processing, pp. 2065–2076 (2018)

CoTE: A Flexible Method for Joint Learning of Topic and Embedding Models

Bo Zhao[1,2](✉) ⓘ, Chunfeng Yuan[1,2], and Yihua Huang[1,2]

[1] National Key Laboratory for Novel Software Technology, Nanjing University, Nanjing 210023, China
chawbhoppi@smail.nju.edu.cn
[2] Collaborative Innovation Center of Novel Software Technology and Industrialization, Nanjing 210023, China

Abstract. The topic and embedding models are two of the most popular categories of techniques to learn the latent semantics from text. In the topic models, each word is generated according to its global context; while in the embedding models, each word occurrence is measured by surrounding words. Thus it is expected to train the topic and embedding models jointly by utilizing multi-context information to learn better representations. In this paper, we propose a flexible method named CoTE to achieve this goal, which can integrate a variety of the topic and embedding models together. And we design a general 3-stage learning procedure to optimize the parameters of CoTE, which adopts a rotation optimization scheme. We chose and combined two groups of the de-facto topic and embedding models to implement the CoTE-PD and CoTE-LW algorithms. Experimental results show that CoTE achieves accuracy improvements in both individual components.

Keywords: Representation learning · Topic model · Embedding model

1 Introduction

Representation learning has become an important research topic in text analysis for discovering latent semantic features from text. There are many techniques of text representation learning, among which the topic models and embedding models are two most popular categories. The topic model uses the statistical distributions to discover the hidden "topics" shared across the collections of documents. The most common topic models include pLSA [7], LDA [3], etc. The embedding model represents words, sentences, etc. as low-dimensional continuous vectors in the same semantic space [2]. For example, word2vec [16], doc2vec [11] and deep neural models like BERT [5] are the most widely-used embedding models in many natural language processing (NLP) tasks.

Both topic and embedding models are natural extensions to the traditional bag-of-words model, as they provide better semantic encodings than the traditional one-hot encoding. The semantics originate from the context that the model based on to generate each words. In the topic models, each word is generated according to its global context such as a document or a sentence, while in the embedding models, each word occurrence is measured by its local context such as surrounding words. Thus, it is expected to combine the topic and embedding model by utilizing multi-context information to make them complement each other to learn better representations.

Some previous works have attempted to combine the topic model with embeddings. The key of the combination is to represent topics as low-dimensional continuous vectors that is generally called "topic embedding". We will discuss these works in detail in the following sections. There are some deficiencies in these works: The first is that they usually take one model to improve the other, rather than the two models improving each other. Further, these works are all newly proposed and customized algorithms, and they lack a general way to cooperate with existing models easily.

To integrate a variety of the topic and embedding models together, and at the same time to train them jointly to improve each other, in this paper we propose a flexible method named CoTE with the following contributions: Firstly, based on summarizing the unified form of the topic models as well as the embedding models, we proposed a general model framework that combines various topic and embedding models with constraints; Next, we designed a genaral and effective 3-stage learning procedure to optimize the parameters of CoTE, which adopts a rotation optimization scheme; Then we chose and combined two groups of the de-facto topic and embedding models to implement the CoTE-PD (pLSA with doc2vec) and CoTE-LW (LDA with word2vec) algorithms. We evaluated CoTE-PD and CoTE-LW and the experimental results show that they can achieve improvements for both individual components.

2 Background

The existing works about combining the topic model with embeddings can be briefly classified into two ways based on how they combine the two models.

One way is to use existing pre-trained embeddings to help learn a better topic model. In the Gaussian LDA [4] approach, a topic is no longer a distribution over words in the vocabulary, but a Gaussian distribution in the word embedding space. MvTM [14] improves Gaussian LDA by replacing the Gaussian distributions with the von Mises-Fisher distributions to achieve better similarity measurement. Xun [23] uses word embeddings to help capture correlation structure in the topic model. These methods build new topic models on top of existing pretrained embeddings, and thus they did not take advantage of the topic models to improve embeddings.

Another way is to simulate the topic model by a link function which models the joint probability of words and topics based on embeddings. With the popularity of deep learning, this type of method has been studied and surged the most, such as [6,8,9,13,15,19,22], among which two remarkable works that need to be mentioned are TopicVec [13] and Embedded Topic Model (ETM) [6]. The main idea of TopicVec is that the topics are drawn by the doc-topic proportions as in LDA, while a link function is employed to output the probability of each word based on embeddings of topics and context words. The ETM models each word with a categorical distribution whose natural parameter is the inner product between the word's embedding and an topic embedding to discover interpretable topics even with large vocabularies and rare words.

As can be seen, most of previous works tried to seek new customized algorithms to combine the topic and embedding models. In this paper, we achieve this goal by taking another approach that finds a general way to cooperate with existing various topic and embedding models, to train them jointly to improve each other.

3   Symbol Definition

Before the introduction of our work, the common notations used throughout the paper are listed in Table 1. To be uniform, we use uppercase bold letters to denote matrices, and lowercase bold letters to denote vectors. The uppercase normal letters that are listed in Table 1 such as D, W denote sets, and the corresponding lowercase normal letters like d, w denotes an element of that set. The rest letters represent scalar constants. In the following description, each document $d \in D$ is a sequence of N_d terms $(w_1, ..., w_{N_d})$ from the vocabulary W.

Table 1. Table of notations

Sym.	Description	Sym.	Description
D	Document set $\{d_1, d_2, ..., d_{|D|}\}$	$\boldsymbol{\theta}_d$	Doc-topic distribution of doc d
W	Vocabulary $\{w_1, w_2, ..., w_{|W|}\}$	$\boldsymbol{\phi}_t$	Topic-word distribution of topic t
T	Topic collection $\{t_1, t_2, ..., t_{|T|}\}$	$\mathbf{V}$	Word representation embeddings
Ψ	Param set of topic model	$\mathbf{U}$	Word context embeddings
Ω	Param set of embedding model	$\mathbf{G}$	Document embeddings
C_w	Context of word w	$\mathbf{Y}$	Topic embeddings

4   CoTE Model Architecture

In order to solve the problems existing in the above related work, in this section, we propose a novel method named CoTE, which can CO-train various Topic and Embedding Models in a unified way. Initially, we introduce several widely-used topic and embedding models in brief and summarize the unified form of them. Next, we build the simulation distribution to map the embedding model into the same metric space as the topic model, and measure its distance from the topic distribution. After that, we formulate the general CoTE model framework that combines various topic and embedding models with constraints. Finally, bringing in widely-used topic and embedding models, we get two concrete CoTE models, namely CoTE-PD and CoTE-LW.

4.1 General Form of Topic and Embedding Models

The topic model assumes that the occurrence of each word in a document is related to the global information, such as the document. The probability of generating each word is equal to the product of the probability of the document generating topic and topic generating words, and summed over all topics. Thus, in topic model, the probability of the word w occurred in the document d can be expressed in the form of Formula 1:

$$p(w|d, \Psi) = \sum_{t \in T} \theta_{dt} \phi_{tw} \tag{1}$$

For different topic models, the content of the parameter Ψ varies. For example, in pLSA [7] and LDA [3] models, their main parameters are two matrices θ and ϕ, which represent the distribution of "document-topic" and "topic-word" respectively. The only difference is that, in pLSA, these two distributions are frequency statistics, while in LDA, these two distributions are instead sampled from Dirichlet priors parameterized by α and β respectively.

The embedding model assumes that the probability of occurrence of each word is related to the words or other information within its context window. Thus, in embedding model, given the word w and its context C_w, the probability of generating w can be expressed in the form of Formula 2:

$$p(w|C_w, \Omega) = \frac{1}{|C_w|} \sum_{c \in C_w} \frac{\exp(\mathbf{v}_w \cdot \mathbf{u}_c)}{\sum_{w'} \exp(\mathbf{v}_{w'} \cdot \mathbf{u}_c)} \tag{2}$$

Similarly, for different embedding models, the content of the parameter Ω varies. For example, in word2vec [16], the parameters are mainly the matrices $\mathbf{V}$ and $\mathbf{U}$, and C_w is all the words in the context window of w. In particular, in doc2vec [11], in addition to the $\mathbf{V}$ and $\mathbf{U}$ matrices, its parameters also include the document vector matrix $\mathbf{G}$. Since documents can be interpreted as a kind of context for words, C_w has one more context vector $\mathbf{g}_d$ than word2vec.

4.2 Measuring Distance Between Topic and Embedding Models

Topic and embedding models have different parameter properties. The parameters in the topic model are mainly probabilities, while the parameters in the embedding model are all dense real numbers. In order to measure the distance between topic and embedding models, we need to map the two models into the same metric space. The approach taken in this paper is to construct a "simulation distribution" to map the embedding model parameters into the probability distribution space.

Assuming that all document vectors $\mathbf{g}_d$ and context word vectors $\mathbf{u}_w$ are given, and extra topic vectors $\mathbf{y}_t$ are introduced. Then $softmax$ function can be applied over the dot product of embeddings to construct two simulation distributions: "document-topic" simulation distribution $\mathbf{m}_d$, "topic-word" simulation distribution $\mathbf{m}_t$, to respectively simulate the "document-topic" distribution θ_d

and "topic-word" distribution ϕ_t in the topic model. The simulation distributions are

$$m_{dt} = softmax(\frac{\mathbf{g}_d \cdot \mathbf{y}_t}{\sqrt{k}}), m_{tw} = softmax(\frac{\mathbf{y}_t \cdot \mathbf{u}_w}{\sqrt{k}}) \tag{3}$$

where k is the dimension of the embedding vectors. The scaling factor $\frac{1}{\sqrt{k}}$ is adopted to reduce the products in magnitude when the dimension value k is large [5].

Divergence can be used to measure the distance between the simulation distribution and the topic distribution. Commonly used in practice is the KL (Kullback-Leibler) divergence [10]. In order to improve the generality of the model, this paper chooses the Rényi Divergence to extend the KL divergence. The Rényi Divergence $D_\alpha(\boldsymbol{p}\|\boldsymbol{q})$ of the distribution $\boldsymbol{p}$ and $\boldsymbol{q}$ is defined as:

$$D_\alpha(\boldsymbol{p}\|\boldsymbol{q}) = \frac{1}{\alpha - 1} \ln \left(\sum_{i=1}^{n} \frac{p_i^\alpha}{q_i^{\alpha-1}} \right) \tag{4}$$

The more similar (overlapping) $\boldsymbol{p}$ and $\boldsymbol{q}$ are, the smaller of $D_\alpha(\boldsymbol{p}\|\boldsymbol{q})$ will be. α is a positive real number, and when $\alpha = 1$, the formula degenerates into the KL divergence formula.

4.3   The General CoTE Framework

CoTE is based on the following natural idea: for the same corpus, the topic model and the embedding model are trained, then there should be a certain semantic similarity between the topic distributions and the simulation distributions constructed by the embedding model; and the richer the two models mined semantics, the greater the similarity should be. Therefore, the training goal of the CoTE model is: for a corpus, simultaneously train the topic model and the embedding model, and hope the differences between the two kinds of distributions as small as possible.

Based on the above discussions, it motivates us to introduce the general CoTE model by combining Formula 1 and 2, with the additional regularizers that maximize the similarity between topic and simulation distributions. Thus, given a collection of documents D and assuming the parameter set of the topic and embedding models be Ψ and Ω respectively, we can derive the log-likelihood objective function $L(D; \Psi, \Omega)$ as:

$$L = \sum_{d \in D} \sum_{w \in d} \left[\ln P_t(w|d, \Psi) + \ln P_e(w|C_w, \Omega) \right]$$

$$-\eta \sum_{d \in D} D_\alpha(\boldsymbol{m}_d \| \boldsymbol{\theta}_d) - \mu \sum_{t \in T} D_\alpha(\boldsymbol{m}_t \| \boldsymbol{\phi}_t) \tag{5}$$

where η and μ are hyper-parameters of the model that are both positive real numbers. $P_t(w|d, \Psi)$ is the probability of a word w occurred in the document d based on the topic model with the parameter Ψ, which satisfies Formula 1. $P_e(w|C_w, \Omega)$

is the probability of the same token generated via the embedding model with the parameter Ω and the context of the token C_w, which satisfies Formula 2. θ_d and ϕ_t are the topic proportions represented by the topic model, while m_d and m_t denote the simulation distributions calculated using Formula 3.

To describe the objective function in detail, we divide the Formula 5 into four parts.

1) The first part is $L_T = \sum_d \sum_w \ln P_t(w|d, \Psi)$, which is the standard log-likelihood function of the topic model.
2) The second part is $L_E = \sum_d \sum_w \ln P_e(w|C_w, \Omega)$, which is the standard log-likelihood function of the embedding model.
3) The third part is $R_D = -\eta \sum_d D_\alpha(m_d \| \theta_d)$, which is a regularizer to constrain m_d and θ_d for all of the document d that maximize the similarity between them.
4) The last part is $R_W = -\mu \sum_t D_\alpha(m_t \| \phi_t)$, which is similar to R_D as a regularizer to constrain m_t and ϕ_t for all of the topic t that maximize the similarity between them.

The Formula 5 is a general form and it depends on the actual chosen topic and embedding models to decide what Ψ and Ω are and how to compute θ_d, ϕ_t, C_w and m_d, m_t. In the next section, we depict two concrete models both under the framework of CoTE.

4.4   Concrete Examples: CoTE-PD and CoTE-LW

Based on the CoTE framework, and choosing pLSA as the topic model component and PV-DBOW (one type of doc2vec) as the embedding model component, and α of Rényi divergence as 0.5 to make the model symmetric, we get the CoTE-PD model. CoTE-PD has a simple objective function form, as follows:

$$L = \sum_{d \in D} \sum_{w \in d} \left[\ln \sum_{t \in T} \theta_{dt} \phi_{tw} + \ln \frac{1}{|C_w|} \sum_{c \in C_w} \frac{e^{\mathbf{v}_w \mathbf{u}_c}}{\sum_{w'} e^{\mathbf{v}_{w'} \mathbf{u}_c}} \right] +$$

$$2\eta \sum_{d \in D} \ln \sum_{t \in T} \sqrt{\theta_{dt} m_{dt}} + 2\mu \sum_{t \in T} \ln \sum_{w \in W} \sqrt{\phi_{tw} m_{tw}} \quad (6)$$

with the parameter set $\{\Theta, \Phi, \mathbf{V}, \mathbf{U}, \mathbf{T}, \mathbf{G}\}$.

LDA and word2vec are two of the most commonly-used text anylasis models. We choose them as the components of CoTE, and set $\alpha = 1$ (as done in [20]), to get the CoTE-LW model. The objective function of CoTE-LW is:

$$L = \sum_{d \in D} \sum_{w \in d} \left[\ln \sum_{t \in T} \theta_{dt} \phi_{tw} + \ln \frac{1}{|C_w|} \sum_{c \in C_w} \frac{e^{\mathbf{v}_w \mathbf{u}_c}}{\sum_{w'} e^{\mathbf{v}_{w'} \mathbf{u}_c}} \right]$$

$$+ \sum_{d \in D} \sum_{t \in T} \left(\alpha' \ln \theta_{dt} - \eta m_{dt} \ln \frac{m_{dt}}{\theta_{dt}} \right) + \sum_{t \in T} \sum_{w \in W} \left(\beta' \ln \phi_{tw} - \mu m_{tw} \ln \frac{m_{tw}}{\phi_{tw}} \right) (7)$$

with the parameter set $\{\Theta, \Phi, \mathbf{V}, \mathbf{U}, \mathbf{T}, \alpha', \beta'\}$. Compared with CoTE-PD, its parameters have Dirichlet priors parameter α' and β', but miss the document vector matrix $\mathbf{G}$. In word2vec, the embedding g_d for document d is not stored and is calculated by the average of all word vectors in document d.

5 Learning Parameters of CoTE

After inferring the CoTE objective, what to do next is to make optimization to maximize the log-likelihood to learn the parameters. It is difficult to learn all the parameters at the same time using a single optimization method, because different parameters have various attributes. For example, parameters of Θ, Φ have inner constraints that are probabilities (i.e. $\forall_{d,t} \theta_{dt} \geq 0, \sum_{t \in T} \theta_{dt} = 1$). Moreover, L_T and L_E are summed over each word tokens, while R_D is summed over all documents and R_W is summed over all the topics.

To solve this problem, we design a general and effective 3-stage learning procedure to optimize the parameters of CoTE, which adopts a rotation optimization scheme that iteratively optimizes a part of parameters while fixing the others.

5.1 The General 3-Stage Learning Procedure

The general 3-stage learning procedure of the CoTE model is depicted as follows.

Topic Model Fitting Stage. The first, we fix embedding parameters to optimize Θ, Φ. Note that in this situation L_E contains no variables and can be ignored, so that the log-likelihood becomes $L_T + R_D + R_W$, which is a general form of ARTM [20]. Due to page limit, we ignored the proof of the inference procedure (please refer to ARTM), then we can get an Expectation-Maximization algorithm to optimize Θ, Φ iteratively: for each word w in each document d, we update the parameters as follows:

$$\gamma_{dwt} = \frac{p(w,t|d,\boldsymbol{\theta}_d,\boldsymbol{\phi}_w)}{p(w|d,\boldsymbol{\theta}_d,\boldsymbol{\phi}_w)}; \tag{8}$$

$$\theta_{dt} \propto \left(n_{dt} + \theta_{dt} \frac{\partial R_D}{\partial \theta_{dt}} \right)_+; \tag{9}$$

$$\phi_{tw} \propto \left(n_{tw} + \phi_{tw} \frac{\partial R_W}{\partial \phi_{tw}} \right)_+. \tag{10}$$

where $n_{dt} = \sum_{w \in d} \gamma_{dwt}$ means the counts of topic t assigned in document d, and $n_{tw} = \sum_d \gamma_{dwt}$ equals the counts of word w assigned with topic t. The simplified notation $p_i \propto q_i$ means $p_i = q_i / \sum_j q_j$, and $(z)_+ = \max(z, 0)$.

Embedding Model Fitting Stage. The second stage is to optimize L_E while fixing topic models and regularizers as constants. The optimization method of L_E is to add the gradients of the embedding parameters Ω multiplied by the learning rate ξ to corresponding embedding vectors:

$$\Omega \leftarrow \Omega + \xi \frac{\partial L_E}{\partial \Omega} \tag{11}$$

Because in this stage the parameters are optimized token-by-token, it is general to use Stochastic Gradient Descent (SGD) to optimize the parameters by calculating gradient on each terms.

Regularizer Fitting Stage. Similar to the second stage, here we fix $L_T + L_E$ and maximize regularizers to optimize Ω.

$$\Omega \leftarrow \Omega + \xi \frac{\partial(R_D + R_W)}{\partial \Omega} \qquad (12)$$

In this stage the full-batch Gradient Descent (GD) is usually used.

5.2 Optimization of CoTE-PD and CoTE-LW

Applying the above learning procedure to CoTE-PD and CoTE-LW, we can get the actual training algorithm of the two models. Due to page limit, We omit the derivation process and directly give the results.

Before inferring the optimization equations, we first introduce two matrices $\mathbf{O}_{DT}$ and $\mathbf{O}_{TW}$, the elements of which are:

$$o_{dt} = \frac{\sqrt{\theta_{dt} m_{dt}}}{\sum_{t'} \sqrt{\theta_{dt'} m_{dt'}}}, o_{tw} = \frac{\sqrt{\phi_{tw} m_{tw}}}{\sum_{w'} \sqrt{\phi_{tw'} m_{tw'}}} \qquad (13)$$

We can see that each row of $\mathbf{O}_{DT}$ and each column of $\mathbf{O}_{TW}$ are all distributions, which we refer to as "bridge distributions", because these distributions are intermediate distributions that connect the topic distributions and the simulation distributions.

The concrete 3-stage learning procedure of CoTE-PD is as follows.

Topic Model Fitting Stage. Plugging the Formula 6 into the Formula 8, 9 and 10, we can get the learning equation for each token:

$$\gamma_{dwt} = \frac{\theta_{dt} \phi_{tw}}{\sum_{t'} \theta_{dt'} \phi_{t'w}}; \qquad (14)$$

$$\theta_{dt} \propto n_{dt} + \eta o_{dt}; \qquad (15)$$

$$\phi_{tw} \propto n_{tw} + \mu o_{tw}. \qquad (16)$$

Embedding Model Fitting Stage. The same as the doc2vec training.

Regularizer Fitting Stage. We write the update gradients in matrix form to optimize all the embedding parameters in batch:

$$\mathbf{G} \leftarrow \mathbf{G} + \xi \eta \mathbf{A} \mathbf{Y} \qquad (17)$$

$$\mathbf{U} \leftarrow \mathbf{U} + \xi \mu \mathbf{B}' \mathbf{Y} \qquad (18)$$

$$\mathbf{Y} \leftarrow \mathbf{Y} + \xi \eta \mathbf{A}' \mathbf{G} + \xi \mu \mathbf{B} \mathbf{U} \qquad (19)$$

where $\mathbf{A}, \mathbf{B}$ are matrices with dimension $(|D| * |T|)$ and $(|T| * |W|)$ respectively, and their corresponding units are $a_{dt} = o_{dt} - m_{dt}, b_{tw} = o_{tw} - m_{tw}$. $\mathbf{A}'$ is the matrix transpose of $\mathbf{A}$.

For the CoTE-LW model, the learning procedure is:

Topic Model Fitting Stage. Substitute the corresponding items of Formula 8, 9 and 10 by the ones in Formula 7, we can get the learning equation for each token:

$$\gamma_{dwt} = \frac{\theta_{dt}\phi_{tw}}{\sum_{t'}\theta_{dt'}\phi_{t'w}};$$ (20)

$$\theta_{dt} \propto n_{dt} + \alpha' + \eta m_{dt};$$ (21)

$$\phi_{tw} \propto n_{tw} + \beta' + \mu m_{tw}.$$ (22)

Embedding Model Fitting Stage. The same as the word2vec training.

Regularizer Fitting Stage. We write the update gradients in matrix form:

$$\mathbf{U} \leftarrow \mathbf{U} + \xi(\eta\mathbf{P}^g\mathbf{A} + \mu\mathbf{B}')\mathbf{Y}$$ (23)

$$\mathbf{Y} \leftarrow \mathbf{Y} + \xi(\eta\mathbf{A}'\mathbf{P} + \mu\mathbf{B})\mathbf{U}$$ (24)

where $\mathbf{P}$ is the index matrix that each element p_{dw} indicates whether the word w is in the document d. $\mathbf{P}^g$ means the pseudo-inverse of the matrix $\mathbf{P}$.

6   Experiments

In order to demonstrate the accuracy improvements of the CoTE joint model framework relative to the individual embedding and topic model, this section evaluates the CoTE-PD and CoTE-LW models proposed above in two categories: the embedding vector quality assessment of the embedding models; the accuracy and interpretability assessment of topic models.

By comparing with the best results in related work, and the reference models selected by the joint model, we demonstrated the accuracy improvements of CoTE-PD and CoTE-LW, so as to indirectly verify the effectiveness of the general CoTE model.

6.1   Document Classification

In this experiment, we first train the text semantic vectors generated by different models, then use the vectors as features to classify documents using the SVM algorithm. We compare CoTE with the best reported performers like TopicVec, ETM. Some well-known methods such as GloVE [12], as well as some related works mentioned in Sect. 2 such as Gaussian LDA and TWE, are not tested because they are already beaten by the methods we compared (see [13]). Also we show the results of bag-of-words model, pLSA, LDA, word2vec, and doc2vec as baselines.

The experiments were conducted on two standard corpora: the 20 Newsgroups and the Reuters-21578, which refer to as 20News and Reuters respectively in the following. After preprocessing, the 20News corpus has 18,821 documents with 93,864 words, while Reuters contains 8,008 documents with 24,056 words.

Experimental Settings. The hyper-parameters of the experiments are applied as followings: the dimension of the embeddings of topics, words and documents is set to 500; The number of topics is specified to 300 and 111 for 20News and Reuters respectively. η and μ are kept as the "vanilla" 1.0 without tuning; and the initial learning rate is set to 0.025.

After 15 iterations on each dataset, CoTE-PD generates three vectors to represent each document d: the document vector $\mathbf{g}_d$; the average vector of all observed words of the document; and a topic vector that equals to $\theta_d \mathbf{Y}$, which is the mapping of the doc-topic proportion θ_d in the embedding space. We refer to these vectors as **DV**, **MeanWV(MWV)**, and **TV** respectively, and assemble them as the classification features. Meanwhile, CoTE-LW has two similar features of **MeanWV** and **TV**, while no **DV** presented.

Classification Results. We used the document representations of the training set as features to feed into the *LinearSVC* classifier from the Python scikit-learn [17] library, then predicted new classes on the holdout test set. We will report accuracy, precision, recall, and F1 score as the measurements. The results are shown in Table 2 and 3. Two best results are in boldface.

Table 2. Reuters classification Results.

(Reuters)	Acc	Prec	Rec	F1
BoW	94.5	92.1	90.3	91.2
pLSA	91.0	80.2	72.1	74.0
LDA	91.1	80.4	72.0	74.1
word2vec	95.1	92.4	89.8	90.6
doc2vec	86.7	85.1	71.1	70.7
TopicVec	96.1	93.2	91.2	92.0
ETM	96.6	93.8	91.6	92.7
CoTE-PD	**97.9**	**95.6**	**93.0**	**93.8**
CoTE-LW	**97.3**	**95.0**	**92.5**	**93.7**

Table 3. 20News classification Results.

(20News)	Acc	Prec	Rec	F1
BoW	78.7	78.5	78.1	78.1
pLSA	71.9	69.9	70.4	70.0
LDA	72.2	70.3	70.6	70.4
word2vec	77.6	75.8	75.6	75.7
doc2vec	72.4	69.3	69.7	69.5
TopicVec	81.3	80.3	80.0	80.1
ETM	81.7	80.9	80.4	80.6
CoTE-PD	**82.4**	**81.9**	**81.4**	**81.4**
CoTE-LW	**81.9**	**81.4**	**81.0**	**81.2**

In many times of experiments, it can be seen that CoTE-PD and CoTE-LW achieved best results on both two corpora in the evaluation criteria. What's more, between the two CoTE algorithms, CoTE-PD reached slightly better than CoTE-LW, and it achieved the best results among all the methods.

Performance of Vector Composition. We performed further experiments to analyze how the various embedding vectors can help with document classification. For this, we first conducted additional experiments for CoTE-PD with

Table 4. The vector combination forms of CoTE-PD on 20News.

(20News)	Acc	Prec	Rec	F1
MeanWV	81.8	81.4	80.8	81.1
TV	81.9	81.5	81.0	81.2
DV	70.4	68.8	68.1	68.4
MWV+TV	82.1	81.5	81.2	81.2
MWV+TV+DV	**82.4**	**81.9**	**81.4**	**81.4**

all vectors (MWV, TV, and DV) composed in several ways, and the results are shown in Table 4.

Compared between each line and with Table 3, we can see that:

1) The word embedding produced by CoTE-PD yielded an improvement of 5% over word2vec.
2) The quality of topic embedding from CoTE-PD was higher than the state-of-art topic embedding discussed in previous works.
3) The TV alone was better than MWV and DV, and TV combined with MWV showed the better performance.
4) The DV alone achieved the worst of all, but it could improve the other embeddings. The combination of the three vectors achieved the best performance.

The combinations of vectors from CoTE-LW (MWV, TV) shared the similar conclusions. The results are shown in Table 5. In addition, comparing between CoTE-PD and CoTE-LW, we can also infer that:

1) In the same combination form, CoTE-PD always performed slightly better than CoTE-LW. Maybe it is because CoTE-PD contains free document embedding parameters **G**, which allows CoTE-PD to learn better models.
2) The best results of CoTE-PD is relatively much better than the best of CoTE-LW, owing to the extra DV features.

Table 5. The vector combination forms of CoTE-LW on 20News.

(20News)	Acc	Prec	Rec	F1
MeanWV	81.6	81.2	80.6	80.9
TV	81.7	81.3	80.9	81.1
MWV+TV	**81.9**	**81.4**	**81.0**	**81.2**

6.2 Word Analogy

Word analogy is another common task to measure embedding models. Mikolov [21] found that word embedding can learn relationships between words automatically using large amounts of unannotated plain text. The linear relationships allow us to do things like $v(king) - v(man) + v(woman) \approx v(queen)$.

In this task, we trained the embedding models on a large external corpus, which was English Wikipedia dump, pre-processed by main text extraction, sentence segmentation, and tokenization. Words that appeared less than 50 times in the corpus were ignored, resulting in vocabularies of 209,533 terms and 2 billion tokens. Then we conducted the analogy experiment in Google's analogy dataset. The dataset presents questions of the form "a is to a* as b is to b*", where "b*" is masked and need be guessed. After filtering questions involving out-of-vocabulary words, we remain with 19258 instances in the dataset.

We compared CoTE (CoTE-PD and CoTE-LW as examples) with some state-of-the-art embedding methods, such as word2vec, GloVe and TWE. Some other methods for comparison in document classification task, such as doc2vec and TopicVec, are ignored here, because they require a huge amount of memory when traing on a large corpus that a normal machine cannot afford. Further, we also tested the results of SVD based on the bag-of-words model as the baseline.

The hyper-parameters of these models are as follows: the dimension of the embeddings of topics, words and documents is set to 500, and the context window is 10; The number of topics is set to 400. The results of guess accuracy are shown in Table 6.

We can see that all the embedding models performed much better than SVD, among which CoTE achieved the best performance. This demonstrates CoTE improves its embedding model part in capturing the semantics of words.

Table 6. The accuracy of word analogy task on Google's analogy dataset.

SVD	word2vec	GloVe	TWE	CoTE-PD	CoTE-LW
35.8	68.3	67.1	70.5	**71.8**	71.5

6.3 Topic Model

In this section, we compared the topic model part of CoTE with its corresponding topic model to prove the improvement of CoTE algorithm. These experiments shared the same datasets and hyper-parameter settings with the classification task.

Model Perplexity. The first measurement of the quality of topic model is the *Perplexity*. We compared the original pLSA with itself as the part of CoTE-PD (CoTE-pLSA for short), as well as LDA with itself as the part of CoTE-LW (CoTE-LDA for short). The perplexity comparison of the above models along with the first 15 iterations is shown in Fig. 1. In the figure, we can see that:

(1) CoTE-pLSA was converged better than pLSA, and CoTE-LDA better than LDA, in these comparable iterations. These demonstrates that CoTE improves its topic model part.
(2) The perplexity of the 4 models are CoTE-LDA $\approx$ CoTE-pLSA $<$ LDA $<$ pLSA.

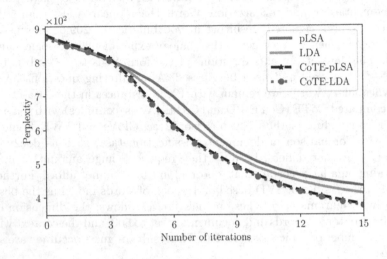

Fig. 1. The perplexity comparison of the four topic models. Note that the curve of the CoTE-pLSA is almost covered by that of CoTE-LDA.

We know that LDA adds the Dirichlet priors to pLSA, while CoTE-PD adds the "bridge distributions" as the priors (Formula 15 and 16), and CoTE-LW adds both kinds of priors (Formula 21 and 22). Since the perplexity of CoTE-pLSA was much better than that of pLSA, while CoTE-LDA only slightly improved CoTE-pLSA, we can infer that the bridge distributions learned by CoTE could be considered as better priors than Dirichlet priors.

Topic Coherence. We also adopted the widely-used Topic Coherence to measure the quality of topic models. We used the C_V metric from [18], which had the highest correlation with human annotators ($r_{Pearson} = 0.731$). To evaluate the topic coherence effectively, only the N top words for each topic were considered for the metric calculation. We chose $N = 10$ because it is a common choice in the literatures [1]. As C_V outputs one coherence value for each topic, we need to average them again to obtain a metric for the entire topic model, and multiply the result value by the factor 100 to achieve a value between 0 and 100. We call this the average topic coherence (ATC) score in the following.

We directly used the authors' implementation of C_V in their Palmetto[1] topic quality measuring toolkit. We evaluated the average topic coherence of the above models on the 20NEWS corpus, as seen in the Table 7.

We can see that the ATC of the topic model component of CoTE is much better than its corresponding original algorithm, which shows the improvement of CoTE.

Table 7. The ATC of the four topic models on 20News dataset.

pLSA	LDA	CoTE-pLSA	CoTE-LDA
39.4	41.9	**47.2**	**47.5**

7   Conclusions and Future Work

In this paper, we proposed CoTE, a flexible method for joint learning of topic and embedding models to improve both models mutually. Experiments show that CoTE can learn high-quality representations of texts such as document-topic distributions, topic embeddings and word embeddings.

In this paper, we choose two groups of de-facto topic and embedding models combined to get the simple CoTE-PD (pLSA with doc2vec) and CoTE-LW (LDA with word2vec) algorithms. In fact, CoTE can combine with a variety of models like neural networks, and either component of CoTE can be supervised or un-supervised. This is worth trying and remains future work, especially when deep topic networks and deep embedding networks are joint together. If integrating other various models, specific tasks in each stage will change accordingly but the overall 3-stage learning precedure keeps the same.

Acknowledgement. This work has been supported by the National Natural Science Foundation of China (No. U181461).

References

1. Aletras, N., Stevenson, M.: Evaluating topic coherence using distributional semantics. In: Proceedings of the 10th International Conference on Computational Semantics (IWCS 2013), pp. 13–22 (2013)
2. Bengio, Y., Ducharme, R., Vincent, P., Jauvin, C.: A neural probabilistic language model. J. Mach. Learn. Res. **3**, 1137–1155 (2003)
3. Blei, D.M., Ng, A.Y., Jordan, M.I.: Latent dirichlet allocation. J. Mach. Learn. Res. **3**, 993–1022 (2003)

---

[1] https://github.com/AKSW/Palmetto.

4. Das, R., Zaheer, M., Dyer, C.: Gaussian LDA for topic models with word embeddings. In: Proceedings of the 53nd Annual Meeting of the Association for Computational Linguistics (2015)
5. Devlin, J., Chang, M.W., Lee, K., Toutanova, K.: Bert: pre-training of deep bidirectional transformers for language understanding. arXiv preprint arXiv:1810.04805 (2018)
6. Dieng, A.B., Ruiz, F.J., Blei, D.M.: Topic modeling in embedding spaces. Trans. Assoc. Comput. Linguist. **8**, 439–453 (2020)
7. Hofmann, T.: Probabilistic latent semantic indexing. In: Proceedings of the 22nd Annual International ACM SIGIR Conference on Research and Development in Information Retrieval, SIGIR 1999, pp. 50–57. ACM, New York (1999)
8. Jiang, D., Shi, L., Lian, R., Wu, H.: Latent topic embedding. In: Proceedings of COLING 2016, the 26th International Conference on Computational Linguistics, pp. 2689–2698. The COLING 2016 Organizing Committee (2016)
9. Keya, K.N., Papanikolaou, Y., Foulds, J.R.: Neural embedding allocation: distributed representations of topic models. Comput. Linguist. **48**(4), 1021–1052 (2022)
10. Kullback, S., Leibler, R.A.: On information and sufficiency. Ann. Math. Stat. **22**(1), 79–86 (1951)
11. Le, Q.V., Mikolov, T.: Distributed representations of sentences and documents. In: ICML, vol. 14, pp. 1188–1196 (2014)
12. Levy, O., Goldberg, Y.: Neural word embedding as implicit matrix factorization. In: Ghahramani, Z., Welling, M., Cortes, C., Lawrence, N.D., Weinberger, K.Q. (eds.) Advances in Neural Information Processing Systems 27, pp. 2177–2185. Curran Associates, Inc. (2014)
13. Li, S., Chua, T.S., Zhu, J., Miao, C.: Generative topic embedding: a continuous representation of documents. In: Proceedings of the 54th Annual Meeting of the Association for Computational Linguistics, pp. 666–675. Association for Computational Linguistics (2016)
14. Li, X., Chi, J., Li, C., Ouyang, J., Fu, B.: Integrating topic modeling with word embeddings by mixtures of vMFs. In: Proceedings of COLING 2016, the 26th International Conference on Computational Linguistics: Technical Papers, pp. 151–160. The COLING 2016 Organizing Committee (2016)
15. Liu, Y., Liu, Z., Chua, T.S., Sun, M.: Topical word embeddings. In: AAAI, pp. 2418–2424 (2015)
16. Mikolov, T., Chen, K., Corrado, G., Dean, J.: Efficient estimation of word representations in vector space. CoRR abs/1301.3781 (2013)
17. Pedregosa, F., et al.: Scikit-learn: machine learning in Python. J. Mach. Learn. Res. **12** (2012)
18. Röder, M., Both, A., Hinneburg, A.: Exploring the space of topic coherence measures. In: Proceedings of the Eighth ACM International Conference on Web Search and Data Mining (WSDM 2015), pp. 399–408 (2015). https://doi.org/10.1145/2684822.2685324
19. Shi, B., Lam, W., Jameel, S., Schockaert, S., Lai, K.P.: Jointly learning word embeddings and latent topics. In: Proceedings of the 40th International ACM SIGIR Conference on Research and Development in Information Retrieval, pp. 375–384 (2017)
20. Vorontsov, K., Potapenko, A.: Additive regularization of topic models. Mach. Learn. (2014)
21. word2vec (2013). https://code.google.com/archive/p/word2vec/

22. Xu, H., Wang, W., Liu, W., Carin, L.: Distilled Wasserstein learning for word embedding and topic modeling. In: Advances in Neural Information Processing Systems, vol. 31 (2018)
23. Xun, G., Li, Y., Zhao, W.X., Gao, J., Zhang, A.: A correlated topic model using word embeddings. In: IJCAI, vol. 17, pp. 4207–4213 (2017)

Construction of Multimodal Dialog System via Knowledge Graph in Travel Domain

Jing Wan[1], Minghui Yuan[1], Zhenhao Dong[1], Lei Hou[2(✉)], Jiawang Xie[2], Hongyin Zhu[2], and Qinghua Wen[2]

[1] Beijing University of Chemical Technology, Beijing 100029, China
[2] Tsinghua University, Beijing 100084, China
`houlei@tsinghua.edu.cn`

Abstract. When traveling to a foreign city, we often find ourselves in dire need of an intelligent agent that can provide instant and informative responses to our various queries. Such an agent should have the ability to understand our queries and possess the knowledge to generate helpful responses. Furthermore, if the agent can comprehend image information, it can provide solutions from multiple perspectives. Knowledge graph-based multimodal dialog systems offer a promising approach to fulfill these requirements. In this paper, we present a solution for efficiently constructing a multimodal dialog system in the travel domain without large-scale datasets. The system's main objective is to assist users in completing various travel-related tasks, specifically attraction recommendation and route planning, which are frequently requested by users while traveling. We introduce the Multimodal Chinese Tourism Knowledge Graph (MCTKG) and integrate image processing and recommendation technology into a dialog system. Specifically, our approach utilizes modular design to construct the dialog system, and leverages the rich information available in the knowledge graph to enhance the performance of each module. To the best of our knowledge, this is the first multimodal travel dialog system that provides users with personalized travel route recommendations. Multiple experiments have proven that our dialog system can effectively enhance the user's travel experience.

Keywords: Multimodal dialog system · Conversational recommender · Multimodal knowledge graph

1 Introduction

In recent years, dialog systems have become ubiquitous in various aspects of our lives and have gained increasing attention [1]. Dialog systems can be broadly classified as open-domain and task-oriented. Open-domain systems can converse on various topics without domain restrictions, while task-oriented systems assist

X. Song et al. (Eds.): APWeb-WAIM 2023, LNCS 14334, pp. 422–437, 2024.
https://doi.org/10.1007/978-981-97-2421-5_28

users in completing specific tasks in vertical fields to reduce manual workload. Leading travel industry companies such as Expedia.com, KLM, and Booking.com are racing to introduce their own online chatbots. Researchers and industry experts agree that a strong and effective task-oriented dialog system can greatly improve the user experience.

Although dialog systems in the travel domain offer significant business potential, building a satisfactory dialog system that can fully serve the domain remains an arduous task. The tourism dialog system faces the following challenges.

1. Multimodal and fine-grained understanding for tourism. Parsing user's input and obtaining semantic information are essential requirements for any dialog system. However, in the field of tourism, most researches has overlooked the utilization of visual information from images. Using visual input of attractions can enhance a dialog system's ability to identify the intent of a person who lacks sufficient knowledge to describe their desired attractions in words. This requires achieving a multimodal understanding that extends beyond natural language understanding. Furthermore, traditional natural language comprehension parses a user's natural language input into predefined semantic slots, which may vary for specific domains. In tourism, users may pay particular attention to specific aspects such as price, distance, attraction type, etc. Mining domain terms and building fine-grained semantic slots automatically is another challenge. Both multimodal information and fine-grained slots can enhance the system.
2. Travel route recommendation. The recommendation task in the tourism domain differs from other domains. Unlike recommending a single item, users seek a satisfactory travel route consisting of multiple target attractions through multiple rounds of dialog interaction. This requires the dialog system to fully consider the explicit and implicit constraints involved in the user's dialog process and recommend suitable attractions. The dialog system needs to have sufficient background knowledge to assist with these tasks.

To address the above problems, we propose using the Multimodal Chinese Tourism Knowledge Graph (MCTKG) as an external knowledge supplement to construct a multimodal dialog system for the tourism domain. In general, MCTKG extracts a large number of attraction entities, attributes, and images from various heterogeneous data sources, such as visitbeijing.com.cn, Baidu Encyclopedia, and tripadvisor.com. The main contributions of our work are as follows.

First, we improve the conventional understanding module in three ways, providing a comprehensive understanding of multimodal and domain-specific inputs. (1) To facilitate visual comprehension, we introduce pre-trained visual models and devise user's intent related to images. (2) Through the analysis of attribute values of entities extracted from MCTKG, we construct a fine-grained slot ontology in the tourism domain to acquire more comprehensive semantic information. (3) To recognize explicit or implicit user requirements, we add the ability to identify domain-related constraints to the dialog system.

Second, we leverage the multimodal information and the knowledge graph to enhance the effect of travel route recommendation. We use an automatic route generation algorithm to generates initial travel routes from user's utterances, which are further refined by visual information. The recommendation module utilizes the user's historical dialog information to identify the most similar candidate entities in the knowledge graph, which are then filtered using explicit and implicit conditional constraints. Users can also modify the initial route based on recommended results. During the dialog, the system provides information assistance to the user through MCTKG.

2   Related Work

2.1   Task-Oriented Dialog Systems

Task-oriented dialog system design has always been a critical area of concern for the research community and industry. It has achieved apparent success in enterprise commerce and education, and significantly reduced the manual workload. Most traditional task-oriented dialog systems are based on text and are constructed in a modular form [14,25], comprising the following components: (1) natural language understanding, usually joint modeling to complete intent extraction and slot filling tasks [13,16]; (2) dialog state tracking, used to track the user's goals and constraints in the dialog process, and determine the value of the predefined slot; (3) policy network, the policy network component receives the information from the dialog state tracking module, and decides what action to take next; (4) a Natural Language Generation (NLG) model, used to convert the action taken by the strategy network into natural language and output, can be achieved technically through predefined sentence templates [6] or generation-based methods [4]. In addition, with the development of deep learning, some end-to-end task-oriented dialog systems have emerged [18,23].

Recently, some dialog systems have introduced external knowledge bases to better answer user's questions and significantly improve the system's performance [9,27]. In regard to complex tasks with multimodal information, these text-based task-oriented systems are greatly restricted and cannot effectively meet user needs.

The demand for multimodal travel dialog systems is increasing due to the rich visual semantics in attraction images. However, limited research exists due to the lack of large-scale multimodal dialog datasets. Saha et al. [17] constructed a multimodal dialog (MMD) dataset. Later, Liao et al. [11] proposed a knowledge-aware multimodal dialog (KMD) model. Nevertheless, a need of pertinent large-scale multimodal dialog datasets still exists in tourism field. We adopt a modular design method to build the travel dialog system.

2.2   Conversational Recommendation

The goal of the recommendation system is to select a subset that meets user's needs from the entire collection of items. In particular, the knowledge graph is

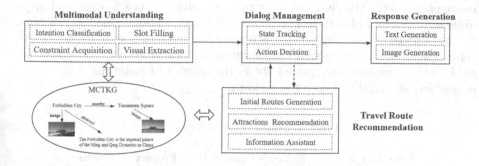

Fig. 1. Framework of multimodal travel dialog system based on modular design.

widely used to improve recommendation performance and interpretability [8, 21]. A conversational recommendation system provides a more natural method of interaction for recommendation services. Instinctively, integrating recommendation technology into the dialog system will significantly enhance user's experience, especially in the tourism industry. For dialog systems, the introduction of recommendation technology can better meet user's needs, such as recommending suitable tourist attractions. For recommendation systems, focus on historical dialog information instead of historical interactive data, which can effectively solve the cold start problem [28].

In our work, we make full use of text and image data, and introduce knowledge graphs as external knowledge supplements. First, we use an automatic route generation algorithm to generate the initial travel route. Then the recommendation module and the information assistant module help users to choose their favorite attractions to complete route planning tasks.

3 Multimodal Travel Dialog System

As shown in Fig. 1, our framework has four modules. The multimodal understanding module recognizes user's intent and fills semantic slots. The dialog management module stores historical conversation information and decides which actions to take. The route planning module is responsible for helping users solve some travel-related tasks. The response generation module generates text or image responses. MCTKG serves as an external knowledge base to improve the performance of the entire dialog system.

3.1 Multimodal Chinese Tourism Knowledge Graph

Multimodal Chinese Tourism Knowledge Graph obtains and integrates heterogeneous Beijing tourist attractions data from major Chinese tourism websites, including visitbeijing.com.cn, Baidu Encyclopedia, tripadvisor.com,

meituan.com, etc. MCTKG [24] construct an ontology classification and orga-
nizes concepts in a hierarchical structure. MCTKG contains nine top concepts.
At the same time, each top concept is further divided into a set of low-level concepts. For example, *Attraction* can be further divided into *natural landscape*,
park, *religiousplaces*, etc. Table 1 shows the number of instances, image links,
properties, and triples of each top concept.

Table 1. Statistics of each concept

Top Concept	Instances	ImageLinks	Properties	Triples
Attraction	2,106	44,539	1,069	96,493
Time-honored shop	1,771	10,798	25	52,127
Brand	60	1,799	3	4,145
Route	232	0	6	1,392
Travel style	22	0	2	22
Person	37,290	194,564	4,102	584,477
Building	3,501	99,850	2,813	161,557
Relic	705	20,810	860	30,436
Organization	10,643	315,698	5,742	442,062

3.2 Multimodal Understanding

The correct understanding of user's intention is an integral part of the dialog system. As a multimodal dialog system, the user's input may contain text, images or
both. Given a multimodal utterance u, the multimodal utterance understanding
module should map u to the user's intention and the corresponding slot value.
This process is followed by generating the corresponding representation, denoted
as U, for the multimodal utterance:

$$U = <I, A, C, V> \tag{1}$$

where I denotes the user's intention of the utterance (Table 2 shows some intentions), A denotes the attributes or slot values extracted from u, C denotes constraints contained in u, and V indicates the visual representation of the images
in u.

Slot Ontology Construction. Developing structured ontologies for language
understanding is challenging in the cold start phase because domain experts must
define the slots and possible values. We have extended traditional slot ontology
to make the dialog system more applicable to tourism field.

First, we extract the different attribute names $\{a_1, \cdots, a_n\}$ of all entities in
MCTKG, and the number of attributes is 8,883. Our experience has shown that
attributes sharing a common suffix typically have similar properties. We merged

the attributes with the same suffix into the same set, treating the suffix as the central word of all the attributes in the set. Finally we obtained 4,490 central words, and recorded the number of attributes in the corresponding set.

Second, we analyzed central words with set length less than k to determine important tourism attributes. We found that most central words have set length less than 10, so k was set to 15, resulting in identification of 240 central words. After manual review, we merged identical attributes and obtained 548 attributes to expand slot ontology, including *Opening hours, Floor space*, and uncommon words such as *Snack provided, Premium Steak* for attractions.

Table 2. Intention types and examples

Intention Type	Example
Information service	How old is the Forbidden City?
Location recognition	Can you tell me which attraction this is?
Chitchat	The weather is nice today
Recommendation	What are the attractions like The Summer Palace?
Route planning	Plan an one-day tour of Beijing for me
Negation	I don't want to go to this place
Add attraction	This is a nice place. Please add it to my tour route

Intent Classification and Slot Filling. Intent classification is a classification problem and slot filling is a sequence labeling task. Recent studies have highlighted the effectiveness of joint learning methods in addressing both tasks simultaneously [5]. We adopt a joint intent classification and slot filling model based on Bidirectional Encoder Representations from Transformers (BERT) as in [2]. Additionally, we employed tourism-specific large-scale text data to fine-tune a BERT model that is optimized for the tourism domain. During the fine-tuning process, for intent classification, we use the output corresponding to the [CLS] and the Softmax classifier.

$$y^i = softmax(W^i h_1 + b^i) \qquad (2)$$

where h_1 is the output corresponding to the special identifier [CLS], W^i and b^i represent the weight coefficient and bias in the feedforward neural network respectively. For slot filling, the other hidden states $\{h_2, \cdots, h_N\}$ are used to feed into the feedforward neural network, and passed through the softmax classifier to obtain the probability distribution of each token on the slot label:

$$y^s_n = softmax(W^s h_n + b^s), n \in 2, \cdots, N \qquad (3)$$

To model the two tasks of intent classification and slot filling together, the training objectives of the model can be formalized as follows:

$$p(y^i, y^s | x) = p(y^i | x) \prod_{n=2}^{N} p(y^s_n | x) \qquad (4)$$

The training goal of the model is to maximize the conditional probability $p(y_n^s|x)$ in the above formula, and the model performs end-to-end fine-tuning by minimizing the cross-entropy loss function.

Constraints Acquisition. Constraints play a crucial role as they allow the system to quickly filter out a large number of potential recommendations. Our system presently incorporates location, price, attraction level, attraction type, distance, time, and travel style as constraints. We employed two methods to get user constraints from user's input: option boxes for fixed categories, and regular expressions or template matching for numeric constraints like price or time. The system ultimately filters out unqualified attractions based on the given constraints.

Visual Extraction. Since every entity in MCTKG is accompanied by a precise multilevel classification and high-quality images, the user's image inputs can be effectively utilized to perform location recognition and image-based attraction recommendation tasks. First, to improve the calculation speed, we use ResNet [7] in advance to extract the visual features $V_{all} = \{v_1, \cdots, v_n\}$ of the existing pictures in the graph and save them offline, where n is the number of images. When the user's input u_i contains an image, we use ResNet to extract its visual feature V_i, and use the cosine similarity calculation method to select the 5 images closest to V_i. If $I_i = Location\ Recognition$, the system selects the image with the highest score and its corresponding entities as the response. If $I_i = Image\ Recommend$, system selects the 5 images and their corresponding entities, entity categories as the response. Users can freely choose the type of attractions they are interested in for further inquiry. Moreover, visual features are also used later in the recommendation module.

3.3   Travel Route Recommendation

The route planning module consists of three parts: initial route generation, attraction recommendation and information assistance. The initial route generation part obtains the user's initial needs and uses the automatic route generation algorithm to plan the initial route for the user.

The attraction recommendation module leverages contextual dialogue information to suggest relevant attractions to users. We also use a multimodal approach combining textual and image information to improve recommendation effectiveness.

The information assistant incorporates a knowledge graph to provide users with a comprehensive understanding of attractions and employs diverse query methods to offer informed responses. Users can modify their initial travel route based on recommendations and relevant information provided by the assistant.

Initial Route Generation. To fulfill user's requirements and generate customized travel route, we have predefined four initial needs slots, which include travel style, starting location, expected duration of stay, and attraction preferences. Our system employs active questioning techniques to gather the necessary values for each slot. Then the automatic route generation algorithm is utilized to create the initial travel route based on the user's input. The pseudocode of the algorithm is shown in Algorithm 1.

Algorithm 1. Automatic generation algorithm of initial routes

---

Input: Travel style S; Starting location L; Expected stay time T; Attraction of interest I; All attractions entity E; Number of routes N
Output: Initial Routes R
1: **function** GENERATEROUTES(S, L, T, I, E, N)
2:     $R \leftarrow [\varnothing]$
3:     $candidates \leftarrow [\varnothing]$
4:     **for** $attraction$ in E **do**
5:         **if** $GetStyle(attraction) == S$ **then**
6:             $candidates.append(attraction)$
7:         **end if**
8:     **end for**
9:     **for** $i = 0 \rightarrow N - 1$ **do**
10:         $route \leftarrow [\varnothing]$
11:         $route.append(L)$
12:         $route.append(I)$
13:         **while** $CalculateTime(route) <= T$ **do**
14:             $route.append(random.sample(candidates, 1))$
15:         **end while**
16:         $route \leftarrow GetShortestRoute(route)$
17:         $R.append(route)$
18:     **end for**
19:     **return** R
20: **end function**

Attractions Recommendation. The traditional search method of constructing a query for recommendation is limited by natural language expression. To make the interaction between users and the dialog system more natural and solve the above problems, we introduce conversational recommendation techniques [28] to indicate the effect of recommendation. Given an attraction entity e_n, e_n^{text} represents its valuable text content, and e_n^{img} represents its corresponding picture. We use BERT and ResNet to extract its text feature t_n and image feature v_n respectively and introduce an MFB layer to obtain their fusion representation f_n, whose effectiveness in combining multimodal features has been proven in many visual tasks [26].

$$t_n = BERT(e_n^{text}) \tag{5}$$

$$v_n = ResNet(e_n^{img}) \tag{6}$$

$$f_n = SumPooling(M_1^T t_n \circ M_2^T v_n, k) \qquad (7)$$

where M_1^T and M_2^T are two transform matrices used to map t_n and v_n to the common high-dimensional space, $\circ$ denotes the element-wise product. For the user's input $U = <I, A, C, V>$, when the system detects that the user's intention is to recommend, it uses the included A, C, and V to make a recommendation. A represents the user's text content, V represents the user's image content. We use the same method to fuse the two content to obtain the user's fusion representation f_{user}. Then, f_{user} and f_n are used to calculate a matching score for ranking. C represents the constraints that filter recommended candidates.

Information Assistant. Recognizing that users may require additional information about specific attractions before modifying their route, we have developed an information assistant to cater to user's needs. This assistant incorporates three information retrieval methods, including the following:

(1) Knowledge graph query based on SPARQL language. The attraction name and attribute name extracted from the user's input are mapped to the corresponding SPARQL sentence, and searched in MCTKG;
(2) Information query based on open information extraction results. We use open relation extraction techniques based on dependency syntactic analysis to generate triples from texts of attractions, such as the summary of attractions, referring to the work of Wen [22]. We save all triples and the corresponding text, and when the answer is not retrieved in the knowledge graph, we search for the relevant answer in these triples and return the original text to the user together;
(3) Information retrieval (IR) based on BERT. A large number of experiments have proven that BERT can also perform well on information retrieval tasks [3]. We convert the input of BERT into the prompt of ([CLS] Query [SEP] Paragraph), where Query represents the user's question and Paragraph represents the paragraph of the attractions. The model calculates the correlation between two paragraphs and selects the one with the highest correlation as the answer. To improve efficiency and reduce computation load, we use the term frequency-inverse document frequency (TF-IDF) algorithm to retrieve 20 most relevant paragraphs and then use the BERT model to determine the correct answer from those paragraphs.

The three methods are executed sequentially, and if one method retrieves an answer, the subsequent methods are not called.

3.4 Dialog Management

Given the absence of large-scale dialog datasets for specific tasks in the tourism field, supervised methods are not feasible to train dialog management models. As an alternative, we designed a set of rules based on different dialog scenarios to ensure a smooth dialog process and enhance the dialog system's performance.

Dialog State Tracking. S_t represents the dialog state at time t, which is jointly determined by the state S_{t-1} at the previous time and the current sentence representation U_t.

$$S_t = \Upsilon(S_{t-1}, <I_t, A_t, C_t, V_t>) \tag{8}$$

where Υ represents our predefined dialog rules, summarized as follows:

(1) if $I_t = Information\ Service$, update S_t based on A_t and V_t;
(2) if $I_t = Location\ Recognition$, update S_t based on V_t;
(3) if $I_t = Chitchat$, $S_t = S_{t-1}$;
(4) if $I_t = Add\ Attraction$, update S_t by A_t and C_t;
(5) if $I_t = Negation$, S_t will inherit information stored in S_{t-1} while updating the parts according to U_t;
(6) if user has not responded for more than *timelimit* minutes, the system will clean S_t and *timelimit* is a predefined constant.

Action Decision. Similar to dialog state tracking, we also use a series of rules to convert the obtained dialog state into the corresponding action.

(1) **Question Answering.** It will be triggered when I_t is $Information\ Service$ and the name of attractions is detected. The system will call the Attractions Information Assistant to return the query result.
(2) **Attractions Recommendation.** It will be triggered when I_t is $Recomm$ $endation$ and A_t, C_t, V_t are not empty. The system will call Attractions Recommendation to return the results.
(3) **Route Planning.** It will be triggered when I_t is $Route\ Planning$ and there are several situations: (a) If the initial route has not been generated, the system will collect information by actively asking questions; (b) If there is an initial route and the user intends to delete or add attractions, the distance and time judgment algorithm will be invoked to determine whether the modified route meets the requirements. For example, if the user exceeds the constraints set for total tour time or travel distance, the system will prompt to adjust constraints or choose different attractions.
(4) **Image Reply.** It will be triggered when I_t is $Location\ Recognition$ or $Image\ Recommend$. The system will return the result of location recognition and the corresponding picture.
(5) **Chitchat.** It will be triggered when I_t is $Chitchat$. if the system cannot respond to questions that are not related to the task, the conversation may not continue. Therefore, it will reply to the chitchat according to the predefined conversation template.
(6) **Route Navigation.** It will be triggered when I_t is $Affirmation$ about the final route, and the Baidu Map API will be called to generate navigation, including walking, cycling, public transportation, etc.

3.5 Response Generation

Response generation is the final step of the entire dialogue process. Our system employs two methods: text reply and picture reply. The text method is further divided into IR-based response generation and template-based response generation.

Text Response Generation. Most of the responses are generated by the template-based method, as it is easier to control. For example, if the following detailed action is required ($TravelStyle = S$) in initial routes generation, The system may respond with the question "What type of attractions do you want to visit?". When the information retrieval module is invoked, we directly retrieve the paragraph that is most relevant to the user's query and use it as a response. For the question "颐和园和哪些园林并称为中国四大名园?" (Which other gardens are commonly known as the Four Famous Gardens of China alongside the Summer Palace?), The system's response may be "颐和园、承德避暑山庄、拙政园、留园并称为中国四大名园." (The Summer Palace, Chengde Summer Resort, Humble Administration Garden and Liuyuan Garden are known as the four famous gardens in China.).

Image Response Generation. Image response generation is used for location recognition and recommendation. When the user inputs a picture, we typically search for similar pictures in the knowledge graph and return them, or we return the corresponding pictures based on the generated list of recommended attractions.

4 Experiments

4.1 Evaluation of Language Understanding

To have a sufficient amount of data for training our BERT model, we initially use a templated method to create different questions for each intent. Placeholders are used to replace specific positions in the questions. When generating questions, the entities, attributes, or concepts in the MCTKG are used to replace the placeholders. Then, we merge our data with CrossWoz [29] data. CrossWoz is a large-scale Chinese dialog dataset. We divide the dataset into a training set, validation set and test set at ratios of 0.6, 0.2 and 0.2. The evaluation metric of intent classification is accuracy, while the metric for slot filling is F1 score. We employ Adam as our optimizer with a learning rate of 5e−5 and gradually reduce it to maintain the stability of training. The number of training epochs is 50, and the batch size is 64.

We compare some baselines: Boosting [20], Boosting+Simplified sentences [19], RNN-EM [15], Encoder-labeler Deep LSTM [10], Attn.BiRNN [12], Slot-Gated [5]. As seen in Table 3, our joint model generally outperforms the independent training model. Compared to standard RNN models, the BERT model has stronger semantic capture and generalization ability due to its pre-training on large-scale corpuses. The joint BERT model significantly outperforms the baseline joint models on both tasks, demonstrating the model's strong capabilities. Additionally, using a joint model for both intent classification and slot filling simplifies the dialog system as only one model needs to be trained and deployed.

Table 3. NLU performance

Model	Intent	Slot
Boosting	0.823	-
Boosting+Simplified sentences	0.848	-
RNN-EM	-	0.927
Encoder-labeler Deep LSTM	-	0.934
Attn.BiRNN	0.887	0.941
Slot-Gated	0.892	0.947
Joint BERT	**0.958**	**0.975**

Table 4. Recommendation performance

Method	HR@5	HR@10
Text only	0.5430	0.6327
Image only	0.5592	0.6504
Text + Image	**0.6948**	**0.7975**

4.2   Evaluation of Attraction Recommendations

To evaluate the efficacy of our recommendation approach, we retrieved popular tourist attractions and their recommended attractions from professional travel websites and applied our method to generate recommended results for these attractions, which were then compared with the actual recommended attractions. We use hit radio@5 and hit radio@10 for evaluation. As shown in Table 4, The results of text recommendation and image recommendation are similar in the field of tourism as attractions can be related based on their background stories or appearances. Combining both methods can improve recommendation performance. Furthermore, constraints collected in the dialog can help users filter candidate results, improving the recommendation performance of the dialog system.

4.3 Evaluation of the Information Assistant

In tourism, providing users with adequate knowledge information is crucial for informed decision-making about attractions. The Information Assistant uses three methods to provide knowledgeable answers. We conducted ablation experiments to analyze the effectiveness of these methods for knowledge questioning. We generated 100 questions from users, used the Information Assistant to provide responses, and evaluated the accuracy of the answers using user feedback.

Table 5 presents the results of Information Assistant for knowledge question answering. We found that only using SPARQL for knowledge graph query has the worst effect, and the combination of the three methods has the best effect, which is precisely in line with our expected results. Open information extraction and information retrieval techniques make full use of the large amount of unstructured text available in the knowledge graph to generate more answers for users. Compared with the two of them, information retrieval has a more noticeable impact as most of the relevant triples are already integrated into the knowledge graph.

Table 5. Ablation study on knowledge question answering

Method	Accuracy
Full AIA	**0.75**
AIA-IR	0.61
AIA-IR-OpenIE	0.56

Table 6. Human evaluation results

	Fluency	Informativeness	Planning satisfaction
Participant1	2.7	3.6	3.8
Participant2	2.9	3.8	4.2
Participant3	2.7	3.8	4.1
Participant4	2.4	3.5	3.9
Participant5	3.0	3.9	4.2
Avg	2.74	3.72	4.04

4.4 Evaluation of the System

We developed a demo of a dialog system for human evaluation. Five participants were recruited to evaluate the system independently, in which the system performed route planning tasks and provided knowledge questions and answers. Participants were asked to rate the conversation's fluency, informativeness, and planning satisfaction. To reduce personal bias, we administered 50 sets of conversation tests to each participant based on different travel goals. They were asked to rate each quality on a Likert scale ranging from 1 (low quality) to 5 (high quality).

Table 6 presents the result of human evaluation for the dialog system. We found that when compared to other aspects, planning satisfaction with the system's recommendations exhibits a more favorable performance. That's probably because the system's dialog contains a significant amount of informative content from various resources, such as textual information, image data, and external APIs. We utilize these information to construct a more comprehensive multimodal travel dialog system.

5   Discussion and Conclusions

In this work, our objective was to construct an intelligent dialog system to address diverse user requirements in the field of tourism. To achieve this, we adopted a modular approach and integrated MCTKG as a knowledge supplement in the system. Facts have proven that the modular construction approach addresses the scarcity of multimodal dialog datasets in the tourism domain. Furthermore, the dialog process can be effectively controlled through a set of rules. Additionally, incorporating image information and recommendation technology has substantially enhanced the dialog system's performance. Our framework can be used as a template to build a new dialog system in other domains using a modular design approach.

In our future work, we plan to construct a large-scale multimodal dialog dataset in the tourism field and develop an end-to-end multimodal dialog system. The lack of large-scale domain-specific conversation datasets make it difficult for deep learning models to improve the dialog management and response generation modules. Additionally, enhancing the interpretability of recommended results is an important task that we aim to focus on.

References

1. Chen, H., Liu, X., Yin, D., Tang, J.: A survey on dialogue systems: recent advances and new frontiers. ACM SIGKDD Explor. Newsl. **19**(2), 25–35 (2017)
2. Chen, Q., Zhuo, Z., Wang, W.: Bert for joint intent classification and slot filling. arXiv preprint arXiv:1902.10909 (2019)
3. Dai, Z., Callan, J.: Deeper text understanding for IR with contextual neural language modeling. In: Proceedings of SIGIR, pp. 985–988. Association for Computing Machinery, New York (2019)
4. Dhingra, B., et al.: Towards end-to-end reinforcement learning of dialogue agents for information access. In: Proceedings of ACL, Vancouver, Canada, pp. 484–495. Association for Computational Linguistics (2017)
5. Goo, C.W., et al.: Slot-gated modeling for joint slot filling and intent prediction. In: Proceedings of NAACL-HLT, New Orleans, Louisiana, pp. 753–757. Association for Computational Linguistics (2018)

6. Han, S., Bang, J., Ryu, S., Lee, G.G.: Exploiting knowledge base to generate responses for natural language dialog listening agents. In: Proceedings of SIGDIAL, Prague, Czech Republic, pp. 129–133. Association for Computational Linguistics (2015)

7. He, K., Zhang, X., Ren, S., Sun, J.: Deep residual learning for image recognition. In: Proceedings of CVPR, New York, USA, pp. 770–778. IEEE (2016)

8. Huang, J., Zhao, W.X., Dou, H., Wen, J.R., Chang, E.Y.: Improving sequential recommendation with knowledge-enhanced memory networks. In: Proceedings of SIGIR, pp. 505–514. Association for Computing Machinery, New York (2018)

9. Jung, J., Son, B., Lyu, S.: AttnIO: knowledge graph exploration with in-and-out attention flow for knowledge-grounded dialogue. In: Proceedings of EMNLP, Stroudsburg, PA, pp. 3484–3497. Association for Computational Linguistics (2020)

10. Kurata, G., Xiang, B., Zhou, B., Yu, M.: Leveraging sentence-level information with encoder LSTM for semantic slot filling. In: Proceedings of EMNLP, Austin, Texas, pp. 2077–2083. Association for Computational Linguistics (2016)

11. Liao, L., Ma, Y., He, X., Hong, R., Chua, T.S.: Knowledge-aware multimodal dialogue systems. In: Proceedings of ACM MM, pp. 801–809. Association for Computing Machinery, New York (2018)

12. Liu, B., Lane, I.: Attention-based recurrent neural network models for joint intent detection and slot filling. In: Proceedings of Interspeech, Baixas, France, pp. 685–689. ISCA-INT Speech Communication Association (2016)

13. Liu, H., Zhang, F., Zhang, X., Zhao, S., Zhang, X.: An explicit-joint and supervised-contrastive learning framework for few-shot intent classification and slot filling. In: Proceedings of EMNLP, Punta Cana, Dominican Republic, pp. 1945–1955. Association for Computational Linguistics (2021)

14. Mrkšić, N., Séaghdha, D.O., Wen, T.H., Thomson, B., Young, S.: Neural belief tracker: data-driven dialogue state tracking. In: Proceedings of ACL, Stroudsburg, PA, pp. 1777–1788. Association for Computational Linguistics (2017)

15. Peng, B., Yao, K., Jing, L., Wong, K.F.: Recurrent neural networks with external memory for spoken language understanding. In: Li, J., Ji, H., Zhao, D., Feng, Y. (eds.) NLPCC 2015. LNCS, vol. 9362, pp. 25–35. Springer, Cham (2015). https://doi.org/10.1007/978-3-319-25207-0_3

16. Qin, L., Xu, X., Che, W., Liu, T.: AGIF: an adaptive graph-interactive framework for joint multiple intent detection and slot filling. In: Proceedings of EMNLP, Stroudsburg, PA, pp. 1807–1816. Association for Computational Linguistics (2020)

17. Saha, A., Khapra, M.M., Sankaranarayanan, K.: Towards building large scale multimodal domain-aware conversation systems. In: Proceedings of AAAI, Palo Alto, CA, pp. 696–704. AAAI Press (2018)

18. Serban, I., Sordoni, A., Bengio, Y., Courville, A., Pineau, J.: Building end-to-end dialogue systems using generative hierarchical neural network models. In: Proceedings of AAAI, Palo Alto, CA, vol. 30, pp. 3776–3783. AAAI Press (2016)

19. Tur, G., Hakkani-Tür, D., Heck, L., Parthasarathy, S.: Sentence simplification for spoken language understanding. In: Proceedings of ICASSP, New York, USA, pp. 5628–5631. IEEE (2011)

20. Tur, G., Hakkani-Tür, D., Heck, L.: What is left to be understood in atis? In: IEEE Spoken Language Technology Workshop, pp. 19–24. IEEE (2010)

21. Wang, X., Wang, D., Xu, C., He, X., Cao, Y., Chua, T.S.: Explainable reasoning over knowledge graphs for recommendation. In: Proceedings of AAAI, Palo Alto, CA, vol. 33, pp. 5329–5336. AAAI Press (2019)

22. Wen, Q., Tian, Y., Zhang, X., Hu, R., Wang, J., Hou, L., Li, J.: Type-aware open information extraction via graph augmentation model. In: Chen, H., Liu, K., Sun, Y., Wang, S., Hou, L. (eds.) CCKS 2020. CCIS, vol. 1356, pp. 119–131. Springer, Singapore (2020). https://doi.org/10.1007/978-981-16-1964-9_10
23. Wen, T.H., et al.: A network-based end-to-end trainable task-oriented dialogue system. In: Proceedings of EACL, Stroudsburg, PA, pp. 438–449. Association for Computational Linguistics (2017)
24. Xie, J., et al.: Construction of multimodal Chinese tourism knowledge graph. In: Zeng, J., Qin, P., Jing, W., Song, X., Lu, Z. (eds.) ICPCSEE 2021. CCIS, vol. 1452, pp. 16–29. Springer, Singapore (2021). https://doi.org/10.1007/978-981-16-5943-0_2
25. Yan, Z., Duan, N., Chen, P., Zhou, M., Zhou, J., Li, Z.: Building task-oriented dialogue systems for online shopping. In: Proceedings of AAAI, Palo Alto, CA, vol. 31, pp. 4618–4625. AAAI Press (2017)
26. Yu, Z., Yu, J., Fan, J., Tao, D.: Multi-modal factorized bilinear pooling with co-attention learning for visual question answering. In: Proceedings of ICCV, New York, USA, pp. 1839–1848. IEEE (2017)
27. Zhang, C., Wang, H., Jiang, F., Yin, H.: Adapting to context-aware knowledge in natural conversation for multi-turn response selection. In: Proceedings of the Web Conference, pp. 1990—2001. Association for Computing Machinery, New York (2021)
28. Zhou, K., Zhao, W.X., Bian, S., Zhou, Y., Wen, J.R., Yu, J.: Improving conversational recommender systems via knowledge graph based semantic fusion. In: Proceedings of KDD, pp. 1006–1014. Association for Computing Machinery, New York (2020)
29. Zhu, Q., Huang, K., Zhang, Z., Zhu, X., Huang, M.: Crosswoz: a large-scale Chinese cross-domain task-oriented dialogue dataset. Trans. Assoc. Comput. Linguist. 8, 281–295 (2020)

DCNS: A Double-Cache Negative Sampling Method for Improving Knowledge Graph Embedding

Hao Zheng, Donghai Guan$^{(\boxtimes)}$, Shuai Xu, and Weiwei Yuan

College of Computer Science and Technology, Nanjing University of Aeronautics and
Astronautics, Nanjing 211106, China
{2016106,dhguan,xushuai7,yuanweiwei}@nuaa.edu.cn

Abstract. Negative sampling plays an important role in knowledge
graph embedding. A high-quality negative sample can push the model
training to the limit. Most negative triples generated by simple uniform
sampling are low-quality negative samples, which will lead to the problem of vanishing gradients in the training process. Generative Adversarial
Network (GAN) has been used in the study of negative sampling methods. However, the training for the GAN-based negative sampling method
is more complicated. To solve these issues, we propose DCNS. DCNS
designs two caches containing high-quality negative triples, samples from
the cache and updates the cache. In addition, in order to generate harder
negative samples that have a greater impact on training, DCNS adopts a
mixing operation. Finally, we evaluated the results of the link prediction
model using DCNS on four standard datasets. The extensive experiments
show that our method can gain significant improvement on various KG
embedding models, and outperform the state-of-the-art negative sampling methods.

Keywords: Mixing · Negative sampling · Link prediction · Knowledge
graph embedding

1 Introduction

Knowledge graph [13] is a classic graph structure, in which all entities are represented in the form of nodes, and all relations are represented in the form of edges.
The knowledge graph contains a large amount of data and has been successfully
applied in many fields, including intelligent question answering [24], information
retrieval [23], and recommendation systems [7], etc. Such importance has also
inspired many famous knowledge graph projects, e.g., WorldNet [11], FreeBase
[2], NELL [12], and YAGO [15]. Knowledge graph is composed of a large number
of fact triples in the form of $(Entity, Relation, Entity)$, which is simplified to
(h, r, t), where h represents the head entity, t represents the tail entity, and r
represents the relation between h and t.

X. Song et al. (Eds.): APWeb-WAIM 2023, LNCS 14334, pp. 438–450, 2024.
https://doi.org/10.1007/978-981-97-2421-5_29

At present, knowledge graph embedding methods [14,18] that emb entities and relations in knowledge graphs into low-dimensional continuous vector spaces have been widely used. And there is a trend to build scoring functions that improve model performance based on interactions between entities and relations. However, researchers have focused more on building good scoring functions to train better low-dimensional embedding of entities and relations, another very important aspect of knowledge graph embedding: negative sampling [9], has not received sufficient attention. A good negative sampling method can also improve the performance of the model.

Due to the simplicity and efficiency of uniform sampling [20], it is widely used in a large number of knowledge graph embedding. Head entities or tail entities in the negative samples are sampled from the random distribution of all entities, the quality of the obtained negative samples cannot be guaranteed. Low-quality negative samples will lead to 0 loss in the margin-based loss function, resulting in the problem of vanishing gradients [19]. In addition to uniform sampling, a better sampling strategy has been proposed, Bernoulli sampling [21]. Bernoulli sampling considers the one-to-many, many-to-one, and many-to-many relations between entities based on uniform sampling, but this method is also based on sampling from a fixed distribution, which will suffer from the problem of vanishing gradients. A high-quality negative sample can push the model training to the limit. To obtain high-quality negative samples more efficiently, in addition to the above method of sampling from a fixed distribution, recently, the Generative Adversarial Network(GAN) have also been used in negative sampling methods. For example, IGAN [19] and KBGAN [4] are introduced. GAN is used to select high-quality negative samples. However, IGAN and KBGAN also have time-consuming problems.

In this paper, to solve the above-mentioned difficulties encountered in the sampling process, we propose a novel negative sampling method, DCNS. In DCNS, we use double-cache to store samples, and sample and update the negative samples from the cache in each epoch, which greatly avoids the difficulties caused by GAN, and use mixing method to generate harder negative samples. In order to prevent the impact of false negative samples on the model, we extract the K samples with the highest scores from the cache for training, and conduct relevant experiments on the impact of K value on the model. Furthermore, we analyze how to balance *exploration* (i.e., explore all the possible high-quality negative samples) and *exploitation* ((i.e., sample the largest score negative triplet in cache) to fine-tune our model parameters. Contributions of our work are summarized as follows:

- How to design a more efficient and effective negative sampling method is a difficult problem. We propose a new double-cache negative sampling method, DCNS. DCNS is outperform baseline negative sampling methods.
- We propose DCNS which has fewer parameters and uses mixing to generate harder negative samples. We also study the effect of exploration and exploitation on experiments, and fine-tune model parameters to optimize the model by balancing exploration and exploitation.

- We conduct experiments on DCNS on four datasets. Experimental results demonstrate that our method is very efficient and is more effective than the state-of-the-art.

2   Related Works

The main task of knowledge graph embedding is to embed entities and relations in triples into a low-dimensional continuous vector space, and then use the embedded entity and relation vectors to perform various downstream tasks. In the knowledge graph embedding process, a good negative sampling method can push the training of the embedding model to the limit.

2.1   Negative Sampling

Negative sampling methods aim to generate high-quality negative samples, so that the model can maximize the difference between positive samples and negative samples. The negative sampling method can also be divided into two categories: sample from a fixed distribution, and the other is sampling from a dynamic distribution.

Sample from Fixed Distributions: As a classic sampling strategy, the idea of fixed negative sampling is simple and intuitive. The uniform sampling [3] method is widely used in early fixed negative sampling methods. Subsequently, a better method, Bernoulli sampling [21], is proposed, which solves the problem of one-to-many, many-to-one, and many-many relations in uniform sampling. But neither uniform sampling nor Bernoulli sampling fail to form better negative samples and suffer from vanishing gradients.

Sample from Dynamic Distributions: To solve the problems with fixed negative sampling, several GAN-based works [4,19] have been proposed to extend negative sampling from fixed distribution to dynamic distribution. KBGAN [4] and IGAN [19] try to use GAN to generate high-quality negative samples. When GAN is used for negative sample sampling, the generator generates high-quality negative samples through the confusing discriminator. Through this strategy, the generator can finally obtain negative samples from the dynamic distribution of increasingly stable negative samples, and train an efficient knowledge graph embedding model through these high-quality negative samples.

Although the negative sampling method based on GAN has achieved excellent results, it is notoriously unstable and degenerate. Therefore, pre-training is very important in KBGAN and IGAN, but it also increases the burden of model training. Therefore, someone proposed a cache-based method, NSCaching [25], which greatly reduces the burden of model training through the cache.

2.2   Mixing Method

Mixing is a data augmentation method, which generates new data by convex combinations of pairs of samples. Essentially, mixup encourages the model to behave linearly in-between training samples. Mixing methods have shown their superiority in many applications [10]. The core idea of mixing is to construct virtual samples through linear interpolation, so some researchers use mixing for harder negative mining. MixGCF [6] uses positive mixing and hop mixing to obtain high-quality negative samples, which achieves higher performance in the recommended scenarios. MoChi [6] showed that harder negative samples can be obtained by mixing hard negative samples and labels.

In order to deal with the problems existing with the above negative sampling methods, we use cache in the negative sampling method, and combine the mixing method to obtain harder negative samples. Finally, we propose a double-cache negative sampling method, DCNS.

3   Our Method

3.1   The DCNS Framework

Previous negative sampling methods often encounter the following problems: 1. How to involve the dynamic distribution of negative samples. 2. How to generate more efficient negative sampling. To better address these problems, we propose the DCNS, which stores high-quality negative samples in the cache, and the negative samples for each epoch are obtained from the cache instead of sampling from a uniform distribution. Different from the previous model, to generate harder negative samples, we use the mixing method to fuse the acquired negative samples, and mix multiple negative samples per epoch to enhance the training of the model. The framework is shown in Fig. 1.

We denote $\mathcal{H}$ and $\mathcal{T}$ (which store $\bar{h} \in \mathcal{E}$ and $\bar{t} \in \mathcal{E}$ respectively) as the set of negative samples generated by replacing head entities and tails, respectively.

Algorithm 1 represents the knowledge graph embedding framework of the cache-based negative sampling method, and the main difference from the general KG embedding framework lies in steps 5–8. The time complexity of DCNS is $O\left(m\left(N_1 + N_2\right)d\right)$, the main additional cost by introducing cache in step 8.

In our method, different loss functions are used for the translation model and semantic matching modelsemantic matching mode. The margin-based loss function [3] is often used by translation-based models and is often considered to be

$$L_{margin} = \sum_{(h,r,t) \in \mathcal{T}} \left[f(h,r,t) - f\left(\bar{h}, r, \bar{t}\right) + \gamma \right]_+ \tag{1}$$

where γ is the margin, $[\cdot]_+ = \max(0, \cdot)$ is hinge function, and $(\bar{h}, r, \bar{t})$ is negative triple.

Semantic matching models use a similar score-based function to measure the factual similarity of the latent semantics of entities and relations embedded in

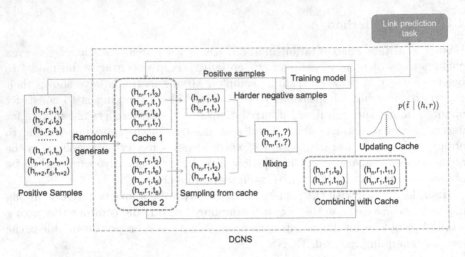

Fig. 1. The DCNS framework. Positive samples are obtained from the training set in each batch, and negative samples are generated by DCNS.

Algorithm 1 DCNS: Double-cache negative sampling method

Input:

training set $\mathcal{S} = \{(h, r, t)\}$, entity set $\mathcal{E}$, relation set $\mathcal{R}$, hidden dimension d, scoring function f, cache size N_1, N_2, the size of epochs E, batch size m, numbers of sample K.

Output: trained model

1: initialize embedding for each $e \in \mathcal{E}$ and $r \in \mathcal{R}$, head cache $\mathcal{H}_1$ $\mathcal{H}_2$ and tail cache $\mathcal{T}_1$ $\mathcal{T}_2$

2: **for** i to E **do**

3:     sample a mini-batch $S_{batch} \in \mathcal{S}$ of size m

4:     **for** each $(h, r, t) \in S_{batch}$ **do**

5:         index the cache $\mathcal{H}_1$ $\mathcal{H}_2$ by (r, t) and $\mathcal{T}_1$ $\mathcal{T}_2$ by (h, r) to get the candidate sets of $\mathcal{H}^1_{(r,t)}$ $\mathcal{H}^2_{(r,t)}$ and $\mathcal{T}^1_{(h,r)}$ $\mathcal{T}^2_{(h,r)}$

6:         *core step* sample $\bar{h}$ from $\mathcal{H}^1_{(r,t)}$ $\mathcal{H}^2_{(r,t)}$ and $\bar{t}$ from $\mathcal{T}^1_{(h,r)}$ $\mathcal{T}^2_{(h,r)}$

7:         *core step* uniformly pick up triplets with K highest scores from $\mathcal{H}^1_{(r,t)}$ $\mathcal{H}^2_{(r,t)}$ and mix head entities using Equation 3 to generate harder negative triple, do the same for $\mathcal{T}^1_{(h,r)}$ $\mathcal{T}^2_{(h,r)}$ again;

8:         *core step* update the cache $\mathcal{H}^1_{(r,t)}$ $\mathcal{H}^2_{(r,t)}$ and $\mathcal{T}^1_{(h,r)}$ $\mathcal{T}^2_{(h,r)}$

9:         calculate loss functions using Equation 1 or 2, then update embedding of entities and relations via gradient descent

10:     **end for**

11: **end for**

---

the vector space. The loss function of the semantic matching model is as follows:

$$L = \sum_{(h,r,t) \in \mathcal{T}} [\ell(+1, f(h,r,t)) + \ell(-1, f(\bar{h}, r, \bar{t}))]_+ \tag{2}$$

where $\ell(\alpha, \beta) = \log(1 + \exp(-\alpha\beta))$ is the logistic loss, other parameters are the same as in Eq. 1.

3.2   Sampling Strategy

The previous sampling methods are all sampling from the uniform distribution, ignoring the influence of a large number of low-quality negative samples, but we use $\mathcal{H}_{(r,t)}$ and $\mathcal{T}_{(h,r)}$ to store high-quality negative samples and draw high-quality negative samples from it, which can effectively avoid the problem of vanishing gradients.

However, as the distribution may change during iterations of the algorithm, the negative triples in the cache may not be accurate enough for sampling in the most recent iteration. In addition, false negative triples exist in the negative sample set, their scores may also be very high. Therefore, we also need to consider other triplets in addition to the highest scoring triplets in the cache. To prevent the influence of high-quality false negative samples, we no longer sample the sample with the highest score, but extract several negative samples with the highest scores of $\mathcal{H}^1_{(r,t)}$ and $\mathcal{H}^2_{(r,t)}$, and the sampling of $\mathcal{T}^1_{(h,r)}$ and $\mathcal{T}^2_{(h,r)}$ is consistent.

3.3   Mixing

In knowledge graphs, many sampled negative triples do not provide discriminative information to help the model learn effective embedding for entities and relations, and only a few negative samples help the model converge in the right direction, and these samples are hard negative samples. Therefore, we use mixing to generate new harder negative samples to improve model training. The mixing formulas of $\mathcal{H}^1_{(r,t)}$ and $\mathcal{H}^2_{(r,t)}$ is as follows:

$$\bar{h}_{1,2} = \alpha * \bar{h}_1 + (1 - \alpha) * \bar{h}_2 \tag{3}$$

where $\bar{h}_1$ from $\mathcal{H}^1_{(r,t)}$ and $\bar{h}_2$ from $\mathcal{H}^2_{(r,t)}$, α denote the adaptive weight.

3.4   Updating Strategy

According to the previous description, the cache needs to be updated in the process of algorithm iteration. Otherwise, because $\mathcal{H}_{(r,t)}$ is fixed, each sampling will eventually become sampling from a fixed distribution, which will affect the experiment. Thus, we need to update the cache in an efficient way during the algorithm iteration progress.

First, we uniformly sample two subsets $\mathcal{R}^1_m, \mathcal{R}^2_m \subset \mathcal{E}$ of size N_2, then union it with $\mathcal{H}^1_{(r,t)}, \mathcal{H}^2_{(r,t)}$ and obtain $\hat{\mathcal{H}}^1_{(r,t)}, \hat{\mathcal{H}}^2_{(r,t)}$. The scores for triplets in $\hat{\mathcal{H}}^1_{(r,t)}, \hat{\mathcal{H}}^2_{(r,t)}$ are evaluated by score function. Then, we construct two subsets $\tilde{\mathcal{H}}^1_{(r,t)}, \tilde{\mathcal{H}}^2_{(r,t)}$

following Eq. 4. Finally, we use $\widetilde{\mathcal{H}}^1_{(r,t)}, \widetilde{\mathcal{H}}^2_{(r,t)}$ as the updated head-cache. The update procedure of tail-cache is also consistent.

$$p(\bar{h} \mid (t,r)) = \frac{\exp(f(\bar{h},r,t))}{\sum_{h_i \in \hat{\mathcal{H}}_{(r,t)}} \exp\left(f\left(\bar{h}_i,r,t\right)\right)} \tag{4}$$

Note that exploration and exploitation also need to be carefully balanced. A bigger N_1 implies more exploitation, while a larger N_2 leads to more exploration. Therefore, if the sizes of N_1 and N_2 can be adjusted appropriately, the exploration and exploration of the model can be better balanced.

4   Experiment

In this section, we conduct the link prediction task to verify the effectiveness of our model with experimental results. In this section, the details of our experiments, such as the settings of relevant parameters, datasets and evaluation metrics, are also introduced.

4.1   Experiment Setup

Dataset: For the experimental datasets, four datasets are used here, WN18 [3], FB15K [3] and their variants WN18RR [5], FB15K-237 [17]. WN18 and FB15K are firstly introduced in TransE, They are widely tested among the most famous knowledge graph embedding learning works TransE, KBGAN, TransR, DistMult, Rescal, and so on. However, WN18 and FB15K have the problem of data leakage, which will have some negative impact on experiments. WN18RR and FB15K-237 remove the inverse-duplicate and near-duplicate relations in WN18 and FB15K, and the stored facts are more real. The statistics are summarized in Table 1.

Table 1. Details of the datasets used, #E is the number of entities. #R is the number of relations

Dataset	#E	#R	#Train/Valid/Test		
WN18	40,943	18	141,442	5,000	5,000
WN18RR	40,943	11	86,835	3,034	3,134
FB15K	14,951	1,345	484,142	50,000	59,071
FB15K-237	14,541	237	272,115	17,535	20,466

Evaluation Metrics: As in previous works [3,21,22], we evaluate different models based on the following metrics:

- Mean reciprocal ranking (MRR) : It is computed by average of the reciprocal ranks $1/|\mathcal{S}| \sum_{i=1}^{|\mathcal{S}|} \frac{1}{\text{rank}_i}$ where $\text{rank}_i, i \in \{1, \ldots, |\mathcal{S}|\}$ is a set of ranking results;

- Mean rank(MR) : It is computed by $\frac{1}{|\mathcal{S}|}\sum_{i=1}^{|\mathcal{S}|}$. Smaller value of MR tends to infer better results;
- Hit@10 : The percentage of appearance in top-10: $1/|\mathcal{S}|\sum_{i=1}^{|\mathcal{S}|}\mathbb{I}(\text{rank}_i < 10)$, where $\mathbb{I}$ is the indicator function.

Hyper-parameter Settings: The embeddings of entities and relations are uniformly initialized. We use grid search to select the following hyper-parameters: hidden dimension $d \in \{50, 100, 200\}$, batch size $m \in \{1024, 2048, 4096\}$, learning rate $\eta \in \{0.0001, 0.001, 0.01, 0.1\}$, cache size $N_1, N_2 \in \{10, 30, 50, 70, 90\}$, the number of negative samples obtained each epoch K. For translational distance models, we tune the margin value $\gamma \in \{1, 2, 3, 4\}$. And for semantic matching models, we tune the penalty value $\lambda \in \{0.001, 0.01, 0.1\}$. The optimizer in our experiments is Adam [8] and we fine-tune the hyper-parameters based on the results of the validation set.

4.2 Baselines

We utilize the following negative sampling algorithms as baselines:

- *Uniform* [16]: Uniform sampling is a type of sampling from a fixed distribution. Uniform sampling is uniformly selects candidate entities from the entire entity set to form negative triplets;
- *Bernoulli* [21]: Bernoulli sampling aims to reduce false negative labels by replacing heads or tails with different probabilities in one-to-many, many-to-one, and many-to-many relations;
- *KBGAN* [4]: KBGAN constructs the generator and discriminator. The generator is used to generate high-quality negative samples, and the discriminator is trained using the positive and negative samples generated by the generator;
- *NSCaching* [25]: NSCaching generates negative samples by sampling from the head cache $\mathcal{H}$ or the tail cache $\mathcal{T}$, and updates the cache after training. NSCaching can delay updating the cache after multiple iterations, which can further save time.
- *SANS* [1]: SANS utilizes the graph structure of a KG to find hard negative examples. Specifically, SANS constructs negative samples using a subset of entities restricted to either the head or tail entity's k-hop neighborhood.

We will evaluate the performance of different negative sampling methods (i.e. Uniform, Bernoulli, KBGAN, NSCaching and KBCache) under different models. In addition, the generator in KBGAN is the TransE model corresponding to the hot start, and we do not consider other models.

4.3 Experiment Results

For the experiments of the proposed negative sampling method, we conducted experiments on the translation-based model and the semantic matching-based model respectively, and the performance on link prediction is compared in Table 2, some experimental results in the table are from NSCaching [25].

Table 2. Results of different negative sampling algorithms for link prediction on five scoring functions and four public datasets. Bold numbers indicate the best performance.

Score function	Datasets	WN18			FB15K			WN18RR			FB15K-237		
	Metrics	MRR	MR	Hit@10	MRR	MR	Hit@10	MRR	MR	Hit@10	MRR	MR	Hit@10
TransE	Uniform	0.4213	**213**	91.50	0.4679	**60**	74.70	0.1753	4038	44.48	0.2262	237	38.64
	Bernoulli	0.5001	249	94.13	0.4951	65	77.37	0.1784	3924	45.09	0.2556	197	41.89
	KBGAN	0.6880	293	94.92	0.4858	83	77.02	0.1808	5356	43.24	0.2938	721	43.42
	NSCaching	0.7867	271	96.62	0.6475	62	81.54	0.2048	2946	45.52	0.3004	218	44.36
	SANS	0.8195	263	95.22	0.6529	65	81.99	0.2133	**2810**	46.19	**0.3094**	201	45.36
	DCNS	**0.8213**	299	**97.11**	**0.6595**	74	**82.11**	**0.2137**	3495	**47.54**	0.3091	**186**	**45.91**
TransH	Uniform	0.4527	**233**	92.71	0.4316	58	73.98	0.1755	5646	43.30	0.2222	223	38.80
	Bernoulli	0.5206	288	94.52	0.4518	60	76.55	0.1862	4113	45.09	0.2329	202	40.10
	KBGAN	0.6167	335	94.84	0.4262	86	75.91	0.1923	4708	45.31	0.2807	401	46.39
	NSCaching	0.8063	286	95.32	0.6520	**54**	81.56	0.2038	4425	48.04	0.2812	**187**	46.48
	SANS	0.8067	271	96.62	0.6575	62	81.54	**0.2048**	4213	48.52	0.2911	218	46.66
	DCNS	**0.8251**	312	**97.02**	**0.6618**	61	**82.56**	0.2012	**3657**	**48.95**	**0.2913**	352	**46.85**
TransD	Uniform	0.4426	**243**	92.69	0.4320	59	73.98	0.1782	4955	42.18	0.2244	215	39.53
	Bernoulli	0.5093	256	94.61	0.4320	63	76.55	0.1901	3555	46.41	0.2451	188	42.89
	KBGAN	0.6168	335	94.84	0.4262	86	75.91	0.1917	3785	46.39	0.2487	798	44.33
	NSCaching	0.8063	286	95.32	0.6520	**54**	81.96	0.2013	2952	48.36	0.2683	**184**	47.85
	SANS	0.8167	1013	94.62	0.6575	62	81.54	0.2048	**2946**	48.52	0.2714	218	48.36
	DCNS	**0.8256**	301	**97.29**	**0.6832**	60	**82.48**	**0.2092**	3289	**48.91**	**0.2871**	312	**49.29**
DistMult	Uniform	0.6340	1174	92.28	0.4985	94	78.28	0.3765	**7405**	44.85	0.2247	408	36.03
	Bernoulli	0.7918	**862**	93.38	0.5376	102	78.69	0.3964	7420	45.25	0.2491	280	42.03
	KBGAN	0.6955	1143	93.11	0.5376	102	78.69	0.3849	7586	44.32	0.2670	370	45.34
	NSCaching	**0.8297**	1038	93.83	**0.7447**	**81**	**84.16**	0.4148	7477	48.80	0.2882	**265**	45.79
	SANS	0.8155	1271	94.61	0.7247	99	82.66	0.4041	7539	48.56	**0.2964**	311	44.81
	DCNS	0.8089	1322	**95.55**	0.7074	153	81.82	0.4018	7666	**48.96**	0.2793	351	43.26
ComplEx	Uniform	0.8046	1106	93.75	0.5191	**85**	78.02	0.3934	8259	41.63	0.2201	418	35.55
	Bernoulli	0.9115	**808**	94.39	0.6253	138	80.72	0.4431	4693	51.77	0.2596	238	43.54
	KBGAN	0.8976	1060	93.73	0.6254	162	80.95	0.4287	6928	47.03	0.2670	370	45.34
	NSCaching	**0.9326**	1079	94.06	0.7994	94	86.32	0.4487	**4861**	51.76	0.3017	**220**	47.75
	SANS	0.9147	1125	95.12	0.8011	91	85.14	0.4513	5210	51.62	0.3004	248	48.36
	DCNS	0.9012	1208	**96.21**	**0.8119**	142	**88.87**	**0.4625**	5357	**52.37**	**0.3125**	358	**49.14**

From the table, we can know that our method DCNS has achieved SOTA results compared with other negative sampling methods. In the translation-based model, we can observe from the experimental results that the performance of KBGAN is not much improved compared with Bernoulli, and the training cost of KBGAN is also relatively high, which means that the negative sampling method based on GAN still has a large room for improvement. NSCaching achieves about two percent improvement over Bernoulli and KBGAN methods. NSCaching is not only in performance, but also higher in efficiency than KBGAN. This result is because NSCaching is based on cache. The DCNS proposed in this paper is based on the advantages of NSCaching. However, unlike NSCaching, DCNS also considers the influence of harder negative samples and false negative samples. DCNS not only uses the mixing method to generate harder negative samples, but also draws multiple negative samples in each round to participate in training in order to prevent the influence of false negative samples. It can also be seen from the experimental results that the DCNS method is not only superior to Bernoulli and KBGAN, but also better than the NScaching method that is also based on cache. In the semantic matching model, it can be observed from the results in the table that NSCaching has achieved better results than Bernoulli and KBGAN methods, and our method DCNS is still better than NSCaching, especially in the model ComplEx. From these experimental results, we can conclude that the effectiveness of DCNS. The better performance of DCNS than NSCaching also proves the effectiveness of the mixing method and multiple sampling in DCNS.

4.4   Experiments on Sampling Ratio

Table 3. Experimental results on datasets WN18RR and FB15K-237 with different sampling rates. Bold numbers indicate the best performance.

Hyper-parameter	Dataset	Metrics/β	0.1	0.2	0.3	0.4	0.5	0.6	0.7	0.8	0.9	1.0
$N_1 = 30\ N_2 = 30$	WN18RR	MRR	**0.2140**	0.2137	0.2125	0.2110	0.2085	0.2077	0.2051	0.2013	0.1986	0.1913
		Hit@10	46.89	**47.54**	46.89	46.67	45.90	45.69	45.10	44.40	43.89	43.12
	FB15K-237	MRR	0.3085	**0.3091**	0.3077	0.3025	0.2985	0.2963	0.2928	0.2897	0.2874	0.2836
		Hit@10	45.02	**45.11**	44.86	44.65	44.33	44.01	43.77	43.35	42.88	42.23
$N_1 = 50\ N_2 = 50$	WN18RR	MRR	0.2210	**0.2235**	0.2207	0.2186	0.2154	0.2132	0.2099	0.2076	0.2035	0.1987
		Hit@10	**47.51**	47.46	47.06	46.87	46.46	46.31	46.10	45.85	44.89	43.22
	FB15K-237	MRR	0.2997	0.3015	**0.3016**	0.3001	0.2983	0.2976	0.2963	0.2946	0.2913	0.2877
		Hit@10	45.79	**45.87**	45.68	45.57	45.34	45.12	44.98	44.68	44.35	44.01

In this section, we take the ratio of the number of negative samples obtained from each epoch to the size of the cache as a hyper-parameter that affects the model performance, which is defined as $\beta = K/N_1$. The influence of this parameter on the model is discussed under the condition that other parameters remain unchanged. We set batch size $m = 1024$, learning rate $\eta = 0.0001$, margin value $\gamma = 4$, hidden dimension $d = 50$, N_1 takes 30 and 50 for comparison, and the value of N_2 is the same as N_1 in this section.

We train TransE on the WN18RR and FB15K-237 datasets, respectively, and observe the changes in the metrics MRR and Hit@10 by adjusting β. As shown in the Table 3, we find that the performance of the model does not increase with the number of negative samples drawn per epoch. We guess that as β increases, the number of low-quality negative samples in the negative samples obtained at each epoch will also increase, thus having a negative impact on the experimental results. Finally, we conclude that the value of K does not need to be too large, which not only reduces the training time of the model, but also improves the performance of the model to a certain extent. We obtain the best performance when β is 0.1 or 0.2.

4.5   Experiments on Exploration and Exploitation

In this part, we analyze steps 6 and 8 in Algorithm 2, and balance exploration and exploitation by adjusting the cache size N_1 and N_2, TransE and WN18RR are used here.

Figure 2(a) shows the model performance when N_2 is fixed. We can find from the results in the figure that when N1 is small, the model converges faster at the beginning than at other values, but the final result is not good. As N_1 increases,

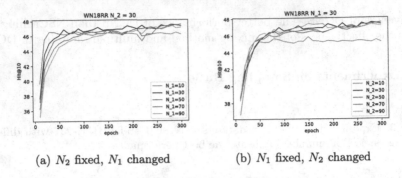

(a) N_2 fixed, N_1 changed   (b) N_1 fixed, N_2 changed

Fig. 2. Balancing on exploration and exploitation of different value of N_1 and N_2. Evaluated by TransE model on WN18RR.

the convergence speed decreases, and the final convergence result becomes better. Figure 2(b) shows the change when N_1 is fixed and the value of N_2 is changed. When $N_2 = 10$, not only the convergence speed is slow, the convergence effect is poor. But when N_2 is other values, the convergence speed and convergence results are improved. Through the results of the experiment, we observe that when the values of N_1 and N_2 are 30, the model performs the best, and can well balance the relation between exploration and exploitation.

5   Conclusions

We propose DCNS as a novel negative sampling method for knowledge graph embedding learning. DCNS obtains multiple high-quality negative samples from the double-cache, and generates harder negative samples by mixing. Experimental results show that the method can generalize well under various settings and achieves state-of-the-arts performance on benchmark datasets. In future work, we can explore the interpretability of the mixed method in the paper and study its feasibility in other knowledge graph embedding models. In addition, we will consider more research on cache-based negative sampling methods and consider how to improve the scalability of DCNS.

References

1. Ahrabian, K., Feizi, A., Salehi, Y., Hamilton, W.L., Bose, A.J.: Structure aware negative sampling in knowledge graphs. arXiv preprint arXiv:2009.11355 (2020)
2. Bollacker, K., Evans, C., Paritosh, P., Sturge, T., Taylor, J.: Freebase: a collaboratively created graph database for structuring human knowledge. In: Proceedings of the 2008 ACM SIGMOD International Conference on Management of Data, pp. 1247–1250 (2008)

3. Bordes, A., Usunier, N., Garcia-Duran, A., Weston, J., Yakhnenko, O.: Translating embeddings for modeling multi-relational data. In: Advances in Neural Information Processing Systems, vol. 26 (2013)
4. Cai, L., Wang, W.Y.: KBGAN: adversarial learning for knowledge graph embeddings. In: Proceedings of NAACL (2018)
5. Dettmers, T., Minervini, P., Stenetorp, P., Riedel, S.: Convolutional 2d knowledge graph embeddings. In: Proceedings of the AAAI Conference on Artificial Intelligence, vol. 32 (2018)
6. Huang, T., et al.: MixGCF: an improved training method for graph neural network-based recommender systems. In: Proceedings of the 27th ACM SIGKDD Conference on Knowledge Discovery & Data Mining, pp. 665–674 (2021)
7. Isinkaye, F.O., Folajimi, Y.O., Ojokoh, B.A.: Recommendation systems: principles, methods and evaluation. Egyptian Inf. J. **16**(3), 261–273 (2015)
8. Kingma, D.P., Ba, J.: Adam: a method for stochastic optimization. In: Proceedings of the 3rd International Conference on Learning Representations (2014)
9. Koller, D., et al.: Introduction to Statistical Relational Learning. MIT Press, Cambridge (2007)
10. Lee, K., Zhu, Y., Sohn, K., Li, C.L., Shin, J., Lee, H.: i-mix: a domain-agnostic strategy for contrastive representation learning. In: International Conference on Learning Representations (2021)
11. Miller, G.A.: Wordnet: a lexical database for English. Commun. ACM **38**(11), 39–41 (1995)
12. Mitchell, T., et al.: Never-ending learning. Commun. ACM **61**(5), 103–115 (2018)
13. Song, X., Li, J., Cai, T., Yang, S., Yang, T., Liu, C.: A survey on deep learning based knowledge tracing. Knowl.-Based Syst. **258**, 110036 (2022)
14. Song, X., Li, J., Lei, Q., Zhao, W., Chen, Y., Mian, A.: Bi-clkt: Bi-graph contrastive learning based knowledge tracing. Knowl.-Based Syst. **241**, 108274 (2022)
15. Suchanek, F.M., Kasneci, G., Weikum, G.: Yago: a core of semantic knowledge. In: Proceedings of WWW, pp. 697–706 (2007)
16. Sun, Z., Deng, Z.H., Nie, J.Y., Tang, J.: Rotate: knowledge graph embedding by relational rotation in complex space. In: ICLR (2019)
17. Toutanova, K., Chen, D.: Observed versus latent features for knowledge base and text inference. In: Proceedings of the 3rd workshop on Continuous Vector Space Models and Their Compositionality, pp. 57–66 (2015)
18. Wang, C., Wang, X., Li, Z., Chen, Z., Li, J.: Hyconve: a novel embedding model for knowledge hypergraph link prediction with convolutional neural networks. In: Proceedings of the ACM Web Conference 2023, pp. 188–198 (2023)
19. Wang, P., Li, S., Pan, R.: Incorporating GAN for negative sampling in knowledge representation learning. In: Proceedings of the AAAI Conference on Artificial Intelligence, vol. 32 (2018)
20. Wang, Q., Mao, Z., Wang, B., Guo, L.: Knowledge graph embedding: a survey of approaches and applications. IEEE Trans. Knowl. Data Eng. **29**(12), 2724–2743 (2017)
21. Wang, Z., Zhang, J., Feng, J., Chen, Z.: Knowledge graph embedding by translating on hyperplanes. In: Proceedings of the AAAI Conference on Artificial Intelligence, vol. 28 (2014)
22. Xiao, H., Huang, M., Hao, Y., Zhu, X.: TransG: a generative mixture model for knowledge graph embedding. arXiv preprint arXiv:1509.05488 (2015)
23. Xiong, C., Power, R., Callan, J.: Explicit semantic ranking for academic search via knowledge graph embedding. In: Proceedings of the 26th International Conference on World Wide Web, pp. 1271–1279 (2017)

24. Yao, X., Van Durme, B.: Information extraction over structured data: Question answering with freebase. In: Proceedings of the 52nd Annual Meeting of the Association for Computational Linguistics, pp. 956–966 (2014)
25. Zhang, Y., Yao, Q., Shao, Y., Chen, L.: NSCaching: simple and efficient negative sampling for knowledge graph embedding. In: 2019 IEEE 35th International Conference on Data Engineering, pp. 614–625. IEEE (2019)

Influence Maximization in Attributed Social Network Based on Susceptibility Cascade Model

Jinyi Chen[1], Junchang Xin[1,3(✉)], Shengnan Lei[1], Keqi Zhou[1], Baoting Li[1], and Zhiqiong Wang[2]

[1] School of Computer Science and Engineering, Northeastern University, Shenyang 110819, China
{chenjinyi,zhoukeqi}@stumail.neu.edu.cn, xinjunchang@mail.neu.edu.cn, leishengnan1997@163.com, lbt1209@163.com
[2] College of Medicine and Biological Information Engineering, Northeastern University, Shenyang 110819, China
wangzq@bmie.neu.edu.cn
[3] Key Laboratory of Big Data Management and Analytics, Northeastern University, Shenyang 110819, Liaoning, China

Abstract. Influence maximization is the problem of finding a small subset of seed nodes in a social network that can effectively maximize the spread of influence using a specific information diffusion model. Traditional methods do not often overlook the attributes of source information and the preference of users in the network, which can result in significant deviations. In this paper, we design a susceptibility cascade model to address the trade-off between influence maximization and the diffusion process. More specifically, we extend the original independent cascade model and propose a more realistic susceptibility cascade model to simulate the information diffusion process. Considering diverse source information attributes and user preferences, we propose an influence maximization algorithm based on the susceptibility cascade model and reverse reachable sampling, and its improvement, that incorporates the community structure. Comprehensive experiments on two real-life social networks obtained from publicly available datasets demonstrate the effectiveness of our algorithms in maximizing the spread of influence.

Keywords: Attributed social network · Influence maximization · Susceptibility cascade model · Source information attributes · User preference · Community structure

1 Introduction

Recently, the success of social networks has enabled information and ideas to rapidly reach and impact a vast number of users within a short span of time [3,12]. Motivated by applications such as viral marketing [20], business location

X. Song et al. (Eds.): APWeb-WAIM 2023, LNCS 14334, pp. 451–466, 2024.
https://doi.org/10.1007/978-981-97-2421-5_30

planning [24] and social network analysis [22,23], the study of maximizing the spread of influence has emerged as a significant and timely research topic in recent years. Consider a practical example. A small company has developed an application for a social network and intends to promote it within this network. At the initial stage of promotion, the company plans to encourage users to like and use the application by offering incentives or gifts. However, due to cost constraints, only a limited number of users can be selected. The company's objective is for these selected users to influence their friends to adopt the application, and subsequently, for their friends to influence their own social circles. Through the word-of-mouth effect [7], a number of users will eventually start using the application. Therefore, the key challenge lies in determining which users should be chosen initially in order to maximize their influence within the network. This is formalized as the influence maximization problem, as presented by [13].

How to accurately model the diffusion process for spreading information from seeds in a social network is crucial for the influence maximization problem [7, 14], as it significantly impacts the adoption of non-seed users. In the diffusion process, it is possible for user u to propagate information to user v, and it is also possible for user v to receive the information. The diffusion process comprises two distinct processes: the influence process and the susceptible process. The success of information diffusion is dependent on the probability of success for both processes, and this probability is influenced by various factors, including the user's preferences and other attributes. Unfortunately, due to the challenges in obtaining real-action diffusion traces for determining edge probabilities, prior research often neglects these intricate details that exist in reality [17].

In addition, existing methods primarily focus on maximizing the spread of influence without considering user preferences, such as entertainment preferences, taste preferences, brand preferences, and so on. This oversight often results in an uneven distribution of influence among different preference groups [8]. Moreover, the information being diffused often possesses various attributes, and the diffusion process differs significantly depending on these attributes. For instance, consider the case of using songs as the source information, where songs have different attributes, such as rock, classical, or instrumental. Users also have distinct preferences when selecting songs, which can lead to substantial deviations in the diffusion process. Therefore, it becomes necessary to incorporate user preference adaptation and source information attributes into the model to accurately describe the process and develop an influence maximization algorithm that better aligns with achieving maximum influence spread.

Aiming to address the limitations of existing research on influence maximization, we analyse the information diffusion process by incorporating diverse source information attributes and user preferences. To quantify the communicator's tendency to influence others and the receiver's susceptibility to being influenced, we introduce the influence factor and the susceptibility factor. Specifically, we enhance the classical independent cascade (IC) model by integrating the susceptibility process and diffusion process, resulting in the proposed susceptibility cascade model. Further, we consider both user preferences and source information attributes, and develop an influence maximization algorithm based on the susceptibility cascade model to find seed nodes. We propose an attributed-based reverse reachable sampling method to identify seed nodes from the network. To avoid full network traversal, we also consider the size of communities and the preferences of community users, and incorporate a community-oriented heuristic screening strategy for iterative seed selection, thereby integrating community screening into the influence maximization algorithm. This integration of micro-user and macro-community considerations enhances the efficiency of problem-solving. The contributions of this paper can be summarized as follows:

- We thoroughly consider both the influence factors and susceptibility factors and propose a more realistic susceptibility cascade model to simulate information diffusion. By combining source information attributes and user preferences, we propose an influence maximization algorithm based on the susceptibility cascade model (SCIM).
- Considering the influences of community size and community preferences, we incorporate community-level considerations to reduce redundant calculations and propose an enhanced community-based SCIM algorithm (CB-SCIM).
- Extensive experiments are conducted on two real-life social networks. The results demonstrate the effectiveness of the proposed algorithms in the task of maximizing the spread of influence.

The rest of this paper is arranged as follows. In Sect. 2, we discuss the related work. In Sect. 3, we provide preliminaries and formally define influence maximization problem. Then, we design the algorithm SCIM in Sect. 4 and its improvement CB-SCIM in Sect. 5. In Sect. 6, the proposed methods are experimentally verified and discussed. The paper is summarized in Sect. 7.

2   Related Work

The concept of influence maximization as a learning problem was initially introduced by Domingos and Richardson [6] in 2001, drawing inspiration from viral marketing. In 2003, Kempe et al. [13] formulated it as an optimization problem. Developing an appropriate diffusion model to propagate information, innovation, and other content in social networks is a crucial aspect in addressing the influence maximization problem. The IC (Independent Cascade) model was first introduced by Goldenberg et al. [7] in the field of marketing. This model assumes that a node has only one chance to activate its inactive neighbor with a certain activation probability, and once a node becomes active, it remains so in the future. Granovetter et al. [10] introduced the threshold model, where each node possesses a random activation threshold. The cumulative influence from active neighbors determines the overall influence of a node, and the node becomes active only if the cumulative influence exceeds its activation threshold. Furthermore, Kim et al. [14] improved upon the IC model by introducing the discrete-time-aware model, which considers time-critical demand.

454   J. Chen et al.

Evaluating the influence between nodes is another focal point in research on the influence maximization problem. Saito et al. [18] proposed a probabilistic graphical model and a maximum likelihood objective function for the diffusion process. They utilized the EM algorithm to estimate the activation probabilities between different nodes. Goyal et al. [8] addressed the problem of learning influence probabilities. They specifically explored the time-varying nature of influence and factors such as the propensity for a certain action to influence others. Based on diffusion models, researchers have extensively studied the influence maximization problem and proposed numerous algorithms. Many of these algorithms employ a greedy approach to select seed nodes, leveraging the fact that the seed set exhibits a monotonic submodularity property [13]. Leskovec et al. [15] developed an optimized greedy algorithm CELF, which reduced the number of influence gain computations by utilizing upper bound constraints and exploiting the submodular nature of the problem. Goyal et al. [9] further enhanced the CELF algorithm with CELF++, which reduced the computational time by optimizing Monte Carlo simulations, but it still required traversing all nodes in the initial iteration. Chen et al. [4] proposed a degree-based approach DegDis that considered the phenomenon of influence overlap by diminishing the influence of selected neighbors in the seed set. Borgs et al. [2] made a theoretical breakthrough by introducing a near-linear time algorithm that employed the reachable sampling method and the IC model to maximize influence.

3   Problem Definition

Let $G = (V, E, W)$ be a weighted and directed social network, where $V(G)$ (resp. $E(G)$) represents the set of nodes (resp. edges) in G (due to the correspondence between social network and graph, the concepts of node and user are not distinguished in this paper), W represents the set of edge weights. An edge $(u, v) \in E$ exists if node u and node v are friends in the social network. Let $N(v, G)$ be the neighbor set of v in G, and $deg(v, G) = |N(v, G)|$ denotes the number of v's neighbors. Each edge (u, v) is associated with a weight p, which represents the probability of u diffusing information to user v. Note that we may omit the input graph in the notions when the context is clear.

Definition 1 *Seed Nodes. Given a social network G, seed nodes $S(G) \in V(G)$ are defined as a set of nodes that participate in the diffusion process of G and act as the source for diffusion.*

Definition 2 *Influenced Nodes. Given a social network G, seed nodes S and a diffusion process C, the influenced nodes $I(G)$ of G are the nodes activated by S with process C.*

Definition 3 *Active Node. An active node $a \in G$ is one that belongs to either the seed nodes $S(G)$ or the influenced nodes $I(G)$.*

Definition 4 *Influence Gain. Given a social network G and seed nodes S, the influence gain of G regarding S, denoted by $f(S, G)$, is the total number of influenced nodes in G.*

Problem Definition. Given a social network G and a parameter K, the influence maximization problem aims to find a set S containing K nodes in G such that the influence gain regarding S is maximized, i.e., $f(S, G)$ is maximized.

4   SCIM Algorithm

4.1   Susceptibility Cascade Model

The diffusion process consists of the influence and the susceptibility process. The influence process involves information spreading from an active node to its neighbors, while the susceptibility process involves inactive nodes being influenced by activated nodes. Below, we formally define the influence factor and the susceptibility factor to quantify the success probabilities of these two processes.

Definition 5 *Influence Factor. In a social network, the influence factor represents the probability of a successful information diffusion from an active node to its inactive neighbors, indicating the node's influence on other nodes.*

Definition 6 *Susceptibility Factor. In a social network, the susceptibility factor represents the probability of an inactive node successfully receiving the information spread by its active neighbors, indicating the degree to which a node is influenced by others.*

Based on the analysis above, we propose an extension of the classic IC model called the susceptibility cascade model. The influence factor and susceptibility factor of a node are influenced by the node's own preferences, which are quantified as the fundamental probabilities used in the calculation of these two factors. The preference value of user u for information t can be computed using Eq. (1).

$$p_u^t = \frac{count_u(t)}{\sum_i count_u(i)} \tag{1}$$

where, p_u^t is the preference value of u regarding t, $count_u(t)$ is the frequency of node u participating in activity t.

Obviously, the degree of a node can serve as an indicator of the closeness of a user's social relationships. The node's influence factor is positively correlated with the rank of its degree, while the susceptibility factor is negatively correlated. We define the node with the largest degree as the core user, and denote the distance between user u and the core user as $dis(u)$. For a core user, its distance from itself is 1. It is evident that the user's influence factor is negatively correlated with $dis(u)$, whereas the susceptibility factor is positively correlated. We provide the calculation method for the influence factor $p_{inf}^t(u)$ and the susceptibility factor $p_{sus}^t(u)$ of node u based on the diffusion process C concerning information t in the network G, as shown in Eq. (2) and Eq. (3):

$$p_{inf}^t(u) = \gamma \times p_u^t \times D_{desc}(u) \times \frac{1}{dis(u)} \tag{2}$$

$$p_{sus}^t(u) = \gamma \times p_u^t \times \frac{1}{D_{desc}(u)} \times dis(u) \tag{3}$$

Here, γ is a parameter. Unless otherwise specified, γ is set to 1 in this paper. $D_{desc}(u)$ represents the degree rank of node u in the network.

In the susceptibility cascade model, the diffused information itself possesses attribute features. For a social network G, let S_i denote the set of nodes affected in the i-th iteration. For any edge $(u, v) \in E$, if u is already in S_i and its neighbor v has not been activated, then v is activated by u in the $(i+1)$-th iteration. The probability $p_{u,v}$ that v joins S_{i+1} can be calculated using Eq. (4).

$$p_{u,v} = 1 - \prod_{u \in S_i} \left(1 - p_{inf}^t(u)p_{sus}^t(v)\right) \tag{4}$$

4.2   SCIM Algorithm

Compared to the simple Monte Carlo simulation method, the reverse reachable sampling method [2] primarily obtains sampling results through reverse simulation acquisition. This approach avoids multiple rounds of simulation calculations during execution, greatly improving computational efficiency without sacrificing accuracy. Inspired by reverse reachable sampling, we provide a specific equation for calculating the influence gain.

Definition 7 Reverse Reachable Set [11] (RRS). *The generation process of a random reverse reachable set is as follows:*

(1) Randomly select a node $v \in V$;
(2) Starting from node v, perform random breadth-first search of node u pointing to node v (incoming neighbors of node v), and add the traversed nodes to set R with activation probability $p_{u,v}$;
(3) Continue the traversal until there are no more nodes to join, and obtain a reverse reachable set R.

Lemma 1. *Let $S \subset V$ be the seed nodes, R be the random reverse reachable set, and the calculation equation for the influence gain $f(S)$ is given by:*

$$f(S) = n \cdot \Pr(S \cap R \neq \emptyset)$$

where n represents the total number of nodes in the network G. If $S \cap R \neq \emptyset$, it means that S covers an RRS R. Assuming that R is the set of generated random RRS, the coverage of set S on R is defined as the number of RRS covered by S in R. The influence gain of set S can be estimated based on the coverage of S on R. Furthermore, we denote the random RRS covered by node u as cover$[u]$.

According to the existing research [13], the set of nodes exhibits a monotonic effect on the influence gain. This means that the function $f_u(S)$ satisfies the condition: when $S \cap T = S$, we have $f_u(S) \leq f_u(T)$. Additionally, the function should also be submodular, which implies that when $S \cap T = S$, we have $f_u(S \cup$

$w) - f_u(S) \geq f_u(T \cup w) - f_u(T)$. To select the seed nodes, we introduce a greedy strategy. In each iteration, we choose the node that maximizes the increase in influence gain from the set of seed nodes until K nodes are selected as the final result. The pseudo-code of the SCIM algorithm is presented in Algorithm 1.

Algorithm 1: SCIM

Input: $G(V, E)$, β, K, t and user preference information T
Output: seed set S

1   $R \leftarrow \emptyset, S \leftarrow \emptyset, S' \leftarrow \emptyset$;
2   initialize $p_{sus}^t(u)$ for each $u \in V$, $p_{u,v}$ for each $(u, v) \in E$;
3   initialize a flag array $visit$;
4   **for** *each* $u \in V$ **do**
5   **if** $p_{sus}^t(u) \geq \beta$ **then**
6         $S' = S' \cup \{u\}$;
7         $visit[u] = false$;

8   $num = K * |S'|$;
9   **for** $i = 0$ *to* num **do**
10      $u \leftarrow$ randomly sample a node from set S' ;
11   Queue $Q.push(u)$ and mark u as active ;
12   **while** Q *is not empty* **do**
13         $u \leftarrow Q.pop()$;
14   **for** *each* v *in* $neighbor(u)$ **do**
15   **if** v *is inactive and active* v *with probability* $p_{u,v}$ **then**
16               $Q.push(v)$;
17               $RRS_u = RRS_u \cup \{v\}$;

18      $R \leftarrow R \cup \{RRS_u\}$;
19   **for** $i = 1$ *to* K **do**
20      $u \leftarrow arg \max_u(f(S \cup u) - f(S))$;
21      $S \leftarrow S \cup \{u\}$;
22   **for** *each* $x \subset cover[u]$ **do**
23   **if** $visit'[x] = false$ **then**
24            $visit'[x] = true$;
25   **for** *each node* $v \in RRS_x$ **do**
26   remove every x in $cover[v]$;

27 **return** S

The SCIM consists of three main stages: 1) network reduction, 2) reverse reachable sampling, and 3) greedy selection of seed nodes. Initially, we initialize relevant variables and flag arrays, calculate node attributes and edge weights in the network (Lines 1–3). Since certain nodes have small susceptibility factors and are less likely to be influenced by seed nodes, it is necessary to filter them before generating the RRS. To achieve this, we set a threshold β and perform reverse reachable sampling only on users whose susceptibility factor exceeds the threshold (Lines 4–7). Next, we determine the number of sampling iterations (Line 8) and conduct random reverse reachable sampling on the potential seed set S' (Lines 9–18). In each iteration, a node from S' is randomly selected to join the queue Q and its state is set to active (Lines 10–11). Then, we activate the inactive neighbors of node u in Q with a probability $p_{u,v}$ one by one (Lines 12–15), adding the activated neighbor v to Q and the corresponding RRS (Lines 16–17). The RRSs obtained from each iteration together form the final RRS (Line 18). In the third stage, we employ a greedy strategy to sequentially select K seed nodes (Lines 19–27). During each selection, we choose the node that yields the largest increase in influence gain and add it to the seed set (Lines 20–21). The RRS is updated accordingly to ensure the algorithm obtains the latest influence gain evaluation in the subsequent selection rounds (Lines 22–26). Finally, the set of seed nodes maximizing influence gain is obtained (Line 27).

Time Complexity. The time complexity of the algorithm is $O(n + K \cdot |S'| \cdot E(|RRS_u|) + K \cdot E(|cover[u]|))$. Among them, the network reduction can be done in linear time, $E(|RRS_u|)$ is the expected of the RRS's size, and $E(|cover[u]|)$ is the expected of the cover set's size.

5   CB-SCIM Algorithm

The long tail effect is ubiquitous in social networks, where only a small number of users contribute the majority of behavioral records. The same holds true for communities within networks, where only a few are active while the majority remain inactive. Furthermore, users within a community exhibit similarities in their behavioral preferences, which in turn influence the diffusion process. Handling the influence maximization for all communities results in computational redundancy. Diffusion between communities is more challenging and time-consuming due to fewer connections, while within-community diffusion is easier and faster. To address this, we leverage communities as units of analysis and transform the problem into sub-problems within each community, approximating the influence at the community level rather than the entire network.

Intuitively, selecting seed nodes from larger communities can activate a greater number of nodes. Typically, the number of selected seed nodes is significantly smaller than the number of communities. Seeding all communities would introduce computational redundancy, aligning with our experimental analysis. Thus, prior to selecting node seeds, we employ a heuristic strategy to filter communities and reduce the number of candidate communities. Similar to other studies considering community structure in influence maximization algorithms, we adopt the non-overlapping community division and utilize the *Louvain* algorithm [1] for division. This algorithm iteratively divides the communities based on modularity until the community assignments of all nodes stabilize. The modularity is defined as shown in Eq. (5).

$$Q = \frac{1}{2m} \sum_{i,j} \left[A_{ij} - \frac{k_i k_j}{2m} \right] \delta\left(c_i, c_j\right) \tag{5}$$

where, A_{ij} represents the weight between node i and j, m is the sum of all edge weights, k_i is the sum of edge weights pointing to node i, C_i is the community where node i is located. $\delta(\cdot)$ is an indicative function. $\delta(C_i, C_j)$ is 1 if node i and j are in the same community, and 0 otherwise.

Algorithm 2: CB-SCIM

Input: $G(V, E)$, α, β, K, t and user preference information T
Output: seed set S

1 $S \leftarrow \emptyset, V_S \leftarrow \emptyset, C' \leftarrow \emptyset$;
2 use the *Louvain* algorithm to divide the community $C = \{C_1, C_2, \cdots, C_n\}$;
3 **if** $|C| > 10 * K$ **then**
4      $avg(C) \leftarrow \frac{\sum_{i=1}^{n} C_i}{n}$;
5   **if** $|C_i| \geq avg(C)$ **then**
6         $C' \leftarrow C' \cup C_i$;

7 **for** *each* $C_i \in C'$ **do**
8   calculate community preferences $p_{C_i}^t$;
9   **if** $p_{C_i}^t < \alpha$ **then**
10         $C' \leftarrow C' \backslash C_i$;
11   **for** *each* $u \in C'$ **do**
12            $V_S \leftarrow V_S \cup u$;

13 $V_S \leftarrow$ filter nodes using $p_{sus}^t(u)$;
14 randomly reverse reachable sampling for each $u \in V_S$;
15 **for** $i = 1$ *to* K **do**
16      $u \leftarrow arg\ \max_u (f(S \cup u) - f(S))$;
17      $S \leftarrow S \cup \{u\}$;
18   update RRS ;

19 **return** S

Based on the analysis above, we propose an enhanced algorithm called CB-SCIM. The pseudo-code of the algorithm is outlined in Algorithm 2. Given a social network $G(V, E)$, the number of seed nodes K, source information t, user preference information T, and two threshold parameters α and β, the algorithm aims to find a candidate seed node set S that maximizes the influence gain. To begin, we initialize necessary variables (Line 1). Next, we employ the *Louvain* algorithm to partition the network into multiple communities (Line 2) and filter out smaller communities (Lines 3–6). For each community, we calculate its community preference $p_{C_i}^t$ (Line 8), which represents the average preference of all nodes within the community. Subsequently, we remove communities with preference values below α (Lines 9–10) and consider the nodes within the remaining communities as the candidate seed set V_S (Lines 11–12). We then combine this approach with the SCIM algorithm to iteratively select seed nodes (Lines 13–18). Finally, we obtain the seed nodes that maximize the influence gain (Line 19).

Time Complexity. The time complexity of CB-SCIM algorithm is further reduced than that of algorithm SCIM, which is $O\big((n_{c'} \setminus n) \cdot (n + num \cdot E\left(|RR_i|\right) + K \cdot E(|cover[u]|)) \big)$. Among them, n_{C_i} is the number of nodes in the pre-sampled community.

6   Result and Discussion

6.1   Experimental Settings

Datasets. To evaluate the proposed algorithms, we utilize two real-life attribute social networks obtained from publicly available datasets: Last.FM and YELP. These networks contain attribute information. Table 1 presents detailed information regarding the datasets, including the number of nodes (n) and edges (m), and the number of available source information (i).

Table 1. Details of dataset

Datasets	n	m	i
Last.FM	2,109	25,434	32
YELP	399,773	2,950,218	1,336

Parameters Setting. There are three parameters in our models: α, β, and γ. Unless stated otherwise, we set the value of γ to 1 throughout the text. The parameters α and β are selected from the range of 0.05 to 0.45. Note that the default values of α and β differ across datasets. For the Last.FM dataset, we use $\alpha = 0.05$ and $\beta = 0.1$. For the YELP dataset, we set $\alpha = 0.2$ and $\beta = 0.15$.

Performance Metrics. We employ two metrics to evaluate the performance of the proposed algorithms: the diffusion range of seed nodes and the running time of the algorithm. To determine the diffusion range, we conduct 10 diffusion simulations on the seed nodes and average the number of influenced nodes. A higher number of influenced nodes indicates better algorithm performance.

6.2   Result Analysis

To assess the effectiveness of SCIM, we compare it with two contrasting methods: Influence Maximization via Martingales (IMM) [21] and Conformity-Aware Influence Maximization (CINEMA) [16]. IMM utilizes the evaluation method of martingales, which offers higher efficiency in practical scenarios. In our experiment, we set the precision parameter ε of IMM to 0.1 to ensure optimal results.

CINEMA considers the influence of node consistency and calculates node influence based on the submodular property. Due to space limitations, we only present results for a subset of source information attributes. The results obtained for other attributes are similar to those shown in the following sections.

When comparing the performance of SCIM with the aforementioned comparison algorithms, we primarily modify the attributes of the source information and the selection of the seed set. We measure the diffusion range and running time of each algorithm. For the Last.FM dataset, we examine the source information attributes t = "Rock" and t = "Metal". The obtained experimental results are presented in Fig. 1.

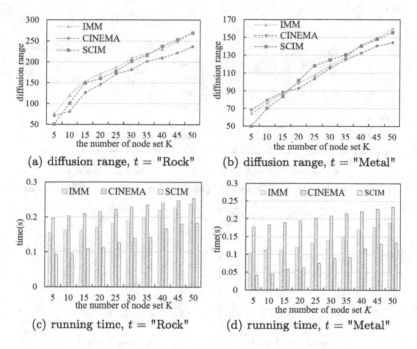

(a) diffusion range, t = "Rock"   (b) diffusion range, t = "Metal"

(c) running time, t = "Rock"   (d) running time, t = "Metal"

Fig. 1. Experimental results on Last.FM (SCIM)

From Fig. 1 (a) and (c), it can be observed that when t = "Rock" and K ranges from 5 to 20, the difference in diffusion range between SCIM and IMM is within 5%, which is significantly better than CINEMA with an average difference of 10.4%. When K increases to [25, 50], the difference in diffusion range between SCIM and IMM decreases to less than 2.5%, while CINEMA performs the worst. In terms of running time, as K increases, SCIM's running time is, on average, 0.056 s lower than that of IMM. However, this difference is not substantial due to the small dataset size. In Fig. 1 (b) and (d), when t = "Metal", the difference in diffusion range between SCIM and IMM is within 5%, while CINEMA performs poorly without a significant lag. Regarding running time, as K increases, SCIM's running time increases noticeably, while IMM's running time increases slightly with no significant change. SCIM's running time is shorter than that of the comparison algorithms. CINEMA has a longer running time due to the Monte Carlo Simulation method. When K is small, SCIM exhibits a greater advantage in running time. As K increases, the running time gap narrows, and SCIM's running time is, on average, 0.74s ahead of IMM. The experimental results are similar across different source information attributes. SCIM performs well and outperforms the comparison algorithms on Last.FM.

To thoroughly verify the efficiency of SCIM, we conduct similar comparative experiments on the YELP dataset. We set the source information attribute t as "Food" or "Toy Stores," $\alpha = 0.05$, $\beta = 0.1$, and the experimental results are shown in Fig. 2.

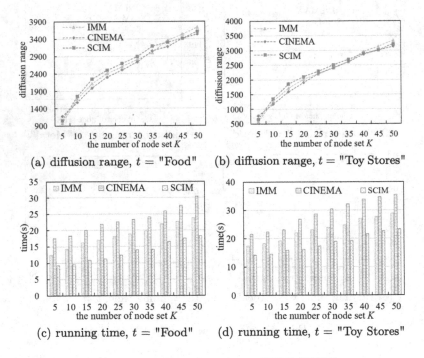

(a) diffusion range, $t =$ "Food"   (b) diffusion range, $t =$ "Toy Stores"

(c) running time, $t =$ "Food"   (d) running time, $t =$ "Toy Stores"

Fig. 2. Experimental results on YELP (SCIM)

As shown in Fig. 2 (a) and (c), when $t =$ "Food", the difference in diffusion range between SCIM and IMM remains within 4.2% as K increases, and CINEMA performs slightly worse. Furthermore, SCIM exhibits a significant advantage in terms of running time, with a reduction of 4.35s compared to IMM. In (b) and (d), when $t =$ "Toy Stores", the diffusion range of SCIM is less than 5% worse than that of IMM, but its running time is significantly better, at least 8.9% higher. Due to the larger scale of the YELP dataset, the pruning optimization strategy is more prominent. In contrast, because the Last.FM dataset is small, SCIM performs relatively similarly to the comparison algorithms. Overall, SCIM shows similar performance to the best comparison algorithms in terms of diffusion range, but it exhibits distinct advantages in running time.

To evaluate the effectiveness of CB-SCIM, we compare it with the IMM algorithm without considering the community, the CIM algorithm (conformity-aware influence maximization) that considers the community [5], and the community-based framework for influence maximization (CoFIM) [19]. IMM was introduced in the previous section. CIM avoids information overlap through the community structure. CoFIM consists of two stages: seed expansion and intra-community diffusion, with seed nodes selected through approximate calculations of the Bernoulli distribution. Similar to the experiments in the previous section, we analyze the variations in diffusion range and running time. For the YELP dataset, we set $t =$ "Food", $\alpha = 0.2$, $\beta = 0.15$. The results are presented in Fig. 3.

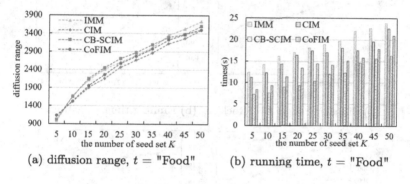

(a) diffusion range, $t =$ "Food"   (b) running time, $t =$ "Food"

Fig. 3. Experimental results on YELP (CB-SCIM)

From Fig. 3, it is evident that CB-SCIM exhibits more pronounced advantages in efficiency, indicating that the community screening strategy has played a crucial role. Among the three comparison methods, IMM achieves the best influence range. As the dataset scale increases, the seed selection strategy becomes more effective. CB-SCIM outperforms the three comparison algorithms in running time, with a lead of at least 1.24 s, as K increases. When K is small, the difference in running time between CoFIM and CB-SCIM is small. However, as K increases, both CIM and CoFIM experience significant increases in running time, and IMM exhibits the longest running time, illustrating the sensitivity of the comparison algorithms to K. Furthermore, we observe that IMM, without considering the community structure, achieves the best performance in terms of diffusion range across different attributes, but its running time is the longest, especially in large social networks. On the other hand, CIM and CoFIM, which consider the community structure, outperform IMM in running time, demonstrating the advantages of considering the community structure.

6.3 Influence of Parameters

In this section, we examine the influence of parameters. To evaluate the impact of parameter β on the influence maximization task, we conduct experiments to measure the performance of SCIM with different values of β. On the Last.FM dataset, with the source information attribute set as "Rock" and the number of seed nodes set to $K = 50$, we analyze the diffusion range and running time of the algorithm, as shown in Fig. 4 (a) and (b). Similarly, on the YELP dataset, with $t =$ "Food" and $K = 50$, we investigate the effect of parameter α on CB-SCIM, examining the diffusion range and running time, as depicted in Fig. 4 (c) and (d).

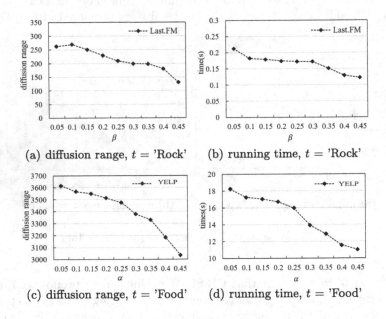

(a) diffusion range, $t = $ 'Rock'   (b) running time, $t = $ 'Rock'

(c) diffusion range, $t = $ 'Food'   (d) running time, $t = $ 'Food'

Fig. 4. Influence of parameters

It is evident from Fig. 4 (a) and (b) that the parameter β significantly impacts the performance of the algorithms. The highest diffusion range is achieved when $\beta = 0.1$. As β increases, the overall running time of the algorithm shows a downward trend. β serves as a threshold parameter, where larger values of β result in more nodes being filtered out and a reduced running time of the algorithm. This observation aligns with the theoretical analysis. As depicted in Fig. 4 (c) and (d), increasing the threshold α leads to a decrease in the influence range of the seed nodes overall. This is due to more communities being filtered out, resulting in lower accuracy in node influence evaluation. When α ranges from 0.05 to 0.2, the decline in the influence range is within 5%. However, when $\alpha > 0.2$, the decline rate of the influence range becomes more pronounced.

7 Conclusions

In this paper, we examine the disparity between conventional influence maximization algorithms and the diffusion patterns in attribute-based social networks. To address this issue, we enhance the traditional IC model and introduce a more realistic susceptibility cascade model to emulate the information diffusion process. By incorporating both the source information attributes and user preferences, we propose the SCIM algorithm for influence maximization, along with its enhanced version CB-SCIM that accounts for community structure. Through extensive experiments conducted on real-life social networks, we demonstrate the effectiveness and efficiency of the proposed algorithms. These findings suggest that our approaches offer scalable solutions to the influence maximization problem in large-scale real-life social networks.

Acknowledgements. This work was supported by the National Natural Science Foundation of China (62072089); Fundamental Research Funds for the Central Universities of China (N2116016, N2104001).

References

1. Blondel, V.D., Guillaume, J.L., Lambiotte, R., et al.: Fast unfolding of communities in large networks. J. Stat. Mech. Theory Exp. **30**(2), 155–168 (2008)
2. Borgs, C., Brautbar, M., Chayes, J., et al.: Maximizing social influence in nearly optimal time. In: SIAM, pp. 946–957 (2014)
3. Cai, T., Li, J., Mian, A., et al.: Target-aware holistic influence maximization in spatial social networks. TKDE **34**(4), 1993–2007 (2020)
4. Chen, W., Wang, Y., Yang, S.: Efficient influence maximization in social networks. In: SIGKDD, pp. 199–208 (2009)
5. Chen, Y.C., Zhu, W.Y., Peng, W.C., et al.: CIM: community-based influence maximization in social networks. TIST **5**(2), 1–31 (2014)
6. Domingos, P., Richardson, M.: Mining the network value of customers. In: SIGKDD, pp. 57–66 (2001). https://doi.org/10.1145/502512.502525
7. Goldenberg, J., Libai, B., Muller, E.: Talk of the network: a complex systems look at the underlying process of word-of-mouth. Mark. Lett. **12**(3), 211–223 (2001). https://doi.org/10.1023/A:1011122126881
8. Goyal, A., Bonchi, F., Lakshmanan, L.V.: Learning influence probabilities in social networks. In: WSDM, pp. 241–250 (2010)
9. Goyal, A., Lu, W., Lakshmanan, L.V.: Celf++ optimizing the greedy algorithm for influence maximization in social networks. In: WWW, pp. 47–48 (2011)
10. Granovetter, M.: Threshold models of collective behavior. Am. J. Sociol. **83**(6), 1420–1443 (1978). https://doi.org/10.1086/226707
11. Guo, Q., Wang, S., Wei, Z., et al.: Influence maximization revisited: efficient reverse reachable set generation with bound tightened. In: SIGMOD, pp. 2167–2181 (2020)
12. Haldar, N.A.H., Reynolds, M., Shao, Q., et al.: Activity location inference of users based on social relationship. World Wide Web **24**(4), 1165–1183 (2021)
13. Kempe, D., Kleinberg, J., Tardos, É.: Maximizing the spread of influence through a social network. In: SIGKDD, pp. 137–146 (2003)

466 J. Chen et al.

14. Kim, J., Lee, W., Yu, H.: CT-IC: continuously activated and time-restricted independent cascade model for viral marketing. KBS **62**, 57–68 (2014)
15. Leskovec, J., Krause, A., Guestrin, C., et al.: Cost-effective outbreak detection in networks. In: SIGKDD, pp. 420–429 (2007)
16. Li, H., Bhowmick, S.S., Sun, A., et al.: Conformity-aware influence maximization in online social networks. VLDB J. **24**(1), 117–141 (2015)
17. Li, J., Cai, T., Deng, K., et al.: Community-diversified influence maximization in social networks. Inf. Syst. **92**, 101522 (2020)
18. Saito, K., Nakano, R., Kimura, M.: Prediction of information diffusion probabilities for independent cascade model. In: Lovrek, I., Howlett, R.J., Jain, L.C. (eds.) KES 2008. LNCS (LNAI), vol. 5179, pp. 67–75. Springer, Heidelberg (2008). https://doi.org/10.1007/978-3-540-85567-5_9
19. Shang, J., Zhou, S., Li, X., et al.: CoFIM: a community-based framework for influence maximization on large-scale networks. KBS **117**, 88–100 (2017)
20. Singh, S.S., Singh, K., Kumar, A., et al.: ACO-IM: maximizing influence in social networks using ant colony optimization. Soft. Comput. **24**(13), 181–203 (2020)
21. Tang, Y., Shi, Y., Xiao, X.: Influence maximization in near-linear time: a martingale approach. In: SIGMOD, pp. 1539–1554 (2015)
22. Yin, H., Song, X., Yang, S., et al.: Sentiment analysis and topic modeling for covid-19 vaccine discussions. World Wide Web **25**(3), 1067–1083 (2022)
23. Yin, H., Yang, S., Song, X., et al.: Deep fusion of multimodal features for social media retweet time prediction. World Wide Web **24**, 1027–1044 (2021)
24. Zeng, Q., Zhong, M., Zhu, Y., et al.: Business location planning based on a novel geo-social influence diffusion model. Inf. Sci. **559**, 61–74 (2021)

Density Ratio Peak Clustering

Shuliang Wang[1]([✉]), Xiaojia Liu[1], Qi Li[1], Hanning Yuan[1], Ye Yuan[1],
Ziwen Feng[1], and Fan Zhang[2]

[1] School of Computer Science and Technology, Beijing Institute of Technology,
Beijing 100081, China
{slwang2011,yhn6,yuan-ye}@bit.edu.cn
[2] China Mobile Information Technology Center, Beijing 102211, China

Abstract. Clustering is an important means of obtaining hidden information, and is widely used in economics, biomedicine and other disciplines. Data imbalance widely exists in real-world datasets. For example, when fraud detection is performs in transaction data, only a very small amount of transaction data has fraudulent behavior. Therefore clustering on density-imbalanced datasets has practical implications. Various clustering algorithms have been proposed in recent years, but most clustering algorithms cannot correctly identify low-density clusters on density-imbalanced datasets, resulting in clustering failure. To this end, we propose a density ratio peak clustering (DRPC) algorithm, which solves the problem that the original density peak clustering (DPC) algorithm cannot correctly identify low-density clusters and non-center points allocation error linkage problem on density-imbalanced datasets. We conduct experiments on shape datasets, density-imbalanced datasets, and UCI real-world datasets, using normalized mutual information NMI as the evaluation metric, comparing with SNN-DPC, DPC-KNN, DPC, DBSCAN, K-Means algorithms. Experiment results show that DRPC not only inherits the advantages of DPC, but also can more accurately cluster density-imbalanced datasets, and the NMI of the clustering results has increased by 1.5% on average.

Keywords: Clustering · Density-imbalanced datasets · Density
peaks · Density ratio peaks

1 Introduction

Data mining is the process of analyzing data to extract hidden knowledge. Clustering is one of the most important means of data mining. Clustering technology has undergone tremendous changes [1–3], and various clustering methods have been proposed, including partition-based clustering methods [4], hierarchy-based clustering methods [5], grid-based clustering methods [6], density-based clustering methods [7,8], model-based clustering methods [9,10], etc. Among them, density-based clustering methods have been extensively studied [11–18] due to their insusceptibility to noise and their ability to identify clusters of arbitrary shapes.

X. Song et al. (Eds.): APWeb-WAIM 2023, LNCS 14334, pp. 467–482, 2024.
https://doi.org/10.1007/978-981-97-2421-5_31

An imbalanced dataset means that the number of data of certain classes in the dataset is far more than that of other classes. The class with a large number of data is called the majority class, and vice versa is called the minority class. Density-imbalanced dataset belongs to the category of imbalanced dataset, in which clusters with high density are called high-density clusters, and vice versa are called low-density clusters. When clustering imbalanced datasets, many algorithms fail to capture enough information from low-density clusters to make accurate predictions. Density-imbalanced datasets widely exist in real life. For example, in fraud detection problems, fraudulent data can be clustered into low-density class. It has strong practical significance to solve the problem of poor clustering on density-imbalanced datasets.

DPC [19] is a density-based clustering algorithm published in science and has gained extensive attention [20–27] in recent years. DPC is based on two assumptions: Cluster centers are surrounded by neighbors with lower local density; Cluster centers are at a relatively large distance from any points with higher local density. DPC first calculates the local density ρ and the distance to the point with higher local density δ as the horizontal and vertical axis, then draws the decision graph and selects the points in the decision graph where both ρ, δ are relatively large as the cluster centers, and finally assigns the non-center points to the cluster to which the nearest higher local density point belongs. DPC is simple and efficient, capable of correctly identifying clusters of arbitrary shape including non-spherical clusters, and can automatically find the correct number of clusters in the decision graph.

However, the clustering results of DPC on density-imbalanced datasets are not satisfactory. 1. The calculation of the local density in DPC is based on the number of points within the d_c neighborhood. Under DPC's local density calculation method, density peaks tend to appear in high-density clusters and the cluster centers of low-density clusters may be overlooked. When clustering the density imbalance dataset, density peaks obtained by this calculation method may not reflect true cluster centers. Figure 1(A) plots the cluster centers (marked by red dots) of the Jain dataset when performing DPC. Jain dataset consists of two crescent-shaped clusters. As can be seen from Fig. 1(A), the two density peaks calculated by the DPC both appear in high-density cluster, which is obviously inconsistent with the real situation.

2. The non-center point allocation strategy of DPC may cause cascading errors. One point being misassigned will cause all points that regard it as the nearest point of higher density being misassigned, resulting in a significant reduction in clustering accuracy. Figure 1(B) shows the clustering results of DPC on Path based dataset, which contains a ring-shaped cluster wrapping two spherical clusters. Due to the chain error in the allocation strategy of DPC, a large number of points on the left and right sides of the ring-shaped cluster are misassigned, leading to a sharp drop in clustering accuracy.

In view of the deficiencies of the original DPC, we propose density ratio peak clustering (DRPC) for density-imbalanced datasets. The contributions of our work are as follows:

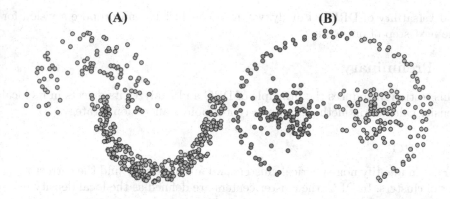

Fig. 1. Cluster Center Identification Error of DPC on Jain Dataset

1. **We augment neighbor information of points and balance the potential density difference between clusters for more accurate selection of cluster centers.** In the selection of cluster centers, first the concept of neighbor propagation is introduced to expand neighbor information of all points, then we define local density ratio to replace local density based on the full neighbor information obtained after neighbor propagation. Local density ratio balances the density differences between clusters in density imbalanced datasets while highlighting the density maxima in all clusters. The newly defined local density ratio can solve the problem that under DPC's local density calculation method density peaks tend to appear in high-density clusters and the cluster centers of low-density clusters may be overlooked.

2. **We make improvements in the allocation strategy of non-center points.** We propose the concept of shared neighbor ratio, using the idea that the greater the proportion of two data points with the same neighbor points in all neighbors, the greater the probability that the two points belong to the same cluster. Therefore, label assignment of each point utilizes label information of the point with the highest similarity. This allocation strategy can overcome the chain error that may occur in the original DPC.

3. **We have conducted extensive experiments.** We compare DRPC with other five clustering algorithms, using the normalized mutual information NMI as the evaluation metric. DRPC achieved optimal results on 8 of the 9 density-imbalanced datasets, among which NMI = 1 is achieved on 7 datasets. Optimal clustering results are also achieved on 4 of the 6 shape datasets, and the NMI reach 1 on two datasets, and sub-optimal clustering results are obtained on the remaining 2 shape datasets. DRPC achieves optimal clustering results on 5 of the 7 UCI real-world datasets and achieves suboptimal clustering results on the remaining two datasets.

The rest of the paper is organized as follows: Sect. 2 introduces the principle of DPC and analyze its deficiencies. In Sect. 3, we give a detailed description of DRPC. In Sect. 4 we present experimental results to demonstrate the superiority and versatility of DRPC. Finally, we draw conclusions and propose a vision for the next step of our work.

2 Preliminary

This section introduces the principle of DPC and analyzes its defects in its local density calculation method and non-center points allocation strategy.

2.1 DPC

DPC can identify non-spherical clusters and automatically find the correct number of clusters. In DPC, the cluster centers are defined as the local density maximum. DPC is based on the assumption that cluster centers are surrounded by points with lower local density and the distance between the cluster center and a points with high local density is relatively large. For each data point i, DPC calculates its local density ρ_i and its distance from the nearest higher density point δ_i, d_{ij} is the distance between data points i and j, local density of data point i is defined as in Eq. 1.

$$\rho_i = \sum_j \chi(d_{ij} - d_c) \tag{1}$$

where $\chi(x) = \begin{cases} 1, x < 0 \\ 0, x \geq 0 \end{cases}$, d_c is the cutoff distance. DPC's local density calculation also has a kernel version as shown in Eq. 2. The local density of a point calculated by both calculation methods is based on the number of points in its d_c neighborhood. The distance between data point i and a data point with higher local density is defined as in Eq. 3:

$$\rho_i = \sum_{i \neq j} exp[-(\frac{d_{ij}}{d_c})^2] \tag{2}$$

$$\delta_i = \min_{j:\rho_j > \rho_i} (d_{ij}) \tag{3}$$

For point i with highest density, $\delta_i = max_j(d_{ij})$. Then DPC plots the decision graph of ρ and δ to identify cluster centers. After finding all cluster centers, remaining points are assigned to the same cluster as its nearest neighbor of higher local density.

2.2 Defects of DPC in Local Density Calculation

The local density of DPC is based on the number of points in the d_c neighborhood. In a density-imbalanced dataset, the number of points in the d_c neighborhood of non-center points in high-density clusters is probably higher than that of points in low-density clusters, or even higher than that of the cluster centers of low-density clusters, that is, **cluster centers in low-density clusters has lower ρ than non-center points in high-density clusters**, which can lead to confusion when selecting cluster centers in the decision graph later.

2.3   Defect of DPC in Allocation Strategy

After all cluster centers being selected, DPC assigns remaining points to the cluster as its nearest higher-density point. Although this one-step allocation strategy seems efficient, when a point is misassigned, the points regarding it as the nearest point of higher density will also be misassigned, that is, **a chain reaction of allocation errors occurs.**

3   Density Ratio Peak Clustering

Aiming at the deficiencies of the original DPC, we propose density ratio peak clustering(DRPC) for more accurate clustering on density-imbalanced datasets. The algorithm is divided into two stages, 1: Identification of cluster centers; 2: Non-center points allocation. To identify cluster centers, the concept of neighbor propagation is introduced to expand the neighbor information of each point, and the newly defined local density ratio is used to replace the original local density to make the identification of cluster centers easier. In the allocation of non-center points, in order to overcome the chain allocation error of density peak clustering, the concept of shared nearest neighbor ratio is introduced, and the k nearest neighbor information is used to assign labels. The detailed introduction of DRPC is as follows:

3.1   Cluster Center Identification

First calculate distance matrix D, $D^{n \times n} = \{d_{ij}\}^{n \times n}$, where d_{ij} is the Euclidean distance of points i, and j. Based on the distance matrix D, here we define the direct neighbor set to represent the direct adjacency between points.

Definition 1. *(Direct Neighbor Set). For any point i, the direct neighbor set is defined as the set of point i's adjacent points within d_c neighborhood denoted by $Adj_0(i)$, expressed as*

$$Adj_0(i) = \{j | d_{ij} < d_c\} \tag{4}$$

The local density in the DPC is based on the number of data points in the d_c neighborhood, that is $|Adj_0(i)|$, resulting in underutilization of the neighbor information. In order to expand the adjacency information of points, the concept of neighbor propagation is introduced. Naturally speaking, a point is most similar to its direct k nearest neighbors, the possibility of being in the same cluster is the highest, and the neighbor information of one's k nearest points is likely to convey extended information of the point. Denote the k nearest points of point i by $\gamma(i)$. Let point i propagate its neighbor information with its k nearest neighbors to expand neighbor information, and the extended information is in point i's complete neighbor set.

Definition 2. *(Complete Neighbor Set). For any point i and j belonging to $\gamma(i)$, the complete neighbor set of i is denoted by $Adj(i)$ defined as*

$$Adj(i) = Adj_0(i) \cup Adj_0(j), j \in \gamma(i) \tag{5}$$

With the complete neighbor information, we define the local density of DRPC.

Definition 3. *(DRPC Local Density). The local density is defined as the number of points in the complete neighbor set $Adj(i)$, denoted by ρ_i.*

$$\rho_i = |Adj(i)| \tag{6}$$

In order to solve the problem that the local density calculation method based on the number of points cannot correctly identify the cluster centers of low-density clusters in density-imbalanced datasets. Based on the assumption that the cluster centers in clusters of different densities are all density maximas, we define local density ratio instead of the local density in DRPC to assist the identification of cluster centers.

Definition 4. *(DRPC Local Density Ratio). For any point i, its local density ratio ρ_i^* is defined as the ratio of its local density to the average local densities of its complete neighbor set.*

$$\rho_i^* = \frac{\rho_i}{\bar{\rho_j}}, j \in Adj(i) \tag{7}$$

After replacing local density with local density ratio, the local density ratio of the non-center points of the high-density clusters shall be lower than the local density ratio of the cluster centers in the low-density clusters, so that the correct cluster centers can be more accurately selected in the decision graph.

Compared with local density, local density ratio can better reflect cluster centers, making it easier to select the cluster centers in the subsequent decision graph.

After using the local density ratio to replace the local density, the distance to the higher local density point in original DPC is correspondingly changed to the distance to the nearest higher local density ratio point in DRPC.

Definition 5. *(Distance From Nearest Higher Local Density Ratio Point). For any point i, the distance from its nearest point of higher local density ratio is denoted as δ_i^*, expressed as in*

$$\delta_i^* = \min_{j:\rho_j^* > \rho_i^*}(d_{ij}) \tag{8}$$

For point i of highest local density ratio, $\delta_i^* = max_j(d_{ij})$. Draw decision graph with ρ^*, δ^* as the horizontal and vertical axis. Select points with relatively large ρ^*, δ^* as cluster centers.

3.2   Allocation Strategy of Non-center Points

After all the cluster centers being identified, the remaining points need to be assigned class labels. Some previous methods use shared nearest neighbors(SNN) in the label assigning process, which is defined as the size of their common neighbor set. However, the number of common neighbors of two points is not enough to express their similarity. Assuming that the $SNN(i,j) = 5$, if point i only has 6 neighbor points including point j, then the label of j is very important for the label assignment of i. But if i has 100 neighbor points in total making the common neighbors of i and j a small fraction of $i's$ neighbors, so the importance of point j's label to i will decrease accordingly. Based on the limitation of SNN, the idea behind the allocation strategy of DRPC is as follows: The greater the proportion of two points with the same neighbors in all neighbors, the higher the similarity of the two points, and the greater the possibility that the two points belong to the same cluster.

Here we define the shared neighbor ratio (SNR) between point i and point j as the ratio of the shared neighbors of point i and point j to the number of point i's neighbors.

Definition 6. *(Shared Neighbor Ratio). According to the above idea, for any point i and point j belonging to $\gamma(i)$, the shared neighbor ratio is defined as:*

$$SNR(i,j) = \frac{|Adj(i) \cap Adj(j)|}{|Adj(i)|} \tag{9}$$

The shared neighbor ratio of two points is more representative of the similarity than just the number of shared neighbors. In the allocation of non-center points, in order to overcome the problem of chain allocation errors that may be caused by the allocation strategy of the DPC, for each non-center point, calculate the SNR with its k nearest neighbors. A point is assigned to the cluster with the most frequent label among its k nearest neighbors whose shared neighbor ratio with it is greater than $\frac{1}{2}$.

4   Experiment

4.1   Experiment Settings

We compare DRPC with SNN-DPC, DPC-KNN, DPC, DBSCAN, K-Means to verify the effectiveness of DRPC, and select the best results to present. In order to prove the effectiveness of DRPC, we select 6 shape datasets, 9 density-imbalanced datasets, and 7 real-world UCI datasets [28] for experiment, which are presented in Table 1. Column N stands for Number of Records.

We use normalized mutual information (NMI) as the evaluation metric, and the maximum value of NMI is 1. The higher the NMI of the clustering result is, the better the performance of the algorithm.

4.2   Shape Datasets

DPC achieves good clustering results on shape datasets. In order to prove that DRPC also retains the advantages of DPC, experiments are carried out on shape datasets. Table 2 shows the clustering results of DRPC, SNN-DPC, DPC, DBSCAN, DPC-KNN, and K-Means on shape datasets. On shape dataset, DRPC achieves the best clustering results on Aggregation, Jain, PathBased, and Spiral. On Compound and Flame datasets, DRPC achieves the suboptimal clustering result, and DBSCAN and DPCs achieves optimal clustering result respectively.

Table 1. Datasets Description

Type	Dataset	Dimension	N	Class	Dataset	Dimension	N	Class
Shape	Aggregation	2	788	7	Jain	2	373	2
	Compound	2	399	6	Pathbased	2	300	3
	Flame	2	240	2	Spiral	2	312	3
Imbalanced	Un_Spiral	2	462	4	4-columns	2	1200	4
	Un_Aggregation	2	758	7	2-columns	2	950	2
	Un_Flame	2	920	4	3-columns	2	1200	3
	Un_Jain	2	925	2	5-columns	2	1200	5
	5-eps	2	771	5				
Real-World	seeds	7	210	3	Wine	13	178	3
	Iris	4	150	3	Dermatology	34	358	6
	Ecoli	7	336	9	Glass	9	214	6
	segmentation	19	210	7				

Table 2. Performances on shape datasets.

	DRPC	SNN-DPC	DPC	DBSCAN	DPC-KNN	K-Means
Aggregation	**0.9958**	0.9555	0.9924	0.9829	0.8685	0.8792
Compound	0.9172	0.8526	0.7371	**0.9382**	0.7491	0.7185
Flame	0.9335	0.8994	**0.9359**	0.9259	0.4476	0.3988
Jain	**1.0**	**1.0**	0.542	**1.0**	0.4859	0.369
Pathbased	**0.9359**	0.9013	0.5491	0.8721	0.5347	0.5458
Spiral	**1.0**	**1.0**	0.9409	**1.0**	0.0004	0.0013

4.3   Density-Imbalanced Datasets

The focus of DRPC is accurate clustering on density-imbalanced dataset. Previous sections have proved that DRPC inherits the advantages of DPC, so in this section, we conduct sufficient experiments on density-imbalanced datasets to demonstrate the advantages of DRPC on density-imbalanced datasets. The

NMI of clustering results of each algorithm on nine density-imbalanced datasets are shown in Table 3. Of the nine density-imbalanced datasets, DRPC achieves best clustering results on eight datasets. On Un_Aggregation dataset, DRPC obtains suboptimal clustering result, and DBSCAN obtains optimal clustering result. NMI of DRPC on Un_Spiral, Un_Jain, 2-column, 3-column, 4-column, 5column, and 5-eps is 1. The clustering results of each algorithm are visualized and analyzed as follows. The six subgraphs are the clustering result obtained by DRPC, SNN-DPC, DPC, DBSCAN, DPC-KNN, and K-Means algorithms. Points clustered into different clusters are represented by different colors. Cluster centers are marked by stars.

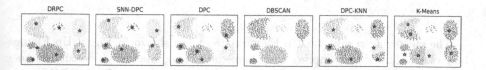

Fig. 2. The clustering results on Un_Aggregation by 6 algorithms.

Table 3. Performances on density-imbalanced datasets.

	DRPC	SNN-DPC	DPC	DBSCAN	DPC-KNN	K-Means
Un_Aggregation	0.9522	0.9519	0.903	**0.9747**	0.8235	0.8143
Un_Flame	**0.98**	0.98	0.8248	0.8627	0.7589	0.7363
Un_Spiral	**1.0**	0.8284	0.7355	0.94	0.545	0.46
Un_Jain	**1.0**	0.5011	0.3675	**1.0**	0.3198	0.1356
2-columns	**1.0**	**1.0**	0.04	**1.0**	0.8442	0.7516
3-columns	**1.0**	**1.0**	0.673	**1.0**	**1.0**	**1.0**
4-columns	**1.0**	0.8201	0.5042	**1.0**	0.8687	**1.0**
5-columns	**1.0**	0.7984	0.5115	**1.0**	0.717	0.7971
5-eps	**1.0**	0.8914	0.8628	**1.0**	0.9458	**1.0**

The clustering results on Un_Aggregation dataset are shown in Fig. 2. This is an imbalanced dataset obtained by sampling transformation of the Aggregation dataset. In result of DRPC on this dataset, the two spherical clusters in the pink area on the lower left are connected together, and the points in the larger spherical cluster are not assigned to their correct cluster centers (blue point circled by black star), indicating that DRPC still has certain limitations and can be improved in future work. SNN-DPC misassigns the points located on the upper left border of the blue spherical cluster to the adjacent pink cluster, and the points in the area connecting the two spherical clusters on the right are misassigned to the lower cluster. Although DBSCAN can achieve the highest

NMI, it can be found that some points (gray) that should belong to the upper left pink cluster are misassigned to the middle spherical cluster. DPC, DPC-KNN, and K-Means all appear to recognize one cluster as multiple clusters, and the results are not ideal.

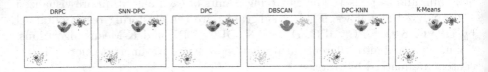

Fig. 3. The clustering results on Un_Flame by 6 algorithms.

The clustering results on Un_Flame dataset are shown in Fig. 3. Un_Flame dataset is obtained by sampling the Flame dataset to obtain high-density flame-shape clusters and adding two low-density spherical clusters. Both DRPC and SNN-DPC achieve satisfactory clustering results. DPC failed to identify the cluster center of the upper right spherical cluster. Its local density calculate method mistakenly identifies two cluster centers in one high-density cluster. DPC also has a chain error in the allocation of non-central points. The points on the left side of the flame-shaped dataset are closer to the higher-density points in other clusters, therefore they are assigned to the wrong cluster. DBSCAN fails to correctly identify the number of clusters, and identifies two spherical clusters as one cluster, and the result is poor. DPC-KNN has an error in the allocation of non-central points, and its allocation strategy misassigns the points assigns the left and right ends of the lower flame-shaped cluster to the upper cluster (gray cluster). K-Means can correctly identify two spherical clusters, but fails in flame-shaped clusters.

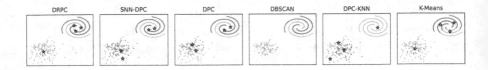

Fig. 4. The clustering results on Un_ Spiral by 6 algorithms.

The clustering results on Un_Spiral dataset are shown in Fig. 4. Un_Spiral dataset consists of three spiral-shaped clusters and a low-density spherical cluster at the lower left, a total of 4 clusters and is obtained by adding a low-density spherical cluster to the Spiral dataset. On the Un_Spiral dataset, only DRPC can cluster correctly. SNN-DPC, DPC, DBSCAN, and DPC-KNN all misidentified the low-density cluster located in the lower left as multiple clusters. DBSCAN can identify the three spiral-shaped clusters but fails to identify the number of clusters. It mistakenly identifies the low-density clusters in the lower left corner as two clusters. K-Means is capable of finding spherical clusters, so it can correctly identify the low-density spherical clusters in Un_Spiral, but cannot identify the upper right spiral clusters.

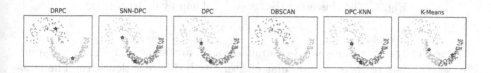

Fig. 5. The clustering results on Un_Jain by 6 algorithms.

The clustering results on Un_Jain dataset are shown in Fig. 5. Un_Jain dataset contains two crescent-shaped clusters. The density of the crescent-shaped cluster at top is low, and the density of the crescent-shaped clusters located below is high, which is obtained by sampling the Jain dataset to further expand the level of density imbalance of the dataset. Only DRPC and DBSCAN algorithms can get the correct results. SNN-DPC, DPC, and DPC-KNN all have errors in identifying the high-density clusters below as two clusters. The allocation strategy of DPC and DPC-KNN leads to the allocation failure of a large number of points. K-Means algorithm cannot identify non-spherical clusters, therefore its clustering result is bad.

Fig. 6. The clustering results on 2-column by 6 algorithms.

The clustering results on 2-column dataset are shown in Fig. 6 2-column consists of two clusters with different densities. The left cluster has low density and the right cluster has high density. On 2-column dataset, DRPC, SNN-DPC, DBSCAN can obtain accurate clustering results. The local density calculation method of the DPC is invalid for identifying the cluster centers of the 2-column dataset. As a result, both cluster centers are identified in the high-density cluster on the right by mistake. The allocation strategies of DPC and DPC-KNN algorithms all lead to some points of the low-density cluster being assigned to the high-density cluster. The K-Means algorithm is not suitable for non-spherical clusters, and its optimization strategy of optimizing the sum of the distances from all points to the cluster centers leads to the wrong assignment of the left cluster.

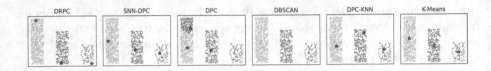

Fig. 7. The clustering results on 3-column by 6 algorithms.

The clustering results on 3-column dataset are shown in Fig. 7 3-column dataset consists of three rectangular clusters with different densities. The cluster density decreases from left to right. It can be found that DRPC, SNN-DPC, DBSCAN, DPC-KNN, and K-Means algorithms can all perform clustering correctly. The local density calculation method of DPC causes two local density maxima in the high-density cluster to be misidentified as cluster centers, thereby identifying the leftmost high-density cluster as two clusters. The chain errors caused by the allocation strategy lead to all points in the rightmost cluster being misassigned.

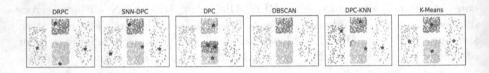

Fig. 8. The clustering results on 4-column by 6 algorithms.

The clustering results on 4-column dataset are shown in Fig. 8, 4-column consists of four rectangular clusters, the two clusters in the middle have high density, and the clusters on both sides have low density. DRPC, SNN-DPC, DBSCAN, K-Means can all get accurate clustering results. DPC's local density calculation method based on the number of points in the d_c neighborhood cannot correctly identify the cluster centers of the low-density clusters. Consequently, the high-density cluster in the middle is identified as three clusters. In addition, its allocation strategy also leads to misallocation of a large number of points. DPC-KNN assigns points to the cluster where the nearest cluster center is located, causing the points on the border of two clusters in the middle being assigned to the wrong clusters.

The clustering results on 5-column dataset are shown in Fig. 9. 5-column consists of 3 rectangular clusters and two triangular clusters. The triangular clusters on both sides are of lowest density, followed by the middle two rectangular clusters, and the upper rectangular cluster with the highest density. Both DRPC and DBSCAN algorithms achieve correct clustering results. SNN-DPC, DPC, DPC-KNN, and K-Means all have errors of identifying one cluster as two clusters. Among them, DPC and DPC-KNN also have errors in the allocation

Fig. 9. The clustering results on 5-column by 6 algorithms.

of non-center points, which should belong to low-density clusters. Points of low-density clusters are assigned to high-density clusters that are closer to them, and the optimization strategy of K-Means algorithm to optimize the sum of distances from all points to cluster centers leads to massive misassignment of clusters.

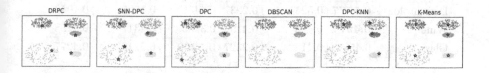

Fig. 10. The clustering results on 5-eps by 6 algorithms.

The clustering results on 5-eps dataset are shown in Fig. 10. 5-eps consists of 5 spherical clusters with different densities, the lower left spherical cluster has the lowest density, followed by the upper two clusters, and the right two clusters have the highest density. It can be seen that DRPC, DBSCAN, and K-Means all achieve correct clustering results. SNN-DPC failed to fully obtain the information in the low-density cluster, and misidentified the lower left low-density cluster as two clusters. The local density calculation method of DPC recognizes the upper left high-density cluster as two clusters, and the allocation strategy of DPC-KNN causes points of the upper right spherical cluster to be misassigned to the high-density below cluster which is closer to them.

4.4 Real-World Datasets

In this section, seven UCI real-world datasets are selected to further verify the generality of DRPC's performance. The experimental results are shown in Table 4. DRPC achieves the best results on five datasets of Iris, Ecoli, segmentation, Dermatology, and Wine, and the NMI of the clustering results has increased by 8.62% on average. DRPC achieves sub-optimal results on the seeds and Glass datasets. SNN-DPC achieves optimal clustering results on the seeds dataset, suboptimal clustering results on Iris, Ecoli, segmentation, Dermatology, Wine, and slightly inferior to DBSCAN and DRPCs on the Glass dataset. DPC performs well on Iris, seeds, Ecoli, segmentation, and Wine, but performs poorly on Dermatology and Glass, and the results is only 0.223, 0.2327 respectively.

DBSCAN achieves the optimal clustering result on the Glass dataset, and the clustering results on Iris, seeds, Ecoli, segmentation, Dermatology, and Wine are average. DPC-KNN performs poorly on all seven real-world dataset. K-Means performs well on Iris, seeds, Ecoli, segmentation, Wine, and Glass datasets, but performs poorly on Dermatology, and the NMI is only 0.1049.

Table 4. Performances on real-world UCI datasets.

	DRPC	SNN-DPC	DPC	DBSCAN	DPC-KNN	K-Means
Iris	**0.9192**	0.8851	0.8057	0.7337	0.4698	0.7582
seeds	0.7379	**0.7565**	0.6751	0.5554	0.4669	0.6949
Ecoli	**0.6853**	0.6604	0.5234	0.5456	0.4680	0.5896
segmentation	**0.680**	0.6227	0.5392	0.5336	0.3778	0.5733
Dermatology	**0.5777**	0.5241	0.223	0.4451	0.1725	0.1049
Wine	**0.4977**	0.423	0.4193	0.3465	0.4317	0.4288
Glass	0.4586	0.4132	0.2327	**0.5269**	0.2218	0.4277

5   Conclusion

We propose density ratio peak clustering algorithm (DRPC), which solves the problem that original DPC has an unsatisfactory clustering effect on the density imbalanced datasets. DRPC utilizes the local density ratio to find the cluster centers and its improved allocation strategy using the shared neighbor ratio overcomes the chain error that may occur in the non-center points allocation process. Experimental results on the density-imbalanced datasets show that the algorithm can obtain highly accurate clustering results on density imbalanced datasets, which is considered as great improvement over DPC, SNN-DPC, KNN-DPC, K-Means, and DBSCAN. Satisfactory results also are achieved on shape and real-world datasets, and the algorithm is versatile. Future work will focus on developing a parameter adaptive density-based clustering algorithm for density-imbalanced datasets.

Acknowledgements. This research is supported by National Natural Science Foundation of China (62306033, 62076027).

References

1. Fang, U., Li, J., Akhtar, N., Li, M., Jia, Y.: Gomic: multi-view image clustering via self-supervised contrastive heterogeneous graph co-learning. World Wide Web-Internet Web Inf. Syst. (2023)

2. Wang, J., Shi, Y., Li, D., Zhang, K., Chen, Z., Li, H.: MCHA: a multistage clustering-based hierarchical attention model for knowledge graph-aware recommendation. World Wide Web-Internet and Web Inf. Syst. **25**(3, SI), 1103–1127 (2022)

3. Yuan, C., Zhu, Y., Zhong, Z., Zheng, W., Zhu, X.: Robust self-tuning multi-view clustering. WORLD WIDE WEB-INTERNET AND WEB INFORMATION SYSTEMS **25**(2, SI), 489–512 (MAR 2022)

4. Wang, H.Z.: Corrigendum to 'a fuzzy k-prototype clustering algorithm for mixed numeric and categorical data. Knowl.-Based Syst. **30**, 129–135 (2012)

5. Tian, Z., Ramakrishnan, R.: Miron livny: birch: an efficient data clustering method for very large databases. ACM SIGMOD Rec. **25**(2) (1999)

6. Wang, W., Yang, J., Muntz, R.: Sting: A statistical information grid approach to spatial data mining. In: VLDB'97, Proceedings of 23rd International Conference on Very Large Data Bases, 25–29 August 1997, Athens, Greece (1997)

7. Cai, J., Luo, J., Wang, S., Yang, S.: Feature selection in machine learning: a new perspective. Neurocomputing **300**, 70–79 (2018)

8. Du, M., Ding, S., Jia, H.: Study on density peaks clustering based on k-nearest neighbors and principal component analysis. Knowl.-Based Syst. **99**, 135–145 (2016)

9. Xiang, L.Y.H..: Dynamic resource allocation algorithm based on big data stream characteristic and improved SOM clustering. Comput. Appl. Softw. (2019)

10. Hu, T., Sung, S.Y.: A Hybrid EM Approach to Spatial Clustering. Elsevier Science Publishers B. V., Amsterdam (2006)

11. Pourbahrami, S., Hashemzadeh, M.: A geometric-based clustering method using natural neighbors. Inf. Sci. **610**, 694–706 (2022)

12. Pourbahrami, S., Khanli, L.M., Azimpour, S.: Improving neighborhood construction with apollonius region algorithm based on density for clustering. Inf. Sci. **522**, 227–240 (2020)

13. Xu, X., Ding, S., Wang, L., Wang, Y.: A robust density peaks clustering algorithm with density-sensitive similarity. Knowl.-Based Syst. **200** (2020)

14. Cheng, D., Zhang, S., Huang, J.: Dense members of local cores-based density peaks clustering algorithm. Knowl.-Based Syst. **193** (2020)

15. Tao, X., et al.: Adaptive weighted over-sampling for imbalanced datasets based on density peaks clustering with heuristic filtering. Inf. Sci. **519**, 43–73 (2020)

16. Gong, C., Su, Z.G., Wang, P.H., Wang, Q.: Cumulative belief peaks evidential k-nearest neighbor clustering. Knowl.-Based Syst. **200** (2020)

17. Flores, K.G., Garza, S.E.: Density peaks clustering with gap-based automatic center detection. Knowl.-Based Syst. **206** (2020)

18. Lu, H., Shen, Z., Sang, X., Zhao, Q., Lu, J.: Community detection method using improved density peak clustering and nonnegative matrix factorization. Neurocomputing **415**, 247–257 (2020)

19. Rodriguez, A., Laio, A.: Clustering by fast search and find of density peaks. Science **344**(6191), 1492–1496 (2014)

20. Wang, S., Li, Q., Zhao, C., Zhu, X., Dai, T.: Extreme clustering - a clustering method via density extreme points. Inf. Sci. **542** (2020)

21. Cheng, D., Huang, J., Zhang, S., Liu, H.: Improved density peaks clustering based on shared-neighbors of local cores for manifold data sets. IEEE Access **7**, 151339–151349 (2019)

22. Liu, R., Huang, W., Fei, Z., Wang, K., Liang, J.: Constraint-based clustering by fast search and find of density peaks. Neurocomputing **330**, 223–237 (2019)

23. Liu, L., Yu, D.: Density peaks clustering algorithm based on weighted k-nearest neighbors and geodesic distance. IEEE Access **8**, 168282–168296 (2020)

24. Li, J., Zhu, Q., Wu, Q.: A self-training method based on density peaks and an extended parameter-free local noise filter for k nearest neighbor. Knowl.-Based Syst. **184** (2019)

25. Li, R., Yang, X., Qin, X., Zhu, W.: Local gap density for clustering high-dimensional data with varying densities. Knowl.-Based Syst. **184** (2019)

26. Wang, Y., Wong, K.C., Li, X.: Exploring high-throughput biomolecular data with multiobjective robust continuous clustering. Inf. Sci. **583**, 239–265 (2022)

27. Tao, X., Chen, W., Zhang, X., Guo, W., Qi, L., Fan, Z.: SVDD boundary and DPC clustering technique-based oversampling approach for handling imbalanced and overlapped data. Knowl.-Based Syst. **234** (2021)

28. Dua, D., Newman, D.: UCI machine learning repository, University of California, School of Information and Computer Science (2017)

Feds: A Highly Efficient Keyword Search System Operating on Federated RDF Systems

Mingdao Li[✉], Peng Peng, and Zheng Qin

Hunan University, Changsha, China
{limingdao,hnu16pp,zqin}@hnu.edu.cn

Abstract. In this demonstration, we develop Feds, a federated RDF-based key-word search system. Our approach involves constructing an offline schema graph which represents the structure of the federated RDF system. By mapping user-inputted keywords to the vertices in the schema graph, we establish a connection between the keywords and the underlying RDF data. Once the keyword-to-vertex mapping is established, we leverage the schema graph to transform the user's keywords into SPARQL queries. The constructed queries are then executed over the federated RDF system, allowing us to obtain the desired search results. To measure the efficiency and effectiveness of our system, we propose cost models that evaluate the performance of keyword mapping and query construction. To further optimize the evaluation process, we introduce a multiple query optimiza-tion strategy. Through extensive experimental studies using real-world federated RDF datasets, our demonstration shows remarkable efficiency, effectiveness, and scalability in performing keyword search over federated RDF systems.

1 Introduction

The *Resource Description Framework* (RDF) is a technical standard used to mark resources on the web and has been widely applied in real-life scenarios. RDF uses triples to describe data, with each triple denoted as ⟨subject, property, object⟩. SPARQL is a query language and protocol specifically designed for querying RDF data. It provides a standardized way to query and manipulate RDF graphs, which are used to represent structured information in a machine-readable format. Presently, increasingly data providers are willing to represent data in the RDF model and provide the SPARQL interface on their sites. Federated RDF systems [2,4] are proposed to integrate and provide transparent access over many sites.

Federated RDF systems typically rely on the SPARQL query language as the pri-mary means for data acquisition. in real-world applications, it can be challenging for users to fully understand the schemas and write correct SPARQL queries. This difficulty arises due to several reasons:complexity of RDF schemas and lack of query guidance. These challenges limit the feasibility of federated RDF systems. Unlike SPARQL, key-word search is a more commonly used and user-friendly search method.

In our demonstration, we have developed a keyword search system called Feds, which facilitates keyword search in federated RDF systems. Feds operates in two phases: offline and online. During the offline phase, we merge the schemas from differ-ent SPARQL endpoints to create a schema graph. Moving to the online phase, we first

X. Song et al. (Eds.): APWeb-WAIM 2023, LNCS 14334, pp. 483–488, 2024.
https://doi.org/10.1007/978-981-97-2421-5_32

utilize the full-text search interfaces offered by existing SPARQL endpoints to generate initial queries based on the provided keywords. We then map the user-input keywords to the candidate classes present in the schema graph. Next, we traverse the schema graph to construct additional queries that capture the users' query intentions. These queries are designed to identify relevant substructures within the federated RDF systems, connecting the vertices that match the keywords. Finally, to enhance the effectiveness of our approach, we introduce cost models for both keyword mapping and query construction. These models help us optimize the process and improve the accuracy of the search results. Additionally, we propose an optimization technique that allows us to rewrite multiple queries together, further enhancing the efficiency of query evaluation.

Technical details of Feds have been published in our previous paper [3] and are summarized in Sect. 2 of this paper. In this demonstration paper we present the prototype system architecture and functionality.

2   System Architecture

In this section, we provide an overview of the different steps involved in our specific execution process. Figure 1 shows the architecture of our system. Considering the challenges in a federated RDF system, a solution in which keywords are mapped through SPARQL query syntax to the schema graph is proposed. Subsequently, multiple SPARQL queries are constructed and executed to obtain the final result. Feds consists of three main components: Keyword Mapping, SPARQL Query Construction, Query Execution.

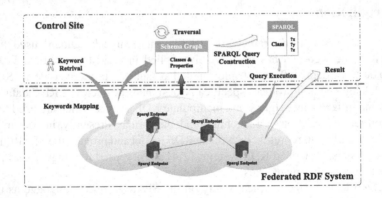

Fig. 1. System Architecture

Keyword Mapping. The graph in the federated RDF system is composed of data from multiple sites. However, it is challenging to establish a global index on the federated RDF graph due to some data providers only offering SPARQL query interfaces and not allowing others to traverse their RDF graphs. To enable the creation of a global index on the federated RDF graph, we build a schema graph during the offline phase. The schema graph includes only class vertices of the RDF graph and edges representing relationships between different classes, capturing the relationships intuitively. Figure 2 shows an example schema graph. After the construction of the schema graph, we look up the authoritative documents of classes and properties in the schema graph and build inverted indices for them. All of the indices are maintained at the control site.

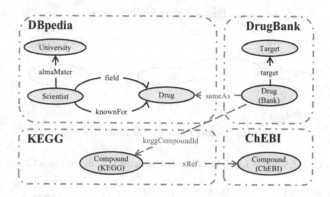

Fig. 2. Example Schema Graph

In the online phase, given a set of keywords $KW=\{w_1, w_2, \ldots, w_n\}$, a candidate class vertex set is generated for each keyword through keyword mapping. In the schema graph, class vertices mapped to keywords are denoted as *keyword elements* in this context. To measure the relevance between keywords and classes, we propose a cost model where the most relevant class vertex mapping is assigned the lowest cost.

SPARQL Query Construction. After finding the keyword element, we explore the schema graph to generate the top-k minimum substructures that connects the keyword element. The top-k substructures is then utilized to generate SPARQL queries. Figure 3 shows the top-2 substructures of a sample query. Vertices are mapped to variables or constants and edges are mapped to properties according to the structure mapping of the query's triple pattern and edges of the schema graph. This process continues until the top-k smallest queries are computed. Then, we construct a set of SPARQL queries $\mathbb{Q}$ to generate results related to the keyword KW over the federated RDF system W. In practice, it is common to retrieve queries of the same size as the top-k smallest queries. To further distinguish them, we propose a cost model.

Query Execution. Once the set of SPARQL queries $\mathbb{Q}$ is constructed, it is dispatched to the corresponding SPARQL endpoints for evaluation. In order to enhance the performance of evaluation, an optimization technique is suggested, which involves rewriting multiple SPARQL queries in $\mathbb{Q}$ simultaneously. This optimization is aimed at improving efficiency by reducing the number of individual query evaluations. Figure 4 illustrates an example of query rewriting. The results of query execution will be obtained and presented as the final results.

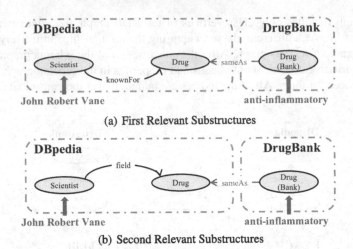

(a) First Relevant Substructures

(b) Second Relevant Substructures

Fig. 3. Example Top-2 Relevant Substructures in the Schema Graph

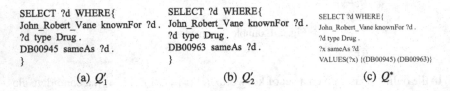

```
SELECT ?d WHERE{
John_Robert_Vane knownFor ?d .
?d type Drug .
DB00945 sameAs ?d .
}
```

(a) Q'_1

```
SELECT ?d WHERE{
John_Robert_Vane knownFor ?d .
?d type Drug .
DB00963 sameAs ?d .
}
```

(b) Q'_2

```
SELECT ?d WHERE{
John_Robert_Vane knownFor ?d .
?d type Drug .
?x sameAs ?d
VALUES(?x) {(DB00945) (DB00963)}
```

(c) Q^*

Fig. 4. Query Rewrite Example

3 Demonstration

In this demo, we perform keyword searches on five major federated RDF datasets, including LifeScience, CrossDomain, LargeRDFBench, LUBM, and QALD-4. We input eight queries for each dataset, four of which consist of two keywords and the other four consist of three keywords.

3.1 Case Study

In Fig. 5, we sampled a query on CrossDomain, labeled CD_1, for case study. CD_1 is to find out which scientists were born in oakland. For this problem, we input the keyword {*scientist, oakland*}. Next we will show the query page of our system, along with screenshots of the results of CD_1. More demonstration could be referred with "http://47.92.255.34:8080/feds/".

Fig. 5. Result of CD_1

3.2 Effectiveness Study

In this section, we assess the effectiveness of our federated multiple query optimization method on both synthetic and real RDF datasets. To evaluate our system, we compared it with FuhSen [1], a federated keyword search engine that has publicly released its source code on GitHub[1].

The experiment used the mean average precision (MAP) [5] factor to evaluate the effectiveness of the results. Table 1 presents the MAP values for 40 queries, with 8 queries per dataset. Overall, the results indicate that Feds achieved higher MAP values compared to the comparative system FuhSen. This demonstrates that Feds outperforms FuhSen in terms of performance.

Table 1. MAP Values (Our Method vs FuhSen)

	Our Method	FuhSen
FedBench (Life Science)	0.81	0.25
FedBench (Cross Domain)	0.68	0.22
LargeRDFBench	0.73	0.28
LUBMft	1	0.69
QALD	0.70	0.32

4 Conclusion

In this demo, we design a federated RDF query system called Feds to solve the keyword search problem in federated RDF systems. This system utilizes the full-text search interface provided by the RDF repository to map keywords to a schema graph and generates SPARQL queries by exploring the schema graph. Subsequently, the generated queries are sent to the SPARQL endpoint and evaluated. Furthermore, to improve the efficiency

---

[1] https://github.com/LiDaKrA/FuhSen-reactive.

of query evaluation, a rewrite-based query optimization method is introduced, which rewrites and merges the generated queries. The query tests demonstrate that this system effectively enhances the keyword search efficiency in federated RDF systems.

References

1. Collarana, D., Lange, C., Auer, S.: FuhSen: a platform for federated, RDF-based hybrid search. In: WWW(Companion Volume), WWW '16 Companion, pp. 171–174, Republic and Canton of Geneva, CHE (2016). International World Wide Web Conferences Steering Committee
2. Görlitz, O., Staab, S.: SPLENDID: SPARQL endpoint federation exploiting void descriptions. In: COLD, pp. 13–24, Aachen, DEU (2011). CEUR-WS.org
3. Li, M., Peng, P., Tian, Z., Qin, Z., Huang, Z., Liu, Y.: Optimizing keyword search over federated RDF systems. IEEE Trans. Big Data 9(3), 918–935 (2023)
4. Quilitz, B., Leser, U.: Querying distributed RDF data sources with SPARQL. In: Bechhofer, S., Hauswirth, M., Hoffmann, J., Koubarakis, M. (eds.) ESWC 2008. LNCS, vol. 5021, pp. 524–538. Springer, Heidelberg (2008). https://doi.org/10.1007/978-3-540-68234-9_39
5. Turpin, A., Scholer, F.: User performance versus precision measures for simple search tasks. In: SIGIR, SIGIR '06, pp. 11–18, New York, NY, USA (2006). Association for Computing Machinery

CY-Apollo: A Multi-view Profile System for Complicated Network Attacks

Zhaoquan Gu[1,2], Haiyan Wang[1]([✉]), Xiayu Xiang[1], Ke Zhou[1], Wenying Feng[1], and Jianxin Li[1]

[1] Peng Cheng Laboratory, Shenzhen 518000, China
{guzhq,wanghy01,xiangxy,zhouk,fengwy,lijx01}@pcl.ac.cn
[2] Harbin Institute of Technology, Shenzhen 518000, China

Abstract. With the fast development of information technologies, many new variants of complicated network attacks, characterized by high covert, persistence, and diffusion, have made identifying and detecting such attacks increasingly difficult. Although many intrusion detection methods or products can help identify possible attack behaviors, it is still a challenging problem to detect complicated network attacks, such as the Advanced Persistent Threat (APT) attacks, which are composed of many single-step attacks. A portrait-based analysis method can provide a multi-view of the complicated attacks, such as the attack path and commonly adopted tools, thus can assist in improving the efficiency and accuracy of network attack detection. However, to the best of our knowledge, there does not exist such a profile system for these attacks. In our work, we first construct the attack profile model and define the expression specifications of attack profile data, then we establish the attack graph based on various types of attack element data, and ultimately generate the profile for different complicated attacks. Furthermore, we develop a multi-view profile system called CY-Apollo for different types of attacks. This system comprises four integral functional modules: data collection module, data preprocessing module, attack profile knowledge graph constructing module and attack profile knowledge management module. CY-Apollo gives the visualization of typical complicated network attacks from four different latitudes: attack principle view, panoramic event view, attack organization view and threat intelligence view, which can help industry professionals quickly understand, learn, and analyze existing complex and cross-domain attacks from multiple perspectives.

Keywords: Attack profile analysis · Attack data · Knowledge graph

1 Introduction

With the rapid development of information technologies, cyberspace security has become a global challenge. Traditional network threats exploit vulnerabilities, defects, or weaknesses in the target object, using methods such as detection,

X. Song et al. (Eds.): APWeb-WAIM 2023, LNCS 14334, pp. 489–495, 2024.
https://doi.org/10.1007/978-981-97-2421-5_33

penetration, intrusion, privilege escalation, theft, tampering, etc., to compromise the confidentiality, integrity, availability, and other security attributes of the target object. In the last decades, a large number of new network attacks or the variants of existing attacks are evolving towards more covert and persistent with high proliferation, leading to a sharp increase in the difficulty of attack identification and detection.

Many network intrusion detection methods are developed to identify possible attack behaviors according to the rules or characteristics of the existing attacks, and many security products have shown good performance in detecting many single-step attacks, such as SQL injection, etc. However, identifying and detecting complicated network attacks, such as the APT attacks, have still been a severe problem in both academic and industrial fields. In order to better detect such attacks, understanding the attackers' intentions and predicting possible further attacks can achieve comprehensive, accurate, timely and effective processing of attacks, assisting in improving detection efficiency and accuracy. Attack profile data analysis technology [1–4] is a promising method that constructs the profile model for the attack to provide a comprehensive understanding of the attacks.

In this work, we construct a multi-view profile model for different types of complicated network attacks and define the representation specification of the profile data. Then we utilize attack data that is collected from various methods to establish the attack graphs, and generate the attack profile based on different attack elements. The attack profile data analysis process is divided into three specific steps: first, analyzing the attacker's IP, country, organization, affected industry, and other information to identify the attacker's identity; second, identifying the attack capabilities by analyzing the attack strategies, tactics, techniques, tools, and vulnerabilities used by the attacker; finally, utilizing data analysis capabilities to find the underlying infrastructure behind the attack.

The key to attack profile data analysis is conducting network attack traceability, which consists of three parts: first, capturing network attacks through security device alerts, log and traffic analysis, service resource anomalies, honeypot systems, etc., to detect attacks; second, using existing technologies such as IP geolocation, malicious sample analysis, ID tracking, etc. to trace and counter collect attacker information; finally, drawing the attack path and classifying attacker identity information to form an attack profile, completing the entire network attack traceability. Most research on network attack tracing [5–7] is currently focused on locating and tracking the attacker's IP address, these studies can be broadly categorized into the following types: 1) active querying: confirm the flow path by actively querying all routers that the data flow may pass through; 2) data detection: monitor data flow in the network by constructing monitoring points which can cover the entire network; 3) path reconstruction: reconstruct the attack data packet path using reconstruction algorithms. We combine these methods and can generate a comprehensive profile for the attacks.

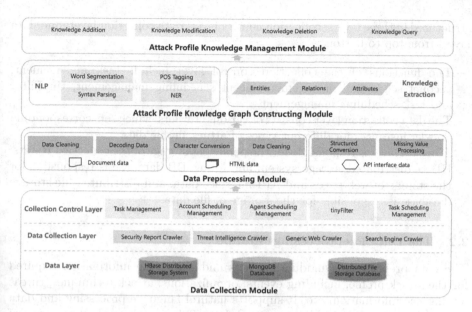

Fig. 1. System architecture of CY-Apollo.

Moreover, we develop CY-Apollo, an innovative multi-view profile system meticulously designed for intricate network attacks. This system consists of four essential functional components: data collection module, data preprocessing module, attack profile knowledge graph constructing module and attack profile knowledge management module. In order to better visualize the typical complicated network attacks, CY-Apollo provides four different latitudes for the attacks, including the attack principle view, the panoramic event view, the attack organization view and the threat intelligence view. CY-Apollo aims to provide a platform for accessing various attack information and assisting in decision-making in the field of cyberspace security.

2   System Architecture of CY-Apollo

In this section, we introduce the system architecture of CY-Apollo. As shown in Fig. 1, CY-Apollo includes four main modules: data collection module, data preprocessing module, attack profile knowledge graph construction module and attack profile knowledge management module, and we introduce the details of each module as follows.

2.1   The Data Collection Module

The data collection module crawls a series of public attack event data through the interface. The data includes threat organizations, security event reports, attack behavior techniques, threat intelligence, etc. The architecture of this module is structured in three layers: collection control layer, data collection layer, and data layer, from top to bottom.

- The functions of the collection control layer include task management, account scheduling management, agent scheduling management, tinyFilter and task scheduling management.
- The data collection layer consists of various crawlers, which sets corresponding crawler instances for different types of attack profile data.
- The data layer is responsible for the persistent storage of the collection results. HBase distributed storage system and MongoDB database is responsible for the storage of web pages and API interface data, and distributed file storage database is responsible for the storage of file data.

2.2   The Data Preprocessing Module

The data preprocessing module extracts and parses the information required for the attack profile, including events, organizations, attack techniques, threat intelligence information, etc. It supports natural language processing and data parsing for structured and unstructured data. This module reads three types of raw data from the data collection: file data, HTML data, and API data, and then executes corresponding cleaning strategies based on different data types.

2.3   The Attack Profile Knowledge Graph Construction Module

This module constructs a model based on typical complex attack behavior types on the Internet. It supports the construction of the entity model for attack profiling, and supports the extraction of attributes and association relation information of entities. The attack profile knowledge graph construction module contains two functions: natural language processing and rule-based knowledge extraction.

- Natural language processing is the basic knowledge extraction and fusion. It performs basic semantic understanding and feature extraction for the input unstructured text data. This process supports the extraction of knowledge such as attributes, relations and events.
- Knowledge extraction is the process of knowledge structuring for unstructured text data. It is defined based on the MDATA model of attack profile, and constructing the extraction models for various entities and relations.

2.4   The Attack Profile Knowledge Management Module

The attack profile knowledge management module queries or enters into the threat intelligence platform or security event data source through an interface. This module is mainly responsible for the management of attack profile knowledge. It provides interfaces for knowledge addition, knowledge modification, knowledge deletion and knowledge query.

3   Visualization of CY-Apollo

CY-Apollo provides a multi-view profile for different types of complicated network attacks. We mainly focus on five types of attacks, APT, botnet, ransomware, worm, and DDoS, and thus CY-Apollo has five different visualizations for these types. Specifically, we present the visualization details as follows:

- **APT attack profile.** This part is responsible for displaying basic information about APT attacks, including the typical attack path, commonly adopted tools, the organization information and security events of the reported APT attacks. The system also shows the regions where APT events occurred by year in the form of maps, and supporting the statistics of attacked regions and industry information.
- **Botnet profile.** This part is responsible for demonstrating the principles of zombie network attacks, displaying basic information about zombie network families, and displaying zombie network events by region and year in map form, supporting statistics on the attacked areas, attack scale, and infected device information.
- **Ransomware profile.** This part is responsible for presenting the attack principles of ransomware, displaying the basic information of ransomware families and the regions where worm viruses have infected in the form of a map by year, also supporting the statistics of infected countries/regions, industries, and the number and scale of infected host systems.
- **Worm profile.** This part is responsible for displaying the infected asset information of worm viruses, demonstrating the attack principles of worm viruses, and displaying the basic information of worm families and the regions where worm virus infections occurred in the form of a map by year. It also supports statistics on infected countries/regions, infected industries, infected host systems and the number of infections.
- **DDOS attack profile.** This part is responsible for displaying the characteristics of DDOS attacks, presenting the latest trends in DDOS attacks, showing the distribution of DDOS attack methods, and displaying the regions where DDOS events occurred in a map format by year. It also supports statistics on the distribution of countries/regions that have been attacked, the industries that have been attacked, the scale of infected devices, and the distribution of DDOS attack types.

The visualization process of CY-Apollo consists of three stages of basic processing, as shown in Fig. 2:

- **Profile command issuance:** based on the query demand from the front end, the attack visualization knowledge management module is used to obtain the required dimensional data for the attack visualization of the five types of complex attacks.
- **Profile knowledge acquisition:** the attack visualization module is responsible for receiving the command issued by the front end, parsing and generating specific knowledge query conditions, and calling the attack visualization knowledge management module to obtain the knowledge.

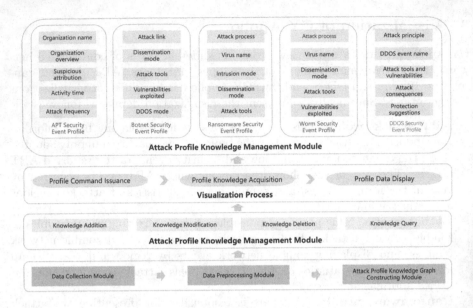

Fig. 2. System's visualization workflow.

– **Profile data display**: after obtaining the attack visualization knowledge, different structured dimensional data of the visualization is generated according to the front-end display request and returned to the end-users.

For different types of complicated network attacks, CY-Apollo mainly provides four high-level reviews as shown in Fig. 3: attack principle view shows the typical attack paths, tools and strategies; the panoramic event view shows the reported security events sorted by different years; the attack organization view

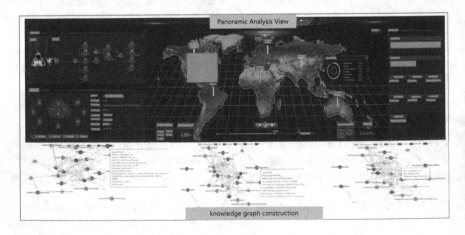

Fig. 3. Visualization architecture of CY-Apollo.

shows the basic information of the behind organization (if exists), such as the attributes, etc.; and threat intelligence view shows the collected events and cyber threat intelligence information.

4 Conclusion

In this paper, we introduced CY-Apollo, a multi-view profile system for complicated network attacks that helps end-users effectively to understand and learn about typical complex attacks from four different aspects: data collection, data preprocessing, construction of attack profile knowledge graph, and management of attack profile knowledge, even if they lack background knowledge of attack profiling. The system also provides the visualizations of attack profiles from four different dimensions: attack principle view, panoramic event view, attack organization view, and threat intelligence view, enabling end-users to have a more intuitive understanding of complex attacks in typical scenarios. The system can provide effective support for end-users to reproduce and analyze typical complex attacks, which can help detect and identify them.

Acknowledgments. This work is supported by the Major Key Project of PCL (Grant No. PCL2022A03).

References

1. Rivers, A.T., Vouk, M.A., Williams, L.A.: On coverage-based attack profiles. In: 2014 IEEE Eighth International Conference on Software Security and Reliability-Companion, San Francisco, CA, USA, pp. 5–6 (2014). https://doi.org/10.1109/SERE-C.2014.15.
2. Maghrebi, H.: Assessment of common side channel countermeasures with respect to deep learning based profiled attacks. In: 2019 31st International Conference on Microelectronics (ICM), Cairo, Egypt, pp. 126–129 (2019). https://doi.org/10.1109/ICM48031.2019.9021728.
3. Barenghi, A., Fornaciari, W., Pelosi, G., Zoni, D.: Scramble suit: a profile differentiation countermeasure to prevent template attacks. IEEE Trans. Comput. Aided Des. Integr. Circuits Syst. **39**(9), 1778–1791 (2020). https://doi.org/10.1109/TCAD.2019.2926389
4. Alotaibi, F., Lisitsa, A.: Matrix profile for DDoS attacks detection. In: 2021 16th Conference on Computer Science and Intelligence Systems (FedCSIS), Sofia, Bulgaria, pp. 357–361 (2021). https://doi.org/10.15439/2021F114.
5. Li, P., Feng, Y., Kawamoto, J., Sakurai, K.: A proposal for cyber-attack traceback using packet marking and logging. In: 2016 10th International Conference on Innovative Mobile and Internet Services in Ubiquitous Computing (IMIS), Fukuoka, Japan, pp. 603–607 (2016). https://doi.org/10.1109/IMIS.2016.89
6. Ling, Y., Yang, C., Li, X., Xie, M., Ming, S.: WEB attack source tracing technology based on genetic algorithm. In: 2022 7th International Conference on Cyber Security and Information Engineering (ICCSIE), Brisbane, Australia, pp. 123–126 (2022). https://doi.org/10.1109/ICCSIE56462.2022.00032
7. Li, Y., Liu, S., Yan, Z., Deng, R.H.: Secure 5G positioning with truth discovery, attack detection, and tracing. IEEE Internet Things J. **9**(22), 22220–22229 (2022). https://doi.org/10.1109/JIOT.2021.3088852

GaussTS: Towards Time Series Data Management in OpenGauss

Lu Li[1], Xu Zhang[1], Feifan Pu[1], Yi Li[1], Jianqiu Xu[1(✉)], and Zhou Zhong[2]

[1] Nanjing University of Aeronautics and Astronautics, Nanjing 210016, China
jianqiu@nuaa.edu.cn
[2] Gauss Database Department, Huawei Company, Beijing, China

Abstract. The rapid advancement of sensor technology presents novel challenges in the efficient management of large scale time series data. In this demo, we demonstrate a time series data management module named GaussTS in database, which provides four key components specifically designed for processing and analysis over time series data application scenarios. Among them, TS Data Query and Analysis mainly focuses similarity search which can be used for enabling data mining and information extraction. TS Data Compression consists of data dimensionality reduction and data partitioning facilitating efficient storage and streamlined processing. TS Data cleaning and TS Data evaluation are capable of accurately handling missing or abnormal data and efficiently assessing data quality. GaussTS has been implemented in a domestic open-source database openGauss and the demonstration showcases the effectiveness and usability of GaussTS in managing and analyzing large scale time series data.

Keywords: Time Series · openGauss · Data Cleaning · Data Evaluation · Similarity Search

1 Introduction

The proliferation of sensors leads to the widespread adoption of time series data across diverse domains, including finance, healthcare, energy, and manufacturing [1]. A time series data management system offers a holistic approach to effectively handle time series data. In order to achieve autonomous management of time series data, we develop a module named GaussTS in openGauss, a domestic enterprise-level open-source relational database built on Linux operating system, to manage time series data. GaussTS serves as a plugin for the database and offers functions that are more in line with the characteristics and application scenarios of time series data.

In recent years, there are a number of time series databases, such as openTSDB [2], InfluxDB [3], and Uqbar [4]. OpenTSDB is an open-source distributed time series database that uses HBase as the storage engine and supports SQL-like query languages and more flexible queries including data aggregation, data grouping, and data filtering. Uqbar, as a domestic time series database, supports complex association queries between time series data and relational data, but lacks specific functionalities tailored for time series data. Despite the extensive research on time series databases, there still exists problems that they cannot meet the requirements of most time series data applications and lack specific data processing and querying functions. GaussTS possesses a wide range of functions applicable to time series data scenarios, such as data evaluation and similarity search which are lacking in existing time series databases. Furthermore, our designed functional algorithms are more general compared to other domain-specific time series databases.

The rest of the paper is organized as follows: The module architecture is discussed in Sect. 2. Section 3 introduces detailed technologies of GaussTS in openGauss and the demonstration is implemented in Sect. 4.

2   GaussTS Architecture

GaussTS consists of four main components: *(i) TS Data Cleaning, (ii) TS Data Evaluation, (iii) TS Data Compression, and (vi) TS Data Query and Analysis*, the framework of which is shown in Fig. 1. Users can utilize TS Data Cleansing and TS Data Evaluation components to preprocess the raw time series data and access data quality reports. Based on the application requirements, users can perform operations such as dimensionality reduction, partitioning, or similarity search in TS Data Compression and TS Data Query and Analysis components on time series data. In order to visualize the processed results, users can make use of third-party visualization software to effectively showcase the impact and outcomes achieved.

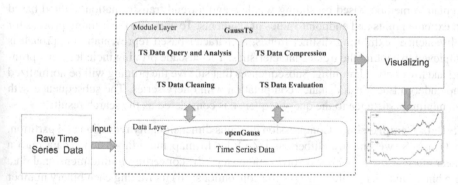

Fig. 1. The framework of GaussTS

TS Data Query and Analysis mainly include similarity search which aims at finding objects in a collection that are close to a given query according to a predefined similarity criterion [5]. The component holds significant prominence in time series analysis and finds applicability in diverse higher-level applications, such as recommendation systems and clustering tasks [6]. Time series similarity search algorithms are typically based on distance measures such as ED (Euclidean Distance) and DTW (Dynamic Time Warping) which is the best method in most fields. TS Data Compression contains dimensionality reduction techniques transforming high-dimensional data into a lower-dimensional representation using specific methodologies which serves to mitigate the complexity of the data. Additionally, partitioning is employed to partition a relational database table into columns, thereby enhancing its suitability for efficient data analysis and visualization purposes. TS Data Cleaning includes identifying and handling missing data, inconsistent attribute values and abnormal values. The cleaning component poses unique data quality challenges due to the presence of auto correlations, trends, seasonality, and gaps in the time series data [8]. The purpose of time series data cleaning is to ensure data quality, reduce the influence of errors, and make analysis more accurate and reliable. TS Data Evaluation is focused on assessing and analyzing the quality of time series data to ascertain its reliability and effectiveness. This process aims to improve forecasting and prediction accuracy, uphold the integrity of time series data, and enhance decision-making. The ultimate goal is to provide a comprehensive evaluation of the data, enabling informed and reliable analysis outcomes.

3   Detailed Techniques of GaussTS

In this section, we mainly introduce the detailed techniques in four components of GaussTS.

TS Data Query and Analysis. In this section, our primary implementation revolves around similarity search, which contains three steps: (i) trend representation, (ii) cascade pruning, and (iii) normalization and distance measurement. The whole search process only needs to scan the time series data once to obtain the result without building any index which greatly reducing the space waste. The trend representation includes segmentation method based on sliding-window and a trend transformation method based on extreme points and endpoint values of segments. To expedite the pruning process for subsequences exhibiting distinct trends, a character-based representation approach is employed to capture the trend characteristics. The cascade pruning includes trend pruning and lower bound pruning. Subsequences that survive the pruning will be normalized and undergo Euclidean distance computation with query series. The subsequence with the minimum distance to the query sequence is considered as the search result.

TS Data Compression. Compression contains dimensionality reduction and partition. In GaussTS, we employ Hilbert and Z-Order to map multidimensional data onto a linear number line. Z-Order involves three steps: (i) converting multidimensional data into binary numbers following specific conventions; (ii) reordering each binary number according to the specified rules to obtain a new binary number; (iii) converting this binary number to a decimal number, and using this integer as the Z-Order encode for the point. For partition function, we utilize the built-in SPI (Server Programming Interface) in openGauss to extract both the time and non-time columns.

TS Data Cleaning. In the data filling and repair stage of cleaning task, we employ the Long Short-Term Memory (LSTM) technique to predict missing values which is capable of uncovering the relationships among variables in multi-variate data and utilizing existing variables to infer the missing values accurately.

TS Data Evaluation. The evaluation follows an incremental design approach, starting with an initial state comprising a user-constructed hierarchy that distinguishes various levels using different classification methods, and a table that monitors the data quality proportion. Subsequently, users have the flexibility to build the hierarchy as required, gradually incorporating additional views such as time distribution and problem severity, thereby forming a comprehensive evaluation function.

4 Demonstration

In this demo, we demonstrate GaussTS on two time series datasets with different tuple numbers and dimensions. Datasets statistics are reported in Table 1. GaussTS support dimensionality redunction, partition, similarity search and data cleaning and evaluation over time series data. We perform operations on TCPC dataset in the command line as an example.

Table 1. Datasets Statistics

Name	#Tuples	#Attributes	Time Interval
Tetuan City Power Consumption (TCPC)	52417	6	10 min
Intel Lab (IL)	2313153	4	31 s

Dimensionality Reduction. The result of dimensionality reduction using Z-order algorithm is illustrated in Fig. 2(a). The function reduces five columns within time series to a single column, decreasing the storage space from 3136 KB to 1880 KB.

Partition. Figure 2(b) shows the result tables of partition which divides time series data into several tables automatically and each table represents a time series data which includes original timestamps and necessary columns.

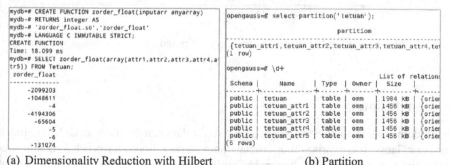

(a) Dimensionality Reduction with Hilbert   (b) Partition

Fig. 2. Demonstration in openGauss

Similarity Search. Upon completion of the similarity search operation, the command line will display the information of the search results in a table format as depicted in Fig. 3. This table encompasses pertinent information such as the names of the query and target series, timestamps, and the computed distance using the Euclidean distance metric. The experiment demonstrates that the process of similarity search can be completed within milliseconds, which is significantly faster than the traditional search process based solely on Euclidean distance. The visualization of the results of similarity search is presented in Fig. 4. It can be observed that both the overall and local trends of the query series align closely with the result sub-sequence and the area between series are also relatively small, indicating a significant similarity.

```
openGauss=# select similaritysearch('query_series','tetuan');
 query_series | target_series |   start_timestamp   |    end_timestamp    | distance
--------------+---------------+---------------------+---------------------+----------
 query_series | tetuan        | 2017-08-11 09:20:00 | 2017-08-14 19:20:00 |  2.3e-07
(1 row)
```

Fig. 3. Demonstration of similarity search

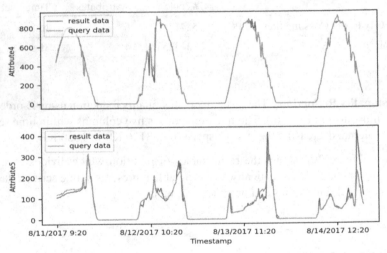

Fig. 4. The similarity between the query series and the result subsequence

5   Conclusion

In this demo, we introduce GaussTS, a time series data management module integrated into openGauss. It comprises four crucial components tailored for processing and analyzing time series data. Our demonstration illustrates the prowess and practicality of GaussTS in efficiently managing and analyzing time series data.

Acknowledgement. The paper is sponsored by National Natural Science Foundation of China (61972198) and CCF-Huawei Populus euphratica Innovation Research Funding.

References

1. Box, G.E.P., Jenkins, G.M., Reinsel, G.C., et al.: Time Series Analysis: Forecasting and Control. Wiley (2015)
2. OpenTSDB (2023). http://opentsdb.net/
3. InfluxDB (2023). https://www.influxdata.com/
4. Uqbar (2023). https://docs.mogdb.io/zh/uqbar/v1.1/overview
5. Echihabi, K., Zoumpatianos, K., Palpanas, T.: High-dimensional similarity search for scalable data science. In: 2021 IEEE 37th International Conference on Data Engineering (ICDE), pp. 2369–2372 (2021)
6. Dai, S., Yanwei, Y., Fan, H., Dong, J.: Spatio-temporal representation learning with social tie for personalized POI recommendation. Data Sci. Eng. **7**(1), 44–56 (2022)
7. Dasu, T., Duan, R., Srivastava, D.: Data quality for temporal streams. IEEE Data Eng. Bull. **39**(2), 78–92 (2016)
8. Fang, C., Wang, F., Yao, B., Xu, J.: GPSClean: A framework for cleaning and repairing GPS data. ACM Trans. Intell. Syst. Technol. **13**(3), 40:1–40:22 (2022)

URPWS: An Urban Road Ponding Monitoring and Warning System Based on Surveillance Video

Rui Xu[1] , Fang Fang[1], Qingyi Hao[1(✉)], Kang Zheng[1], Yi Zhou[1],
Yuting Feng[1], Shengwen Li[1], and Zijing Wan[2]

[1] School of Computer Science, China University of Geosciences, Wuhan
430074, China
{uiirux,fangfang,1202110795,zhengkang,zyi,fengyuting99,swli}@cug.edu.cn
[2] College of Letters and Science, University of California Santa Barbara,
Santa Barbara, CA 93106, USA
zijing_wan@ucsb.edu

Abstract. Efficient and accurate monitoring urban ponding by surveillance video is of great significance to reduce the risk of inundation and urban traffic. The previous work lacks consideration of real-time performance and integration of a unified management platform, which leads to low monitoring efficiency. This demo presents an urban road ponding monitoring and warning system (URPWS) based on surveillance video. URPWS provides a platform that integrates intelligent monitoring, real-time warning and unified management, which realizes the real-time and manageability of monitoring. In this demo, we bring forth the application of URPWS in Nanning of Guangxi Province, which delivers a new feasible solution for urban road ponding monitoring and management.

Keywords: Road ponding monitoring · Real-time warning ·
Management system

1 Introduction

With the increasingly drastic global climate change, the frequency of extreme rainfall is increasing [2]. At the same time, the rapid development of urbanization has greatly changed the conditions of the underlying surface, making the problem of urban ponding and inundation more prominent. Safe, real-time and accurate monitoring of urban ponding is of great significance to reduce the risk of inundation, urban traffic and public safety [1].

As the "eyes" of modern cities, video surveillance contains real-time and useful information [4], which is widely used to road ponding monitoring. However, the traditional method relying on the manual cannot guarantee the accuracy of monitoring and real-time warning. How to monitor the road ponding in real-time and accurately from the video surveillance is a difficult task for the city management department.

Deep learning has preponderance in data processing, which is beneficial to solving urban anomaly detection and disaster management [5]. Recent works have applied deep learning techniques to automatic monitoring of ponding [1,3]. However, they are not only under-optimized for efficiency in practical applications, but also lack a systematic solution for unified monitoring and management.

In this demo, we throw light on the Urban Road Ponding Monitoring and Warning System (URPWS), which integrates intelligent ponding monitoring, real-time warning and unified management. And, an optimized lightweight semantic segmentation network is introduced into the proposed system to improve the efficiency of road ponding monitoring. In addition, the proposed system is capable of flexible configuration and management service integration. Practical applications show that URPWS provides real-time monitoring and automatic warning of road ponding segments, improving the productivity of urban management.

2   System Overview

Figure 1 shows the architecture of URPWS. The system consists of client module, video streaming module, server module, storage module and ponding detection module. Firstly, URPWS uniformly accesses surveillance videos from the video streaming module under the user-specified configuration. Then, the server module schedules the video stream, identifies and segments the road ponding through the ponding detection module. The detection results and their scene photos are stored into the storage module to provide support for the warning service. Finally, URFDWS provides relevant services according to the requirements of users. The detailed description of each module are as follows.

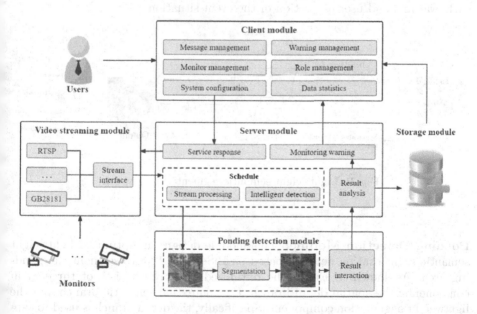

Fig. 1. System architecture

Client Module: In URPWS, the user interacts directly with the system through the client module. The client module provides users with the functions of message management, warning management, monitor management, role management, system configuration and data statistics. The diversity of URPWS configuration and management allows users to respond to different scenarios according to their actual needs.

Video Streaming Module: To manage and schedule video stream uniformly, URPWS designs video stream module. This module not only supports the access of multiple video streams, but also provides a unified interface for the server module to schedule and use.

Server Module: As the hub of URPWS, the server module responds to the requests of the client module to provide services for users, and it schedules the modules in the background of the system to realize monitoring and warning. Specifically, the module collects the video stream and samples the key frames, then sends the frame image to the ponding detection module for intelligent segmentation, and stores the returned result to the storage module. According to the system configuration, the server module performs monitoring and early warning in the form of message queue.

Storage Module: This module is used to store the detection results, including ponding time, ponding road location, event level, scene photos and other information. In addition, all of this stored information is also used as a reference for early warning and user inspection of the event situation.

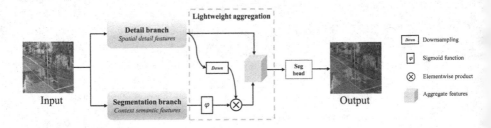

Fig. 2. Lightweight semantic segmentation network

Ponding Detection Module: This module designs an optimized lightweight semantic segmentation network based on BiSeNet V2 [6] to segment the ponding area. As shown in Fig. 2, the optimized network consists of three main components: the spatial detail branch, the contextual semantic branch and the lightweight aggregation component. Specifically, the detail branch is used to capture of low-level details and generate high-resolution feature representations, and the segmentation branch is used to focus on the acquisition of high-level semantic contexts. The lightweight aggregation component guides the aggregation layer to enhance the interconnectivity and merge the above two types of feature representations. The inference speed of the proposed optimized network in practical application is 27.39 FPS, which is an improvement of 1.7 FPS over the origin method. And, the result is only 2.6 FPS lower than the 30 FPS of the original video. The lightweight design of the network ensures efficient performance while maintaining recognition accuracy, which is crucial for the real-time monitoring of the system.

3   Demonstration Scenarios

URPWS is encapsulated well with a friendly interface that users can use with simple click. Take Nanning of Guangxi Province as an example, this demo employs live video from street surveillance cameras in downtown Nanning to illustrate some noteworthy functions and examples of road ponding monitoring and warning. Figure 3 shows the user interface of URPWS.

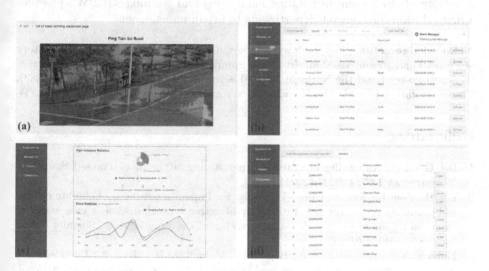

Fig. 3. Demonstration of URPWS

Figure 3(a) shows the road ponding monitoring of surveillance video. This web page provides the road location, event time and detection result image. The ponding area in the image is segmented by red mask. All the information helps users to understand the situation of road ponding in time for further decision-making and prevention.

Figure 3(b) shows the message list of the ponding events. The list displays event information such as the time and location. Users can view detailed pictures of the ponding detection, which is consistent with the segmentation results in

Fig. 3(a). In particular, whenever there is a ponding event occurs, the top right corner of the page will be timely pop-up reminder. In addition, URPWS provides statistical function. As shown in Fig. 3(c), this page provides statistics on the road of frequently occurring events and supports viewing the frequency of events in terms of diverse time dimensions. These assist users to spot water-prone roads and times of day, and make a strategic decision timely to avoid inundation.

Figure 3(d) shows the configuration page. Users can adjust and manage the surveillance video according to their actual needs to cope with different scenarios, such as increasing the frequency of monitoring during the plum rain season or adding surveillance cameras to areas prone to ponding.

4   Conclusion

In this demo, we design an urban road ponding monitoring and warning system based on surveillance video, called URPWS. URPWS performs ponding monitoring through a lightweight semantic segmentation network and integrates warning, and management functions. The demonstration scenarios indicate the feasibility of our system in ponding monitoring and warning. URPWS provides a new solution for relevant departments in urban management.

Acknowledgements. This work was supported by the Open Fund of Key Laboratory of Urban Land Resources Monitoring and Simulation, Ministry of Natural Resources (KF-2021-06-088) and the National Natural Science Foundation of China (42071382). The demo data is provided by Nanning Survey and Design Institute Group Co., LTD.

References

1. Bai, G., et al.: An intelligent water level monitoring method based on ssd algorithm. Measurement **185**, 110047 (2021)
2. Liu, H., Zou, L., Xia, J., Chen, T., Wang, F.: Impact assessment of climate change and urbanization on the nonstationarity of extreme precipitation: a case study in an urban agglomeration in the middle reaches of the yangtze river. Sustain. Urban Areas **85**, 104038 (2022)
3. Muhadi, N.A., Abdullah, A.F., Bejo, S.K., Mahadi, M.R., Mijic, A.: Deep learning semantic segmentation for water level estimation using surveillance camera. Appl. Sci. **11**(20), 9691 (2021)
4. Wang, Y., Li, K., Chen, G., Zhang, Y., Guo, D., Wang, M.: Spatiotemporal contrastive modeling for video moment retrieval. World Wide Web **26**, 1525–1544 (2022)
5. Wu, S., Li, X., Dong, W., Wang, S., Zhang, X., Xu, Z.: Multi-source and heterogeneous marine hydrometeorology spatio-temporal data analysis with machine learning: a survey. World Wide Web **26**(3), 1115–1156 (2023)
6. Yu, C., Gao, C., Wang, J., Yu, G., Shen, C., Sang, N.: Bisenet v2: bilateral network with guided aggregation for real-time semantic segmentation. Int. J. Comput. Vision **129**, 3051–3068 (2021)

RDBlab: An Artificial Simulation System for RDBMSs

Yu Yan, Hongzhi Wang[✉], Junfang Huang, Jian Geng, Zixuan Wang,
and Yuzhuo Wang

Harbin Institute of Technology, Harbin, China
{yuyan0618,wangzh}@hit.edu.cn

Abstract. With the development of cloud database, the simulation system of RDBMSs become increasing important for avoiding database failures. For example, a simulation system could hypothetically collect the database knob performance before implementing knob tuning in RDBMSs. However, existing works have paid less attention to the simulation system of RDBMSs. To fill this gap, we firstly design an artificial simulation system for RDBMSs called RDBlab, which could be utilized for enhancing system tunings, such as index tuning, knob tuning and etc. Our RDBlab provides friendly UI design and artificial simulation model. Users only need to configure some basic information about their database environment. Then they could efficiently utilize RDBlab to gather the synthetic performance of their tuning behaviors. In this paper, we clarify the architecture, core techniques and key scenarios of RDBlab.

Keywords: automatic management · simulation system · database · machine learning

1 Introduction

The simulation system of the complex software is a valuable tool for developers and researchers. For example, it provides the simulation and analysis for the performance of system tunings to reducing the potential performance degradation risks. A simulation system of autonomous driving [5] could simulate the road conditions and improve the performance of self-driving system. Also, a simulation system is significant for RDBMSs to enhance the system tunings.

Specifically, the simulation system for RDBMSs could provide two kinds of service. On the one hand, users could perceive the estimated performance of system tunings before real execution, avoiding the risks of performance degradation. On the other hand, the simulation system could provide sufficient training data for the AI4DB techniques by the synthetic performance labels. Moreover, with the development of cloud databases and self-driving techniques, the simulation system for RDBMSs is urgently in demand [4].

However, existing works have paid less attention to constructing the simulation system for RDBMSs. And some partial simulation model like [2] could

X. Song et al. (Eds.): APWeb-WAIM 2023, LNCS 14334, pp. 507–512, 2024.
https://doi.org/10.1007/978-981-97-2421-5_36

only be used for index selection and fail to work with other database task, like knob tuning, storage changing, etc. To fill this gap, we firstly design an artificial simulation system for comprehensively simulating the performance of RDBMSs. Our system has the following main advantages:

User-Friendly: The important goal of RDBlab is to achieve user-friendliness, so that non-expert users can efficiently implement the performance simulation of complicated database tunings. In order to achieve this goal, we design a simple interactive interface with only a few configurations. Users only need to configure the IP address, the database class, the port of database, the name of database and the workload, then our RDBlab could quickly establish a simulation system for enhancing DBMS tasks, like index revision, knob tunings, storage changes, etc.

Artificial Simulation Model: RDBlab could efficiently construct the simulation system by an artificial hierarchical simulation model. For efficiently simulating the multiple variables of RDBMSs (including the hardware, the storage, the index, the data table, the workload and etc.), our key idea is to divide the various variables of RDBMSs according their characteristics and design different artificial learning model to simulate these variables. Due to the hierarchical design, our RDBlab could largely save time consumption and space consumption by simplifying the simulation model as much as possible.

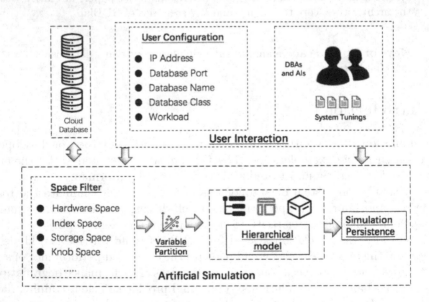

Fig. 1. System Architecture

2   System Architecture and Implementation

In this section, we present the overall architecture of our simulation system, RDBlab. We propose an artificial simulation system to enhance the system tunings, including the *User Interaction, cloud databases* and *Artificial Simulation*. As we can see in Fig. 1, the *User Interaction* has two main functions. One is to deliver the configuration of database environment to the *Artificial Simulation* for obtaining the simulation model, containing the IP address, database name and etc. The other is to transfer the system tuning tasks to the *Artificial Simulation* for calculating the tuning performance hypothetically. The *Artificial Simulation* has four main components to complete the system simulation. (i) The *Space Filter* is responsible for collecting the potential simulation space according to user configuration. (ii) Based on the simulation space, the *Variable Partition* partition these variables according to their characteristics for efficiently simulating these variables. (iii) After gathering the partition results of simulation space, RDBlab design an artificial hierarchical model for the multiple variables. (iv) The *simulation persistence* provides the persistent management of simulation models and interacts with the system tunings by database administrators (DBAs) and AI machines (AIs).

3   Key Technologies

In this section, we introduce the key techniques of RDBlab, including the variable partition princples and the hierarchical simulation model.

The simulation of RDBMSs faces multiple variables, containing hardware, storage, indexes, database knobs, workload and etc. Directly modeling these variables will cost large time and space resource due to their complex structures. Thus, we construct variable partition method to identify these variables according to two kinds metrics. (i) The characteristics of variable itself: We utilize the frequency of variable as the one of the partition metric because the frequency of variables is directly related the complexity of simulation model. (ii) The influential of variables: We utilize the factor (CPU time and IO time) and number (Tuple number) to evaluate the influential of variables because the variables influence the database performance by influencing the factor and the number metrics. Then, based on the above metrics, we construct a Bayes classification model [6] to partition these variables.

After the partition of multiple variables, we construct a hierarchical model to simulate the performance of RDBMSs. Specifically, we utilize the linear regression model [7] to model the infrequent and factor centered variables as a inner vector, because these variables has simpler regular than the frequent variables. Further, for the frequent and number centered variables (complex), we design the deep learning model [3] to simulate combined performance of the inner vector and the complex variables. By designing hierarchical model, our RDBlab could reduce the training overhead of establishing a simulation system while keeping the high simulation accuracy.

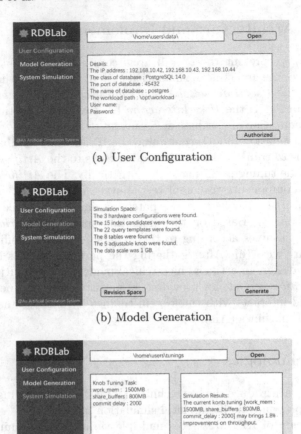

(a) User Configuration

(b) Model Generation

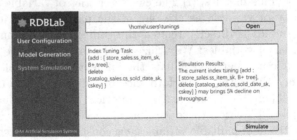

(c) Knob Simulation

(d) Index Simulation

Fig. 2. The demonstrations of RDBlab.

4 Demonstration Scenario

In this section, we utilize the open source benchmark TPC-H [1] to demonstrate our RDBlab, which is a typical and complex OLAP scenario. For database, our RDBlab is compatible with OpenGauss 3.0 and PostgreSQL 15.0 and we utilize the PostgresQL 15.0 as the demonstration. Next, we show some important function of RDBlab, containing the user configuration, the construction of simulation system and the simulation of system tunings.

User Configuration. As shown in Fig. 2a, users need to provide some basic information and authorize us to access its host and database. After that, our RDBlab could automatically collect the potential simulation space.

Model Generation. Figure 2b shows the simulation space we find it based on user configuration. For the current database, we find three kinds of hardwares, 15 index candidates, 22 query templates and etc. Based these automatic simulation space, our RDBlab allows users to further revision the space for efficiently simulating. If users identify the simulation space, they could click the 'generate' botton to construct the simulation model for current simulation space.

Knob Simulation. After obtaining the simulation model, RDBlab provides simulation function for the system tunings. Figure 2c shows the simulation result of a knob tuning task (change work_mem to 1500 MB and etc.). The result shows that this knob tuning may brings 1.8% improvements on throughput.

Index Simulation. Also, RDBlab provides the simulation of index tunings. Figure 2d shows the simulation result of an index tuning task (add index on attribution ss_item_sk and delete index 'cskey'). The result shows that this index tuning may brings 5% decline on throughput.

Acknowledgements. This paper was supported by Huawei (OpenGauss) and NSFC grant (62232005, 62202126).

References

1. Tpc-h (2023). https://www.tpc.org/tpch/
2. Ding, B., Das, S., Marcus, R., Wu, W., Chaudhuri, S., Narasayya, V.R.: AI meets AI: leveraging query executions to improve index recommendations. In: SIGMOD 2019. Association for Computing Machinery (2019)
3. LeCun, Y., Bengio, Y., Hinton, G.: Deep learning. Nature **521**(7553), 436–444 (2015)
4. Lim, W.S., et al.: Database gyms. In: 13th Annual Conference on Innovative Data Systems Research (CIDR 2023), Amsterdam, The Netherlands (2023)
5. Liu, C., Chen, Y., Tai, L., Ye, H., Liu, M., Shi, B.E.: A gaze model improves autonomous driving. In: Proceedings of the 11th ACM Symposium on Eye Tracking Research & Applications. ETRA 2019, Association for Computing Machinery, New York (2019). https://doi.org/10.1145/3314111.3319846

6. Rish, I.: An empirical study of the naive bayes classifier. J. Univ. Comput. Sci. 1(2), 127 (2001)
7. Su, X., Yan, X., Tsai, C.L.: Linear regression. Wiley Interdisc. Rev. Comput. Stat. 4(3), 275–294 (2012)

SeCPlat: A Secure Computation Platform Based on Homomorphic Encryption in Cloud

Fanyou Zhao, Yiping Teng$^{(\boxtimes)}$, Zheng Yang, Yuyang Xie, Jiayv Liu, and Jiawei Qi

School of Computer, Large-Scale Distributed System Laboratory,
Shenyang Aerospace University, Shenyang, China
typ@sau.edu.cn

Abstract. To achieve diversified secure computations on encrypted data, in this paper, we propose and demonstrate a Secure Computation Platform (SeCPlat) based on homomorphic encryption in cloud environments. Using the Paillier cryptosystem as the foundation, we design two main sets of secure computation protocols, i.e., Basic Secure Protocols and Advanced Secure Protocols. Basic Secure Protocols support arithmetic operations, comparison operations, and bit operations in ciphertext, while Advanced Secure Protocols upon the basic protocols can provide secure computational operations (e.g., Euclidean distance) in more complex algorithms. For a better demonstration, we further developed a prototype for secure computation applications, in which several secure algorithms were implemented to show the high supportability to secure computational tasks and the user interfaces were designed to setup and illustrate the parameters, results, and workflows of SeCPlat. Moreover, we provided APIs to facilitate the implementation of secure computational applications.

Keywords: Secure computation protocols · Homomorphic encryption

1 Introduction

To guarantee the confidentiality of sensitive data, a straightforward and effective solution is to encrypt data before storage and processing. In traditional cryptosystems, encrypted data need to be decrypted before computation, while it may consume high costs on encryption and decryption and suffer risks of privacy leakage in data decryption. With the development of homomorphic cryptosystem, data can be directly calculated in the ciphertext.

However, existing partial homomorphic cryptosystems (e.g., Paillier cryptosystem, RSA) can only support limited types of computational methods under encryption, while systems based on fully homomorphic cryptosystems (e.g., FHE compilers [1]) suffer expensive computational overhead, which can hardly facilitate secure arithmetic or comparison calculations and even complex algorithms.

In this paper, we propose and demonstrate a Secure Computation Platform (SeCPlat) based on homomorphic encryption in cloud environments, which mainly supports various complex computational operations assisted by different secure calculation protocols. In the system implementation, using the Paillier cryptosystem as the foundation, we design two main sets of secure computation protocols, i.e., Basic Secure Protocols and Advanced Secure Protocols. Based on the homomorphic properties of the Paillier cryptosystem, Basic Secure Protocols support arithmetic operations, comparison operations, and bit operations in ciphertext, while Advanced Secure Protocols upon the basic protocols can provide more secure computational operations in complex algorithms such as secure Euclidean distance calculation, secure dominance calculation, etc. For a better demonstration, we further developed a prototype in SeCPlat for secure computation applications, which provides high supportability for secure computational tasks. The parameters of the applications can be set, the results of the secure computations can be displayed, and the internal workflow of secure protocols can be illustrated through the user interfaces designed in SeCPlat.

2   System Architecture

2.1   System Design

To better showcase our platform, in the design of our system, we divided our system into three layers: Service Layer, Communication Layer, and User Layer. We embed SeCPlat in the Service Layer to complete the operations for the secure computation requirements. We designed the Web interfaces in the User Layer, where user parameters can be set and the computation results can be visualized through browsers. Between the Service Layer and User Layer, the Communication Layer is designed to transmit user parameters and computation results.

Architecture of SeCPlat. As shown in Fig. 1, using Paillier Homomorphic Encryption [2] as the foundation of SeCPlat, we design two main sets of secure computation protocols, i,e., Basic Secure Protocols and Advanced Secure Protocols. Paillier Homomorphic Encryption supports the homomorphic addition and the homomorphic multiplication and provides data encryption and decryption operations. On the basis of homomorphic encryption, the set of Basic Secure Protocols involves Secure Arithmetic Computation Protocols, Secure Comparison Protocols, and Secure Bit Computation Protocols to support main arithmetic operations, comparison operations, and bit operations in ciphertext. In order to diversify secure operations in more complex algorithms, we further developed Advanced Secure Protocols upon these basic protocols, which can provide secure Euclidean distance calculation, secure dominance calculation and other operations over encrypted data.

2.2   System Model of Secure Computation Protocols

All the secure computation protocols implemented in SeCPlat follow the cloud service framework widely applied in several important studies [3,4], which consists of two non-colluded cloud servers (C_1 and C_2). C_1 maintains the encrypted datasets and the public key of the data owner, and non-colluded C_2 keeps the public/secret key. After receiving encrypted calculation requests from the user, C_1 executes the secure computations cooperated with C_2, and sends the encrypted results to the corresponding user. Due to space limitations, the details of the system model of secure computation protocols can be referred to [3–5].

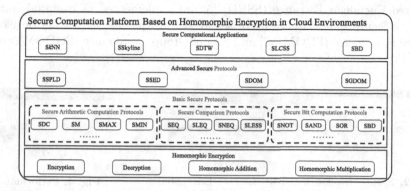

Fig. 1. Architecture of SeCPlat.

3   System Implementation

The user interface in the User Layer of our SeCPlat is implemented through HTML, CSS and JavaScript. Adopting the open-sourced *Django* framework, the Communication Layer is implemented using the JSON package to transfer data to the User Layer. To follow the system model of secure protocols, the two non-colluded cloud servers are simulated in two separated virtual machines deployed in a workstation, which can cooperate with each other through virtual connections. We implement all the algorithms in Python 3.8 and perform experiments on a Tower Server with two 40-core Intel(R) Xeon(R) Bronze 3204 1.90 GHz CPUs and 256 GB RAM running Ubuntu 20.04.

We implement encryption and decryption based on the Paillier cryptosystem [2], which supports the properties of homomorphic additions, homomorphic multiplications, and semantic security.

Implementation of Basic Secure Protocols. Based on the homomorphic properties of the Paillier cryptosystem, we implemented three sets of basic secure protocols supporting secure arithmetic computation, secure comparison, and secure bit computation (shown in Fig. 2). For secure arithmetic computation protocols, we implement Secure Multiplication (SM), Secure Maximum Computation (SMAX), Secure Minimum Computation (SMIN) [6] and Secure Division Computation (SDC) [5]. Secure Equal (SEQ) [7], Secure Less (SLESS) and Secure Less Than or Equal (SLEQ) [6] are implemented as secure comparison protocols. In the secure bit computation protocol set, Secure Bit-Decomposition

(SBD), Secure Bit-OR (SBOR) [8], Secure And (SAND) [6] and Secure NOT (SNOT) [7] are implemented in SeCPlat.

Implementation of Advanced Secure Protocols. On the strength of the basic protocols, we further developed several advanced secure protocols. Within them, Secure Squared Point to Line-segment Distance (SSPLD) and Secure Squared Euclidean Distance (SSED) [5] are implemented as essential components of secure spatial queries. For secure skyline queries, we further developed Secure Dominance (SDOM) [7] and Secure Group Dominance (SGDOM) protocols [6].

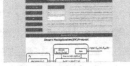

Fig. 2. Secure Protocols   **Fig. 3.**   Encrypted Result   **Fig. 4.** Performance

Fig. 5. Search Steps   **Fig. 6.** SLCSS   **Fig. 7.** SSkyline

4   Demonstration

4.1   Demonstration of Secure Protocols

We demonstrate secure computation processing based on the secure protocols in SeCPlat. After system login, the input data of the protocols can be typed into the user interface, and the parameters of the secret keys can be selected. According to the requirement, the secure computation protocol can be chosen on the menu bar, including the basic secure protocols and advanced secure protocols. Click the *Calculation* button to issue secure computation processing w.r.t the selected protocol and secret keys. The internal workflow of the selected secure protocol is shown at the bottom of the page. After the calculation, the encrypted result will be displayed and that in plaintext as well (shown in Fig. 3), which can be utilized to verify the correctness of the secure computation processing. Furthermore, these protocols can also be utilized by the users to develop their own secure applications.

4.2   Prototype

SeCPlat supports various types of computation tasks over encrypted data, including skyline and kNN queries for multidimensional data, DTW, LCSS and bi-directional similarity computation for sequential data, etc. In the demonstration, we provide five secure computation applications: secure kNN (SkNN) on R-tree, secure longest common subsequence (SLCSS), secure dynamic time warping (SDTW), secure bi-directional (SBD) similarity, and secure skyline (SSkyline) query. In SkNN, the coordinates of the query points and the spatial objects with the R-tree index are encrypted using the encryption tools of SeCPlat. Exploiting SSPLD and SMIN protocols, the nearest node of the R-tree to the query point can be found, which facilitates the best-first retrieval on the encrypted index. To search for nearer objects, SSED and SLESS are employed to calculate and compare the Euclidean distances between them. Finally, the k nearest objects can be retrieved and displayed on the page, with each searching step and its performance illustrated (shown in Fig. 4 and Fig. 5). In SLCSS, SDTW and SBD applications, we apply Bing Maps API to visualize the trajectories, and the performance comparison of the implemented algorithms can be displayed through varying different variables (shown in Fig. 6). In SSkyline query application, based on the dominating relations calculated by SDOM protocol, the skyline results can be computed according to the dataset selected by the user (shown in Fig. 7).

5   Conclusion

In this paper, we propose and demonstrate SeCPlat based on the Paillier cryptosystem in cloud. Basic Secure Protocols and Advanced Secure Protocols are implemented in this platform. For better demonstration, we further developed a prototype including many secure computational applications and user interfaces to show the high supportability and visualization of SeCPlat. Finally, we have also provided APIs on SeCPlat for the users' application development.

Acknowledgement. The work is supported by National Natural Science Foundation of China (61902260) and College Students' Innovative Entrepreneurial Training Project of Shenyang Aerospace University (202210143013). Yiping Teng is the corresponding author.

References

1. Viand, A., Jattke, P., Haller, M., Hithnawi, A.: HECO: fully homomorphic encryption compiler. In: USENIX Security 2023 (2023)
2. Paillier, P.: Public-key cryptosystems based on composite degree residuosity classes. In: Stern, J. (ed.) EUROCRYPT 1999. LNCS, vol. 1592, pp. 223–238. Springer, Heidelberg (1999). https://doi.org/10.1007/3-540-48910-x_16
3. Cui, N., Yang, X., Wang, B., Li, J., et al.: SVKNN: efficient secure and verifiable k-nearest neighbor query on the cloud platform. In: ICDE, pp. 253–264. IEEE (2020)

4. Liu, A., Zheng, K., Li, L., Liu, G., Zhao, L., Zhou, X.: Efficient secure similarity computation on encrypted trajectory data. In: ICDE, pp. 66–77. IEEE (2015)
5. Teng, Y., Shi, Z., Zhao, F., Ding, G., Xu, L., Fan, C.: Signature-based secure trajectory similarity search. In: IEEE TrustCom, pp. 196–206 (2021)
6. Teng, Y., Sun, Y., Shi, Z., Jiang, D., Zhao, L., Fan, C.: Secure skyline groups queries on encrypted data on cloud platform. In: HPCC/DSS/SmartCity/DependSys, pp. 551–560. IEEE (2021)
7. Liu, J., Yang, J., Xiong, L., Pei, J.: Secure and efficient skyline queries on encrypted data. IEEE Trans. Knowl. Data Eng. 31(7), 1397–1411 (2019)
8. Elmehdwi, Y., Samanthula, B.K., Jiang, W.: Secure k-nearest neighbor query over encrypted data in outsourced environments. In: ICDE, pp. 664–675. IEEE (2014)

Author Index

X. Song et al. (Eds.): APWeb-WAIM 2023, LNCS 14334, pp. 519–521, 2024.
https://doi.org/10.1007/978-981-97-2421-5

Printed in the United States
by Baker & Taylor Publisher Services